D1423360

Interdisciplinary Applied Mathematics
Volume 7

Series Preface
Problems in engineering, computational science, and the physical and biological sciences are using increasingly sophisticated mathematical techniques. Thus, the bridge between the Mathematical Sciences and other disciplines is heavily traveled. The correspondingly increased dialog between the disciplines has led to the establishment of the series: *Interdisciplinary Applied Mathematics*

The purpose of this series is to meet the current and future needs for the interaction between various science and technology areas on the one hand and mathematics on the other. This is done, firstly, by encouraging the ways that mathematics may be applied in traditional areas, as well as point towards new and innovative areas of applications; secondly, by encouraging other scientific disciplines to engage in a dialog with mathematicians outlining their problems to both access new methods as well as to suggest innovative developments within mathematics itself.

The series will consist of monographs and high level texts from researchers working on the interplay between mathematics and other fields of science and technology.

Interdisciplinary Applied Mathematics

1. *Gutzwiller:* Chaos in Classical and Quantum Mechanics
2. *Wiggins:* Chaotic Transport in Dynamical Systems
3. *Joseph/Renardy:* Fundamentals of Two-Fluid Dynamics:
   Part I: Mathematical Theory and Applications
4. *Joseph/Renardy:* Fundamentals of Two-Fluid Dynamics:
   Part II: Lubricated Transport, Drops and Miscible Liquids
5. *Seydel:* Practical Bifurcation and Stability Analysis:
   From Equilibrium to Chaos
6. *Hornung:* Homogenization and Porous Media
7. *Simo/Hughes:* Computational Inelasticity
8. *Keener/Sneyd:* Mathematical Physiology
9. *Han/Reddy:* Plasticity: Mathematical Theory and Numerical Analysis
10. *Sastry:* Nonlinear Systems: Analysis, Stability, and Control
11. *McCarthy:* Geometric Design of Linkages
12. *Winfree:* The Geometry of Biological Time

Springer
New York
Berlin
Heidelberg
Barcelona
Hong Kong
London
Milan
Paris
Singapore
Tokyo

J.C. Simo† T.J.R. Hughes

Computational Inelasticity

With 85 Illustrations

Springer

J.C. Simo (deceased)
Formerly, Professor of Mechanical Engineering
Stanford University
Stanford, CA 94305
USA

T.J.R. Hughes
Mechanics and Computation
Durand Building
Stanford University
Stanford, CA 94305
USA

Editors

Mathematics Subject Classification (1991): 73EXX, 73FXX, 65MXX, 73CXX

Library of Congress Cataloging-in-Publication Data
Simo, J.C. (Juan C.), 1952–1994
Computational inelasticity / J.C. Simo, T.J.R. Hughes.
p. cm. — (Interdisciplinary applied mathematics ; 7)
Includes bibliographical references and index.
ISBN 0-387-97520-9 (hardcover : alk. paper)
1. Elasticity. 2. Viscoelasticity. I. Hughes, Thomas J.R.
II. Title. III. Series: Interdisciplinary applied mathematics ; v. 7.
QA931.S576 1997
531′.382—dc21 97-26427

Printed on acid-free paper.

Production managed by Timothy Taylor; manufacturing supervised by Jeffrey Taub.
Photocomposed copy produced from the author's LaTeX files.
Printed and bound by Maple-Vail Book Manufacturing Group, York, PA.
Printed in the United States of America.

9 8 7 6 5 4 3 2 (Corrected second printing, 2000)

ISBN 0-387-97520-9 SPIN 10760898

Springer-Verlag New York Berlin Heidelberg
A member of BertelsmannSpringer Science+Business Media GmbH

Preface

This book goes back a long way. There is a tradition of research and teaching in inelasticity at Stanford that goes back at least to Wilhelm Flügge and Erastus Lee. I joined the faculty in 1980, and shortly thereafter the Chairman of the Applied Mechanics Division, George Herrmann, asked me to present a course in plasticity. I decided to develop a new two-quarter sequence entitled "Theoretical and Computational Plasticity" which combined the basic theory I had learned as a graduate student at the University of California at Berkeley from David Bogy, James Kelly, Jacob Lubliner, and Paul Naghdi with new computational techniques from the finite-element literature and my personal research. I taught the course a couple of times and developed a set of notes that I passed on to Juan Simo when he joined the faculty in 1985. I was Chairman at that time and I asked Juan to further develop the course into a full year covering inelasticity from a more comprehensive perspective. Juan embarked on this path creating what was to become his signature course. He eventually renamed it "Computational and Theoretical Inelasticity" and it covered much of the material that was the basis of his research in material modeling and simulation for which he achieved international recognition. At the outset we decided to write a book that would cover the material in the course. The first draft was written quite expeditiously, and versions of it have been circulated privately among friends, colleagues, and interested members of the research community since 1986. Thereafter progress was intermittent and slow. Some things were changed and some new chapters were added, but we both had become distracted by other activities in the early 1990s. Prior to that, we frequently discussed what would be necessary "to get it out the door," but I do not recall the subject even coming up once in the years immediately preceding Juan's death in 1994. Since that time I have been repeatedly urged to bring the project to completion. Through the efforts of a number of individuals, the task is now completed.

This book describes the theoretical foundations of inelasticity, its numerical formulation, and implementation. It is felt that the subject matter described herein constitutes a representative sample of state-of-the-art methodology currently used in inelastic calculations. On the other hand, no attempt has been made to present a careful account of the historical developments of the subject or to examine in detail important physical aspects underlying inelastic flow in solids. Likewise, the list of references should, by no means, be regarded as a complete literature survey of the field.

Chapter 1 begins with an overview of small deformation plasticity and viscoplasticity in a one-dimensional setting. Notions introduced in Chapter 1 are generalized to multiple dimensions and developed more comprehensively in subsequent chapters. Ideas of convex optimization theory, which are the foundations of the numerical implementation of plasticity, are first introduced in Chapter 1. In Chapter 2 the theory is generalized to multiple dimensions. In addition to the three-dimensional case, plane-strain and plane-stress cases are presented, as well as thermodynamic considerations and the principle of maximal plastic dissipation. Chapter 3 deals with integration algorithms for the constitutive equations of plasticity and viscoplasticity. The two most important classes of return-mapping algorithms are described, namely, the closest-point projection and cutting-plane algorithms. The classical radial return method is also presented. Another important mathematical tool in the construction of numerical methods for inelastic constitutive equations, the operator-splitting methodology, is also introduced in Chapter 3. Chapter 4 deals with the variational setting of boundary-value problems and discretization by finite element methods. Key technologies for successful implementation of inelasticity, such as the assumed strain method and the B-bar approach, are described. The generalization of the theory to nonsmooth yield surfaces is considered in Chapter 5. Mathematical numerical analysis issues of general return-mapping algorithms and, in particular, their nonlinear stability are presented in Chapter 6. The generalization to finite-strain inelasticity theory commences in Chapter 7 with an introduction to nonlinear continuum mechanics, the notion of objectivity, variational formulations of the large-deformation case, and hyperelastic and hypoelastic constitutive equations. The practically important subject of objective integrative algorithms for rate constitutive equations is described in Chapter 8. In Chapter 9 the theory of hyperelastic-based plasticity models is presented. This chapter covers the local multiplicative decomposition of the deformation gradient into elastic and plastic parts and numerical formulations of this concept by way of return-mapping algorithms. Chapter 10 deals with small and large deformation viscoelasticity.

I believe a good, basic course of a semester's or quarter's duration would focus on Chapter 1 to 4. For more advanced students wishing to understand the large deformation theory, Chapters 7 and 8 are essential. Chapter 8, in particular, deals with the types of formulations commonly used in large-scale commercial computer programs. There is more research interest in the hyperelastic-based theories of Chapter 9, which are more satisfying from a theoretical point of view. However, as of this writing, they have not enjoyed similar attention from the developers of most commercial computer programs.

Over the past two years, this text has been used as the basis of courses at Stanford and Berkeley which provided vehicles for readying the manuscript for publication. I wish to sincerely thank the students in these classes for their considerable patience and effort. Present and past graduate students of Juan's and mine were also instrumental in bringing the endeavor to fruition. Among them I wish to thank, in particular, Francisco Armero, Krishnakumar Garikipati, Sanjay Govindjee, John Kennedy, and Steve Rifai. However, without the hard work and devotion of two recent students, I doubt that this project would have been completed: Vinay Rao and Eva Petőcz critically read the manuscript and interacted with the other individuals who provided corrections. Vinay and Eva synthesized the inputs, made changes, and managed the master file containing the manuscript. They searched for and found lost drawings, and when missing figures could not be located, they drew them themselves. They spent many hours in this effort, and I wish to express my sincere thanks and gratitude to them.

Thomas J. R. Hughes
Stanford, March 1998

Contents

1

Motivation. One-Dimensional Plasticity and Viscoplasticity

In this chapter we consider the formulation and numerical implementation of one-dimensional plasticity and viscoplasticity models. Our objective is to motivate our subsequent developments of the theory in the simplest possible context afforded by a one-dimensional model problem. Since the main thrust of this monograph is the numerical analysis and implementation of classical plasticity, viscoplasticity, and viscoelasticity models, an attempt is made to formulate the basic governing equations in a concise form suitable for our subsequent numerical analysis. To this end, once a particular model is discussed, the basic governing equations are summarized in a BOX that highlights the essential mathematical aspects of the theory. Likewise, the corresponding numerical algorithms are also summarized in a BOX that highlights the essential steps involved in the actual numerical implementation. We follow this practice throughout the remaining chapters of this monograph.

1.1 Overview

An outline of the topics covered in this introductory chapter is as follows.

In Section **1.2** we present a detailed formulation of the governing equations for a one-dimensional mechanical device consisting of a linear spring and a Coulomb friction device. This simple model problem exhibits all the basic features underlying classical rate-independent (perfect) plasticity, in particular, the notion of *irreversible response* and its mathematical modeling through the Kuhn–Tucker complementarity conditions. Subsequently, we generalize this model problem to account for hardening effects and discuss the mathematical structure of two classical phenomenological hardening models known as *isotropic* and *kinematic* hardening.

In Section **1.3** we summarize the equations of the one-dimensional elastoplastic boundary-value problem and discuss the weak or variational formulation of these equations. Then we provide an outline of the basic steps involved in a numerical solution procedure. With this motivation at hand, in Section **1.4** we discuss the numerical integration of the constitutive models developed in Section **1.2** and introduce the fundamental concept of *return mapping* or *catching up* algorithm. As shown in Chapter **3** this notion has a straightforward generalization to three-dimensional models and constitutes the single most important concept in computational plasticity. In Section **1.5** we illustrate the role of these integrative algorithms by considering the simplest finite-element formulation of the elastoplastic boundary-value problem. We discuss the incremental form of this problem and introduce the important notion of *consistent* or *algorithmic tangent modulus.*

Finally, Section **1.6** generalizes the preceding ideas to accommodate rate-dependent response within the framework of classical viscoplasticity. We examine two possible formulations of this class of models and discuss their numerical implementation. In particular, emphasis is placed on the significance of viscoplasticity as a *regularization* of rate-independent plasticity. This interpretation is important in the solution of boundary-value problems where hyperbolicity of the equations in the presence of softening can always be attained by suitable choice of the relaxation time.

For further reading on the physical background, and generalizations, see Lemaitre and Chaboche [1990].

1.2 Motivation. One-Dimensional Frictional Models

To motivate the mathematical structure of classical rate-independent plasticity, developed in subsequent sections, we examine the mechanical response of the one-dimensional frictional device illustrated in Figure **1.1**.

We assume that the device initially possesses unit length (and unit area) and consists of a spring, with elastic constant E, and a Coulomb friction element, with constant $\sigma_Y > 0$, arranged as shown in Figure **1.1**. We let σ be the applied stress (force) and ε the total strain (change in length) in the device.

1.2.1 Local Governing Equations

Inspection of Figure **1.1** leads immediately to the following observations:

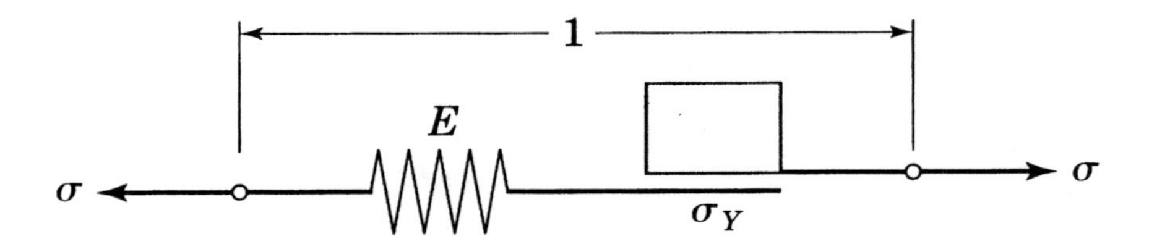

FIGURE 1.1. One-dimensional frictional device illustrating rate–independent plasticity.

a. The total strain ε splits into a part ε^e on the spring with constant E, referred to as the *elastic part*, and a strain ε^p on the friction device referred to as the *plastic part*, that is

$$\boxed{\varepsilon = \varepsilon^e + \varepsilon^p.} \tag{1.2.1}$$

b. By obvious equilibrium considerations, the stress on the spring with constant E is σ, and we have the elastic relationship

$$\boxed{\sigma = E\varepsilon^e \equiv E\left(\varepsilon - \varepsilon^p\right).} \tag{1.2.2}$$

Now we characterize the mechanical response of the friction element as follows.

1.2.1.1 Irreversible frictional response.

Assume that ε, ε^p and σ are functions of time in an interval $\left[0, T\right] \subset \mathbb{R}$. In particular, we let

$$\varepsilon^p : \left[0, T\right] \to \mathbb{R},$$

and

$$\dot{\varepsilon}^p = \frac{\partial}{\partial t}\varepsilon^p. \tag{1.2.3}$$

Change in the configuration of the frictional device is possible only if $\dot{\varepsilon}^p \neq 0$. To characterize this change, we isolate the frictional device as shown in Figure **1.2**. We make the following physical assumptions.

1. The stress σ in the frictional device *cannot be greater in absolute value than* $\sigma_Y > 0$. This means that the *admissible stresses* are constrained to lie in the closed interval $[-\sigma_Y, \sigma_Y] \subset \mathbb{R}$. For future use we introduce the notation

$$\boxed{\mathbb{E}_\sigma = \left\{\tau \in \mathbb{R} \mid f(\tau) := |\tau| - \sigma_Y \leq 0\right\}} \tag{1.2.4}$$

to designate the set of admissible stresses. For reasons explained below, we denote by σ_Y the *flow stress* of the friction device. The function $f : \mathbb{R} \to \mathbb{R}$, defined as

$$f(\tau) := |\tau| - \sigma_Y \leq 0, \tag{1.2.5}$$

then is referred to as the *yield condition.* Note that $\mathbb{E}_\sigma$ is a closed interval and, therefore, it is a *closed convex set.*

2. If the absolute value σ of the applied stress is less than the flow stress σ_Y, *no change* in ε^p takes place, i.e., $\dot{\varepsilon}^p = 0$. This condition implies

$$\dot{\varepsilon}^p = 0 \text{ if } f(\sigma) := |\sigma| - \sigma_Y < 0. \tag{1.2.6}$$

From (1.2.2) and (1.2.6) it follows that

$$f(\sigma) < 0 \Rightarrow \dot{\sigma} = E\dot{\varepsilon}, \tag{1.2.7}$$

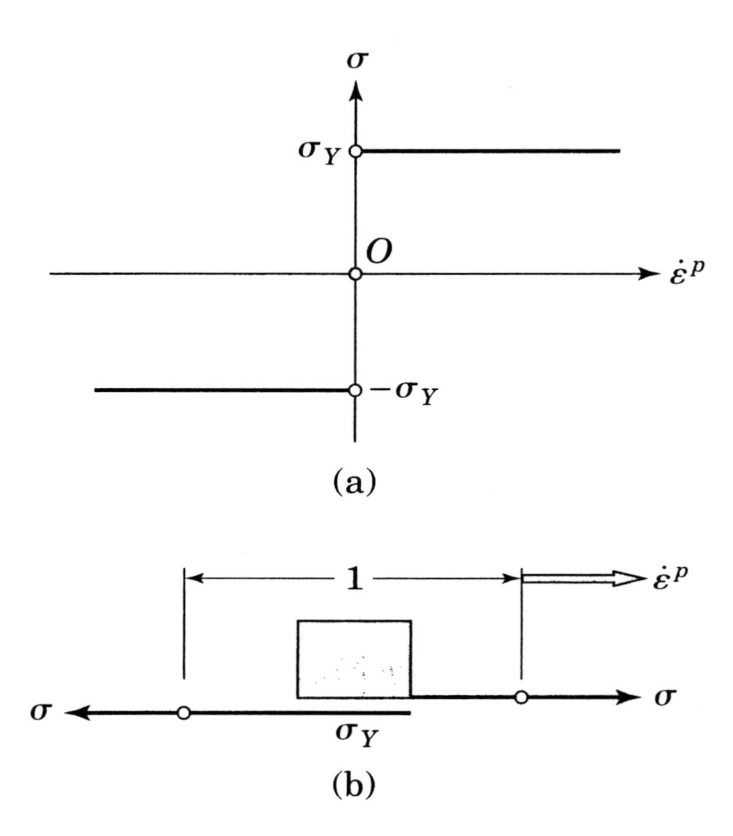

FIGURE 1.2B. Characterization of frictional response for a device with constant $\sigma_Y > 0$.

and the *instantaneous* response of the device is *elastic* with spring constant E. This motivates the denomination of *elastic range* given to the *open* set

$$\operatorname{int}\left(\mathbb{E}_\sigma\right) = \left\{\tau \in \mathbb{R} \;\middle|\; f(\tau) := |\tau| - \sigma_Y < 0\right\}, \tag{1.2.8}$$

since (1.2.6) and (1.2.7) hold for $\sigma \in \operatorname{int}\left(\mathbb{E}_\sigma\right)$.

3. Because, by assumption 1, stress states σ such that $f(\sigma) = |\sigma| - \sigma_Y > 0$ are *inadmissible* and $\dot{\varepsilon}^p = 0$ for $f(\sigma) < 0$ by assumption 2, a *change in* ε^p *can take place only if* $f(\sigma) = |\sigma| - \sigma_Y = 0$. If the latter condition is met, the frictional device experiences *slip in the direction of the applied stress* σ, with *constant slip rate.* Let $\gamma \geq 0$ be the absolute value of the slip rate. Then the preceding physical assumption takes the form

$$\left.\begin{aligned} \dot{\varepsilon}^p &= \gamma \geq 0 \quad \text{if} \quad \sigma = \sigma_Y > 0, \\ \dot{\varepsilon}^p &= -\gamma \leq 0 \quad \text{if} \quad \sigma = -\sigma_Y < 0. \end{aligned}\right\} \tag{1.2.9}$$

Whether $\gamma \geq 0$ is actually positive, i.e., $\gamma > 0$ or zero depends on further conditions involving the applied strain rate $\dot{\varepsilon}$, which are discussed below and are referred to as *loading/unloading conditions*. For now we note that (1.2.9) can be recast into the following single equation

$$\dot{\varepsilon}^p = \gamma \operatorname{sign}(\sigma) \quad \text{iff} \quad f(\sigma) := |\sigma| - \sigma_Y = 0 \tag{1.2.10}$$

where $\gamma \geq 0$, which goes by the name *flow rule*. Here, sign : $\mathbb{R} \to \mathbb{R}$ is the sign function defined as

$$\operatorname{sign}(\sigma) = \begin{cases} +1 & \text{if } \sigma > 0 \\ -1 & \text{if } \sigma < 0. \end{cases} \tag{1.2.11}$$

The boundary $\partial\mathbb{E}_\sigma$ of the convex set $\mathbb{E}_\sigma$, defined by

$$\partial\mathbb{E}_\sigma = \left\{\tau \in \mathbb{R} \mid f(\tau) = |\tau| - \sigma_Y = 0\right\}, \tag{1.2.12}$$

is called the *yield surface*. In the present one-dimensional model, $\partial\mathbb{E}_\sigma = \{-\sigma_Y, \sigma_Y\}$ reduces to two points. Note that

$$\mathbb{E}_\sigma = \operatorname{int}\left(\mathbb{E}_\sigma\right) \cup \partial\mathbb{E}_\sigma, \tag{1.2.13}$$

that is, $\mathbb{E}_\sigma$ is the closure of the elastic range $\operatorname{int}\left(\mathbb{E}_\sigma\right)$.

To complete the description of the model at hand, it remains only to determine the slip rate $\gamma \geq 0$. This involves the following essential conditions that embody the *notion of irreversibility* inherent in the response of the model in Figure **1.1**.

1.2.1.2 Loading/unloading conditions.

With the observations made above in mind, we show that the evaluation of $\varepsilon^p : [0, T] \to \mathbb{R}$ can be completely described, for *any* admissible stress state $\sigma \in \mathbb{E}_\sigma$, with the single evolutionary equation

$$\boxed{\dot{\varepsilon}^p = \gamma \operatorname{sign}(\sigma),} \tag{1.2.14}$$

provided that γ and σ are restricted by certain *unilateral constraints*.

i. First, we note that σ must be admissible, i.e., $\sigma \in \mathbb{E}_\sigma$ by assumption 1, and γ must be nonnegative by assumption 3. Consequently,

$$\boxed{\begin{array}{c} \gamma \geq 0, \\ \text{and} \\ f(\sigma) \leq 0. \end{array}} \tag{1.2.15a}$$

ii. Second, by assumption 2, $\gamma = 0$ if $f(\sigma) < 0$. On the other hand, by assumption 3, $\dot{\varepsilon}^p \neq 0$, and, therefore, $\gamma > 0$ only if $f(\sigma) = 0$. These observations imply the conditions

$$\left.\begin{array}{r} f(\sigma) < 0 \Rightarrow \gamma = 0, \\ \gamma > 0 \Rightarrow f(\sigma) = 0. \end{array}\right\}$$

It follows that we require that

$$\boxed{\gamma f(\sigma) = 0.} \tag{1.2.15b}$$

Conditions (1.2.15) express the physical requirements, elaborated upon above, that the *stress must be admissible* and that the plastic flow, in the sense of *nonzero frictional strain rate* $\dot{\varepsilon}^p \neq 0$, can take place only on the yield surface $\partial\mathbb{E}_\sigma$. These conditions (i.e., (1.2.15a,b)) are classical in the convex mathematical programming literature (see e.g., Luenberger [1984]) and go by the name of Kuhn–Tucker conditions.

The last condition to be described below enables us to determine the actual value of $\gamma \geq 0$ at any given time $t \in [0, T]$ and is referred to as the *consistency requirement.* A precise formulation requires a further observation.

iii. Let $\{\varepsilon(t), \varepsilon^p(t)\}$ be given at time $t \in [0, T]$, so that $\sigma(t)$ is also known at time t by the elastic relationships (1.2.2), i.e., $\sigma(t) = E\left[\varepsilon(t) - \varepsilon^p(t)\right]$.* Assume that we *prescribe the total strain rate* $\dot{\varepsilon}(t)$ at time t. Further, consider the case where

$$\sigma(t) \in \partial\mathbb{E}_\sigma \iff \hat{f}(t) := f\left[\sigma(t)\right] = 0$$

at time t. Then, it is easily shown that $\dot{\hat{f}}(t) \leq 0$, since should $\dot{\hat{f}}(t)$ be positive it would imply that $\hat{f}(t + \Delta t) > 0$ for some $\Delta t > 0$, which violates the admissibility condition $f \leq 0$.† Further, we *specify that* $\gamma > 0$ *only if* $\dot{\hat{f}}(t) = 0$, and set $\gamma = 0$ if $\dot{\hat{f}} < 0$, that is, dropping the hat to simplify the notation, we set

$$\left.\begin{aligned} \gamma > 0 &\Rightarrow \dot{f} = 0, \\ \dot{f} < 0 &\Rightarrow \gamma = 0. \end{aligned}\right\}$$

Therefore, we have the additional condition

$$\boxed{\gamma \dot{f}(\sigma) = 0.} \tag{1.2.15c}$$

Condition (1.2.15c) is alternatively referred to as the *persistency (or consistency) condition,* and corresponds to the physical requirement that for $\dot{\varepsilon}^p$ to be nonzero (i.e., $\gamma > 0$) the stress point $\sigma \in \partial\mathbb{E}_\sigma$ must "persist" on $\partial\mathbb{E}_\sigma$ so that $\dot{f}\left[\sigma(t)\right] = 0$.

1.2.1.3 Frictional slip (plastic flow).

For the model at hand, the expression for $\gamma > 0$ when the consistency condition (1.2.15c) holds, takes a particularly simple form. By the chain rule and conditions

---

*We could alternatively prescribe $\left\{\sigma(t), \varepsilon^p(t)\right\}$ and define $\varepsilon(t)$ as $\varepsilon(t) = \varepsilon^p(t) + E^{-1}\sigma(t)$.

†A formal argument can easily be constructed using a Taylor expansion, as shown in the next chapter; see Lang [1983]

(1.2.2) and (1.2.14),

$$\dot{f} = \frac{\partial f}{\partial \sigma} E\left(\dot{\varepsilon} - \dot{\varepsilon}^p\right)$$
$$= \frac{\partial f}{\partial \sigma} E\dot{\varepsilon} - \gamma \frac{\partial f}{\partial \sigma} E \operatorname{sign}(\sigma). \tag{1.2.16}$$

However,

$$\frac{\partial}{\partial \sigma}|\sigma| = \operatorname{sign}(\sigma) \Rightarrow \frac{\partial f}{\partial \sigma} = \operatorname{sign}(\sigma). \tag{1.2.17}$$

Consequently,* on noting that $[\operatorname{sign}(\sigma)]^2 = 1$, (1.2.16) and (1.2.17) imply

$$\dot{f} = 0 \Rightarrow \gamma = \dot{\varepsilon} \operatorname{sign}(\sigma). \tag{1.2.18}$$

Substitution of (1.2.18) in (1.2.14) yields the result

$$\dot{\varepsilon}^p = \dot{\varepsilon} \text{ for } f(\sigma) = 0, \dot{f}(\sigma) = 0, \tag{1.2.19}$$

which says that "plastic" slip in the frictional device equals the applied strain rate.

The response of the device shown in Figure **1.1** is illustrated in Figure **1.3**. The theory we have presented thus far is called *perfect plasticity*. A summary of the constitutive model is contained in BOX **1.1**.

BOX 1.1. One-Dimensional
Rate-Independent Perfect Plasticity.

i. Elastic stress-strain relationship

$$\sigma = E\left(\varepsilon - \varepsilon^p\right)$$

ii. Flow rule

$$\dot{\varepsilon}^p = \gamma \operatorname{sign}(\sigma)$$

iii. Yield condition

$$f(\sigma) = |\sigma| - \sigma_Y \leq 0.$$

iv. Kuhn–Tucker complementarity conditions

$$\gamma \geq 0, \quad f(\sigma) \leq 0, \quad \gamma f(\sigma) = 0$$

v. Consistency condition

$$\gamma \dot{f}(\sigma) = 0 \quad (\text{if} \quad f(\sigma) = 0)$$

Remarks 1.2.1.

1. The yield condition (1.2.5) and the flow rule (1.2.14) are formulated in terms of stresses. Throughout our discussion, we have considered stress as a *dependent*

*Equation (1.2.17) should be interpreted in a generalized (distributional) sense since $|\sigma|$ is not differentiable at $\sigma = 0$ (see Stakgold [1979]).

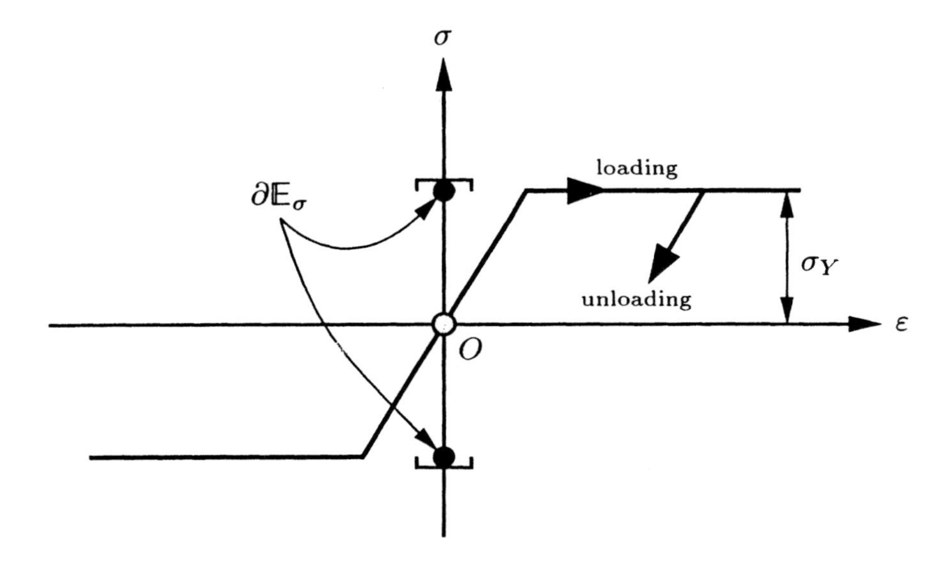

FIGURE 1.3. Schematic representation of the mechanical response of a one-dimensional elastic-friction model.

variable, i.e., a function of $\{\varepsilon, \varepsilon^p\}$ through the elastic relationship (1.2.2). In fact the argument leading to (1.2.19) depends crucially on regarding $\dot{\varepsilon}$ (and not $\dot{\sigma}$) as the "*driving*" variable.

2. Formally, we can recast the entire formulation in *strain space*, in terms of $\{\varepsilon, \varepsilon^p\}$, by eliminating σ with the help of the elastic relationship. For the yield condition (1.2.5), for instance, we find that

$$\bar{f}\left(\varepsilon, \varepsilon^p\right) := f\left[E\left(\varepsilon - \varepsilon^p\right)\right] \equiv E\left|\varepsilon - \varepsilon^p\right| - \sigma_Y \leq 0. \tag{1.2.20}$$

We make use of this formulation in the next chapter.

3. The flow rule given by (1.2.14) is related to the yield condition (1.2.5) through the *potential relationship*

$$\boxed{\dot{\varepsilon}^p = \gamma \frac{\partial f}{\partial \sigma},} \tag{1.2.21}$$

since $\dfrac{\partial f}{\partial \sigma} = \operatorname{sign}(\sigma)$. In the three-dimensional theory, for the case in which (1.2.21) holds, one speaks of an*associative* flow rule.

1.2.2 An Elementary Model for (Isotropic) Hardening Plasticity

As a next step in our motivation of the mathematical theory of plasticity, we examine an enhancement of the model discussed in Section **1.2.1** that illustrates an effect experimentally observed in many metals, called *strain hardening*.

For the model in Section **1.2.1**, slip (i.e., $\dot{\varepsilon}^p \neq 0$) takes place at a constant value of the applied stress σ such that $|\sigma| = \sigma_Y$, leading to the stress-strain response in Figure **1.3**. A *strain-hardening* model, on the other hand, leads to a stress-strain curve of the type idealized in Figure **1.4b**.

The essential difference between the two models illustrated in Figure **1.4** lies in the fact that for perfect plasticity the closure of the elastic range $\mathbb{E}_\sigma$ remains unchanged, whereas for the strain hardening model, $\mathbb{E}_\sigma$ *expands with the amount of slip in the system* (i.e., the amount of plastic flow). A mathematical model that captures this effect is considered ne**xt**

1.2.2.1 The simplest mathematical model.

Our basic assumptions concerning the elastic response of a strain-hardening model remain unchanged. We assume, as in Section **1.2.1**, the additive decomposition

$$\varepsilon = \varepsilon^e + \varepsilon^p. \tag{1.2.22a}$$

In addition, we postulate the elastic stress-strain relationship

$$\sigma = E\left(\varepsilon - \varepsilon^p\right). \tag{1.2.22b}$$

To illustrate the mathematical structure of strain-hardening plasticity we consider the simplest situation illustrated in Figure **1.5**. In this model, the expansion (hardening) experienced by $\mathbb{E}_\sigma$ is assumed to obey two conditions

a. The hardening is isotropic in the sense that at any state of loading, the centeı of $\mathbb{E}_\sigma$ remains at the origin.
b. The hardening is *linear* in the *amount* of plastic flow (i.e., linear in $|\dot{\varepsilon}^p|$) and independent of sign($\dot{\varepsilon}^p$).

The first condition leads to a yield criterion of the form

$$\boxed{f(\sigma, \alpha) = |\sigma| - \left[\sigma_Y + K\alpha\right] \leq 0, \quad \alpha \geq 0,} \tag{1.2.23}$$

where $\sigma_Y > 0$ and $K \geq 0$ are given constants; K is often called the plastic modulus. The variable $\alpha : \left[0, T\right] \to \mathbb{R}$ is a *nonnegative function* of the amount of plastic flow (slip), called an *internal hardening variable*. If $K < 0$, one speaks of a strain-softening response.

With condition **ii** in mind, we consider the simplest evolutionary equation for α, namely,

$$\boxed{\dot{\alpha} = |\dot{\varepsilon}^p|.} \tag{1.2.24}$$

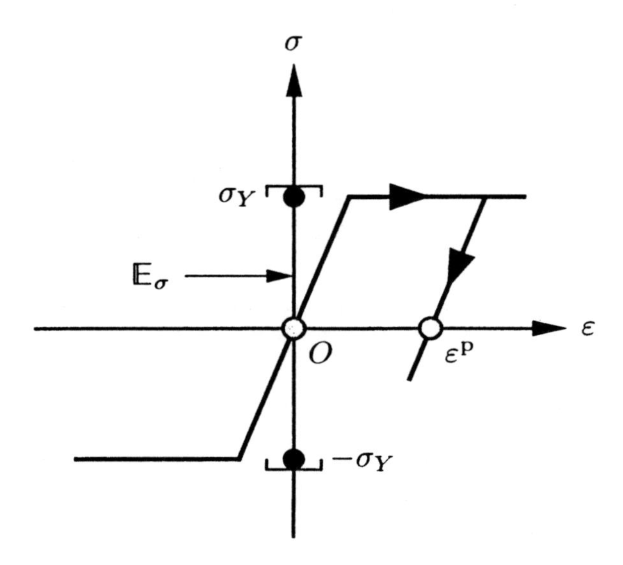

FIGURE 1.4A.

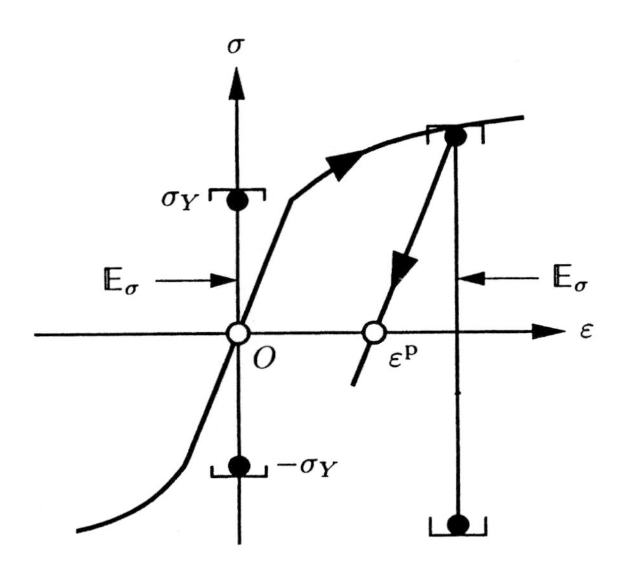

FIGURE 1.4B. Strain-hardening plasticity versus perfect plasticity.

The *irreversible* mechanism that governs the evolution of slip in the system (plastic flow), which is defined by the flow rule, remains unchanged. Consequently, as in Section **1.2.1**, we assume that

$$\dot{\varepsilon}^p = \gamma \ \text{sign}(\sigma), \tag{1.2.25}$$

where $\gamma \geq 0$ is the rate at which slip takes place. The irreversible nature of plastic flow is again captured by means of the Kuhn–Tucker loading/unloading conditions, which in the present context read

$$\gamma \geq 0, \quad f(\sigma, \alpha) \leq 0, \quad \gamma f(\sigma, \alpha) = 0, \tag{1.2.26}$$

where $\gamma \geq 0$ is determined by the consistency condition

$$\gamma \dot{f}(\sigma, \alpha) = 0. \tag{1.2.27}$$

The interpretation of conditions (1.2.26)–(1.2.27) is identical to the one discussed in detail in Section **1.2.1**.

BOX 1.2. One-Dimensional, Rate-Independent Plasticity with Isotropic Hardening.

i. Elastic stress-strain relationship

$$\sigma = E\left(\varepsilon - \varepsilon^p\right).$$

ii. Flow rule and isotropic hardening law

$$\dot{\varepsilon}^p = \gamma \ \text{sign}(\sigma)$$

$$\dot{\alpha} = \gamma.$$

iii. Yield condition

$$f(\sigma, \alpha) = |\sigma| - (\sigma_Y + K\alpha) \leq 0.$$

iv. Kuhn–Tucker complementarity conditions

$$\gamma \geq 0, \quad f(\sigma, \alpha) \leq 0, \quad \gamma f(\sigma, \alpha) = 0.$$

v. Consistency condition

$$\gamma \dot{f}(\sigma, \alpha) = 0 \quad (\text{if} \quad f(\sigma, \alpha) = 0).$$

Remarks 1.2.2.

1. Note from (1.2.24) and (1.2.25) that $\dot{\alpha} = \gamma \geq 0$, an equation which can be used as definition for the evolution of α.
2. In view of expression (1.2.23) which includes α in addition to σ, it is natural to define the closure of the elastic range as the set

$$\boxed{\mathbb{E}_\sigma = \{(\sigma, \alpha) \in \mathbb{R} \times \mathbb{R}_+ \mid f(\sigma, \alpha) \leq 0\}.} \tag{1.2.28}$$

   The intersection of $\mathbb{E}_\sigma$ with lines α = constant defines the elastic range in stress space. See Figure **1.6**.
3. Alternative formulations of the rate equation for $\dot{\alpha}$ are possible. If α is defined by (1.2.24), one speaks (up to numerical factors) of the *equivalent plastic strain*.

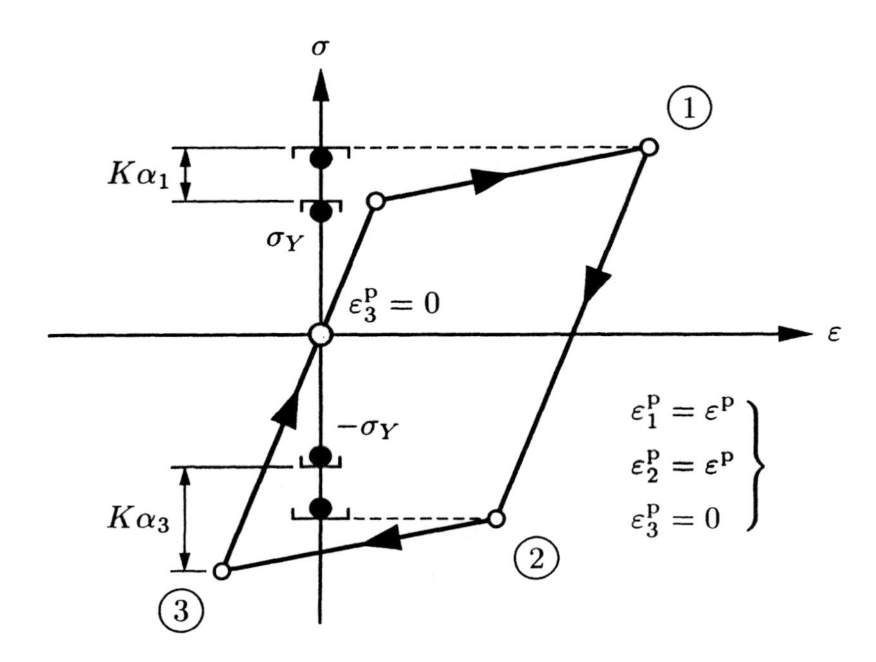

FIGURE 1.5. Response of a *linear isotropic hardening* model in a closed cycle: Although the total plastic strain at the end of the cycle $\varepsilon_3^p = 0$, the value of the hardening variable $\alpha_3 = 2\alpha_1$.

The choice

$$\dot{\alpha} = \sigma \dot{\varepsilon}^p \qquad (1.2.29)$$

goes by the name of *equivalent plastic power*. Note that for this latter case

$$\dot{\alpha} = |\sigma|[\mathrm{sign}(\sigma)]^2 \gamma = |\sigma|\gamma \geq 0. \qquad (1.2.30)$$

Alternative hardening models are discussed in Section **1.2.3**.

1.2.2.2 Tangent elastoplastic modulus.

The consistency condition (1.2.27) enables one to solve explicitly for γ and relate stress rates to strain rates as follows. From (1.2.23), (1.2.24), and (1.2.25), along with the elastic stress-strain relationship,

$$\begin{aligned} \dot{f} &= \frac{\partial f}{\partial \sigma}\dot{\sigma} + \frac{\partial f}{\partial \alpha}\dot{\alpha} \\ &= \mathrm{sign}(\sigma)E(\dot{\varepsilon} - \dot{\varepsilon}^p) - K\dot{\alpha} \\ &= \mathrm{sign}(\sigma)E\dot{\varepsilon} - \gamma[E + K] \leq 0. \end{aligned} \qquad (1.2.31)$$

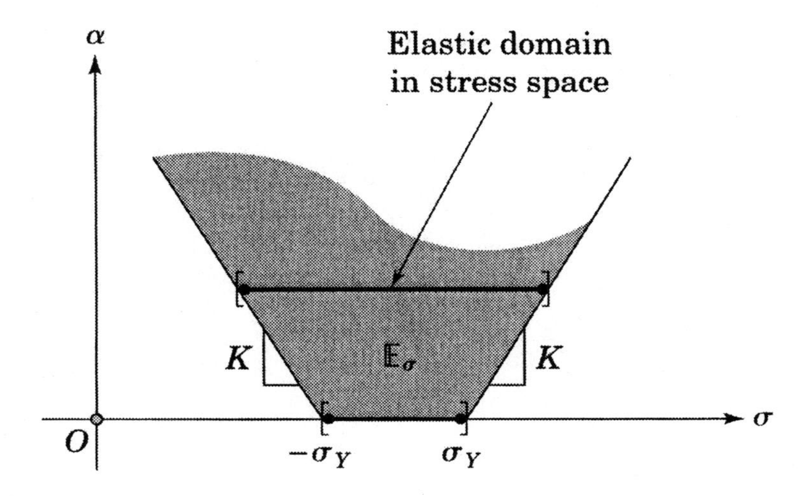

FIGURE 1.6. Elastic range and elastic domain in stress space. Note that $\alpha \geq 0$.

Observe once more that relationship $\dot{f} > 0$ cannot hold. From (1.2.26) and (1.2.27) it follows that γ can be nonzero *only* if

$$\boxed{f = \dot{f} = 0 \Rightarrow \gamma = \frac{\text{sign}(\sigma) E \dot{\varepsilon}}{E + K}.} \tag{1.2.32}$$

Then the rate form of the elastic relationship (1.2.22) along with (1.2.32) yields

$$\boxed{\dot{\sigma} = \begin{cases} E\dot{\varepsilon} & \text{if } \gamma = 0, \\ \dfrac{EK}{E+K}\dot{\varepsilon} & \text{if } \gamma > 0. \end{cases}} \tag{1.2.33}$$

The quantity $EK/(E + K)$ is called the *elastoplastic tangent modulus*. See Figure **1.7a** for an illustration. The interpretation of the plastic modulus is given in Figure **1.7b**. For convenience and subsequent reference, we summarize the constitutive model developed above in BOX **1.2**.

1.2.3 Alternative Form of the Loading/Unloading Conditions

The Kuhn–Tucker unilateral constraints provide the most convenient formulation of the loading/unloading conditions for classical plasticity. To motivate our subsequent algorithmic implementation, we describe a step-by-step procedure within the *strain-driven* format of algorithmic plasticity.

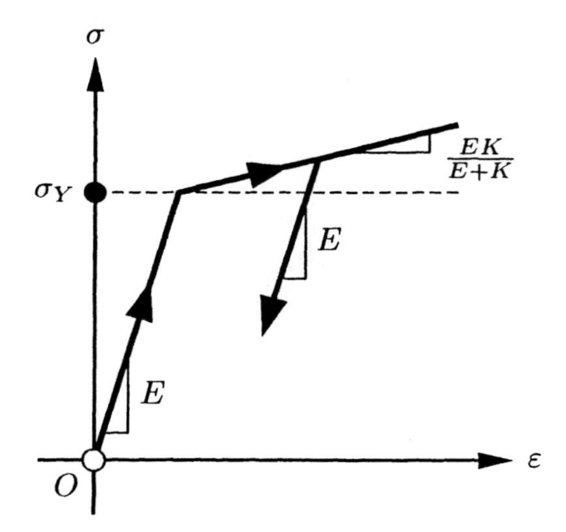

FIGURE 1.7A.

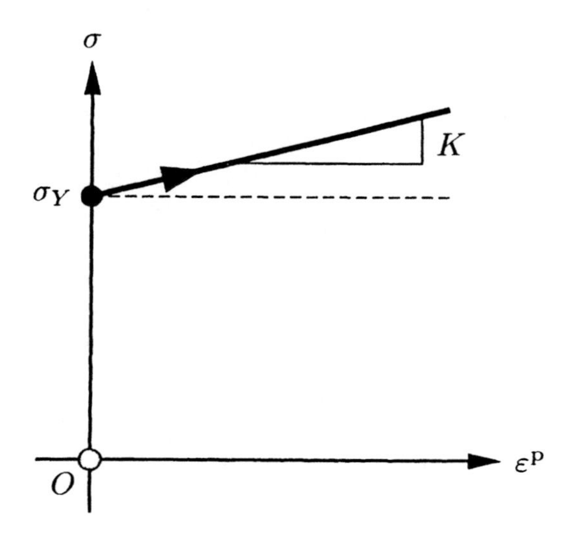

FIGURE 1.7B. **(a)** The tangent modulus and **(b)** the plastic modulus.

(a.) Suppose that we are given an admissible point $(\sigma, \alpha) \in \mathbb{E}$ in the elastic domain and prescribed strain rate $\dot{\varepsilon}$.

(b.) Define the *trial state* as the *rates* computed by freezing plastic flow as

$$\boxed{\begin{aligned} \dot{\sigma}^{\text{trial}} &= E\dot{\varepsilon} \\ \text{and} & \\ \dot{\alpha}^{\text{trial}} &= 0. \end{aligned}} \tag{1.2.34}$$

Note that $\dot{\alpha}^{\text{trial}}$ (and $\dot{\sigma}^{\text{trial}} = 0$) are fictitious rates which do not necessarily coincide with the final rates $\dot{\sigma}$ and $\dot{\alpha}$.

With this definition, since $\dot{\sigma} = E\dot{\varepsilon} - E\dot{\varepsilon}^p$, the final rates are given in terms of the trial state by the formulas

$$\boxed{\begin{aligned} \dot{\sigma} &= \dot{\sigma}^{\text{trial}} - \gamma E \partial f/\partial\sigma, \\ \dot{\alpha} &= \gamma. \end{aligned}} \tag{1.2.35}$$

Observe further that $\dot{\sigma}^{\text{trial}} = E\dot{\varepsilon}$ (and obviously $\dot{\alpha}^{\text{trial}} = 0$) can be computed directly in terms of the prescribed strain rate $\dot{\varepsilon}$. To decide whether $\gamma = 0$ or $\gamma > 0$, we appeal to the Kuhn–Tucker conditions, and proceed as follows:

First, if $f(\sigma, \alpha) < 0$, the condition $\gamma f(\sigma, \alpha) = 0$ immediately gives $\gamma = 0$, and (1.2.35) yields

$$f(\sigma, \alpha) < 0 \Longrightarrow \begin{cases} \dot{\sigma} = \dot{\sigma}^{\text{trial}} = E\dot{\varepsilon}, \\ \dot{\alpha} = \dot{\alpha}^{\text{trial}} = 0. \end{cases} \tag{1.2.36}$$

Thus, if $f(\sigma, \alpha) < 0$, the trial state *is* the actual state, and, as pointed out above, we speak of an instantaneous elastic process.

Second, we examine the case $f(\sigma, \alpha) = 0$. In this situation, the condition $\gamma f(\sigma, \alpha) = 0$ is inconclusive since we can have either $\gamma = 0$ or $\gamma > 0$. To obtain further information, we rewrite expression (1.2.32) $\dot{f}$ in terms of $\dot{\sigma}^{\text{trial}} = E\dot{\varepsilon}$ as

$$\dot{f} = \frac{\partial f}{\partial \sigma}\dot{\sigma}^{\text{trial}} - \gamma(E + K). \tag{1.2.37}$$

Now recall that $\dot{f}(\sigma, \alpha)$ cannot be positive, i.e., $\dot{f}(\sigma, \gamma) \leq 0$, and $\gamma \dot{f}(\sigma, \alpha) = 0$ (consistency condition). By exploiting these conditions, we show that one can immediately conclude whether $\gamma > 0$ or $\gamma = 0$. Remarkably, the criterion involves only the trial elastic stress rate $\dot{\sigma}^{\text{trial}}$.

i. First suppose that $[\partial f/\partial\sigma]\dot{\sigma}^{\text{trial}} > 0$. Then, we require that $\gamma > 0$. To see this we proceed by contradiction and note that if $\gamma = 0$, expression (1.2.37) would imply that $\dot{f} > 0$, which is not allowed. Consequently,

$$\begin{gathered} f(\sigma, \alpha) = 0 \\ \text{and} \\ [\partial f/\partial\sigma]\dot{\sigma}^{\text{trial}} > 0 \Longrightarrow \gamma > 0. \end{gathered} \tag{1.2.38}$$

We say that the process is *instantaneously plastic*. The consistency parameter $\gamma > 0$ is determined from the condition $\dot{f} = 0$ via equation (1.2.32), i.e.,

$$\gamma = \frac{[\partial f/\partial \sigma]\dot{\sigma}^{\text{trial}}}{E + K}. \tag{1.2.39}$$

Then the actual rates $(\dot{\sigma}, \dot{\alpha})$ are then computed from formulas (1.2.35) which gives a correction to the trial state. Thus $\dot{\sigma} \neq \dot{\sigma}^{\text{trial}}$ and $\dot{\alpha} \neq \dot{\alpha}^{\text{trial}}$.

ii. Next, suppose that $[\partial f/\partial \sigma]\dot{\sigma}^{\text{trial}} < 0$. Then, since $\gamma \geq 0$, expression (1.2.37) implies $\dot{f} < 0$ in any case (provided that $E + K > 0$, a condition assumed throughout). From the condition $\gamma \dot{f} = 0$, then we conclude that $\gamma = 0$. Consequently,

$$f(\sigma, \alpha) = 0$$

and (1.2.40)

$$[\partial f/\partial \sigma]\dot{\sigma}^{\text{trial}} < 0 \Longrightarrow \gamma = 0.$$

It follows that $\dot{\sigma} = \dot{\sigma}^{\text{trial}}$ and $\dot{\alpha} = \dot{\alpha}^{\text{trial}} = 0$, and we speak of an *instantaneous elastic unloading process*.

iii. Finally, suppose that $[\partial f/\partial \sigma]\dot{\sigma}^{\text{trial}} = 0$. Condition $\dot{f} < 0$ cannot hold since $\gamma \dot{f} = 0$ would give $\gamma = 0$, which when inserted in (1.2.37) would imply $\dot{f} = 0$, contradicting the hypothesis that $\dot{f} < 0$. Consequently, we require that $\dot{f} = 0$, and expression (1.2.39) gives $\gamma = 0$. Therefore,

$$\dot{f}(\sigma, \alpha) = 0$$

and (1.2.41)

$$[\partial f/\partial \sigma]\dot{\sigma}^{\text{trial}} = 0 \Longrightarrow \gamma = 0.$$

From (1.2.35) we conclude that $\dot{\sigma} = \dot{\alpha}^{\text{trial}}$ and $\dot{\alpha} = \dot{\sigma}^{\text{trial}} = 0$. In this situation, we speak of a process *of neutral loading*.

The preceding analysis shows that instantaneous loading or unloading in the system can be inferred solely in terms of the *trial state*, hence its usefulness. Then the most convenient form of the loading/unloading Kuhn–Tucker conditions take the following form:

Summary. Given $(\sigma, \alpha) \in \mathbb{E}$ and a *prescribed strain rate* $\dot{\varepsilon}$,

a. compute $\dot{\sigma}^{\text{trial}} = E\dot{\varepsilon}$ and $\dot{\alpha}^{\text{trial}} = 0$;

b. if either $f(\sigma, \alpha) < 0$ or $f(\sigma, \alpha) = 0$ and $[\partial f/\partial \sigma(\sigma, \alpha)]\dot{\sigma}^{\text{trial}} \leq 0$, then $\dot{\sigma} = \dot{\sigma}^{\text{trial}}$, and $\dot{\alpha} = \dot{\alpha}^{\text{trial}} = 0$ (*instantaneous elastic process*); and

c. if both $f(\sigma, \alpha) = 0$ and $[\partial f/\partial \sigma(\sigma, \alpha)]\dot{\sigma}^{\text{trial}} > 0$, then

$$\dot{\sigma} = \dot{\sigma}^{\text{trial}} - \gamma E \partial f(\sigma, \alpha)/\partial \sigma \quad \text{and} \quad \dot{\alpha} = \gamma, \tag{1.2.42}$$

where $\gamma = \frac{[\partial f/\partial \sigma]\dot{\sigma}^{\text{trial}}}{E+K} > 0$ (*instantaneous plastic process*).

Remarks 1.2.3.

1. The procedure summarized above motivates the implementation of the so-called *return mapping algorithms* described in detail in the sections that follow, where the rates will be replaced by finite increments.
2. Expression (1.2.35) possesses a compelling geometric interpretation, illustrated in Figure **1.7c** The actual stress rate $\dot{\sigma}$ can be viewed as the projection of the trial state rate $\dot{\sigma}^{\text{trial}}$ on the plane tangent to the yield surface.

1.2.4 Further Refinements of the Hardening Law

The isotropic hardening model discussed above is a gross simplification of the hardening behavior in real materials, in particular, metals. An alternative simple phenomenological mechanism, referred to as *kinematic hardening*, used alone or in conjunction with isotropic hardening, provides an improved means of representing hardening behavior in metals under cyclic loading. The basic phenomenological law is credited to Prager [1956] with further improvements of Ziegler [1959]; see Fung [1965, p.151] for a discussion. Within the present one-dimensional context, the model can be illustrated as follows.

1.2.4.1 Kinematic hardening law.

In many metals subjected to cyclic loading, it is experimentally observed that the center of the yield surface experiences a motion in the direction of the plastic flow. Figure **1.8** gives an idealized illustration of this hardening behavior closely related to a phenomenon known as the Bauschinger effect.

A simple phenomenological model that captures the aforementioned effect is constructed by introducing an additional internal variable, denoted by q and called *back stress*, which defines the location of the center of the yield surface. Then the yield condition is modified to the form

$$\boxed{f(\sigma, q, \alpha) := |\sigma - q| - [\sigma_Y + K\alpha] \leq 0.} \tag{1.2.43}$$

The evolution of the back stress q is defined by Ziegler's rule as

$$\boxed{\dot{q} = H\dot{\varepsilon}^p \equiv \gamma H \operatorname{sign}(\sigma - q),} \tag{1.2.44}$$

where H is called the kinematic hardening modulus. Finally, the evolution of α remains unchanged and given by (1.2.24). The addition of the Kuhn–Tucker conditions of the form (1.2.26) along with a consistency condition analogous to (1.2.27)

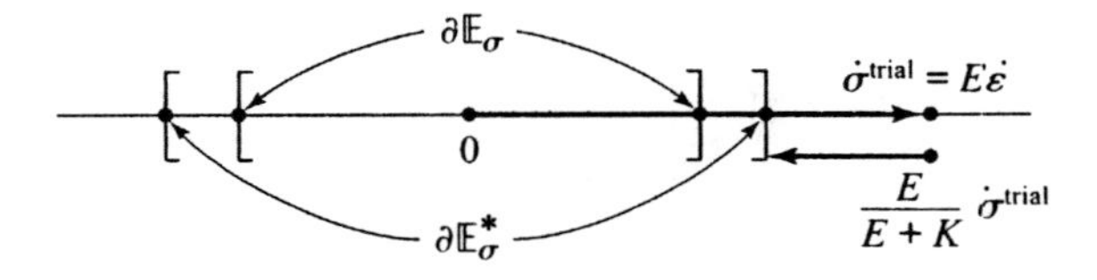

FIGURE 1.7C. Illustration of the trial state correction leading to the final stress rate.

completes the formulation of the model which is summarized for convenience in BOX **1.3**.

1.2.4.2 Tangent elastoplastic modulus.

The computation of the consistency parameter and the elastoplastic tangent modulus proceeds along the same lines discussed in Section **1.2.2.2**. First, one computes the time derivative of the yield criterion by using the chain rule along with the the stress-strain relationship (1.2.22), the flow rule (1.2.25), and the hardening laws (1.2.24) and (1.2.44). Accordingly, from (1.2.43) one finds that

$$\begin{aligned} \dot{f} &= \frac{\partial f}{\partial \sigma}\dot{\sigma} + \frac{\partial f}{\partial q}\dot{q} + \frac{\partial f}{\partial \alpha}\dot{\alpha} \\ &= \operatorname{sign}(\sigma - q)[E(\dot{\varepsilon} - \dot{\varepsilon}^p) - \dot{q}] - K\dot{\alpha} \\ &= \operatorname{sign}(\sigma - q)E\dot{\varepsilon} - \gamma[E + (H + K)] \leq 0. \end{aligned} \tag{1.2.45}$$

Again we recall that the relationship $\dot{f} > 0$ cannot hold in rate-independent plasticity. On the other hand, if γ is nonzero, the Kuhn–Tucker conditions along with the consistency condition require that $f = 0$ and $\dot{f} = 0$. Then the latter

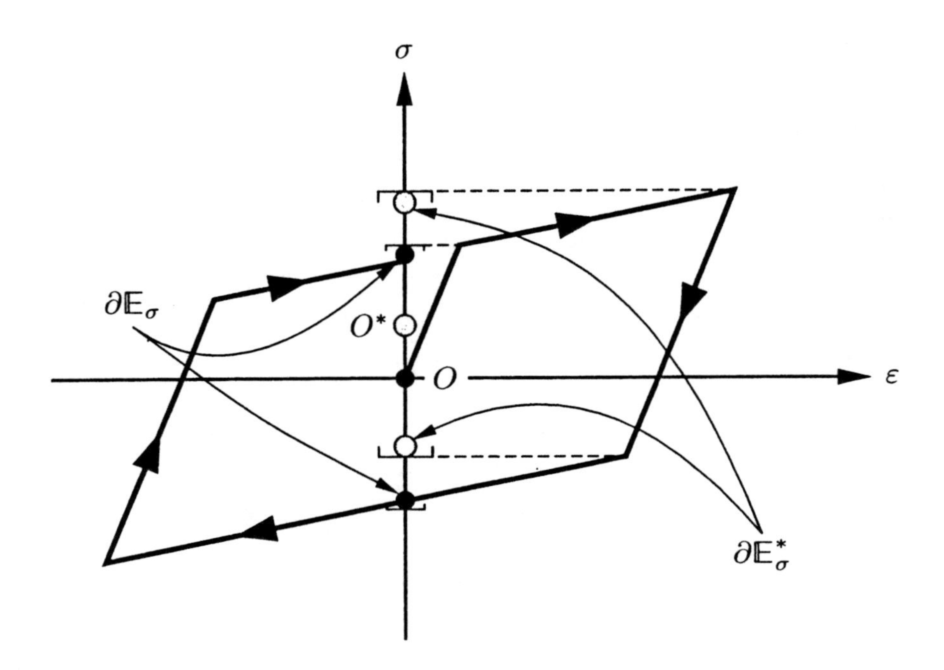

FIGURE 1.8. Idealized illustration of kinematic hardening behavior.

requirement and (1.2.45) yield

$$\boxed{\gamma = \frac{\text{sign}(\sigma - q)E\dot{\varepsilon}}{E + [H + K]}\,.} \tag{1.2.46}$$

Since $\dot{\sigma} = \left[d\sigma/d\varepsilon\right]\dot{\varepsilon}$, the elastic relationship (1.2.22) together with the flow rule (1.2.25) and (1.2.46) result in the expression

$$\boxed{\frac{d\sigma}{d\varepsilon} = \begin{cases} E & \text{if } \gamma = 0\,, \\ \dfrac{E[H + K]}{E + [H + K]} & \text{if } \gamma > 0, \end{cases}} \tag{1.2.47}$$

which defines the elastoplastic tangent modulus for combined isotropic/kinematic hardening.

BOX 1.3. One-Dimensional, Rate-Independent Plasticity. Combined Kinematic and Isotropic Hardening.

i. Elastic stress-strain relationship

$$\sigma = E\left(\varepsilon - \varepsilon^p\right).$$

ii. Flow rule

$$\dot{\varepsilon}^p = \gamma \,\text{sign}(\sigma - q).$$

iii. Isotropic and kinematic hardening laws

$$\dot{q} = \gamma\, H \,\text{sign}(\sigma - q),$$
$$\dot{\alpha} = \gamma.$$

iv. Yield condition and closure of the elastic range

$$f\left(\sigma, q, \alpha\right) := \left|\sigma - q\right| - \left[\sigma_Y + K\alpha\right] \leq 0,$$
$$\mathbb{E}_\sigma = \{\left(\sigma, q, \alpha\right) \in \mathbb{R} \times \mathbb{R}_+ \times \mathbb{R} \mid f\left(\sigma, q, \alpha\right) \leq 0\}.$$

v. Kuhn–Tucker complementarity conditions

$$\gamma \geq 0, \quad f\left(\sigma, q, \alpha\right) \leq 0, \quad \gamma f\left(\sigma, q, \alpha\right) = 0.$$

vi. Consistency condition

$$\gamma \dot{f}\left(\sigma, q, \alpha\right) = 0 \quad (\text{if} \quad f(\sigma, q, \alpha) = 0).$$

1.2.5 Geometric Properties of the Elastic Domain

All the examples of yield criteria described in the preceding sections incorporate two geometric properties which play an important role in the mathematical analysis of classical plasticity and in the numerical analysis of the algorithms described below. These two properties are **(i)** *convexity* of the yield surface and hence of the elastic domain and **(ii)** *degree-one homogeneity* of the yield criterion, in the following sense:

1.2.5.1 Convexity.

A closed subset $\Omega \subset \mathbb{R}^N$ ($N \geq 1$) is said to be convex if the closed segment connecting two arbitrary points in Ω lies entirely within the set Ω. This definition corresponds to our intuitive notion of convexity and is concisely expressed mathematically as the condition

$$\xi \boldsymbol{x}_1 + (1-\xi)\boldsymbol{x}_2 \in \Omega \quad \text{for any } \xi \in [0,1] \quad \text{if } \boldsymbol{x}_1, \boldsymbol{x}_2 \in \Omega. \tag{1.2.48}$$

Similarly, a function $f(\cdot)$ with domain $\Omega \subset \mathbb{R}^N$ is said to be convex if for any two point $\boldsymbol{x}_1, \boldsymbol{x}_2 \in \Omega$ the following property holds

$$f\big(\xi \boldsymbol{x}_1 + (1-\xi)\boldsymbol{x}_2\big) \leq \xi f(\boldsymbol{x}_1) + (1-\xi)\, f(\boldsymbol{x}_2), \quad \text{for all } \xi \in [0,1]. \tag{1.2.49}$$

One can easily use the preceding definitions to show that both the yield criterion and the elastic domain are convex for the models of plasticity described above. Let $\mathbb{E}_\sigma \subset \mathbb{R}^3$ be an elastic domain for plasticity with combined isotropic/kinematic hardening defined, therefore, as

$$\boxed{\mathbb{E}_\sigma = \{\boldsymbol{x} := (\sigma, q, \alpha) \in \mathbb{R} \times \mathbb{R} \times \mathbb{R}_+ \mid f(\boldsymbol{x}) \leq 0\},} \tag{1.2.50}$$

where $f(\boldsymbol{x}) := |\sigma - q| - K\alpha - \sigma_Y$ is the yield criterion defined by (1.2.43) and for simplicity we have used the notation $\boldsymbol{x} = (\sigma, q, \alpha)$. First, we use definition (1.2.49) for the problem at hand to show that the yield condition is convex. Consider two arbitrary points $\boldsymbol{x}_1, \boldsymbol{x}_2 \in \mathbb{R} \times \mathbb{R} \times \mathbb{R}_+$, use the elementary inequality $|a+b| \leq |a| + |b|$, and note that both $\xi \geq 0$ and $(1-\xi) \geq 0$ if $\xi \in [0,1]$ to conclude that

$$\begin{aligned} f(\xi \boldsymbol{x}_1 + (1-\xi)\boldsymbol{x}_2) :&= \big|\, \xi[\sigma_1 - q_1] + (1-\xi)[\sigma_2 - q_2] \,\big| \\ &\quad - K\left[\, \xi\alpha_1 + (1-\xi)\alpha_2 \,\right] - \sigma_Y \\ &\leq \xi \left[|\sigma_1 - q_1| - K\alpha_1 - \sigma_Y\right] \\ &\quad + (1-\xi)\big[|\sigma_2 - q_2| - K\alpha_2 - \sigma_Y\big] \\ &= \xi f(\boldsymbol{x}_1) + (1-\xi) f(\boldsymbol{x}_2), \end{aligned} \tag{1.2.51}$$

i.e., convexity of $f(\boldsymbol{x})$. Note that we have set $\sigma_Y \equiv \xi\sigma_Y + (1-\xi)\sigma_Y$. The convexity of the elastic domain $\mathbb{E}_\sigma$ follows immediately from (1.2.51) since $f(\boldsymbol{x}_1) \leq 0$ and $f(\boldsymbol{x}_2) \leq 0$ if $\boldsymbol{x}_1, \boldsymbol{x}_2 \in \mathbb{E}_\sigma$, so that

$$f\left[\xi \boldsymbol{x}_1 + (1-\xi)\boldsymbol{x}_2\right] \leq \xi f(\boldsymbol{x}_1) + (1-\xi) f(\boldsymbol{x}_2) \leq 0 \quad \text{for } \xi \in [0,1]. \tag{1.2.52}$$

Consequently, $\xi \boldsymbol{x}_1 + (1-\xi)\boldsymbol{x}_2 \in \mathbb{E}_\sigma$. Note that the convexity property holds for both $K \geq 0$ (hardening) and $K < 0$ (softening). A characterization of convexity equivalent to (1.2.49) in the case where $f(\boldsymbol{x})$ is *smooth* is given by the condition

$$f(\boldsymbol{x}_2) - f(\boldsymbol{x}_1) \geq \nabla f(\boldsymbol{x}_1) \cdot (\boldsymbol{x}_2 - \boldsymbol{x}_1) \qquad \text{for all } \boldsymbol{x}_1, \boldsymbol{x}_2 \in \Omega, \tag{1.2.53}$$

which is easily verified by a direct calculation (see Chapter **2**). Convexity of the elastic domain is a classical property that will be postulated at the outset in formulating the general three-dimensional constitutive models described in the following chapters.

1.2.5.2 Degree-one homogeneity.

A function $\phi(\cdot)$ defined on $\Omega \subset \mathbb{R}^N$ $(N \geq 1)$ will be said to be *homogeneous of degree m* if the following condition holds

$$\phi(\xi \boldsymbol{x}) = \xi^m \, \phi(\boldsymbol{x}) \quad \text{for any } \xi > 0 \quad \text{and } \boldsymbol{x} \in \Omega. \tag{1.2.54}$$

The interest in this property lies in the following result that will be exploited in our subsequent analysis:

$$\nabla\phi(\boldsymbol{x}) \cdot \boldsymbol{x} := \sum_{i=1}^{N} \frac{\partial \phi(x_i)}{\partial x_i} x_i = \phi(\boldsymbol{x}). \tag{1.2.55}$$

This result is a direct consequence of a classical theorem going back to Euler, which is easily verified by differentiating both sides of the defining condition (1.2.54) with respect to ξ and particularizing the result at $\xi = 1$. For one-dimensional classical plasticity the yield criterion for combined isotropic-kinematic hardening is written as

$$f(\sigma, q, \alpha) = \phi(\sigma, q, \alpha) - \sigma_Y$$

where (1.2.56)

$$\phi(\sigma, q, \alpha) := |\sigma - q| - K\alpha.$$

By inspection, we immediately conclude that

$$\phi(\xi\sigma, \xi q, \xi\alpha) = \xi\phi(\sigma, q, \alpha) \quad \text{for any } \xi > 0. \tag{1.2.57}$$

Therefore, the part $\phi(\sigma, q, \alpha)$ of the yield criterion is a (convex) function *homogeneous of degree one*.

1.3 The Initial Boundary-Value Problem

To set the stage for our general algorithmic treatment of elastoplasticity and viscoplasticity, in this section we outline the basic structure of the initial boundary-value problem within the framework of a one-dimensional problem. First, we summarize the strong or local form and the weak or variational form of the momentum equation. These are basic principles of mechanics which hold with independence of the specific form adopted for the constitutive model. Next, we provide a precise notion of *dissipativity* in a continuum system and derive an apriori stability estimate based on the key assumption of *positive internal dissipation*. Then we apply these ideas to the elastoplastic problem, with constitutive equations summarized in Boxes **1.1** to **1.3** for several representative examples of the hardening law and arrive at the expression for the dissipation of the system.

In addition, we examine key properties of the solution to the IBVP for elastoplasticity, such as uniqueness and contractivity, along with an apriori energy-decay estimate that arises as a result of the property of positive internal dissipation in the mechanical system. As will be shown subsequently, these properties play a central rule in the stability analysis of the numerical algorithms described in the following sections. Finally, we remark that the structure of these algorithms is motivated by and tailored to the specific form taken by the weak formulation of the momentum equation described below.

1.3.1 The Local Form of the IBVP

We consider a one-dimensional body occupying an interval $\bar{\mathcal{B}} = [0, L]$, with particles labeled by their position $x \in \bar{\mathcal{B}}$. We restrict our attention to an interval of time $[0, T]$. Then the *displacement field*, is a mapping

$$u : \bar{\mathcal{B}} \times [0, T] \to \mathbb{R}. \tag{1.3.1}$$

We denote by $u(x, t)$ the displacement at $x \in \bar{\mathcal{B}}$ and at time $t \in [0, T]$. We let

$$\varepsilon(x, t) := \frac{\partial u(x, t)}{\partial x}$$

and (1.3.2)

$$v(x, t) := \frac{\partial u(x, t)}{\partial t}$$

be the *strain* and *velocity* fields at $(x, t) \in \bar{\mathcal{B}} \times [0, T]$ and denote by $\sigma(x, t)$ the stress field. In this illustrative discussion, we assume that all of the fields involved are as smooth as needed.

The boundary of $\bar{\mathcal{B}}$, which is denoted by $\partial\mathcal{B}$, consists of its two end points. We set

$$\mathcal{B} =]0, L[,$$
$$\partial\mathcal{B} = \{0, L\}, \tag{1.3.3a}$$

and

$$\bar{\mathcal{B}} = \partial\mathcal{B} \cup \mathcal{B}.$$

Typically, one considers *boundary conditions* of the form

$$u\big|_{\partial_u \mathcal{B}} = \bar{u} \quad \text{(prescribed)}, \tag{1.3.3b}$$

and

$$\sigma\big|_{\partial_\sigma \mathcal{B}} = \bar{\sigma} \quad \text{(prescribed)}, \tag{1.3.3c}$$

where $\overline{\partial_\sigma \mathcal{B} \cup \partial_u \mathcal{B}} = \overline{\partial \mathcal{B}}$ and $\partial_\sigma \mathcal{B} \cap \partial_u \mathcal{B} = \emptyset$. An example of conditions (1.3.3) is $\partial_u \mathcal{B} = \{0\}$, $\partial_\sigma \mathcal{B} = \{L\}$,

$$u(0, t) = 0,$$

and

$$\sigma(L, t) = 0,$$

for all $t \in [0, T]$. In addition, for dynamic problems one specifies *initial conditions* of the form

$$u(x, 0) = u_0(x)$$

and (1.3.4)

$$v(x, 0) = v_0(x)$$

in $\bar{\mathcal{B}}$. Of course, (1.3.3b) and (1.3.4) must be compatible in the sense that

$$u_0\big|_{\partial_u \mathcal{B}} = \bar{u}\big|_{t=0}$$

and (1.3.5)

$$v_0\big|_{\partial_u \mathcal{B}} = \frac{\partial}{\partial t}\bar{u}\big|_{t=0}.$$

Finally, one prescribes loading in $\mathcal{B}$ by body forces defined by the function $b : \mathcal{B} \times [0, T] \to \mathbb{R}$.

1.3.1.1 Local momentum equation.

The balance of momentum localized about any point in the one-dimensional body yields the equation

$$\boxed{\frac{\partial}{\partial x}\sigma + \rho b = \rho \frac{\partial v}{\partial t} \quad \text{in} \quad \mathcal{B}\times]0, T[,} \tag{1.3.6}$$

where $\rho : \mathcal{B} \to \mathbb{R}$ is the density of the bar. This equation, along with boundary conditions (1.3.3) and initial conditions (1.3.4)–(1.3.5), defines an *initial boundary-value problem* provided that the stress, $\sigma(x, t)$, is a *known function of the displacement field* (through the strains). Two cases are of interest:

i. The simplest situation is afforded by a *linear elastic* body for which the stress field is defined by the equation

$$\sigma(x, t) = E\varepsilon(x, t). \tag{1.3.7}$$

By virtue of (1.3.2) and (1.3.7), the balance of momentum equation (1.3.6) then reduces to the wave equation in one dimension; i.e., a one-dimensional symmetric hyperbolic linear system of conservation laws.

ii. Here, on the other hand, we shall be concerned where $\sigma(x, t)$ is defined *locally* at each $(x, t) \in \bar{\mathcal{B}} \times [0, T]$ by an inelastic constitutive model, often nonlinear. The example to keep in mind is the *elastoplastic rate-independent models* summarized in Boxes **1.1–1.3**. Then note that (1.3.7) is replaced by the *incremental*, highly nonlinear, *rate form* of (1.2.47), i.e.,

$$\dot{\sigma} = \begin{cases} \dfrac{E[K + H]}{E + [K + H]}\dot{\varepsilon} & \text{iff } f = 0,\ \dot{f} = 0 \text{ at } (x, t) \in \mathcal{B} \times [0, \mathrm{T}], \\ E\dot{\varepsilon} & \text{otherwise.} \end{cases} \tag{1.3.8}$$

Observe that this equation defines only the stress rate (in terms of the strain rate). The problem is nonlinear and highly nontrivial for two reasons: (**i**) integrating

(1.3.8) requires careful consideration of the loading/unloading conditions to decide which tangent modulus applies and **(ii)** the stress (and the internal variables) are subject to an additional constraint defined by the yield criterion. Our ultimate objective is to investigate in detail the numerical solution of the resulting nonlinear initial boundary-value problem.

1.3.2 The Weak Formulation of the IBVP

The local form of the IBVP discussed above is not well suited to a numerical solution by finite-element methods. For this latter purpose it proves more convenient to consider the *weak form* of the IBVP (virtual power principle). In what follows, we shall recall the main ideas needed. (See e.g., Johnson [1987, Chapter **1**], or Hughes [1987, Chapter **1**] for a detailed elaboration).

We denote by $\mathbb{S}_t$ the displacement solution space at time $t \in [0, \mathrm{T}]$ defined as

$$\mathbb{S}_t = \left\{ u(\cdot, t) : \mathcal{B} \to \mathbb{R} \,\middle|\, u(\cdot, t)\big|_{\partial_u \mathcal{B}} = \bar{u}(\cdot, t) \right\}. \tag{1.3.9}$$

(For *hardening* plasticity one takes $\mathbb{S}_t \subset \mathbb{H}^1(\mathcal{B})$ for fixed t, where $\mathbb{H}^1(\mathcal{B})$ denotes the Sobolev space of functions possessing square integrable derivatives. Although important, these technical considerations play no role in our subsequent developments. For present purposes it suffices to note that $\mathbb{H}^1(\mathcal{B})$ contains, in particular, continuous functions (as a direct consequence of the so-called Sobolev embedding theorem). Similarly, associated with $\mathbb{S}_t$ one defines the linear space $\mathbb{V}$ of admissible test functions or kinematically admissible variations, i.e., (virtual) displacements satisfying the homogeneous form of the essential boundary condition (1.3.3b), as

$$\mathbb{V} = \left\{ \eta : \mathcal{B} \to \mathbb{R} \,\middle|\, \eta\big|_{\partial_u \mathcal{B}} = 0 \right\}. \tag{1.3.10}$$

Again, for *hardening* plasticity one takes $\mathbb{V} \subset \mathbb{H}^1(\mathcal{B})$. With these notations in hand, the weak form of the equilibrium equations then reads as follows:

Find the displacement field $u(\cdot, t) \in \mathbb{S}_t$ such that:

$$\int_{\mathcal{B}} \rho \frac{\partial v}{\partial t} \eta \, dx + G(\sigma, \eta) = 0 \text{ for all } \eta \in \mathbb{V} \text{ and all } t \in [0, \mathrm{T}],$$

$$\text{where } G(\sigma, \eta) := \int_{\mathcal{B}} \sigma \, \eta' \, dx - \int_{\mathcal{B}} \rho b \, \eta \, dx - \bar{\sigma} \, \eta\big|_{\partial_\sigma \mathcal{B}} \text{ and } \eta' = \frac{\partial \eta}{\partial x}. \tag{1.3.11}$$

Here, $\sigma(x, t)$ is assumed to satisfy the *local constitutive equations* in BOX **1.3**, and hence it is a function of $u(x, t)$ through the strain $[\varepsilon(x, t) = \frac{\partial u(x,t)}{\partial x}]$ and the internal variables $[\varepsilon^p(x, t), \alpha(x, t), q(x, t)]$. A justification of (1.3.11) is given by elementary considerations as follows.

1.3.2.1 Formal proof of (1.3.11).

First assume that the local momentum equations and boundary conditions hold, that is, the strong form of the IBVP given by

$$\left.\begin{aligned} \rho\frac{\partial v}{\partial t} - \frac{\partial \sigma}{\partial x} - \rho b = 0 \quad &\text{in} \quad \mathcal{B}\times]0,T[,\\ u = \bar{u} \quad &\text{on} \quad \partial_u\mathcal{B}\times]0,T[,\\ \sigma = \bar{\sigma} \quad &\text{on} \quad \partial_\sigma\mathcal{B}\times]0,T[. \end{aligned}\right\} \tag{1.3.12}$$

Now let $\eta \in \mathbb{V}$ be an arbitrary test function, and recall that $\eta|_{\partial_u\mathcal{B}} = 0$ by construction. First, from (1.3.12) we note the relationship

$$\begin{aligned} [\sigma\eta]|_{\partial\mathcal{B}} &= [\sigma\eta]|_{\partial_u\mathcal{B}\cup\partial_\sigma\mathcal{B}}\\ &= [\sigma\eta]|_{\partial_u\mathcal{B}} + [\sigma\eta]|_{\partial_\sigma\mathcal{B}}\\ &= [\sigma\eta]|_{\partial_\sigma\mathcal{B}} = \bar{\sigma}\eta|_{\partial_\sigma\mathcal{B}}. \end{aligned} \tag{1.3.13}$$

Next, multiply $(1.3.12)_1$ by η, and integrate by parts, using the result in (1.3.13), to obtain

$$\begin{aligned} 0 &= \int_{\mathcal{B}}\left(\rho\frac{\partial v}{\partial t}\,\eta - \frac{\partial\sigma}{\partial x}\,\eta - \rho b\,\eta\right)\,dx\\ &= \int_{\mathcal{B}}\left(\rho\frac{\partial v}{\partial t}\,\eta + \sigma\eta' - \rho b\,\eta\right)\,dx - \sigma\eta|_{\partial\mathcal{B}}\\ &= \int_{\mathcal{B}}\left(\rho\frac{\partial v}{\partial t}\,\eta + \sigma\eta' - \rho b\,\eta\right)\,dx - \bar{\sigma}\eta|_{\partial_\sigma\mathcal{B}}\\ &= \int_{\mathcal{B}}\rho\frac{\partial v}{\partial t}\,\eta\,dx + G(\sigma,\eta). \end{aligned} \tag{1.3.14}$$

Therefore, we conclude that (1.3.12) implies the weak form (1.3.14). Conversely, assume that (1.3.11) holds and that σ is smooth. Then, integration by parts of (1.3.11) along with the fact that $\eta|_{\partial_u\mathcal{B}} = 0$ yields

$$\int_{\mathcal{B}}\rho\frac{\partial v}{\partial t}\,\eta\,dx + G(\sigma,\eta) = \int_{\mathcal{B}}\left(\rho\frac{\partial v}{\partial t} - \frac{\partial\sigma}{\partial x} - \rho b\right)\,\eta\,dx + [\sigma - \bar{\sigma}]\eta|_{\partial_\sigma\mathcal{B}} = 0. \tag{1.3.15}$$

Since $\eta \in \mathbb{V}$ can be chosen arbitrarily and $[\rho\partial v/\partial t - \partial\sigma/\partial x - \rho b]$ is assumed continuous, then a standard argument in the calculus of variations implies $(1.3.12)_1$ and $(1.3.12)_3$. Note that boundary condition $(1.3.12)_2$ is enforced at the outset and for this reason is often called an *essential boundary condition*. □

A weak formulation of the constitutive equation for elastoplasticity in either BOX **1.1**, BOX **1.2** or BOX **1.3** is also possible; see e.g., Johnson [1976b,1977] or Simo, Kennedy, and Taylor [1988].

1.3.2.2 The mechanical work identity.

As already pointed out, the weak form (1.3.15) of the momentum equations is independent of the form adopted for the constitutive equations. By specialization of the weak form via a specific choice of test function, we obtain a basic result known as the mechanical work identity.

Consider the case in which the essential boundary conditions are time-independent; i.e., $\partial \bar{u}/\partial t = 0$ on $\partial_u \mathcal{B}$. Under this assumption, for fixed but otherwise arbitrary time $t \in]0, T[$, the velocity field $v(x, t)$ is an admissible test function, i.e., $v(\cdot, t) \in \mathbb{V}$. Setting $\eta(\cdot) = v(\cdot, t)$ in (1.3.15) and making use of the elementary identity

$$\rho \, \frac{\partial v(\cdot, t)}{\partial t} \, v(\cdot, t) = \frac{\partial}{\partial t} \left[\tfrac{1}{2} \, \rho \, |v(\cdot, t)|^2 \right], \tag{1.3.16}$$

yields the following fundamental result:

$$\boxed{\frac{d}{dt} T(v) + P_{\text{int}}(\sigma, v) = P_{\text{ext}}(v) \quad \text{for all} \quad t \in [0, \mathrm{T}],} \tag{1.3.17}$$

where

$$\left.\begin{aligned} T(v) &= \tfrac{1}{2} \int_{\mathcal{B}} \rho \, |v|^2 \, dx \geq 0 && \text{kinetic energy,} \\ P_{\text{int}}(\sigma, v) &= \int_{\mathcal{B}} \sigma \, \frac{\partial v}{\partial x} \, dx && \text{stress power,} \\ \text{and} \qquad P_{\text{ext}}(v) &= \int_{\mathcal{B}} \rho \, b \, v \, dx + \bar{\sigma} \, v\big|_{\partial_\sigma \mathcal{B}} && \text{external power.} \end{aligned}\right\} \tag{1.3.18}$$

Once more, it should be emphasized that the preceding result is independent of the specific form taken by the constitutive equations.

1.3.3 Dissipation. A priori Stability Estimate

We combine the mechanical work identity (1.3.17) and the constitutive equations for classical elastoplasticity, developed in the preceding sections, to arrive at a basic a priori estimate for the elastoplastic IBVP. The importance of this a priori estimate is twofold: **(i)** it provides a natural notion of the dissipative nature of the IBVP and **(ii)** it gives an energy-decay estimate which is used in the nonlinear stability analysis of the algorithms described below.

To begin with, we introduce the notions of *internal energy* and *dissipation* within the specific context of the one-dimensional models of elastoplasticity already discussed. These notions are more general and can be motivated by inspecting the rheological models described above.

For simplicity, we restrict our attention to the model of *linear isotropic hardening plasticity* summarized in BOX **1.2.** and assume dead loading throughout, so that the body force $b(\cdot)$ and the applied traction $\bar{\sigma}$ are time-independent. Then we define the internal energy of the system, denoted by $V_{\text{int}}(\varepsilon, \alpha)$ with $\varepsilon^e := \partial u/\partial x - \varepsilon^p$, and the potential energy of the loading $V_{ext}(u)$ as

$$\left.\begin{aligned} V_{int}(\varepsilon^e, \alpha) &:= \int_{\mathcal{B}} \left(\tfrac{1}{2} E\varepsilon^{e2} + \tfrac{1}{2} K\alpha^2\right) dx \geq 0 \quad (\text{if } K \geq 0 \text{ and } E > 0), \\ V_{ext}(u) &:= -\int_{\mathcal{B}} \rho\, bu\, dx - \bar{\sigma} u\big|_{\partial_\sigma \mathcal{B}} \Rightarrow \frac{dV_{ext}(u)}{dt} = -P_{ext}(v). \end{aligned}\right\} \tag{1.3.19}$$

The quadratic form V_{int} measures the energy stored in the material as a result of deformation, and the linear form $V_{ext}(u)$ measures the potential energy of the applied loads since $dV_{ext}(u)/dt = -P_{ext}(v)$. For the rheological models described above, it is easily argued that V_{int} in fact corresponds to the elastic energy stored in the springs as a result of deformation in the device.

Next, we consider the difference between the (internal) stress power $P_{int}(\sigma, v)$ and the time rate of change of the internal energy, a quantity denoted by $\mathcal{D}_{mech}$. According to a basic principle of mechanics (the Clausius–Duhem version of the second law, see Truesdell and Noll [1965]), $\mathcal{D}_{mech}$ cannot be negative. Therefore,

$$\boxed{\mathcal{D}_{mech} := P_{int}(\sigma, v) - \frac{d}{dt} V_{int}(\varepsilon^e, \alpha) \geq 0 \quad \text{for all } t \in [0, \mathrm{T}].} \tag{1.3.20}$$

We refer to $\mathcal{D}_{mech}$ as the (instantaneous) *mechanical dissipation* in the one-dimensional body $\mathcal{B}$ at time $t \in [0, \mathrm{T}]$, and justify inequality (1.3.20) via simple examples.

i. *Elastic materials.* The stress response is characterized by the constitutive equation $\sigma = E\, \partial u/\partial x$, and the internal energy is given by (1.3.19) with $K \equiv 0$ and $\varepsilon^p \equiv 0$ so that $\varepsilon^e = \varepsilon := \partial u/\partial x$. Consequently, since

$$\frac{d}{dt} V_{int}(\varepsilon) = \int_{\mathcal{B}} E\varepsilon \frac{\partial \varepsilon}{\partial t}\, dx = \int_{\mathcal{B}} \sigma \frac{\partial v}{\partial x}\, dx = P_{int}(\sigma, v), \tag{1.3.21}$$

it follows that $\mathcal{D}_{mech} \equiv 0$ for an elastic material. We remark that no heat conduction effects are considered, otherwise dissipation arises as a result of heat conduction; see Truesdell and Noll [1965].

ii. *Elastoplastic material with isotropic hardening.* The stress response is governed by the constitutive equations in BOX **1.2**, and the internal energy is defined by (1.3.19) with $K \equiv 0$. Since $\sigma = E(\partial u/\partial x - \varepsilon^p)$ and $\partial\varepsilon/\partial t = \partial v/\partial x$, expression (1.3.20) becomes

$$\mathcal{D}_{mech} = \int_{\mathcal{B}} \left[\sigma\, \dot{\varepsilon}^p - K\alpha\, \dot{\alpha}\right] dx. \tag{1.3.22}$$

Inserting the flow rule and hardening law into (1.3.22) and noting that $\sigma\, \mathrm{sign}(\sigma) = |\sigma|$ gives

$$\mathcal{D}_{mech} = \int_{\mathcal{B}} \gamma\, \left[|\sigma| - K\alpha - \sigma_Y + \sigma_Y\right] dx = \int_{\mathcal{B}} \left[\gamma\, f(\sigma, \alpha) + \gamma\, \sigma_Y\right] dx. \tag{1.3.23}$$

Finally, since $\gamma \geq 0$ and $\gamma\, f(\sigma, \alpha) = 0$ as a result of the Kuhn–Tucker conditions, expression (1.3.23) collapses to

$$\mathcal{D}_{\text{mech}} = \int_B \gamma\, \sigma_Y\, dx \geq 0 \quad \text{for all} \quad t \in [0, \mathrm{T}], \tag{1.3.24}$$

which shows that the model of isotropic hardening plasticity conforms to the dissipation inequality (1.3.20). The interpretation of this result within the context of the preceding rheological models is clear. $\mathcal{D}_{\text{mech}}$ is the product of the flow stress times the slip rate in the device, as expected. Note that for $\gamma = 0$ (i.e., instantaneous elastic response) expression (1.3.24) gives $\mathcal{D}_{\text{mech}} = 0$, in agreement with the result obtained for an elastic material. A similar analysis of the model of combined isotropic/kinematic hardening again yields the same result (1.3.24) for a suitably modified internal energy function.

1.3.3.1 The a priori energy-decay estimate.

Accepting inequality (1.3.20) as a basic principle, which certainly holds for the constitutive models described in this monograph, the mechanical work identity (1.3.17) along with (1.3.20) yield the following estimate. Define the function $L(u, v, \varepsilon^e, \alpha)$ by the expression

$$\boxed{L(u, v, \varepsilon^e, \alpha) := \underbrace{V_{\text{ext}}(u) + V_{\text{int}}(\varepsilon^e, \alpha)}_{\text{Potential}} + \underbrace{T(v)}_{\text{Kinetic}}.} \tag{1.3.25}$$

Thus, $L(u, v, \varepsilon^e, \alpha)$ is the sum of the potential energy of the external loading, the internal energy of the system, and the kinetic energy. To compute its rate of change, we use the mechanical work identity (1.3.17) to obtain

$$\begin{aligned} \frac{d}{dt} L(u, v, \varepsilon^e, \alpha) &= \frac{d}{dt}\left[T(v) + V_{\text{ext}}(u)\right] + \frac{d}{dt} V_{\text{int}}(\varepsilon^e, \alpha) \\ &= \left[\frac{d}{dt} T(v) - P_{\text{ext}}(v)\right] + \frac{d}{dt} V_{\text{int}}(\varepsilon^e, \alpha) \\ &= -P_{\text{int}}(\sigma, v) + \frac{d}{dt} V_{\text{int}}(\varepsilon^e, \alpha). \end{aligned} \tag{1.3.26}$$

In view of definition (1.3.20) for the mechanical dissipation, we conclude that

$$\boxed{\frac{d}{dt} L(u, v, \varepsilon^e, \alpha) = -\mathcal{D}_{\text{mech}} \leq 0 \quad \text{for all} \quad t \in [0, \mathrm{T}].} \tag{1.3.27}$$

Therefore, $L(u, v, \varepsilon^e, \alpha)$ is a *nonincreasing function* along the solution of the IBVP. In particular, for an elastic material, $\mathcal{D}_{\text{mech}} \equiv 0$, and (1.3.27) reduces to the familiar law of conservation of the total energy for elastodynamics. Inequality (1.3.27) is an *a priori energy estimate* on the solution of the IBVP that captures its intrinsic dissipative character. As will be discussed below and further elaborated upon in a subsequent chapter, algorithms for the solution of the IBVP that inherit the estimate (1.3.27) incorporate a strong notion of numerical stability.

1.3.4 Uniqueness of the Solution to the IBVP. Contractivity

In this section we exploit the structure of the a priori estimates derived above to show that for hardening plasticity ($K > 0$), the solution to IBVP is *unique*. In addition, we identify a property of the IBVP called *contractivity* which also plays a central role in the numerical analysis of time approximations to the IBVP.

1.3.4.1 Uniqueness.

For simplicity consider isotropic hardening and suppose that $\{u, v, \varepsilon^p, \alpha\}$ and $\{\tilde{u}, \tilde{v}, \tilde{\varepsilon}^p, \tilde{\alpha}\}$ are two solutions of the IBVP for prescribed initial data $\{u_0, v_0\}$ and prescribed boundary conditions $\{b, \bar{u}, \bar{\sigma}\}$ at a fixed but arbitrary time t. To show that these two solutions must coincide (for $K > 0$), we first observe that

$$T(v - \tilde{v}) + V_{\text{int}}(\varepsilon^e - \tilde{\varepsilon}^e) \geq 0, \tag{1.3.28}$$

since both $E > 0$ and $K > 0$. Next, we compute the rate of change of this quadratic form to obtain

$$\begin{aligned}
&\frac{d}{dt}\left[T(v - \tilde{v}) + V_{\text{int}}(\varepsilon^e - \tilde{\varepsilon}^e)\right] \\
&\quad = \int_B \left[\rho \frac{\partial v}{\partial t}(v - \tilde{v}) + \sigma \frac{\partial (v - \tilde{v})}{\partial x} - \rho b(v - \tilde{v})\right] dx - \bar{\sigma}(v - \tilde{v})\big|_{\partial_\sigma B} \\
&\quad + \int_B \left[\rho \frac{\partial \tilde{v}}{\partial t}(\tilde{v} - v) + \tilde{\sigma} \frac{\partial (\tilde{v} - v)}{\partial x} - \rho b(\tilde{v} - v)\right] dx + \bar{\sigma}(\tilde{v} - v)\big|_{\partial_\sigma B} \\
&\quad + \int_B \left[(\sigma - \tilde{\sigma})\dot{\tilde{\varepsilon}}^p - (\alpha - \tilde{\alpha})K\dot{\tilde{\alpha}}\right] dx \\
&\quad + \int_B \left[(\tilde{\sigma} - \sigma)\dot{\varepsilon}^p - (\tilde{\alpha} - \alpha)K\dot{\alpha}\right] dx,
\end{aligned} \tag{1.3.29}$$

where $\sigma = E(\partial u/\partial x - \epsilon^p)$, $\tilde{\sigma} = E(\partial \tilde{u}/\partial x - \tilde{\varepsilon}^p)$, and we have added and subtracted the forcing terms $\rho b(v - \tilde{v})$ and $\bar{\sigma}(v - \tilde{v})\big|_{\partial_\sigma B}$. Now observe that $v - \tilde{v}$ is an admissible test function which, therefore, lies in $\mathbb{V}$. Moreover, by assumption, the two solutions satisfy the weak form of the IBVP. Inspection of (1.3.29) reveals that the first two terms on the right-hand side are precisely the weak forms of each of these two solutions with test function $v - \tilde{v} \in \mathbb{V}$ and, therefore, vanish. Then inserting the flow rule and the hardening law into the last two terms of the right-hand side of (1.3.29) yields

$$\begin{aligned}
&\frac{d}{dt}\left[T(v - \tilde{v}) + V_{\text{int}}(\varepsilon^e - \tilde{\varepsilon}^e)\right] \\
&\quad = \int_B \tilde{\gamma}\left[(\sigma - \tilde{\sigma})\partial_\sigma f(\tilde{\sigma}, \tilde{\alpha}) + (\alpha - \tilde{\alpha})\partial_\alpha f(\tilde{\sigma}, \tilde{\alpha})\right] dx \\
&\quad + \int_B \gamma\left[(\tilde{\sigma} - \sigma)\partial_\sigma f(\sigma, \alpha) + (\tilde{\alpha} - \alpha)\partial_\alpha f(\sigma, \alpha)\right] dx.
\end{aligned} \tag{1.3.30}$$

Since the function $f(\sigma, \alpha)$ is convex, using property (1.2.53), we conclude that

$$\frac{d}{dt}\left[T(v-\tilde{v}) + V_{\text{int}}(\varepsilon^e - \tilde{\varepsilon}^e)\right] \le \int_B \tilde{\gamma}\left[f(\sigma,\alpha) - f(\tilde{\sigma},\tilde{\alpha})\right] dx + \int_B \gamma\left[f(\tilde{\sigma},\tilde{\alpha}) - f(\sigma,\alpha)\right] dx. \quad (1.3.31)$$

From the Kuhn–Tucker conditions, it follows that $\gamma f(\sigma, \alpha) = 0$ and $\tilde{\gamma} f(\tilde{\sigma}, \tilde{\alpha}) = 0$. Moreover, since $\gamma \ge 0$, $f(\sigma, \alpha) \le 0$, and $\tilde{\gamma} \ge 0$, and $f(\tilde{\sigma}, \tilde{\alpha}) \le 0$,

$$\frac{d}{dt}\left[T(v-\tilde{v}) + V_{\text{int}}(\varepsilon^e - \tilde{\varepsilon}^e)\right] \le \int_B \left[\tilde{\gamma} f(\sigma,\alpha) + \gamma f(\tilde{\sigma},\tilde{\alpha})\right] dx \le 0. \quad (1.3.32)$$

Then integrating in time (1.3.32) and noting that $\varepsilon^e - \tilde{\varepsilon}^e\big|_{t=0} = 0$ yields

$$T(v-\tilde{v}) + V_{\text{int}}(\varepsilon^e - \tilde{\varepsilon}^e) \le 0. \quad (1.3.33)$$

By comparing (1.3.33) and (1.3.28), we conclude that $K(v-\tilde{v}) + V_{\text{int}}(\varepsilon^e - \tilde{\varepsilon}^e) \equiv 0$. Finally, since this quadratic form is positive definite by virtue of the assumptions $K > 0$ and $E > 0$, it follows that

$$\begin{aligned} v &= \tilde{v}, \\ \varepsilon^e &= \tilde{\varepsilon}^e \end{aligned} \quad (1.3.34)$$

and

$$\alpha = \tilde{\alpha}.$$

In addition, $\sigma = \tilde{\sigma}$ since the elastic modulus $E > 0$. Moreover, since the elasto-plastic modulus $E^{ep} = EK/(E+K) > 0$, we conclude that $\dot{\varepsilon}^p = \dot{\tilde{\varepsilon}}^p$. Uniqueness of the displacement field; i.e., $u = \tilde{u}$ then follows from this condition along with $v = \tilde{v}$ and the fact that the two solutions are computed with the same initial data.

1.3.4.2 Contractivity.

A question closely related to the uniqueness of solution to the IBVP for elasto-plasticity concerns the effect of a change in the initial conditions. More precisely, consider two initial conditions $\{u_0, v_0\}$ and $\{\tilde{u}_0, \tilde{v}_0\}$ and the same forcing functions $\{b, \bar{u}, \bar{\sigma}\}$. Let $\{u, v, \varepsilon^p, \alpha\}$ and $\{\tilde{u}, \tilde{v}, \tilde{\varepsilon}^p, \tilde{\alpha}\}$ denote the two solutions of the IBVP associated with these two sets of initial data which, as was shown above, must be unique. By proceeding exactly in the same manner as in the uniqueness argument given above, we arrive at the inequality (1.3.32), namely,

$$\frac{d}{dt}\left[T(v-\tilde{v}) + V_{\text{int}}(\varepsilon^e - \tilde{\varepsilon}^e)\right] \le 0. \quad (1.3.35)$$

We remark that, since $E > 0$ and $K > 0$, the quadratic form $T(v) + V_{\text{int}}(\varepsilon^e, \alpha)$ is positive-definite and defines an *energy* norm which is the natural norm for the problem at hand. Then integration in time reveals that

$$\boxed{T(v-\tilde{v}) + V_{\text{int}}(\varepsilon^e - \tilde{\varepsilon}^e) \le T(v_0 - \tilde{v}_0) + V_{\text{int}}(\varepsilon^e_0 - \tilde{\varepsilon}^e_0).} \quad (1.3.36)$$

Therefore, if we view $u_0 - \tilde{u}_0$ and $v_0 - \tilde{v}_0$ as an *error in the initial data*, this inequality shows that the subsequent error $u - \tilde{u}$ and $v - \tilde{v}$ in the solutions, measured in terms of the kinetic and internal energies, *decreases* at any subsequent time. We say that the elastoplastic problem is *contractive* relative to the norm defined by the internal and kinetic energies.

1.3.5 Outline of the Numerical Solution of the IBVP

The numerical solution of problem (1.3.11), which constitutes the central theme of this monograph, involves the following two steps.

i. Time discretization of the interval of interest $[0, \mathrm{T}] = \bigcup_{n=1}^{N} [t_n, t_{n+1}]$. We shall see that the relevant problem within a typical time interval $[t_n, t_{n+1}]$ can be posed as follows.

- **i.a.** For the dynamic problem, the time derivatives arising in the weak form are replaced by suitable algorithmic approximations. This step leads to the formulation of a global time-stepping algorithm and, to a large extent, is independent of the specific constitutive model.
- **i.b.** Attention is now restricted to a *particular point* $x \in \mathcal{B}$ predetermined by the spatial discretization discussed below (in fact, a *quadrature point* of a typical finite element). The goal is to compute an approximation to the stress appearing in the weak form.
- **i.c.** At the point $x \in \mathcal{B}$ of interest and at time t_n, the (*incremental*) displacement (leading to t_{n+1}), denoted by $\Delta u_{n+1}(x)$, is *regarded as given*.
- **i.d.** At time t_n the state at $x \in \mathcal{B}$ characterized by $\{\varepsilon_n(x), \sigma_n(x), \varepsilon_n^p(x), \alpha_n(x)\}$ is *given and is assumed to be equilibrated*, i.e., it satisfies (1.3.11).
- **i.e.** Then, the problem at this stage is to *update* the state variables *at* $x \in \mathcal{B}$ to the values $\{\varepsilon_{n+1}(x), \sigma_{n+1}(x), \varepsilon_{n+1}^p(x), \alpha_{n+1}(x)\}$ in a manner consistent with the model BOX **1.2**.

ii. Spatial discretization of the domain of interest $\mathcal{B} = [0, L]$ to arrive at the *discrete counterpart* of the weak form of the equilibrium equations. At this stage the problem can be posed as follows

- **ii.a.** Attention is focused on a typical element $\mathcal{B}_e$. For *given stress field* $\sigma_{n+1}(x)$ at predetermined points, one evaluates the weak form $G(\sigma_{n+1}, \eta)$ restricted to $\mathcal{B}_e$.
- **ii.b.** Assemble the contributions of all elements and determine whether the system is in equilibrium under the state $\{u_{n+1}, \sigma_{n+1}\}$ by testing whether $G(\sigma_{n+1}, \eta)$ is zero.
- **ii.c.** Determine a correction to the displacement field and return to step *i* to evaluate the associated state $\{\varepsilon_{n+1}, \sigma_{n+1}, \varepsilon_{n+1}^p, \alpha_{n+1}\}$.

Step **ii** is accomplished by the standard finite-element method (see e.g., Hughes [1987]) and typically remains unchanged even if a model different from the ones summarized in Boxes **1.1–1.3** is employed. To illustrate the essential aspects of the method, a simple example is discussed in Section **1.5**.

On the other hand, step **i** depends crucially on the particular constitutive model chosen, such as the one in BOX **1.2**, and constitutes the central objective of this monograph. Step **i.a** refers to the global time-discretization of the dynamic weak form and will not be a point of central interest in our investigation. In fact, in what follows we shall restrict our attention to the quasi-static problem obtained by neglecting the effect of inertial forces. Our goal is to address in detail the implementation of the subsequent steps which are specific to the elastoplastic problem. The next section discusses these and related aspects within the context of the model problem in BOX **1.2**.

1.4 Integration Algorithms for Rate-Independent Plasticity

The problem to be addressed in this section is purely local and is stated as follows:

i. Let $x \in \bar{\mathcal{B}} = [0, L]$ be a *given* point of interest in the body obeying the rate-independent constitutive model in BOX **1.1**, BOX **1.2**, or BOX **1.3**.

ii. Assume that the local state of the body at point $x \in \mathcal{B}$ and current time, say t_n, is completely defined. By this statement we mean that the value of

$$\{\varepsilon_n(x), \varepsilon_n^p(x), \alpha_n(x)\} \tag{1.4.1a}$$

is known and, therefore, the stress state

$$\sigma_n(x) = E\left[\varepsilon_n(x) - \varepsilon_n^p(x)\right] \tag{1.4.1b}$$

is also known. For the model in BOX **1.3** the additional internal variable $q_n(x)$ must be included in (1.4.1a).

iii. Suppose that one is *given* an "increment" in total strain at $x \in \mathcal{B}$, say $\Delta\varepsilon(x)$, which drives the state to time $t_{n+1} = t_n + \Delta t$. The basic problem we shall be concerned with is the update of the basic variables (1.4.1) to time t_{n+1} in a manner consistent with the constitutive model in BOX **1.2** (or BOX **1.3**).

Thus, within the context outlined above, the "*incremental*" integration of the rate-independent elastoplastic model in either BOX **1.2** or **1.3** over a time step $[t_n, t_n + \Delta t]$ is regarded as a *strain-driven process* in which the total strain $\varepsilon = \partial u/\partial x$ is the basic independent variable. As we shall illustrate in the following section, this is precisely the appropriate framework for the numerical solution of the elastoplastic boundary value problem by the finite-element method. Once more, note that this integration process is local in space, that is, it takes place at specific points of the body (as shown below, these points correspond, typically, to quadrature points of a finite element). An illustration of the basic problem in computational plasticity is given in Figure **1.9**.

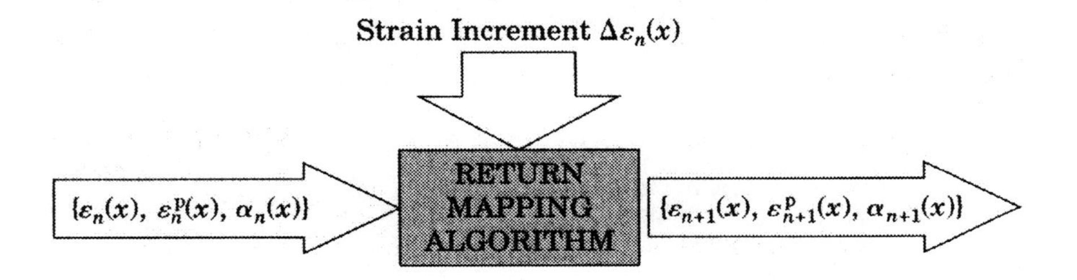

FIGURE 1.9. The role of elastoplastic return-mapping algorithms. Note that σ is a dependent variable computed by the relationship $\sigma(x) = E\left[\varepsilon(x) - \varepsilon^p(x)\right]$.

1.4.1 The Incremental Form of Rate-Independent Plasticity

To motivate the integration scheme adopted in the formulation of return-mapping algorithms, first we review a classical family of schemes for the numerical integration of ordinary differential equations.

1.4.1.1 The generalized midpoint rule.

Let $f : \mathbb{R} \to \mathbb{R}$ be a smooth function, and consider the initial-value problem

$$\left.\begin{aligned} \dot{x}(t) &= f(x(t)) \\ x(0) &= x_n \end{aligned}\right\} \quad \text{in } [0, \mathrm{T}]. \tag{1.4.2}$$

We shall be concerned with the following one-parameter class of integration algorithms called the generalized midpoint rule

$$\boxed{\begin{aligned} x_{n+1} &= x_n + \Delta t f(x_{n+\vartheta}) \\ x_{n+\vartheta} &= \vartheta x_{n+1} + (1-\vartheta)\, x_n; \qquad \vartheta \in [0, 1]\,. \end{aligned}} \tag{1.4.3}$$

Here, $x_{n+1} \cong x(t_{n+1})$ denotes the algorithmic approximation to the exact value $x(t_{n+1})$ at time $t_{n+1} = t_n + \Delta t$. We note that this family of algorithms contains well-known integrative schemes, in particular,

$$\left.\begin{aligned} \vartheta &= 0 \Rightarrow && \text{forward (explicit) Euler} \\ \vartheta &= \frac{1}{2} \Rightarrow && \text{midpoint rule} \\ \vartheta &= 1 \Rightarrow && \text{backward (implicit) Euler.} \end{aligned}\right\} \tag{1.4.4}$$

We refer to standard textbooks, e.g., Gear [1971] or Hairer, Norsett and Wanner [1987], for a discussion of this class of algorithms and, in particular, for relevant notions of *consistency, stability*, and *accuracy*. A complete numerical analysis of the class of methods discussed below for the general three-dimensional problem is deferred to Chapter 6. Here, we merely recall that second-order accuracy is obtained only for $\vartheta = \frac{1}{2}$ whereas unconditional (linearized) stability requires $\vartheta \geq \frac{1}{2}$.

Our objective here is to illustrate the application of the class of algorithms (1.4.3) to the integration of the elastoplastic initial-value problem. First we shall consider a detailed analysis of the isotropic hardening model in BOX **1.2**. The extension of the analysis to the combined isotropic/kinematic hardening model in BOX **1.3** is examined in Section **1.4.4**.

At this stage it is worth pointing out a fundamental difference between (1.4.2) and the model in BOX **1.2**. In (1.4.2), the curve $t \in \mathbb{R} \mapsto x(t) \in \mathbb{R}$ (the "*flow*") is *unconstrained* whereas in the model in BOX **1.2**, the curve

$$t \in \mathbb{R} \mapsto (\sigma(t), \alpha(t)) \in \mathbb{E}_\sigma \tag{1.4.5}$$

is *constrained* to lie within the closure of the elastic domain $\mathbb{E}_\sigma$. Thus, one speaks of a *constrained problem of evolution*. Note that by using the stress-strain relationship we can equivalently, consider the curve

$$t \in \mathbb{R} \mapsto [\varepsilon(t), \varepsilon^p(t), \alpha(t)] \in \mathbb{E}_\varepsilon, \tag{1.4.6}$$

where $\mathbb{E}_\varepsilon$ is the closure of the elastic domain in *strain space*, which is defined from (1.2.28) and the stress-strain relationship as

$$\mathbb{E}_\varepsilon = \left\{(\varepsilon, \varepsilon^p, \alpha) \in \mathbb{R}^2 \times \mathbb{R}_+ \mid f\left(E(\varepsilon - \varepsilon^p), \alpha\right) \leq 0\right\}. \tag{1.4.7}$$

The presence of the constraint condition is precisely the essential feature that characterizes plasticity.

1.4.1.2 Incremental elastoplastic initial value problem. Isotropic hardening.

As motivation, we start our analysis by considering the simplest case for which $\vartheta = 1$ in (1.4.3). This choice of ϑ corresponds to the backward-Euler method, and leads to the classical return-mapping algorithms. From BOX **1.2**, by application of (1.4.3) with $\vartheta = 1$, we obtain

$$\left.\begin{aligned} \varepsilon^p_{n+1} &= \varepsilon^p_n + \Delta\gamma \operatorname{sign}(\sigma_{n+1}), \\ \alpha_{n+1} &= \alpha_n + \Delta\gamma, \end{aligned}\right\} \tag{1.4.8a}$$

where $\Delta\gamma = \gamma_{n+1}\Delta t \geq 0$ is a Lagrange multiplier (the algorithmic counterpart of the consistency parameter $\gamma \geq 0$) and

$$\left.\begin{aligned} \sigma_{n+1} &= E(\varepsilon_{n+1} - \varepsilon^p_{n+1}), \\ \varepsilon_{n+1} &:= \varepsilon_n + \Delta\varepsilon_n. \end{aligned}\right\} \tag{1.4.8b}$$

The variables $(\sigma_{n+1}, \alpha_{n+1})$ along with $\Delta\gamma$ are constrained by the following *discrete version* of the Kuhn–Tucker conditions:

$$\left.\begin{aligned} f_{n+1} &:= \left|\sigma_{n+1}\right| - (\sigma_Y + K\alpha_{n+1}) \leq 0, \\ \Delta\gamma &\geq 0, \\ \Delta\gamma f_{n+1} &= 0. \end{aligned}\right\} \tag{1.4.9}$$

We observe that $\Delta\varepsilon_n$ is *given*, and therefore equation (1.4.8*b*) is regarded merely as the definition for ε_{n+1}. Further, we note that, by applying the implicit backward Euler algorithm, we have transformed the initial constrained problem of evolution into a *discrete constrained* algebraic problem for the variable $\{\varepsilon^p_{n+1}, \alpha_{n+1}\}$. Remarkably, there is an underlying variational structure explained below which renders equations (1.4.8)–(1.4.9) the optimality conditions of a *discrete constrained optimization* problem. We postpone the discussion of these ideas to Section **1.4.3** and focus our attention next on the direct solution of problem (1.4.8)–(1.4.9).

1.4.2 Return-Mapping Algorithms. Isotropic Hardening

For now we assume that the solution of problem (1.4.8)–(1.4.9) is *unique.* A justification of this assumption is given in Section **1.4.3**. An essential step in the solution of (1.4.8)–(1.4.9) is the introduction of the following auxiliary problem.

1.4.2.1 The trial elastic state.

We consider an auxiliary state, which as shown below need not correspond to an actual state, and is obtained by *freezing plastic flow*. In other words, first we consider a purely elastic (trial) step defined by the formulas

$$\boxed{\begin{aligned} \sigma^{\text{trial}}_{n+1} &:= E\left(\varepsilon_{n+1} - \varepsilon^p_n\right) \equiv \sigma_n + E\Delta\varepsilon_n \\ \varepsilon^{p^{\text{trial}}}_{n+1} &:= \varepsilon^p_n \\ \alpha^{\text{trial}}_{n+1} &:= \alpha_n \\ f^{\text{trial}}_{n+1} &:= \left|\sigma^{\text{trial}}_{n+1}\right| - [\sigma_Y + K\alpha_n]. \end{aligned}} \tag{1.4.10}$$

We observe that the trial state is determined solely in terms of the initial conditions $\{\varepsilon_n, \varepsilon^p_n, \alpha_n\}$ and the *given* incremental strain $\Delta\varepsilon_n$. Once more, we remark that this state may not, and in general *will not*, correspond to any actual, physically admissible state unless the incremental process is elastic in the sense described below.

1.4.2.2 Algorithmic form of the loading conditions.

Once the trial state is computed by (1.4.10), first we consider the case for which

$$f^{\text{trial}}_{n+1} \leq 0. \tag{1.4.11}$$

It follows that the *trial state* is *admissible* in the sense that

$$\begin{aligned} \varepsilon^p_{n+1} &= \varepsilon^p_n, \\ \alpha_{n+1} &= \alpha_n, \end{aligned} \tag{1.4.12a}$$

and

$$\sigma_{n+1} = \sigma^{\text{trial}}_{n+1},$$

satisfy:

1. the stress-strain relationship,
2. the flow rule and the hardening law with, $\Delta\gamma \equiv 0$, and
3. the Kuhn–Tucker conditions, since conditions

$$f_{n+1} \equiv f^{\text{trial}}_{n+1} \leq 0 \quad \text{and} \quad \Delta\gamma = 0 \tag{1.4.12b}$$

are consistent with (1.4.9).

Therefore, since the solution to problem (1.4.8)–(1.4.9) is *unique* (see Section **1.6** below), the trial state in fact *is* the solution to the problem. An illustration of this situation is given in Figure **1.10**. Next, we consider the case for which $f^{\text{trial}}_{n+1} > 0$. Clearly, the trial state *cannot* be a solution to the incremental problem since $(\sigma^{\text{trial}}_{n+1}, \alpha_n)$ *violates* the constraint condition $f(\sigma, \alpha) \leq 0$. Thus, we require that $\Delta\gamma > 0$ so that $\varepsilon^p_{n+1} \neq \varepsilon^p_n$ to obtain $\sigma_{n+1} \neq \sigma^{\text{trial}}_{n+1}$. By the Kuhn–Tucker conditions

$$\Delta\gamma > 0$$

and (1.4.13)

$$\Delta\gamma f_{n+1} = 0 \Rightarrow f_{n+1} = 0,$$

and the process is incrementally plastic. See Figure **1.11** for an illustration

To summarize our results, the conclusion that an incremental process for given incremental strain is elastic or plastic is drawn solely on the basis of the trial state according to the criterion

$$f^{\text{trial}}_{n+1} \begin{cases} \leq 0 \Rightarrow & \text{elastic step} \quad \Delta\gamma = 0, \\ > 0 \Rightarrow & \text{plastic step} \quad \Delta\gamma > 0. \end{cases} \tag{1.4.14}$$

Note that these loading/unloading conditions are the algorithmic counterpart of the alternative form of the Kuhn–Tucker conditions in Section **1.2.3**.

1.4.2.3 The return-mapping algorithm.

Here we examine the algorithmic problem for an incrementally plastic process characterized by the conditions

$$f^{\text{trial}}_{n+1} > 0 \Longleftrightarrow f(\sigma_{n+1}, \alpha_{n+1}) = 0, \tag{1.4.15}$$

and

$$\Delta\gamma > 0.$$

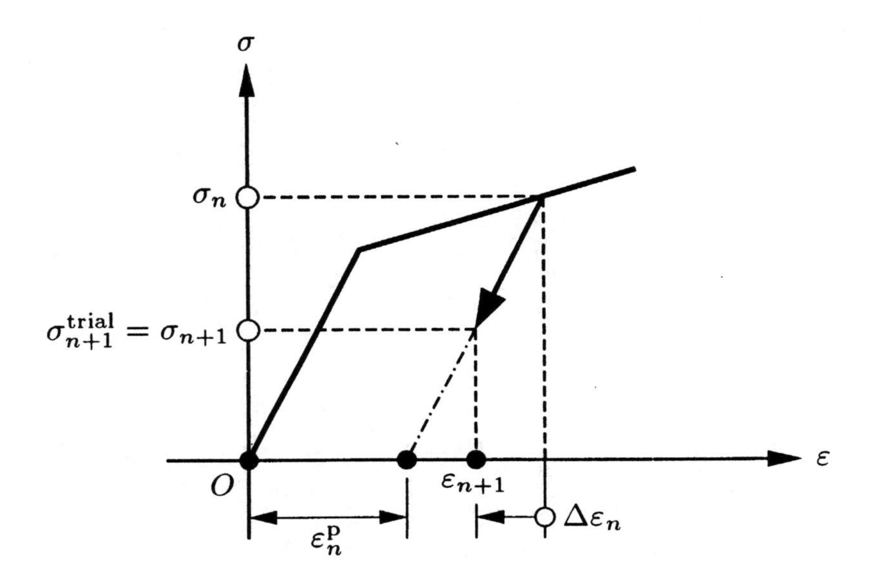

FIGURE 1-10. Example of an incremental elastic step from a plastic state. The final solution coincides with the trial state.

Our objective is to determine the solution $\{\varepsilon^p_{n+1}, \alpha_{n+1}, \sigma_{n+1}, \Delta\gamma\}$ to problem (1.4.8)–(1.4.9). To accomplish this task we first express the final stress σ_{n+1} in terms of $\sigma^{\text{trial}}_{n+1}$ and $\Delta\gamma$ as follows:

$$\begin{aligned}
\sigma_{n+1} &= E\left(\varepsilon_{n+1} - \varepsilon^p_{n+1}\right) \\
&= E\left(\varepsilon_{n+1} - \varepsilon^p_n\right) - E\left(\varepsilon^p_{n+1} - \varepsilon^p_n\right) \\
&= \sigma^{\text{trial}}_{n+1} - E\Delta\gamma \operatorname{sign}\left(\sigma_{n+1}\right).
\end{aligned} \tag{1.4.16}$$

Therefore, since $\Delta\gamma > 0$, (1.4.8)–(1.4.9) is written, in view of (1.4.16), as

$$\left.\begin{aligned}
\sigma_{n+1} &= \sigma^{\text{trial}}_{n+1} - \Delta\gamma E \operatorname{sign}\left(\sigma_{n+1}\right) \\
\varepsilon^p_{n+1} &= \varepsilon^p_n + \Delta\gamma \operatorname{sign}\left(\sigma_{n+1}\right) \\
\alpha_{n+1} &= \alpha_n + \Delta\gamma \\
f_{n+1} &\equiv \left|\sigma_{n+1}\right| - \left[\sigma_Y + K\alpha_{n+1}\right] = 0.
\end{aligned}\right\} \tag{1.4.17}$$

Now Problem (1.4.17) is solved explicitly in terms of the trial elastic state by the following procedure. From $(1.4.17)_1$,

$$\left|\sigma_{n+1}\right| \operatorname{sign}\left(\sigma_{n+1}\right) = \left|\sigma^{\text{trial}}_{n+1}\right| \operatorname{sign}\left(\sigma^{\text{trial}}_{n+1}\right) - \Delta\gamma E \operatorname{sign}\left(\sigma_{n+1}\right). \tag{1.4.18a}$$

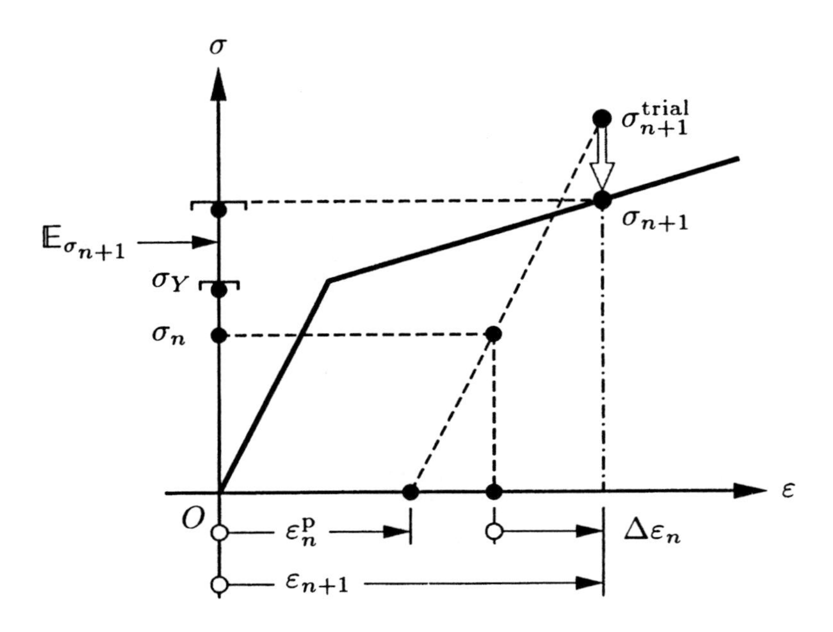

FIGURE 1-11. The trial state violates the constraint condition $f \leq 0$. Consequently, the incremental process must be plastic since $\Delta\gamma > 0$ to achieve $\sigma_{n+1} \neq \sigma_{n+1}^{\text{trial}}$.

Collecting terms in (1.4.18a), we find that

$$\left[\left|\sigma_{n+1}\right| + \Delta\gamma E\right] \operatorname{sign}\left(\sigma_{n+1}\right) = \left|\sigma_{n+1}^{\text{trial}}\right| \operatorname{sign}\left(\sigma_{n+1}^{\text{trial}}\right). \tag{1.4.18b}$$

Since $\Delta\gamma > 0$ and $E > 0$, we observe that the term within brackets in (1.4.18b) is necessarily positive. Therefore we require that

$$\boxed{\operatorname{sign}\left(\sigma_{n+1}\right) = \operatorname{sign}\left(\sigma_{n+1}^{\text{trial}}\right),} \tag{1.4.19}$$

along with the condition

$$\left|\sigma_{n+1}\right| + \Delta\gamma E = \left|\sigma_{n+1}^{\text{trial}}\right|. \tag{1.4.20}$$

Finally, the algorithmic consistency parameter $\Delta\gamma > 0$ is determined from the discrete consistency condition $(1.4.17)_4$ as follows. In view of (1.4.20), the yield criterion f_{n+1} is written as

$$\begin{aligned} f_{n+1} &= \left|\sigma_{n+1}^{\text{trial}}\right| - E\Delta\gamma - \left[\sigma_Y + K\alpha_n\right] - K\left(\alpha_{n+1} - \alpha_n\right) \\ &= f_{n+1}^{\text{trial}} - \Delta\gamma(E + K), \end{aligned} \tag{1.4.21}$$

where we have used (1.4.10) and $(1.4.17)_3$ in obtaining (1.4.21). Hence

$$f_{n+1} = 0 \Rightarrow \boxed{\Delta\gamma = \frac{f_{n+1}^{\text{trial}}}{E + K} > 0.} \tag{1.4.22}$$

Substituting (1.4.19) and (1.4.22) in (1.4.17) yields the desired result:

$$\boxed{\begin{aligned} \sigma_{n+1} &= \sigma_{n+1}^{\text{trial}} - \Delta\gamma E \,\text{sign}\left(\sigma_{n+1}^{\text{trial}}\right) \\ \varepsilon_{n+1}^{p} &= \varepsilon_n^p + \Delta\gamma \,\text{sign}\left(\sigma_{n+1}^{\text{trial}}\right) \\ \alpha_{n+1} &= \alpha_n + \Delta\gamma. \end{aligned}} \tag{1.4.23}$$

Remarks 1.4.1.

1. A compelling interpretation of algorithm (1.4.22)–(1.4.23) illustrated in Figure **1.12** is derived by writing (1.4.23) in a slightly different form. From $(1.4.23)_1$ and (1.4.22) we obtain the alternative expression

$$\begin{aligned} \sigma_{n+1} &= \left(\left|\sigma_{n+1}^{\text{trial}}\right| - \Delta\gamma E\right) \,\text{sign}\left(\sigma_{n+1}^{\text{trial}}\right) \\ &= \left[1 - \frac{\Delta\gamma E}{\left|\sigma_{n+1}^{\text{trial}}\right|}\right] \sigma_{n+1}^{\text{trial}}. \end{aligned} \tag{1.4.24}$$

   Since $f_{n+1} = 0$, in view of (1.4.24), we conclude that *the final stress state is the projection of the trial stress onto the yield surface.* A more fundamental explanation of this simple result is given in the next subsection. Because of this interpretation, the algorithm summarized in BOX **1.4** is called a *return-mapping algorithm.*

2. An almost identical development is carried out for the generalized midpoint rule formula (see Simo and Taylor [1986]), which corresponds to $q = \frac{1}{2}$. A complete numerical analysis of the resulting algorithm is given in Simo and Govindjee [1988], Simo [1991], and summarized in Chapter **6.** An introduction to these topics is given in Section **1.6** below.

1.4.3 Discrete Variational Formulation. Convex Optimization

The algorithm in BOX **1.4** possesses a more fundamental interpretation which is the manifestation of a basic variational structure underlying classical rate-independent plasticity. We show below that this algorithm is interpreted as the *Kuhn–Tucker optimality conditions of a convex-optimization problem* which is, in fact, the discrete counterpart of a classical postulate known as the principle of maximum plastic dissipation (or entropy production). A discussion of the role played by this principle is given in Chapter **2.**

BOX 1.4. Return-Mapping Algorithm for 1-D, Rate-Independent Plasticity. Isotropic Hardening.

1. Database at $x \in \mathcal{B} : \{\varepsilon_n^p, \alpha_n\}$.

2. *Given* strain field at $x \in \mathcal{B} : \varepsilon_{n+1} = \varepsilon_n + \Delta\varepsilon_n$.

3. Compute elastic trial stress and test for plastic loading

$$\sigma_{n+1}^{\text{trial}} := E(\varepsilon_{n+1} - \varepsilon_n^p)$$

$$f_{n+1}^{\text{trial}} := \left|\sigma_{n+1}^{\text{trial}}\right| - [\sigma_Y + K\alpha_n]$$

IF $f_{n+1}^{\text{trial}} \leq 0$ THEN

Elastic step: set $(\bullet)_{n+1} = (\bullet)_{n+1}^{\text{trial}}$ & *EXIT*

ELSE

Plastic step: Proceed to step **4**.

ENDIF

4. Return mapping

$$\Delta\gamma := \frac{f_{n+1}^{\text{trial}}}{(E + K)} > 0$$

$$\sigma_{n+1} := \left[1 - \frac{\Delta\gamma E}{\left|\sigma_{n+1}^{\text{trial}}\right|}\right] \sigma_{n+1}^{\text{trial}}$$

$$\varepsilon_{n+1}^p := \varepsilon_n^p + \Delta\gamma \ \text{sign}\left(\sigma_{n+1}^{\text{trial}}\right)$$

$$\alpha_{n+1} := \alpha_n + \Delta\gamma$$

1.4.3.1 Discrete convex mathematical programming problem.

We consider the following functional depending on the variables (σ, α):

$$\chi(\sigma, \alpha) : = \frac{1}{2}\left(\sigma_{n+1}^{\text{trial}} - \sigma\right) E^{-1} \left(\sigma_{n+1}^{\text{trial}} - \sigma\right) + \frac{1}{2}(\alpha_n - \alpha) K (\alpha_n - \alpha). \tag{1.4.25}$$

A physical interpretation of this function in a general setting is be given in Chapter **2**. Here we simply note that $\chi(\sigma, \alpha)$ is interpreted as the (complementary) energy at the increment between the trial stress and the state (σ, α). We show that by minimizing $\chi(\sigma, \alpha)$ over $\mathbb{E}_\sigma$ one obtains the algorithm in BOX **1.4**. To this end, first we recall that the closure of the elastic domain $\mathbb{E}_\sigma$ is defined by

$$\mathbb{E}_\sigma := \left\{(\sigma, \alpha) \in \mathbb{R} \times \mathbb{R}_+ \mid f(\sigma, \alpha) \leq 0\right\}. \tag{1.4.26}$$

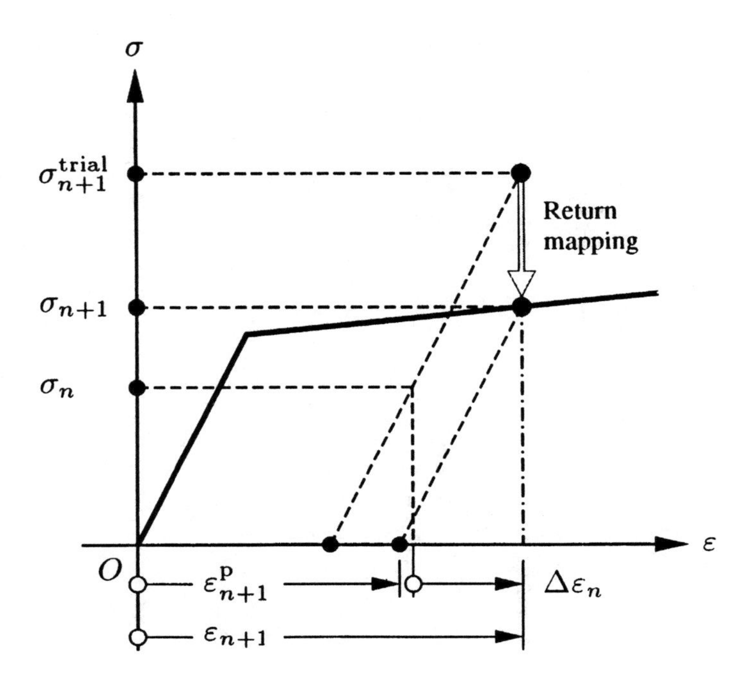

FIGURE 1-12. The final stress is obtained by "*returning*" the trial stress to the yield surface through a scaling, hence, the denomination return mapping.

As remarked above, $\mathbb{E}_\sigma$ is a closed convex set, since $f : \mathbb{R} \times \mathbb{R}_+ \to \mathbb{R}$ is assumed to be convex. Now consider the following minimization problem:

$$\boxed{\begin{aligned} &\text{Find } \left(\sigma_{n+1}, \alpha_{n+1}\right) \in \mathbb{E}_\sigma \text{ such that} \\ &\chi\left(\sigma_{n+1}, \alpha_{n+1}\right) = \underset{(\sigma, \alpha) \in \mathbb{E}_\sigma}{\text{MIN}} \left\{\chi(\sigma, \alpha)\right\}. \end{aligned}} \tag{1.4.27}$$

Since $E > 0$ and, *by assumption*, $K > 0$, it follows that $\chi(\sigma, \alpha)$ is convex. Thus, we have the situation depicted in Figure **1.13**.

Problem (1.4.27) is a *constrained convex minimization problem* in the variables $(\sigma, \alpha) \in \mathbb{E}_\sigma$, with convex constraint $f(\sigma, \alpha) \leq 0$. By standard results in optimization theory (see e.g., Bertsekas [1982]), it is known that this problem has a *unique solution*. Note that $(\sigma_{n+1}^{\text{trial}}, \alpha_n)$ is *given* by (1.4.17) and, therefore, is regarded as known data. In what follows we assume that $f(\sigma_{n+1}^{\text{trial}}, \alpha_n) > 0$.

1.4.3.2 Optimality conditions: Closest point projection.

To characterize the solution of (1.4.27), we use the method of Lagrange multipliers. Accordingly, we consider the Lagrangian

$$\mathcal{L}(\sigma, \alpha, \Delta\gamma) := \chi(\sigma, \alpha) + \Delta\gamma f(\sigma, \alpha). \tag{1.4.28}$$

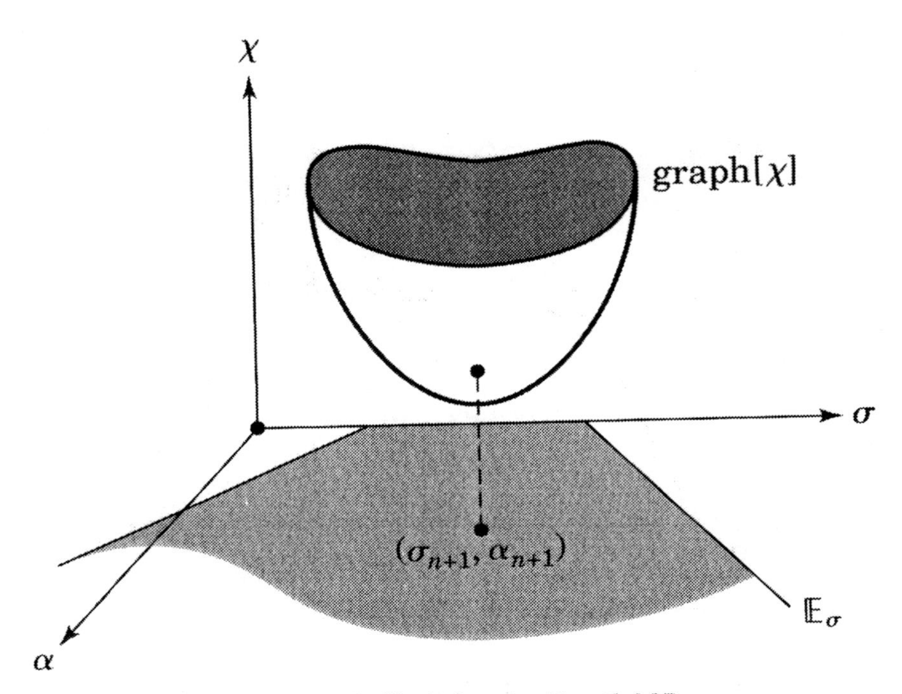

FIGURE 1-13. Illustration of problem (1.4.27).

If $(\sigma_{n+1}, \alpha_{n+1})$ is the minimum of (1.4.27), the standard optimality conditions require that (see e.g., Luenberger [1984])

$$\left.\begin{aligned} \frac{\partial}{\partial \sigma} \mathcal{L}\left(\sigma_{n+1}, \alpha_{n+1}, \Delta\gamma\right) &= 0, \\ \frac{\partial}{\partial \alpha} \mathcal{L}\left(\sigma_{n+1}, \alpha_{n+1}, \Delta\gamma\right) &= 0, \end{aligned}\right\} \tag{1.4.29a}$$

along with the additional requirements

$$\left.\begin{aligned} &\Delta\gamma \geq 0, \quad f(\sigma_{n+1}, \alpha_{n+1}) \leq 0 \\ &\Delta\gamma f(\sigma_{n+1}, \alpha_{n+1}) = 0. \end{aligned}\right\} \tag{1.4.29b}$$

However, since $\dfrac{\partial f}{\partial \sigma} = \operatorname{sign}(\sigma)$, condition (1.4.29a) yields

$$\left.\begin{aligned} \sigma_{n+1} &= \sigma_{n+1}^{\text{trial}} - E\Delta\gamma \operatorname{sign}\left(\sigma_{n+1}\right) \\ \alpha_{n+1} &= \alpha_n + \Delta\gamma, \end{aligned}\right\} \tag{1.4.30}$$

which coincides with the return-mapping equations (1.4.23). Thus, *the return-mapping algorithm (*1.4.28*) characterizes the solution* $(\sigma_{n+1}, \alpha_{n+1})$ *as the closest* *point projection of the trial state* $(\sigma_{n+1}^{\text{trial}}, \alpha_n)$ *onto the yield surface* $f(\sigma, \alpha) = 0$ *in the complementary energy* $\chi(\sigma, \alpha)$.

The interpretation of the algorithm in BOX **1.4** as optimality conditions of a convex minimization problem is of *fundamental significance* in the three-dimensional theory, especially in the presence of complicated yield conditions, as is often the case in practice. This interpretation opens the possibility of applying a number of algorithms well developed in convex mathematical programming to solving the elastoplastic problem.

1.4.4 Extension to the Combined Isotropic/Kinematic Hardening Model

Next, we further illustrate the development of integrative algorithms for rate-independent plasticity by considering the slightly more general situation afforded by a combined isotropic/kinematic hardening mechanism, as summarized in BOX **1.3**. We shall see that the steps involved in formulating the algorithm are identical to those examined in detail in Section **1.4.2**. In fact, as shown in Chapter **3**, the procedure is completely general and applies, essentially without modification, to the most general three-dimensional plasticity model.

To address the numerical implementation of the model in BOX **1.3**, it proves convenient to introduce the auxiliary variable

$$\xi := \sigma - q, \tag{1.4.31}$$

known as the *relative stress*.

1.4.4.1 Trial elastic state.

In the present context, the auxiliary problem (1.4.10) obtained by *freezing plastic flow* in the interval $[t_n, t_{n+1}]$ now takes the form

$$\boxed{\begin{aligned}
\sigma_{n+1}^{\text{trial}} &:= E\left(\varepsilon_{n+1} - \varepsilon_n^p\right) \equiv \sigma_n + E\Delta\varepsilon_n, \\
\xi_{n+1}^{\text{trial}} &:= \sigma_{n+1}^{\text{trial}} - q_n, \\
\varepsilon_{n+1}^{p^{\text{trial}}} &:= \varepsilon_n^p, \\
\alpha_{n+1}^{\text{trial}} &:= \alpha_n, \\
q_{n+1}^{\text{trial}} &:= q_n, \\
f_{n+1}^{\text{trial}} &:= \left|\xi_{n+1}^{\text{trial}}\right| - [\sigma_Y + K\alpha_n].
\end{aligned}} \tag{1.4.32}$$

Again we observe that the trial state is determined solely in terms of the initial conditions $\{\varepsilon_n, \varepsilon_n^p, \alpha_n\}$ and the *given* incremental strain $\Delta\varepsilon_n$. An analysis entirely analogous to that carried out in detail in Section **1.4.2.2** leads to the following statement of the algorithmic counterpart of the Kuhn–Tucker loading/unloading conditions

$$f_{n+1}^{\text{trial}} \begin{cases} \leq 0 \Rightarrow & \text{elastic step} \quad \Delta\gamma = 0, \\ > 0 \Rightarrow & \text{plastic step} \quad \Delta\gamma > 0. \end{cases} \tag{1.4.33}$$

If the step is elastic in the sense of (1.4.33), the trial state is the actual solution of the incremental problem associated with the constitutive model in BOX **1.3**. On the other hand, if the step is plastic, a closed form algorithm is constructed as follows.

1.4.4.2 The return-mapping algorithm.

The discrete algorithmic equations are obtained from the continuum model in BOX **1.3** by applying an implicit backward Euler difference scheme. Proceeding along the same lines leading to equations (1.4.17), the final result expressed as (we invite the reader to supply the necessary details)

$$\left.\begin{aligned} \sigma_{n+1} &= \sigma_{n+1}^{\text{trial}} - \Delta\gamma E \operatorname{sign}(\xi_{n+1}), \\ \varepsilon_{n+1}^{p} &= \varepsilon_n^p + \Delta\gamma \operatorname{sign}(\xi_{n+1}), \\ \alpha_{n+1} &= \alpha_n + \Delta\gamma, \\ q_{n+1} &= q_n + \Delta\gamma H \operatorname{sign}(\xi_{n+1}), \\ f_{n+1} &:= \left|\xi_{n+1}\right| - (\sigma_Y + K\alpha_{n+1}) = 0, \end{aligned}\right\} \tag{1.4.34}$$

where

$$\xi_{n+1} := \sigma_{n+1} - q_{n+1}. \tag{1.4.35}$$

The crucial step involved in the closed-form solution of the discrete problem (1.4.34)-(1.4.35) relies on exploiting an expression for ξ_{n+1} obtained as follows. By subtracting $(1.4.34)_4$ from $(1.4.34)_1$ and using definition (1.4.35),

$$\xi_{n+1} = (\sigma_{n+1}^{\text{trial}} - q_n) - \Delta\gamma(E + H)\operatorname{sign}(\xi_{n+1}). \tag{1.4.36}$$

Now we use the fact that $\xi_{n+1}^{\text{trial}} := \sigma_{n+1}^{\text{trial}} - q_n$ and rearrange terms in (1.4.36) to obtain

$$\left[\left|\xi_{n+1}\right| + \Delta\gamma(E + H)\right] \operatorname{sign}(\xi_{n+1}) = \left|\xi_{n+1}^{\text{trial}}\right| \operatorname{sign}(\xi_{n+1}^{\text{trial}}). \tag{1.4.37}$$

Since $\Delta\gamma > 0$ and $H + E > 0$ by assumption, it necessarily follows that the coefficient of $\operatorname{sign}(\xi_{n+1})$ in (1.4.37) must be positive. Therefore, (1.4.37) implies the result

$$\boxed{\operatorname{sign}(\xi_{n+1}) = \operatorname{sign}(\xi_{n+1}^{\text{trial}}),} \tag{1.4.38}$$

along with the condition

$$\left|\xi_{n+1}\right| + \Delta\gamma[E + H] = \left|\xi_{n+1}^{\text{trial}}\right|. \tag{1.4.39}$$

Now the incremental plastic consistency parameter $\Delta\gamma > 0$ is determined from the consistency requirement $(1.4.34)_5$ by using (1.4.39) and $(1.4.34)_3$:

$$\begin{aligned} f_{n+1} &= \left|\xi_{n+1}^{\text{trial}}\right| - (E+H)\Delta\gamma - [\sigma_Y + K\alpha_{n+1}] \\ &= \left|\xi_{n+1}^{\text{trial}}\right| - (E+H)\Delta\gamma - \left[\sigma_Y + K\alpha_n\right] - K\left(\alpha_{n+1} - \alpha_n\right) \\ &= f_{n+1}^{\text{trial}} - \Delta\gamma[E + (K+H)] = 0. \end{aligned} \tag{1.4.40}$$

Solving this algebraic equation for $\Delta\gamma$ yields the following result:

$$\boxed{\Delta\gamma = \frac{f_{n+1}^{\text{trial}}}{E + [K+H]} > 0.} \tag{1.4.41}$$

Thus, conditions (1.4.38) and (1.4.41) completely determine the algorithm in (1.4.34). For convenience, a step-by-step outline of the overall computational scheme is given in BOX **1.5.**

BOX 1.5. Return-Mapping Algorithm for One-Dimensional, Rate-Independent Plasticity. Combined Isotropic/Kinematic Hardening.

1. Database at $x \in \mathcal{B} : \left\{\varepsilon_n^p, \alpha_n, q_n\right\}$.

2. *Given* strain field at $x \in \mathcal{B} : \varepsilon_{n+1} = \varepsilon_n + \Delta\varepsilon_n$.

3. Compute elastic trial stress and test for plastic loading

$$\begin{aligned} \sigma_{n+1}^{\text{trial}} &:= E\left(\varepsilon_{n+1} - \varepsilon_n^p\right) \\ \xi_{n+1}^{\text{trial}} &:= \sigma_{n+1}^{\text{trial}} - q_n \\ f_{n+1}^{\text{trial}} &:= \left|\xi_{n+1}^{\text{trial}}\right| - \left[\sigma_Y + K\alpha_n\right] \end{aligned}$$

IF $f_{n+1}^{\text{trial}} \leq 0$ THEN

Elastic step: set $(\bullet)_{n+1} = (\bullet)_{n+1}^{\text{trial}}$ & *EXIT*

ELSE

Plastic step: Proceed to step **4.**

ENDIF

4. Return mapping

$$\begin{aligned} \Delta\gamma &:= \frac{f_{n+1}^{\text{trial}}}{E + [K+H]} > 0 \\ \sigma_{n+1} &:= \sigma_{n+1}^{\text{trial}} - \Delta\gamma E \operatorname{sign}(\xi_{n+1}^{\text{trial}}) \\ \varepsilon_{n+1}^p &:= \varepsilon_n^p + \Delta\gamma \operatorname{sign}(\xi_{n+1}^{\text{trial}}) \\ q_{n+1} &:= q_n + \Delta\gamma H \operatorname{sign}(\xi_{n+1}^{\text{trial}}) \\ \alpha_{n+1} &:= \alpha_n + \Delta\gamma \end{aligned}$$

1.5 Finite-Element Solution of the Elastoplastic IBVP. An Illustration

To illustrate the role of the integrative algorithms developed in the preceding section, here we outline a typical numerical solution scheme for the elastoplastic IBVP within the context of the finite-element method.

The point of departure in our developments is the weak form of the IBVP defined by the variational equation (1.3.11). For simplicity, in our discussion we assume that the displacement boundary conditions are homogeneous, i.e.,

$$u\big|_{\partial_u\mathcal{B}} = \bar{u} \equiv 0,$$

and (1.5.1)

$$\partial_u\mathcal{B} = \{0\}.$$

For loading, i.e., for given body forces and boundary tractions given by functions

$$b : \mathcal{B} \times [0, \mathrm{T}] \to \mathbb{R}$$

and (1.5.2)

$$\bar{\sigma} : \partial_\sigma\mathcal{B} \times [0, \mathrm{T}] \to \mathbb{R},$$

the problem is the numerical approximation for the solution $u(x, t)$ to (1.3.11), where $\sigma(x, t)$ is assumed to satisfy the local constitutive equations in BOX **1.3**. Note that (1.3.11) must hold for *all* $t \in [0, \mathrm{T}]$. In addition, because of the simplifying assumption (1.5.1), the displacement field $u : \mathcal{B} \times [0, \mathrm{T}] \to \mathbb{R}$ is such that $u(\cdot, t) \in \mathbb{V}$ for all $t \in [0, \mathrm{T}]$.

A typical algorithmic scheme for the solution of (1.3.11) is briefly discussed below. Subsequently we describe a typical incremental solution procedure restricted, for simplicity, to the quasi-static problem obtained by neglecting inertial forces, namely, $G(\sigma, \eta) = 0$ for all $\eta \in \mathbb{V}$.

1.5.1 Spatial Discretization. Finite-Element Approximation

Conceptually, within the context of the simplest finite-element method, one proceeds as follows (see Hughes [1987, Chapters 1 and 2] for a detailed account of these ideas).

i. The domain $\mathcal{B} = [0, L]$ is discretized into a sequence of nonoverlapping elements

$$\mathcal{B}_e = [x_e, x_{e+1}],$$

and (1.5.3)

$$\mathcal{B} = \bigcup_{e=1}^{n_{\mathrm{el}}} \mathcal{B}_e$$

where $x_1 = 0$ and $x_{n_{el}+1} = L$. We let $h_e := x_{e+1} - x_e$ be the "mesh size" which, for simplicity, is assumed to be uniform.

ii. Then the simplest (conforming) finite-dimensional approximation to $\mathbb{V}$, denoted by $\mathbb{V}^h \subset \mathbb{V}$, is constructed as follows. The restriction w_e^h to a typical element $\mathcal{B}_e$ of a test function $w^h \in \mathbb{V}^h$ is locally interpolated linearly as

$$w_e^h := \sum_{a=1}^{2} N_e^a(x) w_e^a, \tag{1.5.4}$$

where $N_e^a : \mathcal{B}_e \to \mathbb{R}$, $a = 1, 2$, are the linear shape functions given by

$$\begin{aligned} N_e^1 &= \frac{x_{e+1} - x}{h_e}, \\ N_e^2 &= \frac{x - x_e}{h_e}, \\ x &\in \mathcal{B}_e, \end{aligned} \tag{1.5.5}$$

and $\boldsymbol{w}_e = [w_e^1, w_e^2]^T$ is the vector containing the nodal values of the *local* element test functions. Then, a *global, piecewise*, continuous function $w^h \in \mathbb{V}^h$ is obtained from the above element interpolation by *matching* the value of $\boldsymbol{w}_e$ at the nodes:

$$w_e := w_e^1 \equiv w_{e-1}^2,$$

and (1.5.6)

$$w_{e+1} := w_e^2 \equiv w_{e+1}^1 \, .$$

iii. The computation of $G(\sigma^h, w^h)$, given by (1.3.11) with $w^h \in \mathbb{V}^h$ (and u^h also in $\mathbb{V}^h$ since, by assumption $\bar{u} \equiv 0$), is performed in an *element-by-element* fashion by setting, in view of $(1.5.3)_2$,

$$G\left(\sigma^h, w^h\right) = \sum_{e=1}^{n_{el}} G_e\left(\sigma^h, w^h\right). \tag{1.5.7}$$

For a typical element $\mathcal{B}_e$, first one computes

$$\frac{\partial}{\partial x} w_e^h = \left[\frac{\partial}{\partial x} N_e^1 \quad \frac{\partial}{\partial x} N_e^2\right] \boldsymbol{w}_e =: \boldsymbol{B}_e \boldsymbol{w}_e, \tag{1.5.8}$$

where, in the present simple context, from (1.5.5),

$$\boldsymbol{B}_e := \left[\frac{\partial}{\partial x} N_e^1 \quad \frac{\partial}{\partial x} N_e^2\right] = \left[-\frac{1}{h_e} \quad \frac{1}{h_e}\right]. \tag{1.5.9}$$

Therefore, from expression (1.3.11), we obtain

$$G_e\left(\sigma^h, w^h\right) = \boldsymbol{w}_e^T \left[\boldsymbol{f}_e^{\text{int}}\left(\sigma^h\right) - \boldsymbol{f}_e^{\text{ext}}(t)\right], \tag{1.5.10}$$

where

$$f_e^{\text{int}}(\sigma^h) := \int_{\mathcal{B}_e} \boldsymbol{B}_e^T \sigma^h(x, t) dx \tag{1.5.11}$$

is the so-called *element internal force vector*. Note that f_e^{int} is implicitly a function of u^h along with (ε^p, α) through the constitutive equations in BOX **1.2**. In addition,

$$f_e^{\text{ext}} := \int_{\mathcal{B}_e} \begin{Bmatrix} N_e^1 \\ N_e^2 \end{Bmatrix} \rho b(x, t) dx + \left[\bar{\sigma}(t) \begin{Bmatrix} N_e^1 \\ N_e^2 \end{Bmatrix} \right] \Bigg|_{\partial\mathcal{B}_e \cap \partial_\sigma \mathcal{B}} \tag{1.5.12}$$

is referred to as the *element external load vector*. By using condition (1.5.1) and (1.5.6), expression (1.5.7) is *assembled* from the element contributions given by (1.5.11) and (1.5.12) as follows:

$$G^h(u^h, w^h) =: \boldsymbol{w}^T \left[\boldsymbol{F}^{\text{int}}(\sigma^h) - \boldsymbol{F}^{\text{ext}}(t) \right], \tag{1.5.13}$$

where $\boldsymbol{w}^T := [w^2, w^3, \ldots, w^{n_{el}+1}] \in \mathbb{R}^{n_{el}}$. The global force vectors are computed from element contributions, viz.,

$$\left. \begin{aligned} \boldsymbol{F}^{\text{int}}(\sigma^h) &= \mathop{\mathbf{A}}_{e=1}^{n_{el}} f_e^{\text{int}}(\sigma^h) \\ \boldsymbol{F}^{\text{ext}}(t) &= \mathop{\mathbf{A}}_{e=1}^{n_{el}} f_e^{\text{ext}}(t), \end{aligned} \right\} \tag{1.5.14}$$

where $\mathbf{A}$ is the standard, finite-element, assembly operator (see Hughes [1987] for further details).

iv. The computation of the inertial term $\int_{\mathcal{B}} \partial v^h/\partial t \, w^h \, dx$, where $v^h = \partial u^h/\partial t$ is in $\mathbb{V}^h$, is performed in an element-by-element fashion exactly as in evaluating the static term $G(\sigma^h, w^h)$. The final result takes a form entirely analogous to (1.5.13), namely,

$$\int_{\mathcal{B}} \rho \frac{\partial^2 u^h}{\partial t^2} w^h \, dx = \boldsymbol{w}^T \boldsymbol{M} \ddot{\boldsymbol{d}}, \tag{1.5.15}$$

where $\boldsymbol{M}$ is the *mass matrix* computed by assembling the element mass matrices according to the standard expression

$$\boldsymbol{M} = \mathop{\mathbf{A}}_{e=1}^{n_{el}} \boldsymbol{m}_e$$

with (1.5.16)

$$m_e^{ab} := \int_{\mathcal{B}^e} \rho \, N_e^a \, N_e^b \, dx, \quad a, b = 1, 2.$$

This result defines the so-called consistent mass matrix. Diagonal and high-order mass matrices are obtained by a number of techniques, including special quadrature formulas (see e.g., Hughes [1987]).

The finite-element counterpart of the weak form (1.3.11) of the momentum equation takes the form

$$\int_{B} \frac{\partial^2 u^h}{\partial t^2} w^h \, dx + G(\sigma^h, w^h) = 0 \quad \text{for all} \quad w^h \in \mathbb{V}^h. \tag{1.5.17}$$

Since the test function w^h is arbitrary, it follows that $\boldsymbol{w} \in \mathbb{R}^N$ is also arbitrary and from (1.5.13) along with (1.5.15) one arrives at the discrete system of (nonlinear) differential equations:

$$\boldsymbol{M}\ddot{\boldsymbol{d}} + \boldsymbol{F}^{\text{int}}(\sigma^h) - \boldsymbol{F}^{\text{ext}}(t) = 0, \tag{1.5.18}$$

which constitutes the discrete counterpart of the momentum equations.

The crucial step in the outline given above which remains to be addressed concerns the *computation of the stress field* $\sigma^h(x, t)$ *within a typical element for time* $t \in [0, \mathrm{T}]$. First, we make a crucial observation concerning numerical quadrature.

1.5.1.1 The role of numerical quadrature.

The element internal force vector given by (1.5.11) is evaluated by numerical quadrature according to the formula (see e.g. Hughes [1987, Chapter **3**])

$$\boldsymbol{f}_e^{\text{int}}(\sigma^h) = \sum_{\ell=1}^{n_{\text{int}}} \boldsymbol{B}_e^T \sigma^h(x, t)\big|_{x=x_e^\ell} w^\ell h_e, \tag{1.5.19}$$

where $x_e^\ell \in \mathcal{B}_e$ denotes a quadrature point, w^ℓ is the corresponding weight, and n_{int} is the number of quadrature points for element $\mathcal{B}_e$. In a more general context, (1.5.19) is implemented through isoparametric mapping.

The important conclusion to be extracted from expression (1.5.19), also valid in a general finite-element formulation, is that *the stress within an element* $\mathcal{B}_e$ *is required only at discrete points; typically the quadrature points* x_e^ℓ *of the element.*

1.5.2 Incremental Solution Procedure

For simplicity, we restrict our attention to the static problem obtained by neglecting $\boldsymbol{M}\ddot{\boldsymbol{d}}$ in the discrete system (1.5.18). Then an iterative solution procedure for the resulting quasi-static elastoplastic problem proceeds as follows:

1.5.2.1 Incremental loading.

Let $[0, \mathrm{T}]$ be the time interval of interest. As in Section **1.4**, consider a partition

$$[0, \mathrm{T}] = \bigcup_{n=1}^{M} [t_n, t_{n+1}]. \tag{1.5.20}$$

Let $x_e^\ell \in \mathcal{B}_e$ be the quadrature points of a typical finite element $\mathcal{B}_e$, and let $\{\varepsilon_n^p, \alpha_n\}$ be the *internal* variables at x_e^ℓ. We assume that the body is equilibrated at $t = t_n$ under forces $b_n : \mathcal{B} \to \mathbb{R}$ and $\bar{\sigma}_n : \partial_\sigma \mathcal{B} \to \mathbb{R}$, so that the stress field σ_n^h satisfies

$$\boldsymbol{F}^{\text{int}}(\sigma_n) - \boldsymbol{F}_n^{\text{ext}} = \boldsymbol{0}. \tag{1.5.21}$$

The associated *displacement field* at t_n is $u_n^h \in \mathbb{V}^h$. Now consider an incremental load $(\Delta b_n, \Delta\bar{\sigma}_n)$ so that

$$b_{n+1} = b_n + \Delta b_n,$$

and (1.5.22)

$$\bar{\sigma}_{n+1} = \bar{\sigma}_n + \Delta\bar{\sigma}_n$$

is the loading at $t_{n+1} \in [0, \mathrm{T}]$, which defines a discrete *external load vector* $\boldsymbol{F}_{n+1}^{\text{ext}}$. This constitutes the given data. The problem can be stated as follows:

> Find $\Delta u_n^h \in \mathbb{V}^h$, the updated displacement field $u_{n+1}^h = u_n^h + \Delta u_{n+1}^h$, the updated internal variables $\{\varepsilon_{n+1}^p, \alpha_{n+1}, q_{n+1}\}$, and the stress field σ_{n+1}^h (at discrete points $x_e^\ell \in \mathcal{B}_e$) such that
>
> **i.** $\boldsymbol{F}^{\text{int}}(\sigma_{n+1}^h) - \boldsymbol{F}_{n+1}^{\text{ext}} = \boldsymbol{0}$ (equilibrium) and
>
> **ii.** the discrete constitutive equations in BOX **1.5** hold.

(1.5.23)

To simplify the notation in what follows, the superscript "h" is omitted.

1.5.2.2 Iterative solution procedure.

The solution to problem (1.5.23) is obtained by an iterative solution procedure which proceeds as follows. We let $(\bullet)_{n+1}^{(k)}$ be the value of a variable $(\bullet)$ at the kth iteration during the load step in $[t_n, t_{n+1}]$. Accordingly,

i. let $\Delta\boldsymbol{d}_{n+1}^{(k)}$ be the incremental *nodal* displacement at the kth iteration, and let

$$\boldsymbol{d}_{n+1}^{(k)} := \boldsymbol{d}_n + \Delta\boldsymbol{d}_{n+1}^{(k)} \tag{1.5.24}$$

be the total nodal displacement. Since $u^h \in \mathbb{V}^h$, the displacement field over a typical element $\mathcal{B}_e$ is given by an expression having the same form as (1.5.4), and the strain field is computed by an expression analogous to that in (1.5.8):

$$\varepsilon_{n+1}^{(k)}\big|_{\mathcal{B}_e} = \boldsymbol{B}_e \boldsymbol{d}_e\big|_{n+1}^{(k)}. \tag{1.5.25}$$

ii. given the strain field (1.5.25), at each quadrature point $x_e^\ell \in \mathcal{B}_e$, we *compute the stress* $\sigma_{n+1}^{(k)}$ by means of the algorithm in BOX **1.5**;

iii. we evaluate the internal force vector $\boldsymbol{f}_e^{\text{int}}(\sigma_{n+1}^h)$ by (1.5.19) and assemble the contribution of all elements by (1.5.14);

iv. *check convergence*: if (1.5.18) is satisfied for $\sigma = \sigma_{n+1}^{(k)}$ then $(\bullet)_{n+1}^{(k)}$ *is the solution*; otherwise, continue; and

v. *determine* $\Delta \boldsymbol{d}_{n+1}^{(k)} \in \mathbb{R}^N$, set $k \leftarrow k+1$, and go to step **i**.

The only step in the solution procedure outlined above which remains to be addressed is determining $\Delta \boldsymbol{d}_{n+1}^{(k)}$ (step *v*). Although several schemes are possible, here we consider determinating $\Delta \boldsymbol{d}_{n+1}^{(k)}$ by linearizing $\boldsymbol{f}^{\text{int}}(\sigma_{n+1}^{(k)})$ about the current state, defined by $\boldsymbol{d}_{n+1}^{(k)}$.

First, since the assembly operator $\mathbf{A}$ is linear, by the chain rule,

$$\begin{aligned}
\frac{\partial \boldsymbol{F}^{\text{int}}(\sigma_{n+1}^{(k)})}{\partial \boldsymbol{d}_{n+1}^{(k)}} \Delta \boldsymbol{d}_{n+1}^{(k+1)} &= \mathop{\mathbf{A}}_{e=1}^{n_{\text{el}}} \frac{\partial \boldsymbol{f}_e^{\text{int}}(\sigma_{n+1}^{(k)})}{\partial \boldsymbol{d}_e\big|_{n+1}^{(k)}} \Delta \boldsymbol{d}_e\big|_{n+1}^{(k+1)} \\
&= \mathop{\mathbf{A}}_{e=1}^{n_{\text{el}}} \int_{\mathcal{B}_e} \boldsymbol{B}_e^T \left[\frac{\partial \sigma_{n+1}^{(k)}}{\partial \varepsilon_{n+1}^{(k)}}\right] \frac{\partial \varepsilon_{n+1}^{(k)}}{\partial \boldsymbol{d}_e\big|_{n+1}^{(k)}} \Delta \boldsymbol{d}_e\big|_{n+1}^{(k+1)} dx \\
&= \mathop{\mathbf{A}}_{e=1}^{n_{\text{el}}} \left[\int_{\mathcal{B}_e} \boldsymbol{B}_e^T \left[\frac{\partial \sigma_{n+1}^{(k)}}{\partial \varepsilon_{n+1}^{(k)}}\right] \boldsymbol{B}_e dx\right] \Delta \boldsymbol{d}_e\big|_{n+1}^{(k+1)}. \qquad (1.5.26)
\end{aligned}$$

Next, we introduce a matrix $\boldsymbol{k}_e\big|_{n+1}^{(k)} \in \mathbb{R}^{2\times 2}$, called the *element tangent stiffness matrix*, and defined as

$$\boldsymbol{k}_e\big|_{n+1}^{(k)} := \int_{\mathcal{B}_e} \boldsymbol{B}_e^T \left[\frac{\partial \sigma_{n+1}^{(k)}}{\partial \varepsilon_{n+1}^{(k)}}\right] \boldsymbol{B}_e dx. \qquad (1.5.27)$$

By performing an assembly operation similar to that in (1.5.14) (see e.g. Hughes [1987, Chapter **1**] for a detailed description), we arrive at

$$\frac{\partial \boldsymbol{F}^{\text{int}}(\sigma_{n+1}^{(k)})}{\partial \boldsymbol{d}_{n+1}^{(k)}} \Delta \boldsymbol{d}_{n+1}^{(k+1)} = \boldsymbol{K}_{n+1}^{(k)} \Delta \boldsymbol{d}_{n+1}^{(k+1)}; \qquad \boldsymbol{K}_{n+1}^{(k)} = \mathop{\mathbf{A}}_{e=1}^{n_{\text{el}}} \boldsymbol{k}_e\big|_{n+1}^{(k)}, \qquad (1.5.28)$$

where $\boldsymbol{K}_{n+1}^{(k)}$ is called the global stiffness matrix at time t_{n+1} and iteration (k). With expression (1.5.28) in hand, we estimate $\Delta \boldsymbol{d}_{n+1}^{(k+1)}$ by replacing the equilibrium equation (1.5.18) with the *linear* approximation

$$\left[\boldsymbol{F}^{\text{int}}(\sigma_{n+1}^{(k)}) - \boldsymbol{F}_{n+1}^{\text{ext}}\right] + \frac{\partial \boldsymbol{F}^{\text{int}}(\sigma_{n+1}^{(k)})}{\partial \boldsymbol{d}_{n+1}^{(k)}} \Delta \boldsymbol{d}_{n+1}^{(k+1)} = \boldsymbol{0}. \qquad (1.5.29)$$

We note that all the terms in this *linear* equation are known, except for $\Delta \boldsymbol{d}_{n+1}^{(k+1)}$, *provided that one can compute* $\partial \sigma_{n+1}^{(k)} / \partial \varepsilon_{n+1}^{(k)}$. Assuming this to be the case, from (1.5.28) and (1.5.29), we obtain the following expression:

$$\Delta \boldsymbol{d}_{n+1}^{(k+1)} = -\left[\boldsymbol{K}_{n+1}^{(k)}\right]^{-1} \left[\boldsymbol{F}^{\text{int}}(\sigma_{n+1}^{(k)}) - \boldsymbol{F}_{n+1}^{\text{ext}}\right]. \qquad (1.5.30)$$

Using this formula in the solution scheme **i–v**, results in a procedure equivalent to the classical Newtonian method.

1.5.2.3 Algorithmic tangent modulus. Combined isotropic/kinematic hardening.

To complete the algorithmic procedure discussed above, there only remains to be computed an explicit expression for the coefficient

$$\mathbf{C}_{n+1}^{(k)} := \frac{\partial \sigma_{n+1}^{(k)}}{\partial \varepsilon_{n+1}^{(k)}} \tag{1.5.31}$$

in expression (1.5.27). One refers to (1.5.31) as the *algorithmic tangent modulus*, a notion first introduced in Simo and Taylor [1986]. The procedure used to derive a closed-form expression for $\mathbf{C}_{n+1}^{(k)}$ entails differentiating the update formulas in BOX **1.5**, as explained below.

For clarity we omit the superindex k in the following development. First, from step **3** in BOX **1.5** we obtain (note that ε_n^p, α_n and q_n are constants in the derivation that follows)

$$\left.\begin{aligned} \frac{\partial \sigma_{n+1}^{\text{trial}}}{\partial \varepsilon_{n+1}} &= E \\ \frac{\partial \xi_{n+1}^{\text{trial}}}{\partial \varepsilon_{n+1}} &= \frac{\partial \sigma_{n+1}^{\text{trial}}}{\partial \varepsilon_{n+1}} = E. \end{aligned}\right\} \tag{1.5.32}$$

Using these results and differentiating the first formula in step **4**, assuming that $f_{n+1}^{\text{trial}} > 0$, leads to

$$\begin{aligned} \frac{\partial(\Delta\gamma)}{\partial \varepsilon_{n+1}} &= \frac{1}{E + [K + H]} \frac{\partial f_{n+1}^{\text{trial}}}{\partial \varepsilon_{n+1}} \\ &= \frac{1}{E + [K + H]} \frac{\partial |\xi_{n+1}^{\text{trial}}|}{\partial \xi_{n+1}^{\text{trial}}} \frac{\partial \xi_{n+1}^{\text{trial}}}{\partial \varepsilon_{n+1}} \\ &= \frac{E}{E + [K + H]} \operatorname{sign}\left(\xi_{n+1}^{\text{trial}}\right). \end{aligned} \tag{1.5.33}$$

Next, we rearrange the second formula in step **4** of BOX **1.5** as follows:

$$\begin{aligned} \sigma_{n+1} &= (\sigma_{n+1}^{\text{trial}} - q_n) + q_n - \Delta\gamma E \operatorname{sign}(\xi_{n+1}^{\text{trial}}) \\ &= q_n + \xi_{n+1}^{\text{trial}} - \Delta\gamma E \operatorname{sign}(\xi_{n+1}^{\text{trial}}) \\ &= q_n + \left[1 - \frac{\Delta\gamma E}{|\xi_{n+1}^{\text{trial}}|}\right] \xi_{n+1}^{\text{trial}}. \end{aligned} \tag{1.5.34}$$

Finally, we differentiate the algorithmic constitutive equation (1.5.34) with respect to ε_{n+1} by using the chain rule along with relationships $(1.5.32)_2$ and (1.5.33). Then we obtain

$$\frac{\partial \sigma_{n+1}}{\partial \varepsilon_{n+1}} = \left[1 - \frac{\Delta\gamma E}{\left|\xi_{n+1}^{\text{trial}}\right|}\right] E + \frac{\Delta\gamma E}{\left|\xi_{n+1}^{\text{trial}}\right|^2} \xi_{n+1}^{\text{trial}} \frac{\partial \left|\xi_{n+1}^{\text{trial}}\right|}{\partial \varepsilon_{n+1}}$$

$$- \frac{E}{\left|\xi_{n+1}^{\text{trial}}\right|} \frac{E}{E + [K + H]} \xi_{n+1}^{\text{trial}} \operatorname{sign}\left(\xi_{n+1}^{\text{trial}}\right)$$

$$= E\left[1 - \frac{\Delta\gamma E}{\left|\xi_{n+1}^{\text{trial}}\right|}\right] + \frac{\Delta\gamma E^2}{\left|\xi_{n+1}^{\text{trial}}\right|^2} \xi_{n+1}^{\text{trial}} \operatorname{sign}\left(\xi_{n+1}^{\text{trial}}\right) - \frac{E^2}{E + [K + H]}$$

$$= E\left[1 - \frac{\Delta\gamma E}{\left|\xi_{n+1}^{\text{trial}}\right|}\right] + \frac{\Delta\gamma E^2}{\left|\xi_{n+1}^{\text{trial}}\right|} - \frac{E^2}{E + [K + H]}$$

$$= \frac{E[K + H]}{(E + [K + H])}, \quad \text{for} \quad f_{n+1}^{\text{trial}} > 0. \tag{1.5.35}$$

Since $\sigma_{n+1} = \sigma_{n+1}^{\text{trial}}$ for $f_{n+1}^{\text{trial}} \leq 0$, from (1.5.32) and (1.5.35),

$$\boxed{\mathbf{C}_{n+1}^{(k)} := \frac{\partial \sigma_{n+1}^{(k)}}{\partial \varepsilon_{n+1}^{(k)}} = \begin{cases} E & \text{iff} \quad f_{n+1}^{\text{trial}} \leq 0, \\ \dfrac{E[K + H]}{E + [K + H]} & \text{iff} \quad f_{n+1}^{\text{trial}} > 0. \end{cases}} \tag{1.5.36}$$

By comparing (1.5.36) and (1.2.47), we conclude that *for the one-dimensional problem, the algorithmic tangent modulus coincides with the elastoplastic tangent modulus*. However, as shown in subsequent chapters, *this result does not hold in higher dimensions*.

1.6 Stability Analysis of the Algorithmic IBVP

In this section we provide an introduction to the stability analysis of the algorithmic initial boundary-value problem obtained by combining the return mapping algorithms described above with a time-discretization of the algorithmic weak form. The methodology employed is based on an adaptation of the *energy method* of stability analysis. A detailed account of this technique is described in detail in a subsequent chapter. The notion of numerical stability for the problem at hand is directly inspired by the properties of the IBVP described above and is stated for the case of isotropic hardening as follows.

An algorithmic approximation to the IBVP is said to be stable if the numerical solution $\{u_n, v_n, \varepsilon_n^e, \alpha_n\}$ generated via such an approximation within a typical time step $[t_n, t_{n+1}]$ inherits the a priori estimate (1.3.27) on the exact solution to the IBVP, namely, if

$$L(u_{n+1}, v_{n+1}, \varepsilon_{n+1}^e, \alpha_{n+1}) \leq L(u_n, v_n, \varepsilon_n^e, \alpha_n). \tag{1.6.1}$$

If this inequality holds for *any* step-size $\Delta t := t_{n+1} - t_n > 0$, the algorithm is said to be *unconditionally stable*. On the other hand, if (1.6.1) holds only for time steps $\Delta t < \Delta t_{\text{crit}}$, where $\Delta t_{\text{crit}} > 0$ is a finite critical value, the algorithm is said to be *conditionally stable* with stability limit Δt_{crit}. Roughly speaking, this definition ensures that algorithmic approximations do not blow up, a natural condition to be required of the algorithm in view of the property (1.3.27) of the continuum problem. For hardening plasticity, this notion of stability is closely related to the concept of B-stability explained in detail in a subsequent chapter which, in turn, is motivated by the contractivity property of the IBVP.

1.6.1 Algorithmic Approximation to the Dynamic Weak Form.

To present the main ideas in the simplest possible context, we turn our attention to the dynamic problem and consider the following algorithmic time approximation to the dynamic weak form within the time interval $[t_n, t_{n+1}]$:

$$\left.\begin{aligned} \frac{1}{\Delta t}\int_B \rho(v_{n+1} - v_n)\,\eta\,dx + G(\sigma_{n+\vartheta}, \eta) &= 0 \text{ for all } \eta \in \mathbb{V}, \\ \frac{1}{\Delta t}(u_{n+1} - u_n) &= v_{n+\vartheta}, \end{aligned}\right\} \tag{1.6.2}$$

$$G(\sigma_{n+\vartheta}, \eta) := \int_B \sigma_{n+\vartheta}\,\eta'\,dx - \int_B \rho b_{n+\vartheta}\,\eta\,dx - \bar{\sigma}_{n+\vartheta}\,\eta\big|_{\partial_\sigma B}.$$

As before the subscripts n and $n+1$ refer to algorithmic approximations to the exact values at t_n and t_{n+1}, respectively, and the subscript $n+\vartheta$ refers to algorithmic approximations computed via the interpolation formulas

$$\left.\begin{aligned} u_{n+\vartheta} &:= \vartheta u_{n+1} + (1-\vartheta)u_n \\ v_{n+\vartheta} &:= \vartheta v_{n+1} + (1-\vartheta)v_n \\ \sigma_{n+\vartheta} &:= \vartheta\sigma_{n+1} + (1-\vartheta)\sigma_n \end{aligned}\right\} \quad \vartheta \in [0, 1]. \tag{1.6.3}$$

Expression (1.6.2) can be viewed as a generalized midpoint rule approximation to the dynamic weak form in which ϑ is a design parameter that defines the algorithm.

The first step in the stability analysis is deriving the algorithmic counterpart of the mechanical work identity (1.3.17). For simplicity, we shall assume that the forcing terms $\{b, \bar{\sigma}\}$ and the essential boundary condition $\bar{u}$ are time-independent. Choosing as a test function

$$\eta = \Delta t\, v_{n+\vartheta} = u_{n+1} - u_n \in \mathbb{V}, \tag{1.6.4}$$

and using of the algebraic identity

$$v_{n+\vartheta} = v_{n+\frac{1}{2}} + \left[\vartheta - \tfrac{1}{2}\right](v_{n+1} - v_n)$$

where (1.6.5)

$$v_{n+\frac{1}{2}} := \tfrac{1}{2}(v_{n+1} + v_n),$$

the integrand in the first term of the algorithmic weak form (1.6.2) becomes

$$\rho(v_{n+1} - v_n)\,v_{n+\frac{1}{2}} = \tfrac{1}{2}\rho v_{n+1}^2 - \tfrac{1}{2}\rho v_n^2 + (2\vartheta - 1)\tfrac{1}{2}\rho(v_{n+1} - v_n)^2. \tag{1.6.6}$$

Inserting this expression into (1.6.2) and using the definition of $V_{\text{ext}}(u)$ along with the assumption of a time-independent forcing yields

$$\begin{aligned}&\left[T(v_{n+1}) + V_{\text{ext}}(u_{n+1})\right] - \left[T(v_n) + V_{\text{ext}}(u_n)\right] + \int_{\mathcal{B}} \sigma_{n+\vartheta} \frac{\partial (u_{n+1} - u_n)}{\partial x}\, ds \\ &\quad = -(2\vartheta - 1) T(v_{n+1} - v_n).\end{aligned} \tag{1.6.7}$$

To interpret this result and illustrate its role in an analysis of numerical stability, first we consider the case of a linear elastic material.

1.6.1.1 Linear elasticity.

Setting $\varepsilon_{n+\vartheta} = \vartheta \varepsilon_{n+1} + (1 - \vartheta)\vartheta \varepsilon_n$ and using the identity

$$\varepsilon_{n+\vartheta} = \varepsilon_{n+\frac{1}{2}} + \left[\vartheta - \tfrac{1}{2}\right](\varepsilon_{n+1} - \varepsilon_n) \tag{1.6.8}$$

with $\varepsilon_{n+\frac{1}{2}} := \frac{1}{2}(\varepsilon_{n+1} + \varepsilon_n)$, from the elastic constitutive equation $\sigma_{n+\vartheta} = E\varepsilon_{n+\vartheta}$, we obtain

$$\sigma_{n+\vartheta}(\varepsilon_{n+1} - \varepsilon_n) = \tfrac{1}{2} E\varepsilon_{n+1}^2 - \tfrac{1}{2} E\varepsilon_n^2 + (2\vartheta - 1)\tfrac{1}{2} E(\varepsilon_{n+1} - \varepsilon_n)^2. \tag{1.6.9}$$

Inserting this result into (1.6.7) and recalling that $V_{\text{int}}(u) = \frac{1}{2}\int_{\mathcal{B}} E\varepsilon^2\, dx$ and the definition of $L(u, v, \varepsilon)$ for linear elasticity yields the result

$$\begin{aligned}&L(u_{n+1}, v_{n+1}, \varepsilon_{n+1}) - L(u_n, v_n, \varepsilon_n) \\ &\quad = -(2\vartheta - 1) \underbrace{\left[T(v_{n+1} - v_n) + V_{\text{int}}(\varepsilon_{n+1} - \varepsilon_n)\right]}_{\text{Quadratic form } \geq 0}.\end{aligned} \tag{1.6.10}$$

Inspection of this result reveals that the a priori stability estimate (1.6.1) holds, provided that $2\vartheta - 1 \geq 0$. In fact,

a. if $\vartheta < \frac{1}{2}$, then $L(u_{n+1}, v_{n+1}, \varepsilon_{n+1}) - L(u_n, v_n, \varepsilon_n) > 0$, and the algorithm is *unconditionally unstable*;
b. if $\vartheta \geq \frac{1}{2}$, then $L(u_{n+1}, v_{n+1}, \varepsilon_{n+1}) - L(u_n, v_n, \varepsilon_n) \leq 0$, and the algorithm is *unconditionally stable*; and
c. if $\vartheta = \frac{1}{2}$, then $L(u_{n+1}, v_{n+1}, \varepsilon_{n+1}) - L(u_n, v_n, \varepsilon_n) = 0$, and the algorithm *conserves exactly* the total potential energy of the elastic system.

Algorithms for $\vartheta > \frac{1}{2}$ exhibit numerical dissipation and are only first-order accurate. Second-order accuracy holds only for $\vartheta = \frac{1}{2}$. The preceding results demonstrate that a stability analysis based on the stability criterion (1.6.1) reproduces the well-known stability results for the generalized midpoint rule applied to linear elastodynamics.

1.6.1.2 Elastoplasticity with isotropic hardening.

To better illustrate the applicability of the present analysis (nonlinear) of numerical stability, we consider the following generalization of the return-mapping algorithms described in the preceding sections:

$$\left.\begin{aligned} \varepsilon^p_{n+1} - \varepsilon^p_n &= \Delta\gamma \,\mathrm{sign}(\sigma_{n+\vartheta}), \\ \alpha_{n+1} - \alpha_n &= \Delta\gamma, \end{aligned}\right\} \tag{1.6.11}$$

$$\Delta\gamma \geq 0, \quad f(\sigma_{n+\delta}, \alpha_{n+\delta}) \leq 0 \quad \text{and} \quad \Delta\gamma f(\sigma_{n+\delta}, \alpha_{n+\delta}) = 0.$$

where, here, $\Delta\gamma = \gamma_{n+\vartheta}\Delta t$. Here, the algorithmic parameter $\vartheta \in [0, 1]$ has the same meaning as before, whereas $\delta \in [0, 1]$ is an additional algorithmic parameter that establishes the point at which the algorithmic consistency condition is to be enforced. Clearly, if $\vartheta = \delta = 1$, one recovers the standard return-mapping algorithms described above. The choice $\delta = \vartheta \in [0, 1]$ gives the generalized return maps proposed in Simo and Taylor [1986]. The choice of $\delta = 1$ and $\vartheta \in [0, 1]$ yields the class of algorithms proposed in Ortiz and Popov [1985].

To assess the stability properties of the preceding approximation, first we use the stress-strain relationship and the algorithmic equations (1.6.11) to compute

$$\begin{aligned} \varepsilon_{n+1} - \varepsilon_n &= E^{-1}(\sigma_{n+1} - \sigma_n) + \varepsilon^p_{n+1} - \varepsilon^p_n \\ &= E^{-1}(\sigma_{n+1} - \sigma_n) + \Delta\gamma \,\mathrm{sign}(\sigma_{n+\vartheta}). \end{aligned} \tag{1.6.12}$$

Once more using the identity $\sigma_{n+\vartheta} = \sigma_{n+\frac{1}{2}} + [\vartheta - \frac{1}{2}](\sigma_{n+1} - \sigma_n)$ along with the fact that $\mathrm{sign}(\sigma_{n+\vartheta})\sigma_{n+\vartheta} = |\sigma_{n+\vartheta}|$ yields the result

$$\sigma_{n+\vartheta}(\varepsilon_{n+1} - \varepsilon_n) = \frac{1}{2E}\sigma^2_{n+1} - \frac{1}{2E}\sigma^2_n + (2\vartheta - 1)\frac{1}{2E}(\sigma_{n+1} - \sigma_n)^2 + \Delta\gamma|\sigma_{n+\vartheta}|. \tag{1.6.13}$$

Next, we use the definition of the yield criterion and the algorithmic equation $\Delta\gamma = \alpha_{n+1} - \alpha_n$ to rewrite the last term in (1.6.13) as

$$\Delta\gamma|\sigma_{n+\vartheta}| = \Delta\gamma f(\sigma_{n+\vartheta}, \alpha_{n+\vartheta}) + \Delta\gamma\sigma_Y + K\alpha_{n+\vartheta}(\alpha_{n+1} - \alpha_n). \tag{1.6.14}$$

Using the identity $\alpha_{n+\vartheta} = \alpha_{n+\frac{1}{2}} + (\vartheta - \frac{1}{2})(\alpha_{n+1} - \alpha_n)$, we can express this result in the equivalent form

$$\begin{aligned} \Delta\gamma|\sigma_{n+\vartheta}| = {} & \Delta\gamma\sigma_Y + \tfrac{1}{2}K\alpha^2_{n+1} - \tfrac{1}{2}K\alpha^2_n \\ & + \Delta\gamma f(\sigma_{n+\vartheta}, \alpha_{n+\vartheta}) + (2\vartheta - 1)\tfrac{1}{2}K(\alpha_{n+1} - \alpha_n)^2. \end{aligned} \tag{1.6.15}$$

By combining (1.6.13) and (1.6.15) and using the stress-strain relationship along with the definition of V_{int}, we obtain

$$\begin{aligned} \int_{\mathcal{B}} \sigma_{n+\vartheta}(\varepsilon_{n+1} - \varepsilon_n)\,dx = {} & V_{\mathrm{int}}(\varepsilon^e_{n+1}, \alpha_{n+1}) - V_{\mathrm{int}}(\varepsilon^e_n, \alpha_n) + \int_{\mathcal{B}} \Delta\gamma\sigma_Y\,dx \\ & + (2\vartheta - 1)\,V_{\mathrm{int}}(\varepsilon^e_{n+1} - \varepsilon^e_n, \alpha_{n+1} - \alpha_n) \\ & + \int_{\mathcal{B}} \Delta\gamma f(\sigma_{n+\vartheta}, \alpha_{n+\vartheta})\,dx. \end{aligned} \tag{1.6.16}$$

By inserting this result in the algorithm counterpart of the mechanical work identity (1.6.7) and recalling the definition of the function $L(u, v, \varepsilon^e, \alpha)$, we arrive at the expression

$$L(u_{n+1}, v_{n+1}, \varepsilon^e_{n+1}, \alpha_{n+1}) - L(u_n, v_n, \varepsilon^e_n, \alpha_{n+1})$$
$$= \underbrace{-\int_{\mathcal{B}} \Delta\gamma \sigma_Y \, dx}_{\text{Dissipation}\geq 0} - (2\vartheta - 1) \underbrace{\left[T(v_{n+1} - v_n) + V_{\text{int}}(\varepsilon^e_{n+1} - \varepsilon^e_n, \alpha_{n+1} - \alpha_n)\right]}_{\text{Quadratic form}\geq 0}$$
$$- \int_{\mathcal{B}} \Delta\gamma f(\sigma_{n+\vartheta}, \alpha_{n+\vartheta}) \, dx. \tag{1.6.17}$$

Then the stability of the numerical approximation depends on the schemes adopted in enforcing the consistency condition:

a. *Consistency enforced for* $\delta = \vartheta$. Then the term $\Delta\gamma f(\sigma_{n+\vartheta}, \alpha_{n+\vartheta}) = 0$ as a result of the Kuhn–Tucker conditions. Inspection of (1.6.17) reveals that (1.6.1) holds provided that $\vartheta \geq \frac{1}{2}$. Therefore, for this class of algorithms, unconditional stability holds if $\vartheta \in [\frac{1}{2}, 1]$.
b. *Consistency enforced for* $\delta = 1$. Then result (1.6.17) is inconclusive except for $\vartheta = \delta = 1$ which corresponds to the standard return-mapping algorithms. To see this, we observe that convexity of the yield function implies that
$$\Delta\gamma f(\sigma_{n+\vartheta}, \alpha_{n+\vartheta}) \leq \vartheta \Delta\gamma f(\sigma_{n+1}, \alpha_{n+1}) + (1 - \vartheta)\Delta\gamma f(\sigma_n, \alpha_n) \leq 0, \tag{1.6.18}$$
since $\Delta\gamma f(\sigma_{n+1}, \alpha_{n+1}) = 0$, $\Delta\gamma \geq 0$ and $f(\sigma_n, \alpha_n) \leq 0$ as a result of the design condition $\delta = 1$. Therefore, the last term on the right-hand side of equality (1.6.17) is positive and one cannot conclude that the left-hand side is nonpositive.

Clearly, both schemes include the classical return-mapping algorithms which, according to the preceding analysis, are unconditionally stable. Additional topics, such as uniqueness of the solution to the algorithmic problem and contractivity of solutions obtained for different initial data, are addressed in detail in subsequent chapters.

1.7 One-Dimensional Viscoplasticity

In this section, in the spirit of our elementary discussion in Section **1.2**, we illustrate the mathematical structure of the constitutive equations for classical viscoplasticity by a simple rheological model. Our objective here is merely to motivate in the simplest possible context the formulation of the general viscoplastic models undertaken in Chapter **2**.

In addition we examine in some detail the structure of a general class of recently proposed integrative algorithms, again within the context of a simple one-dimensional model problem. We show that these algorithms are obtained from the return-mapping algorithms for rate-independent plasticity examined in Section **1.4**, by an *explicit closed-form expression*.

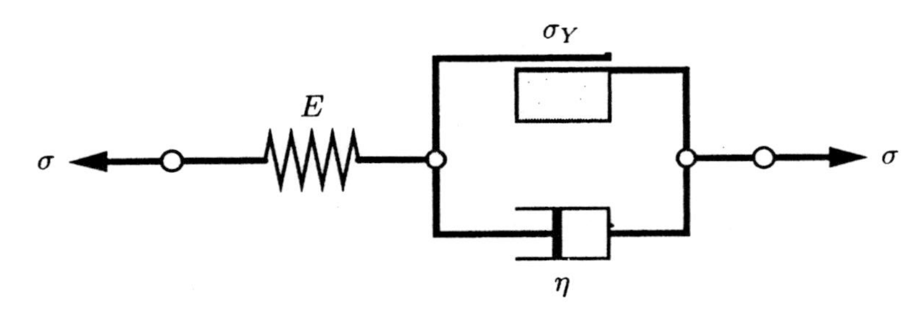

FIGURE 1-14. One-dimensional rheological model illustrating the response of a one-dimensional viscoplastic solid.

1.7.1 One-Dimensional Rheological Model

The mathematical structure underlying classical (rate-dependent) viscoplasticity is motivated by examining the response of the mechanical device arranged as illustrated in Figure **1.14**.

The device possesses unit length (and unit area) and consists of a spring with elastic constant E, which is connected to a dashpot with constant η, in parallel with a coulombic frictional device with constant σ_Y.

Let σ be the applied stress on the device, and let ε be the total strain. As in Section **1.2** we consider the additive decomposition

$$\boxed{\varepsilon = \varepsilon^e + \varepsilon^{\mathrm{vp}},} \tag{1.7.1}$$

where ε^e is the strain in the spring, so that

$$\boxed{\sigma = E\varepsilon^e = E\left(\varepsilon - \varepsilon^{\mathrm{vp}}\right).} \tag{1.7.2}$$

Next, we examine the rate of change of $\varepsilon^{\mathrm{vp}} := \varepsilon - \varepsilon^e$. To this end, consider the set of all possible stresses whose absolute value is less than or equal to the frictional constant σ_Y. This set is the closed interval $[-\sigma_Y, \sigma_Y]$. As before we use the notation

$$\mathbb{E}_\sigma = \left\{\tau \in \mathbb{R} \mid f(\tau) = |\tau| - \sigma_Y \leq 0\right\}, \tag{1.7.3}$$

and call the function $f(\sigma) := |\sigma| - \sigma_Y$ the *loading function*. Further, we recall that $\mathrm{int}(\mathbb{E}_\sigma)$ and $\partial\mathbb{E}_\sigma$ denote the interior and boundary of $\mathbb{E}_\sigma$, respectively, i.e.,

$$\mathrm{int}(\mathbb{E}_\sigma) = (-\sigma_Y, \sigma_Y), \quad \partial\mathbb{E}_\sigma = \{-\sigma_Y, \sigma_Y\}. \tag{1.7.4}$$

Finally we recall that $\mathbb{E}_\sigma$ and $\partial\mathbb{E}_\sigma$ are called the closure and boundary of the *elastic range* $\mathrm{int}(\mathbb{E}_\sigma)$ respectively. With these notations at hand, we consider the following two possibilities.

a. First, let $\sigma \in \mathrm{int}(\mathbb{E}_\sigma)$. Then $f(\sigma) \equiv |\sigma| - \sigma_Y < 0$ and *no instantaneous change should take place in* $\varepsilon^{\mathrm{vp}} = \varepsilon - \varepsilon^e$, that is,

$$\dot{\varepsilon}^{\mathrm{vp}} = 0 \ \text{ if } \ f(\sigma) \equiv |\sigma| - \sigma_Y < 0. \tag{1.7.5}$$

b. Second, assume that $\sigma \notin \mathbb{E}_\sigma$, that is, $f(\sigma) \equiv |\sigma| - \sigma_Y > 0$. Then, the stress in the frictional device is σ_Y and the stress on the dashpot, called the *extra stress* and denoted by σ_{ex}, is given as

$$\sigma_{ex} = \left\{ \begin{array}{ll} \sigma - \sigma_Y & \text{if} \quad \sigma \geq \sigma_Y \\ \sigma + \sigma_Y & \text{if} \quad \sigma \leq -\sigma_Y \end{array} \right\} = (|\sigma| - \sigma_Y)\,\text{sign}(\sigma). \qquad (1.7.6)$$

Using the fact that the stress σ_{ex} on the dashpot is connected to the strain through the viscous relationship $\sigma_{ex} = \eta \dot{\varepsilon}^{vp}$ from (1.7.6), we obtain

$$\dot{\varepsilon}^{vp} = \frac{1}{\eta} f(\sigma)\,\text{sign}(\sigma) \quad \text{if} \quad f(\sigma) = |\sigma| - \sigma_Y \geq 0. \qquad (1.7.7)$$

If we denote the ramp function by $\langle x \rangle = \frac{(x+|x|)}{2}$, (1.7.5) and (1.7.7) combine to yield the expression

$$\boxed{\begin{aligned} \dot{\varepsilon}^{vp} &= \frac{\langle f(\sigma) \rangle}{\eta} \frac{\partial f(\sigma)}{\partial \sigma}, \\ f(\sigma) &:= |\sigma| - \sigma_Y. \end{aligned}} \qquad (1.7.8)$$

We refer to (1.7.8) as a *viscoplastic constitutive equation of the Perzyna type.* An alternative formulation of the rate equation (1.7.8), which is particularly useful in a numerical analysis context, is considered next.

1.7.1.1 Viscoplastic flow rule and closest point projection.

An important interpretation of (1.7.7) is derived by rewriting this evolutionary equation as follows. First introduce a time constant, denoted by τ defined as

$$\boxed{\tau := \frac{\eta}{E}.} \qquad (1.7.9)$$

The ratio τ of the viscosity coefficient in the dashpot to the spring constant in the device in Figure **1.14** is called the *relaxation time* of the device. Its physical significance is illustrated in the example below. Now rewrite (1.7.7) as

$$\begin{aligned} \dot{\varepsilon}^{vp} &= \frac{E^{-1}}{\tau} \left[|\sigma|\,\text{sign}(\sigma) - \sigma_Y \text{sign}(\sigma) \right] \\ &= \frac{E^{-1}}{\tau} \left[\sigma - \sigma_Y \text{sign}(\sigma) \right]. \end{aligned} \qquad (1.7.10)$$

In view of this expression, we set

$$\boxed{\dot{\varepsilon}^{vp} = \frac{E^{-1}}{\tau} \left[\sigma - \mathbf{P}\sigma \right],} \qquad (1.7.10)$$

where $\mathbf{P} : \mathbb{R} \rightarrow \partial \mathbb{E}_\sigma$ is the mapping defined by

$$\boxed{\mathbf{P}\sigma = \sigma_Y\, \text{sign}(\sigma),} \qquad (1.7.11)$$

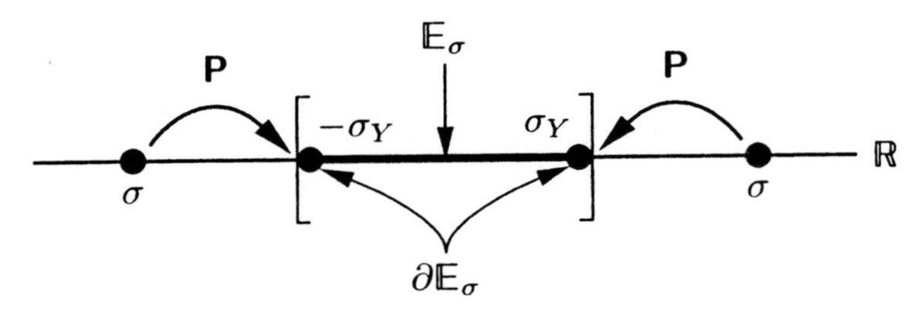

FIGURE 1-15. The map $\mathbf{P} : \mathbb{R} \to \partial\mathbb{E}_\sigma$ "returns" $\sigma \in \mathbb{R}$ to the boundary of $\mathbb{E}_\sigma$.

with a geometric interpretation illustrated in Figure **1.15**. One can easily show that $\mathbf{P} : \mathbb{R} \to \partial\mathbb{E}_\sigma$ is a projection in the sense that

$$\mathbf{P}(\mathbf{P}\sigma) = \mathbf{P}^2\sigma = \mathbf{P}\sigma \iff \mathbf{P}^2 = \mathbf{P}. \tag{1.7.12}$$

The physical significance of (1.7.11) should be clear. $\mathbf{P}$ maps a stress point σ onto the *closest point* of the boundary $\partial\mathbb{E}_\sigma$ of the elastic range. This interpretation of the viscoplastic flow rule, which is the result of the alternative expression (1.7.10) is attributed to Duvaut and Lions [1972].

1.7.1.2 Example: Relaxation test.

To further illustrate the physical significance of the constitutive model just outlined, we consider the following experiment.

At time $t = 0$ to the device in Figure **1.14** we apply an instantaneous strain which is held constant throughout time, that is, we consider the strain history (see Figure **1.16**)

$$\varepsilon(t) = \varepsilon_0 H(t),$$

where

$$H(t) := \begin{cases} 1 \text{ if } t > 0 \\ 0 \text{ otherwise,} \end{cases} \tag{1.7.13}$$

and $\varepsilon_0 > 0$. The discontinuous function $H(t)$ is the Heaviside step function. Let $\sigma_0 := E\varepsilon_0$. Consequently, $\sigma_0 > 0$. Clearly, by (1.7.10), if

$$\sigma_0 := E\varepsilon_0 \begin{cases} < \sigma_Y \Rightarrow \dot{\varepsilon}^{\mathrm{vp}} = 0 & \text{(elastic response),} \\ > \sigma_Y \Rightarrow \dot{\varepsilon}^{\mathrm{vp}} \neq 0 & \text{(viscoplastic response).} \end{cases} \tag{1.7.14}$$

Since the elastic case corresponding to the condition $\sigma_0 - \sigma_Y < 0$ is elementary, we consider the situation illustrated in Figure **1.16** for which $\sigma_0 - \sigma_Y > 0$.

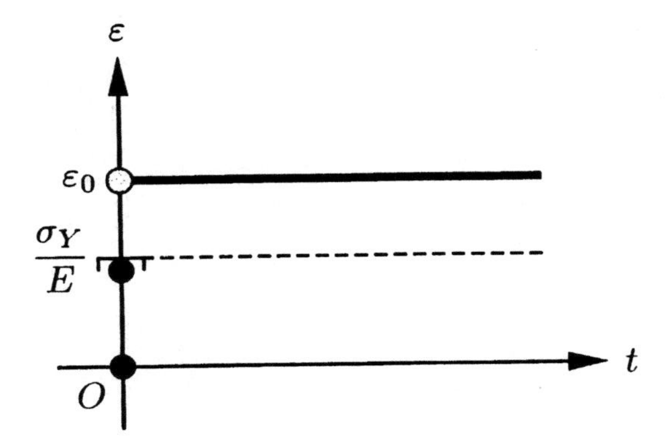

FIGURE 1-16. Strain history for a relaxation test.

To compute the stress history, we need to integrate the constitutive model as follows. From (1.7.2), (1.7.10) and (1.7.9),

$$\left.\begin{aligned} \dot{\sigma} &= E\dot{\varepsilon} - E\dot{\varepsilon}^{\mathrm{vp}}, \\ \dot{\varepsilon}^{\mathrm{vp}} &= \frac{1}{\tau} E^{-1} [\sigma - \sigma_Y] . \end{aligned}\right\} \tag{1.7.15}$$

Combining these equations, we obtain

$$\left.\begin{aligned} \dot{\sigma} + \frac{1}{\tau}\sigma &= E\dot{\varepsilon} + \frac{1}{\tau}\sigma_Y \\ \varepsilon = \varepsilon(0) &> \frac{\sigma_Y}{E} \end{aligned}\right\} \text{ in } (0, \infty). \tag{1.7.16}$$

Equation (1.7.16), integrated in closed form (note that $e^{\frac{t}{\tau}}$ is the integrating factor), yields

$$e^{t/\tau}\sigma - \sigma(0) = \int_0^t e^{s/\tau}\, E\dot{\varepsilon}(s)ds + \sigma_Y \left(e^{t/\tau} - 1\right). \tag{1.7.17}$$

Now, since $\dot{\varepsilon}(t) = 0$ in $(0, \infty)$, it is easily shown that the integral in (1.7.17) vanishes identically. (One needs to be a bit careful with the singularity of $\dot{\varepsilon}(t)$ at $t = 0$.) Consequently, since $\sigma(0) = E\varepsilon(0) = E\varepsilon_0$,

$$\boxed{\sigma(t) = \left[E\varepsilon_0 - \sigma_Y\right] e^{-\frac{t}{\tau}} + \sigma_Y.} \tag{1.7.18}$$

The stress response given by (1.7.18) is shown in Figure **1.17**. Note that the stress decays exponentially with time. In fact, as $t/\tau \to \infty, \sigma(t) \to \sigma_Y$.

From a physical standpoint, it is important to realize that the controlling factor in the *relaxation* process illustrated in Figure **1.17** is the *relative time* t/τ. The absolute time $t \in [0, \infty)$ is regarded as short or long only when compared with $\tau = \eta/E$. Equivalently, what counts is the *ratio* of the viscosity η in the dashpot

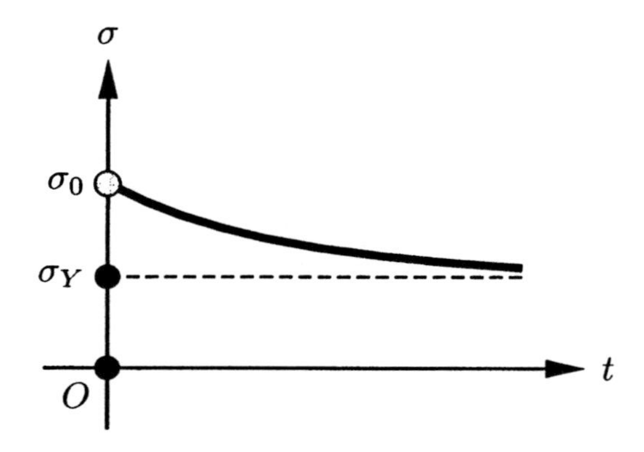

FIGURE 1-17. Stress response in a relaxation test

to the stiffness E in the spring in the device in Figure **1.14**. Because of this, τ is called the *natural relaxation time*.

1.7.1.3 Extensions to account for strain hardening.

Strain-hardening effects are incorporated in the model outlined above by a procedure similar to that discussed in detail in Section **1.2.2.1**. The simplest *linear isotropic hardening* model is obtained by appending an internal variable, denoted by α, to an evolutionary equation given by

$$\boxed{\dot{\alpha} = \left|\dot{\varepsilon}^{vp}\right| \geq 0,} \tag{1.7.19}$$

and modifying the loading function as

$$f(\sigma, \alpha) := |\sigma| - \left[\sigma_Y + K\alpha\right]. \tag{1.7.20}$$

The closure of the elastic range is the time-dependent (*closed convex*) set defined by

$$\boxed{\mathbb{E}_\sigma = \left\{(\sigma, \alpha) \in \mathbb{R} \times \mathbb{R}_+ \mid f(\sigma, \alpha) \leq 0\right\}.} \tag{1.7.21}$$

1.7.1.4 The viscoplastic regularization.

As illustrated in the example 1.7.1.2, within the framework of (rate-dependent) viscoplasticity, the variables (σ, α) are no longer constrained to lie within the closure of the elastic range $\mathbb{E}_\sigma$, in sharp contrast with the situation found in the rate-independent plasticity model.

On the other hand, on physical grounds, by inspecting Figure **1.14**, one concludes that, as $\eta \to 0$, the effect of the dashpot disappears and one recovers the rate–independent model illustrated in Figure **1.1**. In the next chapter we rigorously show that this intuition is correct. This important fact is exploited analytically and numerically and leads to the notion of *viscoplastic regularization* (which is closely related to the Yoshida regularization; see, e.g., Pazy [1983, p.9]) of rate-independent plasticity.

To elaborate further, observe that by setting

$$\gamma := \frac{\langle f(\sigma, \alpha)\rangle}{\eta}, \tag{1.7.22}$$

the equations of evolution (1.7.7) and (1.7.19) are written as

$$\dot{\varepsilon}^{\mathrm{vp}} = \gamma \,\mathrm{sign}(\sigma),$$

and (1.7.23)

$$\dot{\alpha} = \gamma,$$

which is the exact counterpart of the evolutionary equations of classical rate-independent plasticity, but with the Kuhn–Tucker conditions (1.2.26) and the consistency condition (1.2.27) now replaced by (1.7.22).

Because the consistency parameter is no longer determined by the consistency condition but directly through constitutive equation (1.7.22), one speaks of a viscoplastic regularization. For convenience and subsequent reference, the one-dimensional viscoplastic model developed above is summarized in BOX **1.6**.

BOX 1.6. One-Dimensional Classical Viscoplasticity.

1. Elastic stress strain relationship

$$\sigma = E\left(\varepsilon - \varepsilon^{\mathrm{vp}}\right).$$

2. Closure of the elastic range and loading function

$$\mathbb{E}_\sigma := \left\{(\sigma, \alpha) \in \mathbb{R} \times \mathbb{R}_+ \;\mid\; f(\sigma, \alpha) \leq 0\right\}$$
$$f(\sigma, \alpha) := |\sigma| - \left[\sigma_Y + K\alpha\right].$$

3.a. Flow rule and hardening law (Perzyna formulation)

$$\dot{\varepsilon}^{\mathrm{vp}} = \frac{\langle f(\sigma, \alpha)\rangle}{\eta}\,\mathrm{sign}(\sigma)$$

$$\dot{\alpha} = \frac{\langle f(\sigma, \alpha)\rangle}{\eta}.$$

3.b. Flow rule and hardening law (Duvaut–Lions formulation)

$$\dot{\varepsilon}^{\mathrm{vp}} = \begin{cases} \dfrac{E^{-1}}{\tau}\left[\sigma - \mathbf{P}\sigma\right]; & f(\sigma, \alpha) > 0 \\ 0 & \text{otherwise,} \end{cases}$$

$$\dot{\alpha} = \left|\dot{\varepsilon}^{\mathrm{vp}}\right|,$$

where $\mathbf{P} : \mathbb{R} \rightarrow \partial\mathbb{E}_\sigma$ is the closest point projection onto $\partial\mathbb{E}_\sigma$, the boundary of the elastic range.

1.7.2 Dissipation. A Priori Stability Estimate

The viscoplastic IBVP consists of the weak form (1.3.11) of the momentum equations supplemented by the inelastic constitutive equation summarized in BOX **1.6.** This problem possesses an a priori energy-decay estimate analogous to that described in Section **1.3.3** for rate-independent plasticity, which is the direct result of the mechanical work identity (1.3.17) along with the key property of *positive* mechanical dissipation.

To arrive at the property of positive dissipation, first we observe that the internal energy function $V_{\text{int}}(\varepsilon^e, \alpha)$ and the potential energy $V_{\text{ext}}(u)$ for one-dimensional (hardening) viscoplasticty are also given by (1.3.19), assuming dead loading. Again the mechanical dissipation $\mathcal{D}_{\text{mech}}$ is defined by (1.3.20). Then a computation identical to that leading to (1.3.23) gives

$$\begin{aligned} \mathcal{D}_{\text{mech}} &= \int_{\mathcal{B}} \left[\sigma\, \dot{\varepsilon}^{\text{vp}} - K\alpha\, \dot{\alpha}\right] dx \\ &= \int_{\mathcal{B}} \left[\gamma\, f(\sigma, \alpha) + \gamma\, \sigma_Y\right] dx \end{aligned} \tag{1.7.24}$$

with

$$\gamma := \frac{\langle f(\sigma, \alpha)\rangle}{\eta}.$$

This expression is rearranged to conclude that, at any time $t \in [0, \mathrm{T}]$,

$$\boxed{\begin{aligned} &\mathcal{D}_{\text{mech}} = \int_{\mathcal{B}} \underbrace{\gamma\, \sigma_Y}_{\geq 0}\, \underbrace{\left[1 + \eta\, \gamma/\sigma_Y\right]}_{\geq 0}\, dx \geq 0, \\ &\text{since} \\ &\qquad \gamma := \frac{\langle f(\sigma, \alpha)\rangle}{\eta} \geq 0. \end{aligned}} \tag{1.7.25}$$

Therefore, mechanical dissipation is a nondecreasing function in time. It was pointed out above that, as $\eta \to 0$, the model of viscoplasticity reduces to the classical model of rate-independent plasticity. Moreover, it can be shown that the factor $\gamma = \langle f(\sigma, \alpha)\rangle/\eta$ remains bounded as the viscosity $\eta \to 0$ and tends to the plastic multiplier of the rate-independent theory. Then from expression (1.7.25) we conclude that the mechanical dissipation in the viscoplastic model also tends to the dissipation of the rate-independent model since the quadratic term $\eta\, \gamma^2/\sigma_Y \to 0$ as $\eta \to 0$.

1.7.2.1 A priori estimate. Uniqueness and contractivity of the IBVP.

The implications of the dissipation inequality (1.7.25) are the same as those described in Section **1.2.3** for the IBVP of classical rate-dependent plasticity. In particular,

i. *a priori estimate.* The mechanical work identity (1.3.17) along with (1.7.25) yields the estimate:

$$\boxed{\frac{d}{dt} L(u, v, \varepsilon^e, \alpha) = -\mathcal{D}_{\text{mech}} \leq 0 \quad \text{for all} \quad t \in [0, \mathrm{T}],} \tag{1.7.26}$$

where $L(u, v, \varepsilon^e, \alpha)$ is the sum of the potential energy of the external loading, the internal energy of the system, and the kinetic energy, as defined by (1.3.25).

ii. *uniqueness of the solution to the IBVP.* An argument identical to that leading to (1.3.31) shows that the difference of two possible solutions to the IBVP for the same boundary conditions and the same forcing satisfy the identity

$$\begin{aligned}\frac{d}{dt}&\left[T(v - \tilde{v}) + V_{\text{int}}(\varepsilon^e - \tilde{\varepsilon}^e)\right] \\ &\leq \frac{1}{\eta} \int_B \langle f(\tilde{\sigma}, \tilde{\alpha})\rangle \left[f(\sigma, \alpha) - f(\tilde{\sigma}, \tilde{\alpha})\right] dx \\ &\quad + \frac{1}{\eta} \int_B \langle f(\sigma, \alpha)\rangle \left[f(\tilde{\sigma}, \tilde{\alpha}) - f(\sigma, \alpha)\right] dx.\end{aligned} \tag{1.7.27}$$

Now observe that the properties of the function $\langle \cdot \rangle$ imply the relationships

$$f(\sigma, \alpha) \leq \langle f(\sigma, \alpha)\rangle$$

and (1.7.28)

$$f(\sigma, \alpha)\langle f(\sigma, \alpha)\rangle = \left[\langle f(\sigma, \alpha)\rangle\right]^2,$$

which hold for both (σ, α) and $(\tilde{\sigma}, \tilde{\alpha})$. By using these relationships in indentity (1.7.27) and completing a perfect square, we arrive at the inequality

$$\frac{d}{dt}\left[T(v - \tilde{v}) + V_{\text{int}}(\varepsilon^e - \tilde{\varepsilon}^e)\right] \leq -\int_B \frac{1}{\eta} \left[\langle f(\sigma, \alpha)\rangle - \langle f(\tilde{\sigma}, \tilde{\alpha})\rangle\right]^2 dx \leq 0. \tag{1.7.29}$$

If the initial data is the same for both solutions, by integrating (1.7.29) in time, we obtain

$$\boxed{0 \leq T(v - \tilde{v}) + V_{\text{int}}(\varepsilon^e - \tilde{\varepsilon}^e) \leq 0.} \tag{1.7.30}$$

Therefore, $T(v - \tilde{v}) + V_{\text{int}}(\varepsilon^e - \tilde{\varepsilon}^e) \equiv 0$, and we conclude that the two solutions must coincide since this quadratic form is positive-definite by virtue of the assumptions $K > 0$ and $E > 0$.

iii. *Contractivity.* If the two solutions of the IBVP are obtained for the same boundary conditions and the same forcing function but with two different initial conditions $\{u_0, v_0\}$ and $\{\tilde{u}_0, \tilde{v}_0\}$, integrating the inequality (1.7.29) in time yields

$$\boxed{T(v - \tilde{v}) + V_{\text{int}}(\varepsilon^e - \tilde{\varepsilon}^e) \leq T(v_0 - \tilde{v}_0) + V_{\text{int}}(\varepsilon^e_0 - \tilde{\varepsilon}^e_0).} \tag{1.7.31}$$

Therefore, the difference of the two solutions at any time $t \in [0, \mathrm{T}]$, measured in terms of the kinetic and internal energies, is always less or at most equal to the difference of the solutions at time $t = 0$. Equivalently, finite perturbations in the initial data are attenuated in time when measured in terms of energy.

1.7.3 An Integration Algorithm for Viscoplasticity

An algorithm for numerically integrating the viscoplastic constitutive model summarized in BOX **1.6** is developed at once by exploiting the notion of viscoplastic regularization; see Simo, Kennedy, and Govindjee [1988]. An alternative approach was proposed by Hughes and Taylor [1978].

The basic idea is based on the property that *as* $t/\tau \to \infty$, *the viscoplastic solution* relaxes to the rate–independent (inviscid) solution. In other words, by a slight generalization of the argument in Section **1.7.1.1**, we write the viscoplastic evolutionary equations as

$$\left.\begin{aligned} \dot{\varepsilon}^{\mathrm{vp}} &= \frac{E^{-1}}{\tau}\,[\sigma - \sigma_\infty] \\ \dot{\alpha} &= -\frac{1}{\tau}\,[\alpha - \alpha_\infty] \end{aligned}\right\}, \tau = \frac{\eta}{E + K}, \tag{1.7.32}$$

where $\{\sigma_\infty, \alpha_\infty\}$ is the solution of the rate–independent model with constitutive equations summarized in BOX **1.2**. The parameter $\tau = \eta/(E+K)$ is the relaxation time. Thus, now the projection $\mathbf{P}$ is now defined by the relationship

$$\{\sigma, \alpha\} \mapsto \mathbf{P}\{\sigma, \alpha\} := \{\sigma_\infty, \alpha_\infty\}. \tag{1.7.33}$$

We remark that $\{\sigma_\infty, \alpha_\infty\}$ is computed in closed form for given $\{\sigma_n, \alpha_n\}$ and given $\varepsilon_{n+1} = \varepsilon_n + \Delta\varepsilon_n$ by the return-mapping algorithm in BOX **1.4**.

Now, from (1.7.32) and (1.7.2), we obtain the initial value problem

$$\begin{aligned} \dot{\sigma} + \frac{1}{\tau}\sigma &= E\dot{\varepsilon} + \frac{\sigma_\infty}{\tau}, \\ \dot{\alpha} + \frac{1}{\tau}\alpha &= +\frac{\alpha_\infty}{\tau}. \end{aligned} \tag{1.7.34}$$

In a computational context, the objective is to determine the variables $\{\sigma_{n+1}, \varepsilon^{\mathrm{vp}}_{n+1}, \alpha_{n+1}\}$ at the end of the time interval $[t_n, t_{n+1}]$, for given initial data $\{\sigma_n, \varepsilon^{\mathrm{vp}}_n, \alpha_n\}$ and *prescribed strains* $\varepsilon_{n+1} = \varepsilon_n + \Delta\varepsilon_n$. Thus, the initial conditions for (1.7.34) are

$$\begin{aligned} (\sigma, \alpha)\big|_{t=t_n} &= (\sigma_n, \alpha_n) \qquad \text{(given)}, \\ (\varepsilon, \varepsilon^{\mathrm{vp}})\big|_{t=t_n} &= (\varepsilon_n, \varepsilon^{\mathrm{vp}}_n) \qquad \text{(given)}. \end{aligned} \tag{1.7.35}$$

We observe that (1.7.34) is a linear differential equation which can be solved *in closed form for known values* $\{\sigma_\infty, \alpha_\infty\}$. However, $\{\sigma_\infty, \alpha_\infty\}$ can be computed explicitly in *closed form* by the (first-order accurate) algorithm in BOX **1.4**. Therefore, consistent with the first-order accuracy of the return-mapping algorithm for $\{\sigma_\infty, \alpha_\infty\}$, we approximate (1.7.34)–(1.7.35) by an implicit backward-Euler difference scheme as

$$\left.\begin{aligned} \sigma_{n+1}\left[1+\frac{\Delta t}{\tau}\right] &= E\Delta\varepsilon_n + \sigma_n + \frac{\Delta t}{\tau}\sigma_\infty \\ \alpha_{n+1}\left[1+\frac{\Delta t}{\tau}\right] &= \alpha_n + \frac{\Delta t}{\tau}\alpha_\infty . \end{aligned}\right\} \tag{1.7.36}$$

By recalling the definition $\sigma_{n+1}^{\text{trial}} := \sigma_n + E\Delta\varepsilon_n$ of the trial elastic state and solving for $(\sigma_{n+1}, \alpha_{n+1})$, we find the closed-form formulas

$$\boxed{\begin{aligned} \sigma_{n+1} &= \frac{\sigma_{n+1}^{\text{trial}} + \frac{\Delta t}{\tau}\sigma_\infty}{1+\frac{\Delta t}{\tau}} \\ \alpha_{n+1} &= \frac{\alpha_n + \frac{\Delta t}{\tau}\alpha_\infty}{1+\frac{\Delta t}{\tau}} . \end{aligned}} \tag{1.7.37}$$

Remarkably, this approach generalizes without modifying the multidimensional case. A summary of the procedure is outlined in BOX **1.7**.

BOX 1.7. Algorithm for Viscoplasticity Based on the Notion of Viscoplastic Regularization.

1. Database at $x \in \mathcal{B}$: $\left\{\varepsilon_n^{\text{vp}}, \alpha_n\right\}$.
2. Given strain field at $x \in \mathcal{B}$: $\varepsilon_{n+1} = \varepsilon_n + \Delta\varepsilon_n$.
3. Compute rate-independent solution $\{\sigma_\infty, \alpha_\infty\}$ by the return-mapping algorithm in BOX **1.4**.
4. Perform viscoplastic regularization

IF $f_{n+1}^{\text{trial}} < 0$ THEN

Elastic step: *EXIT*

ELSE

$$\tau = \frac{\eta}{E+K}$$

$$\sigma_{n+1} = \frac{\sigma_{n+1}^{\text{trial}} + \frac{\Delta t}{\tau}\sigma_\infty}{1+\frac{\Delta t}{\tau}}$$

$$\alpha_{n+1} = \frac{\alpha_n + \frac{\Delta t}{\tau}\alpha_\infty}{1+\frac{\Delta t}{\tau}}$$

$$\varepsilon_{n+1}^{\text{vp}} = \varepsilon_{n+1} - E^{-1}\sigma_{n+1}$$

ENDIF

1.7.3.1 Algorithmic tangent modulus.

The finite-element implementation of the viscoplastic constitutive model is identical to the one discussed in Section **1.5** for rate–independent plasticity. The only change required is the replacement of BOX **1.5** by BOX **1.7** in computing the stress update. Then the computational procedure is completed by providing a closed-form expression for the algorithmic tangent modulus discussed in Section **1.5.2.3**. To this end, we differentiate expression (1.7.37) and recall that

$$\frac{\partial \sigma_{n+1}^{\text{trial}}}{\partial \varepsilon_{n+1}} = E \quad \text{and} \quad \mathbf{C}_{n+1}^{(k)}\Big|_{\infty} := \frac{\partial \sigma_{n+1}^{(k)}}{\partial \varepsilon_{n+1}^{(k)}}\Bigg|_{\infty}, \tag{1.7.38}$$

where $\mathbf{C}_{n+1}^{(k)}\Big|_{\infty}$ stands for the elastoplastic modulus given by (1.5.36). From (1.7.37) and (1.7.38),

$$\boxed{\mathbf{C}_{n+1}^{(k)} := \frac{\partial \sigma_{n+1}^{(k)}}{\partial \varepsilon_{n+1}^{(k)}} = \begin{cases} E & \text{iff } f_{n+1}^{\text{trial}} \le 0 \\ \dfrac{E}{1+\frac{\Delta t}{\tau}} + \dfrac{\frac{\Delta t}{\tau}}{1+\frac{\Delta t}{\tau}} \mathbf{C}_{n+1}^{(k)}\Big|_{\infty} & \text{otherwise.} \end{cases}} \tag{1.7.39}$$

Again this expression generalizes to the multidimensional case without modification.

Remarks 1.7.1.

1. Our preceding developments generalize immediately to other forms of hardening. In particular, *kinematic hardening* is formulated by introducing the rate equations

$$\boxed{\begin{aligned} \dot{q} &= = \frac{\langle f(\sigma - q, \alpha)\rangle}{\eta} H \operatorname{sign}(\sigma - q), \\ \dot{\alpha} &= \frac{\langle f(\sigma - q, \alpha)\rangle}{\eta}, \\ f(\sigma - q, \alpha) &:= |\sigma - q| - [\sigma_Y + K\alpha], \end{aligned}} \tag{1.7.40}$$

   where H is the kinematic hardening modulus.

2. Alternatively, within the context of a Duvaut–Lions type of formulation, we set

$$\boxed{\begin{aligned} \dot{q} &= -\frac{1}{\tau}[q - q_{\infty}] \\ \dot{\alpha} &= -\frac{1}{\tau}[\alpha - \alpha_{\infty}] \\ \dot{\sigma} &= E\dot{\varepsilon} - \frac{1}{\tau}[\sigma - \sigma_{\infty}] \end{aligned}} \tag{1.7.41a}$$

where τ is the relaxation time now given by

$$\boxed{\tau := \frac{\eta}{E + [H + K]}} \tag{1.7.41b}$$

and $\{\sigma_\infty, \alpha_\infty, q_\infty\}$ is the rate-independent solution which obeys the constitutive model in BOX **1.3**.

3. The algorithmic treatment of (1.7.41a) is identical to that summarized in BOX **1.7**, except for the fact that now step 3 is performed by BOX **1.5**, and one needs to add the following update formula to step 4:

$$\boxed{q_{n+1} = \frac{q_n + \frac{\Delta t}{\tau} q_\infty}{1 + \frac{\Delta t}{\tau}} .} \tag{1.7.42}$$

We remark that now step 4 is performed with the relaxation time defined by (1.7.41b).

4. It can be shown that one arrives at an algorithm identical to that in BOX **1.7** by performing the following steps.
   i. Use a backward Euler difference scheme on the *Perzyna model* given in BOX **1.6**.
   ii. For a viscoplastic process it can be shown that one arrives at update formulas identical to those of the inviscid return-mapping algorithms in BOX **1.4** (or BOX **1.5**) but with $\Delta\gamma$ now given by

$$\boxed{\Delta\gamma = \frac{\Delta t/\tau}{1 + \Delta t/\tau} \Delta\gamma_\infty,} \quad \tau = \frac{\eta}{E + K + H} \tag{1.7.43}$$

   where $\Delta\gamma_\infty$ is given by either (1.4.22) or (1.4.41), depending on the nature of the hardening mechanism.
   iii. By substituting (1.7.43) in the update formulas in BOX **1.4** (or BOX **1.5**), one obtains the algorithm in BOX **1.7**.

   Again, formula (1.7.43) exhibits the fact that, as $\Delta t/\tau \to \infty$, one recovers the rate-independent limit, in agreement with the notion of viscoplastic regularization.

5. From formulas (1.7.39) and expression (1.5.36), since $\tau = \eta/(E + K)$,

$$\begin{aligned} \mathbf{C}_{n+1} &= \frac{1}{1 + \Delta t/\tau} \left[E + \frac{\Delta t}{\tau} \frac{EK}{E + K} \right] \\ &= \frac{E}{1 + \Delta t/\tau} \left[1 + \frac{\Delta t}{\eta} K \right]. \end{aligned} \tag{1.7.44}$$

   Thus, even in the case of a *softening material*, for which $K < 0$,

$$\boxed{\mathbf{C}_{n+1} > 0 \quad \text{if} \quad \Delta t < -\frac{\eta}{K} \quad (K < 0),} \tag{1.7.45}$$

   that is, the (consistent) algorithmic tangent modulus $\mathbf{C}_{n+1}$ is positive for softening materials ($K < 0$) by performing a viscoplastic regularization and choosing

Δt small enough according to the critical limit (1.7.45). Note that this modulus ensures that the incremental (algorithmic) problem is hyperbolic. This observation, which does not appear well known, is made in Simo [1991].

2

Classical Rate-Independent Plasticity and Viscoplasticity

In this chapter we summarize the equations of classical rate-independent plasticity and its viscoplastic regularization. Our presentation is restricted to an outline of the mathematical structure of the governing equations relevant to the numerical solution of boundary-value problems and the analysis of numerical algorithms.

First, for the convenience of the reader, we summarize some basic notation of continuum mechanics with attention restricted to the linearized theory. For further details we refer to standard textbooks, e.g., Sokolnikoff [1956] or Gurtin [1972]. Next, we proceed to outline the basic structure of rate-independent plasticity within the classical framework of response functions formulated in stress space, as in Hill [1950] or Koiter [1960]. Special attention is given to the proper (and *unique*) formulation of *loading/unloading* conditions in the so-called *Kuhn–Tucker* form. These are the *standard* complementarity conditions for problems, such as plasticity, subjected to unilateral constraints. This form of loading/unloading conditions is in fact classical and has been used by several authors, Koiter [1960] and Maier [1970]. Because the algorithmic elastoplastic problem is typically regarded as a *strain-driven* problem, throughout our discussion we adopt the *strain* tensor as the primary (driving) variable. Accordingly, although the response functions are formulated in stress space, the theory is essentially equivalent to a strain-space formulation. This is the standard point of view adopted in the numerical analysis literature, starting from the pioneering work of Wilkins [1964]. Alternative stress-space frameworks have been explored by several authors, e.g., Johnson [1977] and Simo, Kennedy, and Taylor [1988].

We consider the thermodynamic basis of the theory within the context of internal variables. As shown subsequently, this structure is important to understand the algorithmic structure of the discrete problem. Finally, we examine the case of associative plasticity which is intimately connected to the *principle of maximum plastic dissipation*.

Because of the important role played by the principle of maximum dissipation in formulating finite-element approximations, a discussion of this and its equivalence with normality, loading/unloading conditions in Kuhn–Tucker form, and convexity of the yield surface is included. We conclude the chapter with an outline of the so-called viscoplastic regularization leading to the classical viscoplastic constitutive models. As an illustration, we consider in some detail the classical J_2 flow theory. The algorithmic treatment of this important example is considered in subsequent chapters.

2.1 Review of Some Standard Notation

Let $\mathcal{B} \subset \mathbb{R}^{n_{dim}}$ be the reference configuration of the body of interest, where $1 \leq n_{dim} \leq 3$ is the space dimension. We assume that $\mathcal{B}$ is open and bounded with smooth boundary $\partial\mathcal{B}$ and closure $\bar{\mathcal{B}} := \mathcal{B} \cup \partial\mathcal{B}$. Let $[0, \mathrm{T}] \subset \mathbb{R}_+$ be the time interval of interest, and let

$$\boldsymbol{u} : \bar{\mathcal{B}} \times [0, \mathrm{T}] \rightarrow \mathbb{R}^{ndim} \tag{2.1.1}$$

be the *displacement field* of particles with reference position $\boldsymbol{x} \in \mathcal{B}$ at time $t \in [0, \mathrm{T}]$. We write $\boldsymbol{u}(\boldsymbol{x}, t)$ and denote the infinitesimal strain tensor by

$$\boldsymbol{\varepsilon} = \nabla^s \boldsymbol{u} := \frac{1}{2}\left[\nabla \boldsymbol{u} + (\nabla \boldsymbol{u})^T\right]. \tag{2.1.2}$$

Relative to the standard basis $\{\boldsymbol{e}_i\}$ in $\mathbb{R}^{n_{dim}}$,

$$\boldsymbol{u} = u_i \boldsymbol{e}_i,$$

and

$$\nabla^s \boldsymbol{u} = \frac{1}{2}\left(u_{i,j} + u_{j,i}\right) \boldsymbol{e}_i \otimes \boldsymbol{e}_j, \tag{2.1.3}$$

where $\otimes$ denotes a tensor product. Second-order *symmetric* tensors are linear transformations in $\mathbb{S}$, defined as

$$\mathbb{S} := \left\{\boldsymbol{\xi} : \mathbb{R}^{n_{dim}} \rightarrow \mathbb{R}^{n_{dim}} \mid \boldsymbol{\xi} \text{ is linear, and } \boldsymbol{\xi} = \boldsymbol{\xi}^T\right\}. \tag{2.1.4a}$$

This is a vector space with inner product

$$\boldsymbol{\xi} : \boldsymbol{\xi} = \mathrm{tr}\left[\boldsymbol{\xi}^T \boldsymbol{\xi}\right] \equiv \xi_{ij}\xi_{ij}\,. \tag{2.1.4b}$$

With the usual abuse in notation, we often identify $\mathbb{S} = \mathbb{R}^{n(n+1)/2}$ since any $\boldsymbol{\xi} \in \mathbb{S}$ has $n(n+1)/2$ components $\xi_{ij} \in \mathbb{R}$ relative to the standard basis. We denote the stress tensor by

$$\boldsymbol{\sigma} = \sigma_{ij} \boldsymbol{e}_i \otimes \boldsymbol{e}_j. \tag{2.1.5}$$

In what follows, we assume that

$$\overline{\partial\mathcal{B}} = \overline{\partial_u\mathcal{B} \cup \partial_\sigma\mathcal{B}}$$

and (2.1.6)

$$\partial_u\mathcal{B} \cap \partial_\sigma\mathcal{B} = \emptyset,$$

where $\partial_u \mathcal{B}$ is the part of $\partial \mathcal{B}$ where displacements are prescribed as

$$\boldsymbol{u}\big|_{\partial_u \mathcal{B}} = \bar{\boldsymbol{u}} \text{ (given)}, \tag{2.1.7}$$

whereas $\partial_\sigma \mathcal{B}$ is the part of $\partial \mathcal{B}$ where tractions are prescribed as

$$\boldsymbol{\sigma}\big|_{\partial_\sigma \mathcal{B}} \hat{\boldsymbol{n}} = \bar{\boldsymbol{t}} \text{ (given)}. \tag{2.1.8}$$

Here $\hat{\boldsymbol{n}}$ is the field normal to $\partial_\sigma \mathcal{B}$.

2.1.1 The Local Form of the IBVP. Elasticity

Let $\boldsymbol{b}(\boldsymbol{x}, t)$ be the body force per unit of mass, a given vector field defined on $\mathcal{B}\times]0, T[$, and denote the mass density by $\rho : \mathcal{B} \to \mathbb{R}$. The local forms of the momentum equations are

$$\left.\begin{aligned} \frac{\partial^2 \boldsymbol{u}}{\partial t^2} &= \operatorname{div} \boldsymbol{\sigma} + \rho \boldsymbol{b}, \\ \boldsymbol{\sigma} &= \boldsymbol{\sigma}^T, \end{aligned}\right\} \quad \text{in } \mathcal{B}\times]0, T[. \tag{2.1.9a}$$

This system of partial differential equations is supplemented by the boundary conditions specified by (2.1.7) and (2.1.8) subject to the restrictions (2.1.6) and supplemented by the initial data

$$\boldsymbol{u}(\boldsymbol{x}, 0) = \boldsymbol{u}_0(\boldsymbol{x})$$

and (2.1.9b)

$$\frac{\partial}{\partial t} \boldsymbol{u}(\boldsymbol{x}, 0) = \boldsymbol{v}_0(\boldsymbol{x}) \quad \text{in } \mathcal{B},$$

where $\boldsymbol{u}_0(\cdot)$ and $\boldsymbol{v}_0(\cdot)$ are prescribed functions in $\mathcal{B}$. Equations (2.1.9*a*,*b*), together with the boundary conditions (2.1.7) and (2.1.8), yield an initial boundary-valued problem (IBVP) for the displacement field $\boldsymbol{u}(\boldsymbol{x})$, when the stress field $\boldsymbol{\sigma}$ is related to the the displacement field $\boldsymbol{u}$ through a *constitutive equation*.

EXAMPLE: ELASTICITY. The simplest model for a constitutive equation is provided by a *hyperelastic* material, for which the stress response is characterized in terms of a *stored energy* function

$$W : \mathcal{B} \times \mathbb{S} \to \mathbb{R}, \tag{2.1.10}$$

such that

$$\boldsymbol{\sigma}(\boldsymbol{x}) = \frac{\partial W \left[\boldsymbol{x}, \boldsymbol{\varepsilon}(\boldsymbol{x})\right]}{\partial \boldsymbol{\varepsilon}}. \tag{2.1.11}$$

In components, $\sigma_{ij} = \frac{\partial W}{\partial \varepsilon_{ij}}$. One calls

$$\mathbf{C}(\boldsymbol{x}) := \frac{\partial^2 W \left[\boldsymbol{x}, \boldsymbol{\varepsilon}(\boldsymbol{x})\right]}{\partial \boldsymbol{\varepsilon}^2} \tag{2.1.12}$$

the *elasticity tensor*. In components, $\mathsf{C}_{ijkl} = \frac{\partial^2 W}{\partial \varepsilon_{ij} \partial \varepsilon_{kl}}$.

Remarks 2.1.1.

1. Note that $\mathbf{C}$ possesses the symmetries

$$\mathsf{C}_{ijkl} = \mathsf{C}_{klij} = \mathsf{C}_{ijlk} = \mathsf{C}_{jilk} . \qquad (2.1.13)$$

2. For the infinitesimal theory, one assumes that $\mathbf{C}$ is *positive-definite* restricted to $\mathbb{S}$, i.e.,

$$\boldsymbol{\xi} : \mathbf{C} : \boldsymbol{\xi} := \xi_{ij} \mathsf{C}_{ijkl} \xi_{kl} \geq \beta \|\boldsymbol{\xi}\|^2, \qquad (2.1.14)$$

   for some $\beta > 0$ (depending on $\boldsymbol{x} \in \mathcal{B}$), and *any* $\boldsymbol{\xi} \in \mathbb{S}$. Here $\|\boldsymbol{\xi}\|^2 = \boldsymbol{\xi} : \boldsymbol{\xi}$. This condition, also known as pointwise stability (see e.g., Marsden and Hughes [1983, Chapter **3**]) is equivalent to postulating that W is *convex* on $\mathbb{S}$. Convexity is an unacceptable restriction in the nonlinear theory, see e.g. Ciarlet [1988].
3. A weaker condition on W which often holds in the nonlinear theory is the *strong ellipticity* condition:

$$\boldsymbol{a} \otimes \boldsymbol{b} : \mathbf{C} : \boldsymbol{a} \otimes \boldsymbol{b} \geq \alpha \|\boldsymbol{a}\|^2 \|\boldsymbol{b}\|^2, \qquad (2.1.15)$$

   for some $\alpha > 0$ (depending on $\boldsymbol{x} \in \mathcal{B}$) and any $\boldsymbol{a}, \boldsymbol{b} \in \mathbb{R}^n$. It is easily shown that (2.1.14) implies (2.1.15) *but not conversely*; see Marsden and Hughes [1983, Chapter **3**]. On the other hand, condition (2.1.15) is equivalent to the requirement that wave speeds in the material are real, i.e., the so-called Hadamard condition on the acoustic tensor.
4. If W does not depend on $\boldsymbol{x} \in \mathcal{B}$ [that is, $\partial_{\boldsymbol{x}} W = 0$] the material is said to be *homogeneous*. Finally, if W is rotationally invariant, the material is said to be *isotropic*. In addition, if $\mathbf{C}$ is constant, the material is said to be *linearly elastic* and one has the classical result

$$\mathbf{C} = \lambda \mathbf{1} \otimes \mathbf{1} + 2\mu \mathbf{I}, \qquad (2.1.16)$$

   where $\mathbf{1} = \delta_{ij} \boldsymbol{e}_i \otimes \boldsymbol{e}_j$ is the second-order identity tensor, $\mathbf{I} = \frac{1}{2}[\delta_{ik}\delta_{jl} + \delta_{il}\delta_{jk}] \boldsymbol{e}_i \otimes \boldsymbol{e}_j \otimes \boldsymbol{e}_k \otimes \boldsymbol{e}_l$ is the fourth-order symmetric identity tensor, and λ, μ are the Lamé constants.
5. The existence theory for linear elasticity along with the numerical implementation are most easily formulated in terms of the *weak form* of the local equations (2.1.9), (i.e., the virtual work principle). Similarly, the finite numerical solution of this IBVP by finite-element methods relies on the weak formulation of the problem. We postpone the discussion of these ideas to Chapter **4**.

The subject of this monograph is the numerical solution of initial boundary-value problems for constitutive equations other than the hyperelastic model (2.1.11). Our objective is the precise formulation of a particular class of *nonlinear constitutive* equations, known as classical plasticity, and its numerical solution in the context of the finite-element method.

2.2 Classical Rate-Independent Plasticity

Below we summarize the governing equations of classical rate-independent plasticity within the context of the three-dimensional infinitesimal theory. First, we consider the classical formulation in stress space and show that the theory constitutes a straightforward extension of the one-dimensional model motivated in detail in Chapter **1**. Subsequently, we examine the formulation of the theory in strain space. As noted above this is the most suitable framework for computational plasticity, given the fact that the computational problem is always regarded as strain-driven.

Throughout our discussion, if no explicit indication of the arguments in a field is made, it is understood that the fields $\boldsymbol{u}$, $\boldsymbol{\varepsilon}$, $\boldsymbol{\sigma}$ and so on, are *evaluated at a point* $\boldsymbol{x} \in \mathcal{B}$ and *at current time* $t \in [0, \mathrm{T}]$, where $[0, \mathrm{T}]$ is the time interval of interest often taken as the entire $\mathbb{R}_+$ for convenience. In addition, we denote the *strain history at a point* $\boldsymbol{x} \in \mathcal{B}$ *up to current time* $t \in \mathbb{R}_+$ by $\tau \mapsto \boldsymbol{\varepsilon}_\tau(\boldsymbol{x}) = \boldsymbol{\varepsilon}(\boldsymbol{x}, \tau)$, where $\tau \in (-\infty, t]$. Typically, one assumes that this mapping is C°. Frequently, we shall omit explicit indication of the spatial argument and write $\tau \mapsto \boldsymbol{\varepsilon}(\tau)$ or simply use the symbol $\boldsymbol{\varepsilon}_\tau$, for $\tau \in (-\infty, t]$.

2.2.1 Strain-Space and Stress-Space Formulations

Motivated by our elementary discussion in Chapter **1**, from a phenomenological point of view we regard *plastic flow* as an *irreversible process* in a material body, typically a metal, characterized in terms of the history of the strain tensor $\boldsymbol{\varepsilon}$ and two additional variables: the *plastic strain* $\boldsymbol{\varepsilon}^p$ and a suitable set of *internal variables* generically denoted by $\boldsymbol{\alpha}$ and often referred to as *hardening parameters*. Accordingly, in a strain-driven formulation, *plastic flow* at each point $\boldsymbol{x} \in \mathcal{B}$ up to current time $t \in \mathbb{R}_+$ is described in terms of the histories

$$\tau \in (-\infty, t] \mapsto \left\{\boldsymbol{\varepsilon}(\boldsymbol{x}, \tau), \boldsymbol{\varepsilon}^p(\boldsymbol{x}, \tau), \boldsymbol{\alpha}(\boldsymbol{x}, \tau)\right\}. \tag{2.2.1}$$

In this context the stress tensor is a *dependent function* of the variables $\{\boldsymbol{\varepsilon}, \boldsymbol{\varepsilon}^p\}$ through the elastic stress-strain relationships, as discussed below. This leads to a strain-space formulation of plasticity. Even though we regard (2.2.1) as our basic "driving" variables, in classical plasticity the response functions, i.e, the yield condition and the flow rule are formulated in *stress space* in terms of the variables

$$\tau \in (-\infty, t] \mapsto \left\{\boldsymbol{\sigma}(\boldsymbol{x}, \tau), \boldsymbol{q}(\boldsymbol{x}, \tau)\right\}, \tag{2.2.2}$$

where $\boldsymbol{\sigma}$ is the stress tensor (a function of $\{\boldsymbol{\varepsilon}, \boldsymbol{\varepsilon}^p\}$) and $\boldsymbol{q}$ are internal variables which are functions of $\{\boldsymbol{\varepsilon}^p, \boldsymbol{\alpha}\}$.* In the following discussion of classical plasticity, we adopt this point of view and formulate the response functions in stress space. Nevertheless, implicitly we always regard (2.2.1) as the independent variables.

*In the thermodynamic context one thinks of $\boldsymbol{q}$ as "fluxes" conjugate to the affinities $\boldsymbol{\alpha}$.

2.2.2 Stress-Space Governing Equations

We generalize our one-dimensional model problem in Chapter **1** to the three-dimensional setting as follows.

i. *Additive decomposition of the strain tensor.* One assumes that the strain tensor $\boldsymbol{\varepsilon}$ can be decomposed into an elastic and plastic part, denoted by $\boldsymbol{\varepsilon}^e$ and $\boldsymbol{\varepsilon}^p$, respectively, according to the relationship

$$\boxed{\begin{aligned} &\boldsymbol{\varepsilon} = \boldsymbol{\varepsilon}^e + \boldsymbol{\varepsilon}^p, \\ &\text{i.e.,} \\ &\varepsilon_{ij} = \varepsilon^e_{ij} + \varepsilon^p_{ij}. \end{aligned}} \tag{2.2.3}$$

Since $\boldsymbol{\varepsilon}$ is regarded as an independent variable and the evolution of $\boldsymbol{\varepsilon}^p$ is defined through the flow rule (as discussed below), equation (2.2.3) should be viewed as a definition of the elastic strain tensor as $\boldsymbol{\varepsilon}^e := \boldsymbol{\varepsilon} - \boldsymbol{\varepsilon}^p$.

ii. *(Elastic) stress response.* The stress tensor $\boldsymbol{\sigma}$ is related to the elastic strain $\boldsymbol{\varepsilon}^e$ by means of a stored-energy function $W : \mathcal{B} \times \mathbb{S} \to \mathbb{R}$ according to the (hyperelastic) relationship

$$\boldsymbol{\sigma}(\boldsymbol{x}, t) = \frac{\partial W[\boldsymbol{x}, \boldsymbol{\varepsilon}^e(\boldsymbol{x}, t)]}{\partial \boldsymbol{\varepsilon}^e}. \tag{2.2.4a}$$

For linearized elasticity, W is a quadratic form in the elastic strain, i.e., $W = \frac{1}{2}\boldsymbol{\varepsilon}^e : \mathbf{C} : \boldsymbol{\varepsilon}^e$, where $\mathbf{C}$ is the tensor of *elastic moduli* which is assumed constant. Then equations (2.2.4a) and (2.2.3) imply

$$\boxed{\begin{aligned} &\boldsymbol{\sigma} = \mathbf{C} : \left[\boldsymbol{\varepsilon} - \boldsymbol{\varepsilon}^p\right], \\ &\text{i.e.,} \\ &\sigma_{ij} = \mathsf{C}_{ijkl}(\varepsilon_{kl} - \varepsilon^p_{kl}). \end{aligned}} \tag{2.2.4b}$$

We observe that equations (2.2.4) and the decomposition (2.2.3) are *local.* Therefore, although the total strain is the (symmetric) gradient of the displacement field, the elastic strain *is not* in general the gradient of an elastic displacement field. Note further that $\boldsymbol{\varepsilon}^p$ and, consequently, $\boldsymbol{\varepsilon}^e$ are *assumed to be symmetric at the outset,* i.e., $\boldsymbol{\varepsilon}^p \in \mathbb{S}$. Thus, the notion of a plastic spin plays no role in classical plasticity.

2.2.2.1 Irreversible plastic response.

The essential feature that characterizes plastic flow is the notion of *irreversibility.* This basic property is built into the formulation through a straightforward extension of the ideas discussed in Section **1.2.1.1** of Chapter **1**, as follows.

iii. *Elastic domain and yield condition.* We define a function $f : \mathbb{S} \times \mathbb{R}^m \to \mathbb{R}$ called the *yield criterion* and constrain the admissible states $\{\boldsymbol{\sigma}, \boldsymbol{q}\} \in \mathbb{S} \times \mathbb{R}^m$ in *stress space* to lie in the set $\mathbb{E}_\sigma$ defined as

$$\boxed{\mathbb{E}_\sigma := \left\{(\boldsymbol{\sigma}, \boldsymbol{q}) \in \mathbb{S} \times \mathbb{R}^m \mid f(\boldsymbol{\sigma}, \boldsymbol{q}) \leq 0\right\}.} \tag{2.2.5}$$

One refers to the interior of $\mathbb{E}_\sigma$, denoted by int $(\mathbb{E}_\sigma)$ and given by

$$\text{int}\,(\mathbb{E}_\sigma) := \left\{(\boldsymbol{\sigma}, \boldsymbol{q}) \in \mathbb{S} \times \mathbb{R}^m \mid f(\boldsymbol{\sigma}, \boldsymbol{q}) < 0\right\}, \tag{2.2.6}$$

as the *elastic domain*; whereas the boundary of $\mathbb{E}_\sigma$, denoted by $\partial\mathbb{E}_\sigma$ and defined as

$$\partial\mathbb{E}_\sigma := \left\{(\boldsymbol{\sigma}, \boldsymbol{q}) \in \mathbb{S} \times \mathbb{R}^m \mid f(\boldsymbol{\sigma}, \boldsymbol{q}) = 0\right\}, \tag{2.2.7}$$

is called the *yield surface in stress space*. As in the one-dimensional case $\mathbb{E}_\sigma = \text{int}\,(\mathbb{E}_\sigma) \cup \partial\mathbb{E}_\sigma$. Note that states $\{\boldsymbol{\sigma}, \boldsymbol{q}\}$ outside $\mathbb{E}_\sigma$ are *nonadmissible* and are ruled out in classical plasticity.

iv. *Flow rule and hardening law. Loading/unloading conditions.* Now we introduce the notion of irreversibility of plastic flow by the following (nonsmooth) equations of evolution for $\{\boldsymbol{\varepsilon}^p, \boldsymbol{q}\}$, called *flow rule and hardening law*, respectively;

$$\boxed{\begin{aligned} \dot{\boldsymbol{\varepsilon}}^p &= \gamma \boldsymbol{r}(\boldsymbol{\sigma}, \boldsymbol{q}), \\ \dot{\boldsymbol{q}} &= -\gamma \boldsymbol{h}(\boldsymbol{\sigma}, \boldsymbol{q}). \end{aligned}} \tag{2.2.8}$$

Here $\boldsymbol{r} : \mathbb{S} \times \mathbb{R}^m \to \mathbb{S}$ and $\boldsymbol{h} : \mathbb{S} \times \mathbb{R}^m \to \mathbb{R}^m$ are *prescribed* functions which define the direction of plastic flow and the type of hardening. The parameter $\gamma \geq 0$ is a nonnegative function, called the *consistency parameter*, which is assumed to obey the following *Kuhn–Tucker complementarity conditions*:

$$\boxed{\begin{aligned} &\gamma \geq 0, \quad f(\boldsymbol{\sigma}, \boldsymbol{q}) \leq 0, \\ &\text{and} \\ &\gamma f(\boldsymbol{\sigma}, \boldsymbol{q}) = 0. \end{aligned}} \tag{2.2.9}$$

In addition to conditions (2.2.9), $\gamma \geq 0$ satisfies the *consistency requirement*

$$\boxed{\gamma \dot{f}(\boldsymbol{\sigma}, \boldsymbol{q}) = 0.} \tag{2.2.10}$$

In the classical literature, conditions (2.2.9) and (2.2.10) go by the names *loading/un-loading* and *consistency conditions*, respectively. As already discussed in Chapter **1** and further elaborated below, these conditions replicate our intuitive notion of plastic loading and elastic unloading.

2.2.2.2 Interpretation of the Kuhn–Tucker complementarity conditions.

The following alternative situations which give rise to Figure **2.1** occur.

a. First consider the case in which $\{\boldsymbol{\sigma}, \boldsymbol{q}\} \in \text{int}\,(\mathbb{E}_\sigma)$ so that, according to (2.2.6) $f(\boldsymbol{\sigma}, \boldsymbol{q}) < 0$. Therefore, from condition $(2.2.9)_3$ we conclude that

$$\gamma f = 0 \text{ and } f < 0 \Rightarrow \boxed{\gamma = 0.} \tag{2.2.11a}$$

Then from (2.2.8) it follows that $\dot{\boldsymbol{\varepsilon}}^p = \mathbf{0}$ and $\dot{\boldsymbol{q}} = \mathbf{0}$. Thus, (2.2.3) yields $\dot{\boldsymbol{\varepsilon}} = \dot{\boldsymbol{\varepsilon}}^e$, and the rate form of (2.2.4b) leads to

$$\dot{\boldsymbol{\sigma}} = \mathbf{C} : \dot{\boldsymbol{\varepsilon}} \equiv \mathbf{C} : \dot{\boldsymbol{\varepsilon}}^e. \tag{2.2.11b}$$

In view of (2.2.11*b*) we call this type of response *instantaneously elastic*.

b. Now suppose that $\{\boldsymbol{\sigma}, \boldsymbol{q}\} \in \partial\mathbb{E}$ which, in view of (2.2.7), implies that $f(\boldsymbol{\sigma}, \boldsymbol{q}) = 0$. Then condition (2.2.9)$_3$ is automatically satisfied even if $\gamma > 0$. Whether γ is actually positive or zero is concluded from condition (2.2.10). Two situations can arise.
   ii.a. First, if $\dot{f}(\boldsymbol{\sigma}, \boldsymbol{q}) < 0$, from condition (2.2.10) we conclude that

$$\gamma \dot{f} = 0 \text{ and } \dot{f} < 0 \Rightarrow \boxed{\gamma = 0\,.} \tag{2.2.12}$$

   Thus, again from (2.2.8) it follows that $\dot{\boldsymbol{\varepsilon}}^p = \mathbf{0}$ and $\dot{\boldsymbol{q}} = \mathbf{0}$. Since (2.2.11b) holds and $(\boldsymbol{\sigma}, \boldsymbol{q})$ is on $\partial\mathbb{E}_\sigma$, this type of response is called *unloading from a plastic state*.

   ii.b. Second, if $\dot{f}(\boldsymbol{\sigma}, \boldsymbol{q}) = 0$, condition (2.2.10) is automatically satisfied. If $\gamma > 0$, then $\dot{\boldsymbol{\varepsilon}}^p \neq \mathbf{0}$ and $\dot{\boldsymbol{q}} \neq \mathbf{0}$, a situation called *plastic loading*. The case $\gamma = 0$ (and $\dot{f} = 0$) is termed *neutral* loading.

To summarize the preceding discussion we have the following possible situations and corresponding definitions for any $(\boldsymbol{\sigma}, \boldsymbol{q}) \in \mathbb{E}_\sigma$:

$$\left\{\begin{array}{ll} f < 0 \Longleftrightarrow (\boldsymbol{\sigma}, \boldsymbol{q}) \in \operatorname{int}(\mathbb{E}_\sigma) \Longrightarrow & \gamma = 0 \text{ (elastic)} \\ f = 0 \Longleftrightarrow (\boldsymbol{\sigma}, \boldsymbol{q}) \in \partial\mathbb{E}_\sigma & \left\{\begin{array}{l} \dot{f} < 0 \Longrightarrow \gamma = 0 \text{ (elastic unloading)} \\ \dot{f} = 0 \text{ and } \gamma = 0 \text{ (neutral loading)} \\ \dot{f} = 0 \text{ and } \gamma > 0 \text{ (plastic loading).} \end{array}\right. \end{array}\right. \tag{2.2.13}$$

We observe that the possibility $\dot{f} > 0$ has been excluded from the analysis above. Intuitively, it is clear that if $\dot{f}(\boldsymbol{\sigma}, \boldsymbol{q}) > 0$ for some $(\boldsymbol{\sigma}, \boldsymbol{q}) \in \partial\mathbb{E}_\sigma$ at some time $t \in \mathbb{R}_+$, then condition $f \leq 0$ would be violated at a neighboring subsequent time; see Figure **2.2**. A formal argument is given in the following

Lemma 2.2.1. *Let $\tau \mapsto \{\boldsymbol{\sigma}_\tau, \boldsymbol{q}_\tau\}$ for $\tau \in (-\infty, t]$ be the history in stress space up to current time $t \in \mathbb{R}_+$. Set*

$$\hat{f}(t) := f(\boldsymbol{\sigma}_t, \boldsymbol{q}_t), \tag{2.2.14a}$$

and assume that $(\boldsymbol{\sigma}_t, \boldsymbol{q}_t)$ is on $\partial\mathbb{E}_\sigma$ so that $\hat{f}(t) = 0$. Then the time derivative of $\hat{f}(t)$ cannot be positive, i.e.,

$$\boxed{\textit{if } \hat{f}(t) = 0 \textit{ at } t \in \mathbb{R}_+ \textit{ then } \dot{\hat{f}}(t) \leq 0.} \tag{2.2.14b}$$

Proof. Assuming that $\hat{f}(\bullet)$ is smooth, the result follows from elementary considerations. In fact, for $\zeta \geq t$ by Taylor's formula,

$$\hat{f}(\zeta) = \hat{f}(t) + [\zeta - t]\dot{\hat{f}}(t) + \mathcal{O}\left|\zeta - t\right|^2, \tag{2.2.14c}$$

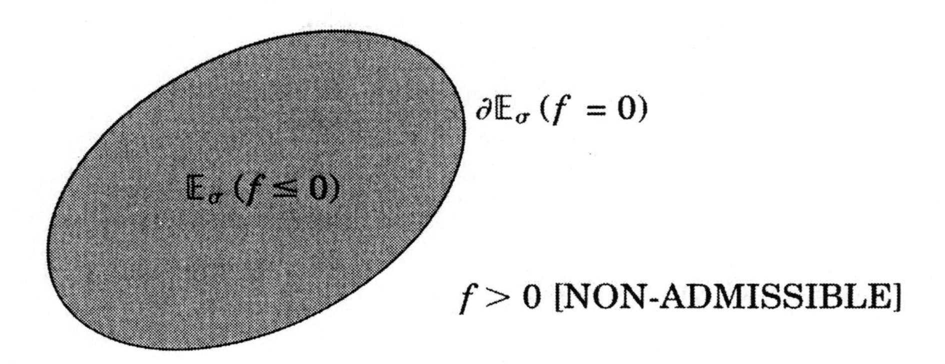

FIGURE 2-1. Illustration of the elastic domain and admissible states in stress space.

where, by definition, $\mathcal{O}|\varsigma - t|^2/[\varsigma - t] \to 0$ as $\varsigma \to t$. Now since $\hat{f}(\varsigma) \leq 0$ and $\hat{f}(t) = 0$, dividing (2.2.14c) by $[\varsigma - t]$ leads to the inequality

$$\dot{\hat{f}}(t) + \frac{\mathcal{O}[\varsigma - t]^2}{[\varsigma - t]} \leq 0. \tag{2.2.15}$$

Then the result follows by taking the limit of (2.2.15) as $[\varsigma - t] \to 0$. □

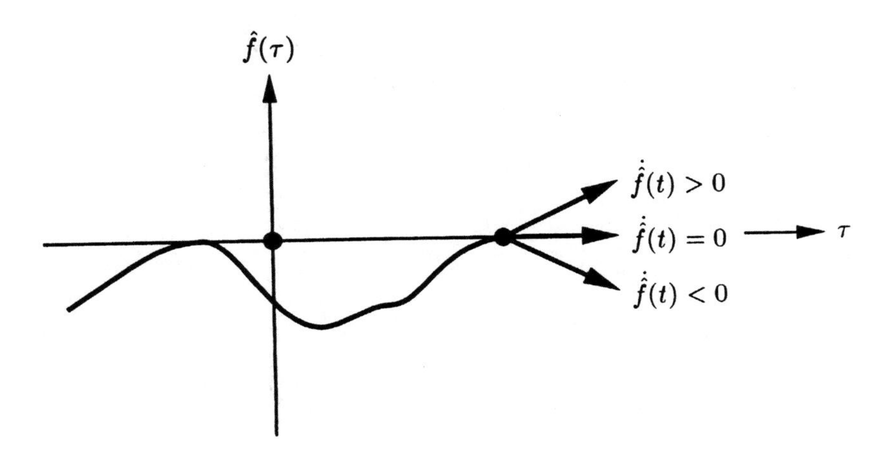

FIGURE 2-2. Illustration of the fact that $\dot{\hat{f}}(t) \leq 0$.

The consistency condition (2.2.10) enables us to relate γ to the *current strain* rate and results in an alternative formulation of the loading/unloading conditions closely related to expressions found in the classical literature.

2.2.2.3 Consistency condition and elastoplastic tangent moduli.

To exploit condition (2.2.10), we start out by evaluating the time derivative of f at $(\boldsymbol{\sigma}, \boldsymbol{q}) \in \mathbb{E}_\sigma$. Using the chain rule, along with the rate forms of the stress-strain relationship (2.2.4), the flow rule, and the hardening law in (2.2.8), we find that

$$\begin{aligned} \dot{f} &= \partial_\sigma f : \dot{\boldsymbol{\sigma}} + \partial_q f \cdot \dot{\boldsymbol{q}} \\ &= \partial_\sigma f : \mathbf{C} : \left[\dot{\boldsymbol{\varepsilon}} - \dot{\boldsymbol{\varepsilon}}^p\right] + \partial_q f \cdot \dot{\boldsymbol{q}} \\ &= \partial_\sigma f : \mathbf{C} : \dot{\boldsymbol{\varepsilon}} - \gamma\left[\partial_\sigma f : \mathbf{C} : \boldsymbol{r} + \partial_q f \cdot \boldsymbol{h}\right] \le 0. \end{aligned} \tag{2.2.16}$$

To carry our analysis further we need to make an additional assumption about the structure of the flow rule and hardening law in (2.2.8). Explicitly, we make the following hypothesis.

Assumption 2.1. The flow rule, hardening law, and yield condition in stress space are such that the following inequality holds:

$$\boxed{\left[\partial_\sigma f : \mathbf{C} : \boldsymbol{r} + \partial_q f \cdot \boldsymbol{h}\right] > 0,} \tag{2.2.17}$$

for all admissible states $\{\boldsymbol{\sigma}, \boldsymbol{q}\} \in \partial\mathbb{E}_\sigma$.

We will see below that this assumption always holds for *associative perfect* plasticity. With such an assumption in hand, it follows from (2.2.10) that

$$\boxed{\dot{f} = 0 \Longleftrightarrow \gamma = \frac{\langle \partial_\sigma f : \mathbf{C} : \dot{\boldsymbol{\varepsilon}}\rangle}{\partial_\sigma f : \mathbf{C} : \boldsymbol{r} + \partial_q f \cdot \boldsymbol{h}},} \tag{2.2.18}$$

where $\langle x\rangle := [x + |x|]/2$ denotes the ramp function. In view of (2.2.17) and (2.2.18), we also conclude that

$$\boxed{\begin{aligned} &\text{for } f = 0 \text{ and } \dot{f} = 0, \\ &\gamma \ge 0 \Longleftrightarrow \partial_\sigma f : \mathbf{C} : \dot{\boldsymbol{\varepsilon}} \ge 0. \end{aligned}} \tag{2.2.19}$$

This relationship provides a useful geometric interpretation of the plastic loading and neutral loading conditions in (2.2.13) which are illustrated in Figure **2.3**. Plastic loading or neutral loading takes place at a point $(\boldsymbol{\sigma}, \boldsymbol{q}) \in \partial\mathbb{E}_\sigma$ if the angle in the *inner product defined by the elasticity tensor* $\mathbf{C}$ between the *normal* $\partial_\sigma f(\boldsymbol{\sigma}, \boldsymbol{q})$ to $\partial\mathbb{E}_\sigma$ at $(\boldsymbol{\sigma}, \boldsymbol{q})$ and the strain rate $\dot{\boldsymbol{\varepsilon}}$ is less or equal than 90°.
Finally, according to (2.2.4) and (2.2.8),

$$\dot{\boldsymbol{\sigma}} = \mathbf{C} : \left[\dot{\boldsymbol{\varepsilon}} - \dot{\boldsymbol{\varepsilon}}^p\right] = \mathbf{C} : \left[\dot{\boldsymbol{\varepsilon}} - \gamma \boldsymbol{r}\right]. \tag{2.2.20}$$

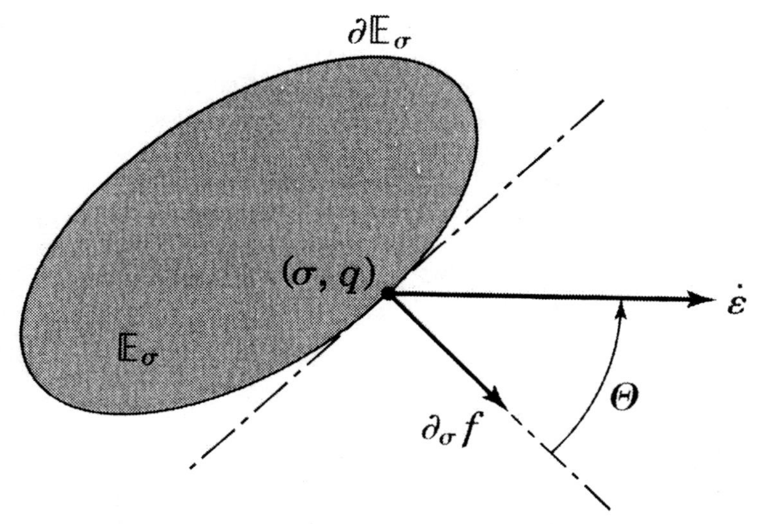

FIGURE 2-3. Plastic loading at $(\sigma, q) \in \partial\mathbb{E}_\sigma$ takes place if the angle Θ defined as $\Theta := \partial_\sigma f : \mathbf{C} : \dot{\varepsilon} \Big/ [\partial_\sigma f : \mathbf{C} : \partial_\sigma f]^{\frac{1}{2}} [\dot{\varepsilon} : \mathbf{C} : \dot{\varepsilon}]^{\frac{1}{2}}$ is such that $\Theta < \pi/2$.

Then substituting (2.2.18) in (2.2.20) then yields the rate of change of $\boldsymbol{\sigma}$ in terms of the total strain rate $\dot{\boldsymbol{\varepsilon}}$ as

$$\boxed{\dot{\sigma} = \mathbf{C}^{\mathrm{ep}} : \dot{\varepsilon},} \tag{2.2.21}$$

where $\mathbf{C}^{\mathrm{ep}}$ is the so-called tensor of tangent elastoplastic moduli given by the expression

$$\boxed{\mathbf{C}^{\mathrm{ep}} = \begin{cases} \mathbf{C} & \text{if } \gamma = 0, \\ \mathbf{C} - \dfrac{\mathbf{C} : \boldsymbol{r} \otimes \mathbf{C} : \partial_\sigma f}{\partial_\sigma f : \mathbf{C} : \boldsymbol{r} + \partial_{\boldsymbol{q}} f \cdot \boldsymbol{h}} & \text{if } \gamma > 0. \end{cases}} \tag{2.2.22}$$

Note that $\mathbf{C}^{\mathrm{ep}}$ is generally *nonsymmetric* for arbitrary $\boldsymbol{r}(\boldsymbol{\sigma}, \boldsymbol{q})$, except in the case for which

$$\boxed{\boldsymbol{r}(\boldsymbol{\sigma}, \boldsymbol{q}) = \partial_\sigma f(\boldsymbol{\sigma}, \boldsymbol{q}),} \tag{2.2.23}$$

which has special significance and is called *an associative flow rule.*

Remarks 2.2.1.

1. The analysis of Section **2.2.2.3**, leading to expressions (2.2.18) and (2.2.19), relies crucially on Assumption 2.1. This assumption is also necessary to establish the equivalence between the Kuhn–Tucker complementarity conditions and the classical loading/unloading conditions in strain space, which are essentially equivalent to (2.2.19). Further discussion on the alternative formulations

of the loading/unloading conditions is deferred to Section **2.2.3.1.** A simple one-dimensional interpretation of Assumption 2.1 is also given there.

2. A more fundamental interpretation of the associative flow rule (2.2.23) is given below where we show that the flow rule (2.2.23) and a *particular form* of the hardening law are the result of a classical hypothesis known as the *principle of maximum plastic dissipation* (or maximum entropy production).
3. For *perfect* plasticity, characterized by $\boldsymbol{h} = \boldsymbol{0}$, Assumption 2.1 always holds provided that the flow rule is associative. This conclusion is the direct result of the positive definiteness (or pointwise stability) of the elasticity tensor $\mathbf{C}$ since, for $\boldsymbol{h} \equiv \boldsymbol{0}$, (2.2.17) reduces to

$$\partial_\sigma f : \mathbf{C} : \partial_\sigma f \geq \beta \|\partial_\sigma f\|^2 > 0, \tag{2.2.24}$$

where $\beta > 0$ is the ellipticity constant. Recall that positive definiteness of $\mathbf{C}$ holds for any symmetric tensor $\boldsymbol{\xi} \in \mathbb{S}$, in particular, for $\boldsymbol{\xi} = \partial_\sigma f$.
4. Expression (2.2.22) clearly exhibits the fact that the formulation outlined above is indeed *rate-independent* in the sense that the stress rate depends linearly on the strain rate. The rate-dependent version of the theory, known as *viscoplasticity*, is considered in Section **2.7**.

For the reader's convenience and subsequent reference, we have summarized the basic governing equations of general rate-independent plasticity in BOX **2.1**.

2.2.3 Strain-Space Formulation

A strain space formulation of the classical model outlined in BOX **2.1** is derived by substituting the elastic stress-strain relationship (2.2.4) in the yield condition, flow rule, and hardening law to obtain

$$\left.\begin{aligned} \bar{f}(\boldsymbol{\varepsilon}, \boldsymbol{\varepsilon}^p, \boldsymbol{q}) &:= f[\partial_\varepsilon W(\boldsymbol{\varepsilon} - \boldsymbol{\varepsilon}^p), \boldsymbol{q}], \\ \bar{\boldsymbol{r}}(\boldsymbol{\varepsilon}, \boldsymbol{\varepsilon}^p, \boldsymbol{q}) &:= \boldsymbol{r}[\partial_\varepsilon W(\boldsymbol{\varepsilon} - \boldsymbol{\varepsilon}^p), \boldsymbol{q}], \\ \bar{\boldsymbol{h}}(\boldsymbol{\varepsilon}, \boldsymbol{\varepsilon}^p, \boldsymbol{q}) &:= \boldsymbol{h}[\partial_\varepsilon W(\boldsymbol{\varepsilon} - \boldsymbol{\varepsilon}^p), \boldsymbol{q}]. \end{aligned}\right\} \tag{2.2.25}$$

Despite some claims in the literature, note that on physical grounds it is difficult to justify the *a priori* formulation of the response functions in strain space. In fact, all the yield conditions typically used in metal plasticity, notably the Mises–Huber yield criterion and the Tresca yield condition or variants thereof, are formulated in stress space.

By using the chain rule and the change of variable formulas (2.2.25), it is possible to recast the developments of the preceding section entirely in strain space. In particular, from (2.2.25) we note the relationship

$$\boxed{\partial_\varepsilon \bar{f} = \mathbf{C} : \partial_\sigma f.} \tag{2.2.26}$$

BOX 2.1. Classical Rate-Independent Plasticity

i. Elastic stress-strain relationships

$$\boldsymbol{\sigma} = \frac{\partial W(\boldsymbol{\varepsilon} - \boldsymbol{\varepsilon}^p)}{\partial \boldsymbol{\varepsilon}} = \mathbf{C} : (\boldsymbol{\varepsilon} - \boldsymbol{\varepsilon}^p)$$

$$\mathbf{C} := \frac{\partial^2 W(\boldsymbol{\varepsilon} - \boldsymbol{\varepsilon}^p)}{\partial \boldsymbol{\varepsilon}^2} = \text{constant (elastic moduli)}.$$

ii. Elastic domain in stress space (single surface)

$$\mathbb{E}_\sigma = \left\{(\boldsymbol{\sigma}, \boldsymbol{q}) \in \mathbb{S} \times \mathbb{R}^m \mid f(\boldsymbol{\sigma}, \boldsymbol{q}) \leq 0\right\}.$$

iii. Flow rule and hardening law

iii.a. General nonassociative model

$$\dot{\boldsymbol{\varepsilon}}^p = \gamma \boldsymbol{r}(\boldsymbol{\sigma}, \boldsymbol{q})$$

$$\dot{\boldsymbol{q}} = -\gamma \boldsymbol{h}(\boldsymbol{\sigma}, \boldsymbol{q}).$$

iii.b. (Particular) associative case

$$\dot{\boldsymbol{\varepsilon}}^p = \gamma \frac{\partial f}{\partial \boldsymbol{\sigma}}$$

$$\dot{\boldsymbol{q}} = -\gamma \mathbf{D} \frac{\partial f}{\partial \boldsymbol{q}}$$

$$\mathbf{D} = \text{matrix of generalized plastic moduli.}$$

iv. Kuhn–Tucker loading/unloading (complementarity) conditions

$$\gamma \geq 0, \;\; f(\boldsymbol{\sigma}, \boldsymbol{q}) \leq 0, \;\; \gamma f(\boldsymbol{\sigma}, \boldsymbol{q}) = 0.$$

v. Consistency condition

$$\gamma \dot{f}(\boldsymbol{\sigma}, \boldsymbol{q}) = 0.$$

By substituting (2.2.26) in (2.2.18), we obtain the strain-space expression

$$\dot{\bar{f}} = 0 \Longleftrightarrow \gamma = \frac{\partial_\varepsilon \bar{f} : \dot{\boldsymbol{\varepsilon}}}{\partial_\varepsilon \bar{f} : \bar{\boldsymbol{r}} + \partial_q \bar{f} \cdot \bar{\boldsymbol{h}}}, \tag{2.2.27}$$

so that the counterpart of (2.2.19) in strain space becomes

$$\boxed{\gamma \geq 0 \Longleftrightarrow \partial_\varepsilon \bar{f} : \dot{\boldsymbol{\varepsilon}} \geq 0, \;\; \text{for } \bar{f} = 0 \text{ and } \dot{\bar{f}} = 0.} \tag{2.2.28}$$

Finally, we remark that the validity of the two expressions above relies on the counterpart in strain space of Assumption 2.1 which now takes the form

Assumption 2.1*. The yield condition, flow rule, and hardening law in strain space are assumed to obey the inequality

$$\boxed{\partial_{\varepsilon} \bar{f} : \bar{\boldsymbol{r}} + \partial_{q} \bar{f} \cdot \bar{\boldsymbol{h}} > 0,} \tag{2.2.29}$$

for all admissible states $\{\boldsymbol{\varepsilon}, \boldsymbol{\varepsilon}^{p}, \boldsymbol{q}\} \in \mathbb{E}_{\varepsilon}$, where

$$\mathbb{E}_{\varepsilon} := \left\{(\boldsymbol{\varepsilon}, \boldsymbol{\varepsilon}^{p}, \boldsymbol{q}) \in \mathbb{S} \times \mathbb{S} \times \mathbb{R}^{m} \mid \bar{f}(\boldsymbol{\varepsilon}, \boldsymbol{\varepsilon}^{p}, \boldsymbol{q}) \leq 0\right\}, \tag{2.2.30}$$

is the elastic domain in strain space.

2.2.3.1 Alternative formulation of the loading/unloading conditions.

In the constitutive theory outline above and summarized in BOX **2.1**, the loading/unloading conditions are formulated as Kuhn–Tucker complementarity conditions, a form which is standard for problems subject to unilateral constraints. We show below that this form of the loading/unloading conditions is equivalent to two alternative characterizations of plastic loadings.

i. *Strain-space loading/unloading conditions*, as discussed in Naghdi and Trapp [1975] or Casey and Naghdi [1981, 1983a,b], are formulated as follows.

$$\begin{cases} \bar{f} < 0 & \text{(elastic)} \\ \bar{f} = 0 \text{ and } \begin{cases} \partial_{\varepsilon} \bar{f} : \dot{\boldsymbol{\varepsilon}} < 0 & \text{(elastic unloading)} \\ \partial_{\varepsilon} \bar{f} : \dot{\boldsymbol{\varepsilon}} = 0 & \text{(neutral loading)} \\ \partial_{\varepsilon} \bar{f} : \dot{\boldsymbol{\varepsilon}} > 0 & \text{(plastic loading).} \end{cases} \end{cases} \tag{2.2.31}$$

In view of relationship (2.2.28), it is apparent that these conditions are equivalent to conditions (2.2.13) and, therefore, equivalent to the classical Kuhn–Tucker conditions.

ii. *The rate-of-trial-stress condition.* Alternatively, loading/unloading conditions are formulated in terms of the so-called *rate-of-trial-stress* defined as

$$\boxed{\dot{\boldsymbol{\sigma}}^{\text{trial}} := \mathbf{C} : \dot{\boldsymbol{\varepsilon}},} \tag{2.2.32}$$

by declaring a process plastic whenever

$$\boxed{\begin{aligned} & f(\boldsymbol{\sigma}, \boldsymbol{q}) = 0, \\ \text{and} & \\ & \partial_{\sigma} f(\boldsymbol{\sigma}, \boldsymbol{q}) : \dot{\boldsymbol{\sigma}}^{\text{trial}} > 0. \end{aligned}} \tag{2.2.33}$$

The fact that this condition is equivalent to the Kuhn–Tucker conditions follows at once from (2.2.32) and (2.2.18) by noting that, for $f(\boldsymbol{\sigma}, \boldsymbol{q}) = 0$,

$$\dot{f} = 0 \Longleftrightarrow \gamma = \frac{\partial_{\sigma} f : \dot{\boldsymbol{\sigma}}^{\text{trial}}}{\partial_{\sigma} f : \mathbf{C} : \boldsymbol{r} + \partial_{q} f \cdot \boldsymbol{h}}. \tag{2.2.34}$$

Consequently, since Assumption 2.1 holds,

$$\boxed{\gamma > 0 \Longleftrightarrow \partial_\sigma f : \dot{\sigma}^{\text{trial}} > 0, \text{ for } f = \dot{f} = 0,} \tag{2.2.35}$$

and the equivalence between (2.2.33) and the Kuhn–Tucker conditions follows. The simple geometric interpretation of $\dot{\sigma}^{\text{trial}} := \mathbf{C} : \dot{\varepsilon}$ should be noted and is illustrated in Figure **2.4**. We observe that $\mathbf{C} : \dot{\varepsilon}$ is the rate of stress obtained by "freezing" the evolution of plastic flow and internal variables (i.e., by setting $\dot{\varepsilon}^p = \mathbf{0}$ and $\dot{\boldsymbol{q}} = 0$) hence the name "rate-of-trial (elastic) stress."

As motivated in Chapter **1** and discussed in detail in Chapter **3**, the notion of trial elastic state arises naturally in the context of the algorithmic treatment of the elasto-plastic problem and can be rigorously justified as a product formula based on an elastic-plastic operator split. In the computational literature, use of the algorithmic counterpart of the rate-of-trial-stress condition goes back to the pioneering work of Wilkins [1964] on the now classical radial return algorithm for J_2 flow theory. The notion was subsequently formalized independently by Moreau [1976,1977] who coined the expression "catching-up-algorithm." The explicit formulation of the loading condition in the form (2.2.33) is found in Hughes [1984].

2.2.4 An Elementary Example: 1-D Plasticity

First we illustrate the general formulation outlined above by returning to our elementary example of Chapter **1** of a one-dimensional bar occupying an interval $\mathcal{B} = [0, L]$. Our objective is to examine in a simple context the significance of

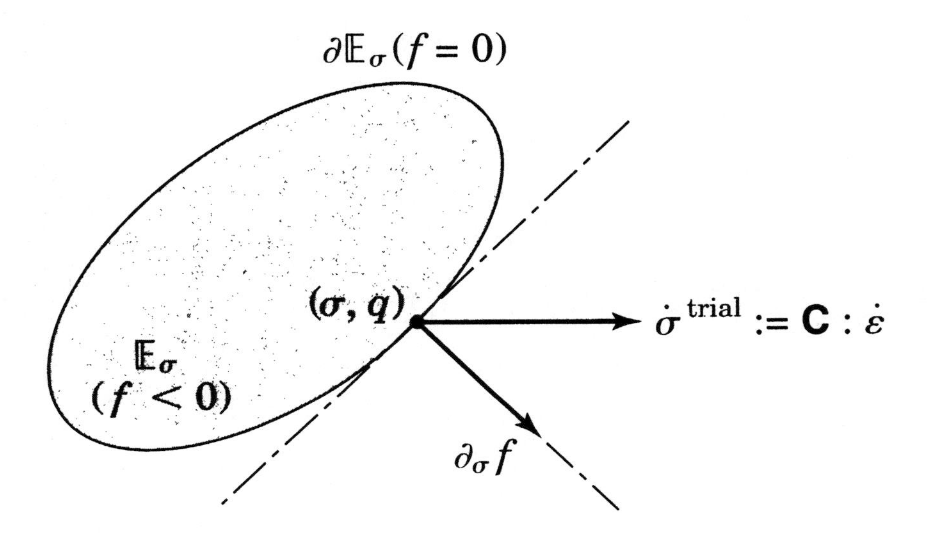

FIGURE 2-4. Interpretation of the loading/unloading conditions in terms of the trial elastic stress $\dot{\sigma}^{\text{trial}}$.

Assumption 2.1 based on which the equivalence between the Kuhn–Tucker conditions and the strain space (or rate-of-trial-stress) loading/unloading conditions was established.

2.2.4.1 Particularization of the general model.

To bring the general model in BOX **2.1** into correspondence with that of BOX **1.3** in Chapter **1**, we let $\{\boldsymbol{\sigma}, \boldsymbol{\varepsilon}, \boldsymbol{\varepsilon}^p\}$ be simply $\{\sigma, \varepsilon, \varepsilon^p\}$. Further, we define a two-dimensional vector of internal variables as

$$\boldsymbol{q} = \begin{Bmatrix} q_1 \\ q_2 \end{Bmatrix} \Rightarrow m = 2. \tag{2.2.36}$$

Now, we specify the yield criterion as

$$\boxed{f(\sigma, \boldsymbol{q}) := |\sigma + q_2| + q_1 - \sigma_Y \leq 0.} \tag{2.2.37}$$

Finally, we specify the flow rule and hardening law by setting

$$\begin{aligned} r(\sigma, \boldsymbol{q}) &:= \text{sign}\,[\sigma + q_2] \\ \boldsymbol{h}(\sigma, \boldsymbol{q}); &= \begin{bmatrix} K & 0 \\ 0 & H \end{bmatrix} \begin{Bmatrix} 1 \\ \text{sign}\,[\sigma + q_2] \end{Bmatrix}. \end{aligned} \tag{2.2.38}$$

Obviously, in the present one-dimensional context the elastic moduli $\mathbf{C}$ reduce to E. We also observe from (2.2.37) that

$$\begin{aligned} \partial_\sigma f &= \text{sign}[\sigma + q_2] \quad ; \\ \text{and} \\ \partial_{\boldsymbol{q}} f &= \begin{Bmatrix} 1 \\ \text{sign}\,[\sigma + q_2] \end{Bmatrix}. \end{aligned} \tag{2.2.39}$$

From the preceding expressions and BOX **1.3** in Chapter **1**, we conclude that the general model in BOX **2.1** reduces to the one-dimensional, kinematic/isotropic model by identifying the back stress with $-q_2$ and the equivalent plastic strain with $-q_1/K$. Note further that, in view of (2.2.38)–(2.2.39), expression (2.2.18) becomes

$$\begin{aligned} \gamma &= \frac{\text{sign}\,[\sigma + q_2] E\dot{\varepsilon}}{E\,(\text{sign}\,[\sigma + q_2])^2 + K + H\,(\text{sign}\,[\sigma + q_2])^2} \\ &= E\,\text{sign}\,[\sigma + q_2] \frac{\dot{\varepsilon}}{[E + H + K]} \end{aligned} \tag{2.2.40}$$

which coincides with (1.2.46) of Chapter **1** under our identification of the back stress q as $-q_2$.

Remarks 2.2.2.

1. From expression (2.2.38) and (2.2.39), it follows that the flow rule and hardening law can be written as

$$\boxed{\begin{aligned} \dot{\varepsilon}^p &= \gamma \frac{\partial f}{\partial \sigma}, \\ \dot{\boldsymbol{q}} &= -\gamma \mathbf{D} \frac{\partial f}{\partial \boldsymbol{q}}, \\ \text{and} & \\ \mathbf{D} &:= \begin{bmatrix} K & 0 \\ 0 & H \end{bmatrix}. \end{aligned}} \tag{2.2.41}$$

   Therefore, the flow rule is associative.

2. Suppose that we define an alternative set of *internal variables in strain space*, denoted by $\boldsymbol{\alpha} := [\alpha_1 \ \alpha_2]^T$, through the relationship

$$\boldsymbol{q} = -\mathbf{D}\boldsymbol{\alpha} \Longleftrightarrow \boldsymbol{\alpha} = -\mathbf{D}^{-1}\boldsymbol{q}. \tag{2.2.42}$$

   Then, the flow rule and hardening law given by $(2.2.41)_{1,2}$, respectively, take the suggestive forms

$$\boxed{\begin{aligned} \dot{\varepsilon}^p &= \gamma \frac{\partial f}{\partial \sigma}, \\ \text{and} & \\ \dot{\boldsymbol{\alpha}} &= \gamma \frac{\partial f}{\partial \boldsymbol{q}}. \end{aligned}} \tag{2.2.43}$$

   It follows from $(2.2.43)_2$ and (2.2.39) that $\dot{\alpha}_1 = \gamma$, i.e., $\alpha_1 = -K^{-1}q_1$ coincides with the equivalent plastic strain as defined by equation (1.2.24) of Chapter **1**. Note that the evolutionary equations for $\{\varepsilon^p, \boldsymbol{\alpha}\}$ are *both* associative in the sense that the yield criterion $f(\sigma, \boldsymbol{q})$ in *stress space* is a *potential* in the sense of (2.2.43) for *both* $\boldsymbol{\varepsilon}^p$ and α. As elaborated in detail in Section **6**, this model is said to *obey the principle of maximum plastic dissipation*.

3. Finally from (2.2.42) and the stress-strain relationship $\sigma = E[\varepsilon - \varepsilon^p]$, we observe that

$$\begin{aligned} \sigma &= \frac{\partial W(\varepsilon - \varepsilon^p)}{\partial \varepsilon} \\ \text{and} & \\ \boldsymbol{q} &= -\frac{\partial \mathcal{H}(\boldsymbol{\alpha})}{\partial \boldsymbol{\alpha}}, \end{aligned} \tag{2.2.44}$$

   where $W(\varepsilon - \varepsilon^p)$ is the elastic stored-energy function and $\mathcal{H}(\boldsymbol{\alpha})$ is a potential defined, respectively, as

$$W := \frac{1}{2}(\varepsilon - \varepsilon^p)E(\varepsilon - \varepsilon^p);$$

and

$$\mathcal{H} = \frac{1}{2}\boldsymbol{\alpha}^T \mathbf{D} \boldsymbol{\alpha}. \tag{2.2.45}$$

The significance of this potential is explored in detail in Section **6**.

2.2.4.2 Significance of Assumption 2.1.

The developments above enable us to provide a simple interpretation of Assumption 2.1. By substituting (2.2.38) and (2.2.39) in (2.2.17), we obtain

$$\begin{aligned} \partial_\sigma f : \mathbf{C} : \boldsymbol{r} + \partial_{\boldsymbol{q}} f \cdot \boldsymbol{h} &= E + \{1 \ \operatorname{sign}[\sigma + q_2]\} \begin{bmatrix} K & 0 \\ 0 & H \end{bmatrix} \begin{Bmatrix} 1 \\ \operatorname{sign}[\sigma + q_2] \end{Bmatrix} \\ &= E + \left\{ K + (\operatorname{sign}[\sigma + q_2])^2 H \right\} \\ &= E + [H + K] > 0. \end{aligned} \tag{2.2.46}$$

The significance of this condition is easily appreciated by inspecting Figure **2.5**. Condition (2.2.46) places a *restriction on the amount of allowable softening* in the sense that

$$\boxed{[H + K] > -E \iff E^{\mathrm{ep}} > -\infty,} \tag{2.2.47}$$

where E^{ep} is defined by (1.2.47) of Chapter **1** or in the general case by (2.2.22).

It should be noted that the *classical uniqueness condition* places a much stronger restriction on the formulation than Assumption 2.1. In effect, a *sufficient condition*

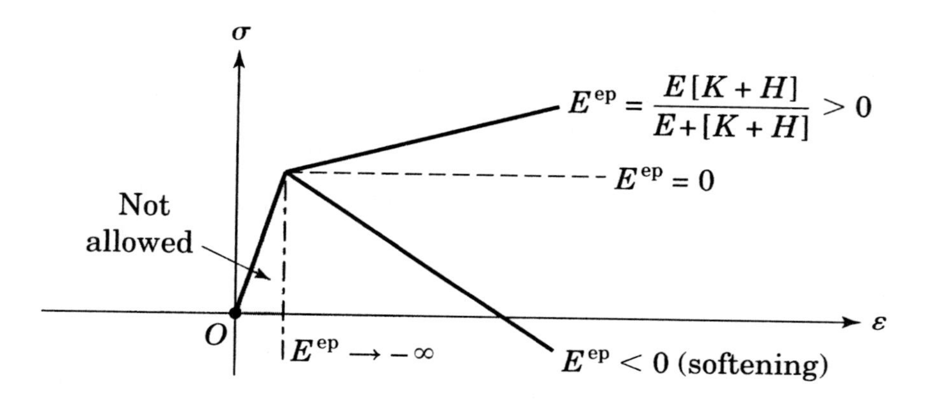

FIGURE 2-5. Assumption 2.1 places a limit on the amount of allowable softening in the model.

for uniqueness of the elastoplastic boundary value problem is that the *second-order work density be nonnegative* (see Hill [1950]), i.e.,

$$\delta^2 W := \frac{1}{2} \dot{\boldsymbol{\sigma}} : \dot{\boldsymbol{\varepsilon}} \geq 0. \tag{2.2.48}$$

This condition admits a simple interpretation in the context of the one-dimensional problem. From (1.2.47) of Chapter **1** and (2.2.48) we find that

$$\dot{\sigma}\dot{\varepsilon} = \frac{E[K + H]}{E + [K + H]} (\dot{\varepsilon})^2 > 0 \Longleftrightarrow \boxed{K + H > 0.} \tag{2.2.49}$$

Thus, the uniqueness condition *precludes the presence of a softening response.* The derivation of the three-dimensional counterpart of this result follows exactly the same lines as above.

2.2.4.3 Remark on sign convention.

Note that the internal variable q_2, which corresponds to the *back stress* in the present context, differs from our definition in Chapter **1** by a minus sign. The reason for this change in sign convention lies in the thermodynamic interpretation of the general theory discussed in Section **5** below. In particular, note that this convention leads to a *positive-definite* matrix of plastic moduli $\mathbf{D}$ defined by $(2.2.41)_3$. Other choices of sign convention, however, are possible.

2.3 Plane Strain and 3-D, Classical J_2 Flow Theory

We summarize below the classical model of metal plasticity for plane-strain or three-dimensional problems. This model is obtained by specializing the framework outlined in Section **2**. The plane-stress case warrants special treatment and is considered in the following section.

2.3.1 Perfect Plasticity

The classical Prandl–Reuss equations of perfect plasticity are obtained by introducing the following assumptions:

1. linear isotropic elastic response,
2. *Huber–von Mises* yield condition, i.e., $f(\boldsymbol{\sigma}) := \sqrt{\|\boldsymbol{\sigma}\|^2 - \frac{1}{3}(\operatorname{tr}[\boldsymbol{\sigma}])^2} - R$, where $R := \sqrt{\frac{2}{3}}\sigma_Y$ is the radius of the yield surface and σ_Y is the flow stress;
3. associative *Levy-Saint Venant* flow rule,; and
4. no hardening, i.e., $\boldsymbol{h} \equiv \mathbf{0}$.

From these assumptions it follows that the plastic-strain rate is given by the evolutionary equation

$$\dot{\varepsilon}^p = \gamma \nabla f(\sigma) = \gamma \frac{\sigma - \frac{1}{3}(\text{tr}[\sigma])\mathbf{1}}{\sqrt{\|\sigma\|^2 - \frac{1}{3}(\text{tr}[\sigma])^2}} \equiv \gamma \frac{\text{dev}[\sigma]}{\|\text{dev}[\sigma]\|}, \tag{2.3.1}$$

where $\text{dev}[\bullet] := (\bullet) - \frac{1}{3}(\text{tr}[\bullet])\mathbf{1}$ denotes the deviator of the indicated argument. Since $\text{tr}[\dot{\varepsilon}^p] \equiv 0$, by particularizing equation (2.2.18), we obtain the following expression for the consistency parameter:

$$\boxed{\gamma = \boldsymbol{n} : \dot{\varepsilon},}$$

where

$$\boxed{\boldsymbol{n} := \frac{\text{dev}[\sigma]}{\|\text{dev}[\sigma]\|}.} \tag{2.3.2}$$

Finally, by particularizing the general expression (2.2.22), we arrive at the following expression for the elastoplastic tangent moduli:

$$\boxed{\mathbf{C}^{ep} = \kappa \mathbf{1} \otimes \mathbf{1} + 2\mu[\mathbf{I} - \tfrac{1}{3}\mathbf{1} \otimes \mathbf{1} - \boldsymbol{n} \otimes \boldsymbol{n}], \text{ for } \gamma > 0,} \tag{2.3.3}$$

where $\kappa := \lambda + \frac{2}{3}\mu > 0$ is the *bulk modulus*, $\mathbf{I}$ is the fourth-order symmetric unit tensor, and $\mathbf{1}$ the second-order symmetric unit tensor.

2.3.2 J_2 Flow Theory with Isotropic/Kinematic Hardening

A choice of internal plastic variables which is typically of metal plasticity is $\boldsymbol{q} := \{\alpha, \bar{\beta}\}$. Here, α is the equivalent plastic strain that defines *isotropic hardening* of the von Mises yield surface, and $\bar{\beta}$ defines the center of the von Mises yield surface in stress deviator space. The resulting J_2-plasticity model has the following yield condition flow rule and hardening law:

$$\boxed{\begin{aligned} \eta &:= \text{dev}[\sigma] - \bar{\beta}, \ \text{tr}[\bar{\beta}] := 0, \\ f(\sigma, \boldsymbol{q}) &= \|\eta\| - \sqrt{\tfrac{2}{3}}\, K(\alpha), \\ \dot{\varepsilon}^p &= \gamma \frac{\eta}{\|\eta\|}, \\ \dot{\bar{\beta}} &= \gamma \tfrac{2}{3} H'(\alpha) \frac{\eta}{\|\eta\|}, \\ \dot{\alpha} &= \gamma \sqrt{\tfrac{2}{3}}. \end{aligned}} \tag{2.3.4}$$

The functions $K'(\alpha)$ and $H'(\alpha)$ are called the isotropic and kinematic hardening modulus, respectively. Since $\|\dot{\varepsilon}^p\| = \gamma$, relationship $(2.3.4)_5$ implies that

$$\alpha(t) := \int_0^t \sqrt{\tfrac{2}{3}}\, \|\dot{\varepsilon}^p(\tau)\| d\tau, \tag{2.3.5}$$

which agrees with the usual definition of equivalent plastic strain. Alternatively, one may use the notion of equivalent plastic work, e.g., see Naghdi [1960], Kachanov [1974] or Malvern [1969] to characterize hardening. In applications to metal plasticity, it is often assumed that the isotropic hardening is linear of the form $K(\alpha) = \sigma_Y + \bar{K}\alpha$, where $\bar{K}$ = constant, and σ_Y is the flow stress. Alternatively, the following form of combined kinematic/isotropic hardening laws is widely used in computational implementations; see, e.g., Hughes [1984],

$$\begin{aligned} H'(\alpha) &= (1-\theta)\bar{H}, \\ K(\alpha) &= [\sigma_Y + \theta\bar{H}\alpha], \quad \theta \in [0,1], \end{aligned} \tag{2.3.6}$$

where $\bar{H}$ = constant. The assumption of a constant kinematic hardening modulus leads to the so-called Prager-Ziegler rule discussed in Chapter **1**. More generally, nonlinear isotropic hardening models are often considered in which a saturation hardening term of the exponential type, as in Voce [1955], is appended to the linear term, i.e.,

$$\boxed{K(\alpha) := \sigma_Y + \theta\bar{H}\alpha + (\bar{K}_\infty - \bar{K}_0)[1 - \exp(-\delta\alpha)]}\,, \tag{2.3.7}$$

where $\bar{H} \geq 0$, $\bar{K}_\infty \geq \bar{K}_0 > 0$, and $\delta \geq 0$ are material constants.

Now, the plastic consistency parameter given by (2.2.18) in the general case takes the explicit form

$$\boxed{\gamma = \frac{\langle \boldsymbol{n} : \dot{\boldsymbol{\varepsilon}} \rangle}{1 + \frac{H' + K'}{3\mu}}}\,,$$

where

$$\boxed{\boldsymbol{n} := \frac{\boldsymbol{\eta}}{\|\boldsymbol{\eta}\|}}\,. \tag{2.3.8}$$

Note that since $\operatorname{tr}[\boldsymbol{n}] = 0$, it follows that $\boldsymbol{n} : \dot{\boldsymbol{\varepsilon}} \equiv \boldsymbol{n} : \operatorname{dev}[\dot{\boldsymbol{\varepsilon}}]$. Finally, for $\langle \gamma \rangle = \gamma \geq 0$, i.e., for plastic loading, the elastoplastic tangent moduli are obtained from (2.2.22) as

$$\boxed{\mathbf{C}^{ep} = \kappa \mathbf{1} \otimes \mathbf{1} + 2\mu \left[\mathbf{I} - \tfrac{1}{3}\mathbf{1} \otimes \mathbf{1} - \frac{\boldsymbol{n} \otimes \boldsymbol{n}}{1 + \frac{H' + K'}{3\mu}} \right], \quad \text{for } \gamma > 0.} \tag{2.3.9}$$

2.4 Plane-Stress J_2 Flow Theory

In this example we cast the basic equations resulting from the plane-stress constraint $\sigma_{3i} \equiv 0$ for $i = 1, 2, 3$ in the general format of Section **2**. This form of the equations plays a crucial role in the algorithmic treatment of the plane-stress problem.

2.4.1 Projection onto the Plane-Stress Subspace

Recall that we denote the vector space of *symmetric* second order tensors by $\mathbb{S}$. This symmetry condition implies that $\dim[\mathbb{S}] = 6$. The *plane-stress* subspace, denoted by $\mathbb{S}_P \subset \mathbb{S}$ in what follows, is obtained from $\mathbb{S}$ by appending *three* additional constraints as

$$\mathbb{S}_P := \{\boldsymbol{\sigma} \in \mathbb{S} \mid \sigma_{13} = \sigma_{23} = \sigma_{33} \equiv 0\}. \tag{2.4.1}$$

Similarly, the subspace of *deviatoric* symmetric second-order tensors, denoted by $\mathbb{S}_D \subset \mathbb{S}$ in what follows, is defined by appending *three* additional constraints on $\mathbb{S}$;

$$\mathbb{S}_D := \{\boldsymbol{s} \in \mathbb{S} \mid s_{13} \equiv s_{23} = 0,\ \operatorname{tr}[\boldsymbol{s}] := s_{kk} \equiv 0\}. \tag{2.4.2}$$

Hence, $\dim[\mathbb{S}_P] = \dim[\mathbb{S}_D] = 3$. Since both $\mathbb{S}_D \subset \mathbb{S}$ and $\mathbb{S}_P \subset \mathbb{S}$ are isomorphic to $\mathbb{R}^3$, it is convenient to introduce vector notation and express $\boldsymbol{\sigma} \in \mathbb{S}_P$ and $\boldsymbol{s} \in \mathbb{S}_D$ as

$$\boldsymbol{\sigma} := \left[\sigma_{11}\ \sigma_{22}\ \sigma_{12}\right]^T,$$

and (2.4.3)

$$\boldsymbol{s} := \left[s_{11}\ s_{22}\ s_{12}\right]^T.$$

The mapping $\bar{\mathbf{P}} : \mathbb{S}_P \rightarrow \mathbb{S}_D$ connecting the *constrained* stress tensor $\boldsymbol{\sigma} \in \mathbb{S}_P$ and its deviator $\boldsymbol{s} := \operatorname{dev}[\boldsymbol{\sigma}] \in \mathbb{S}_D$ plays a crucial role in what follows. In matrix notation

$$\boldsymbol{s} := \operatorname{dev}[\boldsymbol{\sigma}] = \bar{\mathbf{P}}\boldsymbol{\sigma},$$

where

$$\bar{\mathbf{P}} := \frac{1}{3}\begin{bmatrix} 2 & -1 & 0 \\ -1 & 2 & 0 \\ 0 & 0 & 3 \end{bmatrix}. \tag{2.4.4}$$

Observe that although the component s_{33} is *nonzero*, it *need not be explicitly included in* (2.4.3). It should be noted that $\bar{\mathbf{P}}$ *is not* a projection, i.e., $\bar{\mathbf{P}}\bar{\mathbf{P}} \neq \bar{\mathbf{P}}$.

2.4.2 Constrained Plane-Stress Equations

Now we now formulate the basic equations with the plane-stress condition automatically enforced. To this end, in place of the deviatoric *back stress* $\bar{\boldsymbol{\beta}} \in \mathbb{S}_D$ with components $\bar{\beta}_{ij}$ we introduce a vector $\tilde{\boldsymbol{\beta}} \in \mathbb{S}_P$ defined by the relationship

$$\left[\bar{\beta}_{11}\ \bar{\beta}_{22}\ \bar{\beta}_{12}\right]^T =: \bar{\mathbf{P}}\tilde{\boldsymbol{\beta}}, \quad \tilde{\boldsymbol{\beta}} := [\tilde{\beta}_{11}\ \tilde{\beta}_{22}\ \tilde{\beta}_{12}]^T. \tag{2.4.5}$$

In addition, following standard conventions, we collect the components ϵ_{ij} of the strain tensor $\boldsymbol{\varepsilon} \in \mathbb{S}$ in vector form as

$$\boldsymbol{\varepsilon} := \left[\epsilon_{11}\ \epsilon_{22}\ 2\epsilon_{12}\right]^T,$$

and

$$\boldsymbol{\varepsilon}^p := \left[\epsilon_{11}^p \; \epsilon_{22}^p \; 2\epsilon_{12}^p\right]^T, \tag{2.4.6}$$

where we have adopted the convention of multiplying the shear-strain component ϵ_{12} by a factor of two. This convention is motivated as follows. Recall that the stress power is defined by the pairing $\langle \bullet, \bullet \rangle_{\mathbb{S}} : \mathbb{S} \times \mathbb{S} \to \mathbb{R}$ according to the relationship $\langle \boldsymbol{\sigma}, \dot{\boldsymbol{\varepsilon}} \rangle_{\mathbb{S}} := \sigma^{ij}\dot{\epsilon}_{ij}$. In terms of the vector notation introduced above, then the stress power restricted to $\mathbb{S}_P \times \mathbb{S}$ takes the simple form

$$\langle \boldsymbol{\sigma}, \dot{\boldsymbol{\varepsilon}} \rangle_{\mathbb{S}} = \boldsymbol{\sigma}^T \dot{\boldsymbol{\varepsilon}}, \quad \text{for } \boldsymbol{\sigma} \in \mathbb{S}_P \text{ and } \dot{\boldsymbol{\varepsilon}} \in \mathbb{S}. \tag{2.4.7}$$

Again we observe that $\epsilon_{33} \neq 0$ does not appear explicitly since $\sigma_{33} \equiv 0$. To formulate the plane-stress version of J_2 flow theory directly in $\mathbb{S}_D$, we also need to consider strain deviators, which are written in vector notation as

$$\begin{aligned} \operatorname{dev}[\boldsymbol{\varepsilon}] &:= \left[e_{11} \; e_{22} \; 2\epsilon_{12}\right]^T, \\ e_{11} &:= \epsilon_{11} - \tfrac{1}{3}\operatorname{tr}[\boldsymbol{\varepsilon}], \end{aligned} \tag{2.4.8}$$

and

$$e_{22} := \epsilon_{22} - \tfrac{1}{3}\operatorname{tr}[\boldsymbol{\varepsilon}],$$

where, once more, the component e_{33} is omitted. Finally, to conveniently express strain deviators in terms of stress deviators and account for the factor of two in the shear strain component, we modify $\bar{\mathbf{P}}$ in (2.4.4) and set

$$\boxed{\mathbf{P} := \frac{1}{3}\begin{bmatrix} 2 & -1 & 0 \\ -1 & 2 & 0 \\ 0 & 0 & 6 \end{bmatrix}.} \tag{2.4.9}$$

With these notations in hand, the basic equations (2.3.4) of three-dimensional J_2 flow theory are recast in the general format of Section **2** as follows:

$$\boxed{\begin{aligned} \boldsymbol{\varepsilon} &= \boldsymbol{\varepsilon}^e + \boldsymbol{\varepsilon}^p, \\ \boldsymbol{\eta} &:= \boldsymbol{\sigma} - \tilde{\boldsymbol{\beta}}, \\ \boldsymbol{\sigma} &= \mathbf{C}\boldsymbol{\varepsilon}^e, \\ \dot{\boldsymbol{\varepsilon}}^p &= \gamma \mathbf{P}\boldsymbol{\eta}, \\ \dot{\tilde{\boldsymbol{\beta}}} &= \gamma \tfrac{2}{3} H' \boldsymbol{\eta}, \\ f &:= \sqrt{\boldsymbol{\eta}^T \mathbf{P} \boldsymbol{\eta}} - \sqrt{\tfrac{2}{3}}\, K(\alpha) \leq 0. \end{aligned}} \tag{2.4.10}$$

Here, $\mathbf{C}$ is the elastic constitutive matrix for plane stress, and for convenience we have set $\boldsymbol{\eta} := \boldsymbol{\sigma} - \tilde{\boldsymbol{\beta}}$. Further, by using the flow rule $(2.4.10)_4$ and relationship (2.4.4) for the stress deviator, the evolution of the equivalent plastic strain defined by $(2.3.4)_5$ is rephrased as

$$\boxed{\dot{\alpha} = \gamma \sqrt{\tfrac{2}{3}\, \boldsymbol{\eta}^T \mathbf{P} \boldsymbol{\eta}}.} \tag{2.4.11}$$

Now equations (2.4.9)-(2.4.11) are in a form which is ideally suited for applying the general return-mapping algorithms discussed in the next chapter.

2.4.2.1 Simultaneous diagonalization.

For the case of *isotropic* elasticity the constitutive matrix $\mathbf{C}$ and the projection matrix $\mathbf{P}$ have the same characteristic subspaces and, therefore, are easily diagonalized. In fact, their spectral decompositions take the following form

$$\mathbf{P} = Q \Lambda_{\mathbf{P}} Q^T,$$

and (2.4.12a)

$$\mathbf{C} = Q \Lambda_{\mathbf{C}} Q^T,$$

where the orthogonal matrix $Q^{-1} \equiv Q^T$ and the constitutive matrix $\mathbf{C}$ are given by the expressions

$$Q = \frac{1}{\sqrt{2}} \begin{bmatrix} 1 & 1 & 0 \\ -1 & 1 & 0 \\ 0 & 0 & \sqrt{2} \end{bmatrix},$$

and

$$\mathbf{C} := \frac{E}{1-\nu^2} \begin{bmatrix} 1 & \nu & 0 \\ \nu & 1 & 0 \\ 0 & 0 & \frac{1-\nu}{2} \end{bmatrix}, \tag{2.4.12b}$$

where ν is the Poisson ratio. The diagonal matrices $\Lambda_{\mathbf{P}}$ and $\Lambda_{\mathbf{C}}$ are expressed as

$$\Lambda_{\mathbf{P}} = \begin{bmatrix} \frac{1}{3} & 0 & 0 \\ 0 & 1 & 0 \\ 0 & 0 & 2 \end{bmatrix},$$

and

$$\Lambda_{\mathbf{C}} = \begin{bmatrix} \frac{E}{1-\nu} & 0 & 0 \\ 0 & 2\mu & 0 \\ 0 & 0 & \mu \end{bmatrix}. \tag{2.4.12c}$$

Since $\mathbf{P}$ and $\mathbf{C}$ have the same eigenvectors, it follows that $\mathbf{PC} \equiv \mathbf{CP}$, that is, $\mathbf{P}$ and $\mathbf{C}$ commute.

Remarks 2.4.1.

1. For isotropic elasticity, the properties recorded above play a crucial role in the implementing the algorithm discussed in Chapter **3**.
2. It should be noted that the strain components ϵ_{33}, ϵ_{33}^e, and ϵ_{33}^p do not appear explicitly in the formulation. These are *dependent* variables obtained from the basic variables $\{\varepsilon, \varepsilon^p, \sigma\}$, the plane-stress condition, and the condition of isochoric plastic flow. For the case of isotropic elasticity,

$$\epsilon_{33}^e = -\frac{\nu}{1-\nu}(\epsilon_{11}^e + \epsilon_{22}^e)$$

and

$$\epsilon_{33}^p = -\epsilon_{11}^p - \epsilon_{22}^p. \tag{2.4.13}$$

Then the total strain ϵ_{33} follows simply as $\epsilon_{33} \equiv \epsilon_{33}^e + \epsilon_{33}^p$.

We conclude this example by recording the expression for the elastoplastic tangent moduli. From (2.2.18), the plastic consistency parameter is given by

$$\boxed{\gamma = \frac{\langle \boldsymbol{\eta}^T \mathbf{P}\mathbf{C}\dot{\boldsymbol{\varepsilon}} \rangle}{\boldsymbol{\eta}^T \mathbf{P}\mathbf{C}\mathbf{P}\boldsymbol{\eta}(1+\beta)},} \tag{2.4.14}$$

where we have set

$$\boxed{\begin{aligned} \beta &:= \tfrac{2}{3}\frac{(K' + H')\bar{f}^2}{\boldsymbol{\eta}^T \mathbf{P}\mathbf{C}\mathbf{P}\boldsymbol{\eta}}, \\ &\text{and} \\ \bar{f} &:= \sqrt{\boldsymbol{\eta}^T \mathbf{P}\boldsymbol{\eta}}. \end{aligned}} \tag{2.4.15}$$

Finally, expression (2.2.22) for the elastoplastic tangent moduli in the present context takes the following form

$$\boxed{\mathbf{C}^{ep} = \mathbf{C} - \frac{\boldsymbol{n}\otimes\boldsymbol{n}}{1+\beta}, \quad \text{for } \gamma > 0,}$$

where

$$\boxed{\boldsymbol{n} := \frac{\mathbf{C}\mathbf{P}\boldsymbol{\eta}}{\sqrt{\boldsymbol{\eta}^T \mathbf{P}\mathbf{C}\mathbf{P}\boldsymbol{\eta}}}.} \tag{2.4.16}$$

Here $\mathbf{C}$ is the matrix of elastic moduli in plane stress which is given by $(2.4.12b)_2$ for the case of isotropic elasticity.

2.5 General Quadratic Model of Classical Plasticity

The examples considered in the preceding sections are generalized to the case of a yield condition defined by a general quadratic form. This form of classical plasticity includes a special case that most of the rate-independent plasticity models use in practice. In particular, by suitably restricting the quadratic form that defines the yield condition, one can recover the Mises–Huber yield criterion both in plane strain and plane stress, as discussed in the two examples above, and also the anisotropic criterion of Hill [1950], or the general anisotropic yield condition of Tsai and Wu [1971].

2.5.1 The Yield Criterion.

Let $\mathbf{A}$ be a symmetric (positive-definite) fourth-order tensor and, as before, let σ_Y be the flow stress. We consider a yield criterion of the form

$$\left.\begin{aligned} &\boxed{\phi(\boldsymbol{\sigma}) := \sqrt{\boldsymbol{\sigma} : \mathbf{A} : \boldsymbol{\sigma}}} \\ f(\boldsymbol{\sigma}, \alpha) := \phi(\boldsymbol{\sigma}) &- \left[K\alpha + \sigma_Y\right] \leq 0\,. \end{aligned}\right\} \tag{2.5.1}$$

Here, α stands for a hardening variable which models isotropic hardening with evolution defined below, and K is the (isotropic) plastic hardening modulus. If $K < 0$, we speak of a *strain-softening* response. Observe that the function $\phi(\boldsymbol{\sigma})$ satisfies the following property:

i. *Degree-one homogeneity.* Recall that a function $\phi(\boldsymbol{\sigma})$ defined on $\mathbb{S}$ is said to be homogeneous of degree one if the following condition holds:

$$\frac{\partial \phi(\boldsymbol{\sigma})}{\partial \boldsymbol{\sigma}} : \boldsymbol{\sigma} = \phi(\boldsymbol{\sigma}), \quad \text{for all } \boldsymbol{\sigma} \in \mathbb{S}. \tag{2.5.2}$$

For the function $\phi(\cdot)$ defined by (2.5.1),

$$\frac{\partial \phi(\boldsymbol{\sigma})}{\partial \boldsymbol{\sigma}} = \frac{\mathbf{A} : \boldsymbol{\sigma}}{\sqrt{\boldsymbol{\sigma} : \mathbf{A} : \boldsymbol{\sigma}}} \quad \Longrightarrow \quad \frac{\partial \phi(\boldsymbol{\sigma})}{\partial \boldsymbol{\sigma}} : \boldsymbol{\sigma} = \frac{\boldsymbol{\sigma} : \mathbf{A} : \boldsymbol{\sigma}}{\sqrt{\boldsymbol{\sigma} : \mathbf{A} : \boldsymbol{\sigma}}} = \phi(\boldsymbol{\sigma}), \tag{2.5.3}$$

so that condition (2.5.2) holds.

2.5.2 Evolution Equations. Elastoplastic Moduli.

The elastic stress-strain relationships and the expression connecting the stress-like and strain-like hardening variables are given by

$$\left.\begin{aligned} \boldsymbol{\sigma} &= \mathbf{C} : (\boldsymbol{\varepsilon} - \boldsymbol{\varepsilon}^p)\,, \\ q &= -K\alpha\,, \end{aligned}\right\} \tag{2.5.4}$$

where $\mathbf{C}$ denotes the tensor of elastic constants. In terms of the variable q defined by $(2.5.4)_2$, the yield condition $(2.5.1)_2$ then reads

$$f(\boldsymbol{\sigma}, q) := \phi(\boldsymbol{\sigma}) + q - \sigma_Y \leq 0\,. \tag{2.5.5}$$

We define the evolution of $\dot{\boldsymbol{\varepsilon}}^p$ by the associative flow rule

$$\boxed{\dot{\boldsymbol{\varepsilon}}^p = \gamma \frac{\partial f}{\partial \boldsymbol{\sigma}} \equiv \gamma \frac{\mathbf{A} : \boldsymbol{\sigma}}{\sqrt{\boldsymbol{\sigma} : \mathbf{A} : \boldsymbol{\sigma}}}\,,} \tag{2.5.6}$$

where $\gamma \geq 0$. The evolution of the hardening variable α is assumed to be given by the rate equation

$$\dot{\alpha} := \sqrt{\dot{\boldsymbol{\varepsilon}}^p : \mathbf{A}^{-1} : \dot{\boldsymbol{\varepsilon}}^p} = \gamma\,. \tag{2.5.7a}$$

Since $\partial f(\sigma, q)/\partial q = 1$, it follows from (2.5.7*a*) that the hardening law is also associative in the sense that

$$\dot{\alpha} = \gamma \frac{\partial f}{\partial q} . \tag{2.5.7b}$$

Finally, as usual, the loading/unloading conditions are formulated in *Kuhn–Tucker form* as

$$\begin{aligned} \gamma &\geq 0, \\ f(\sigma, q) &\leq 0, \end{aligned}$$

and (2.5.8a)

$$\gamma f(\sigma, q) = 0,$$

where the actual value of γ is determined from the *consistency requirement*

$$\gamma \dot{f}(\sigma, q) = 0 \quad \text{if} \quad f(\sigma, q) = 0 . \tag{2.5.8b}$$

Finally, the elastoplastic moduli are obtained by specializing the general expression given in (2.2.22). Using the preceding relationships, an easy manipulation yields the result

$$\mathbf{C}^{\text{ep}} = \begin{cases} \mathbf{C} & \text{if } \gamma = 0, \\ \mathbf{C} - \dfrac{\mathbf{C} : [\mathbf{A} : \sigma] \otimes \mathbf{C} : [\mathbf{A} : \sigma]}{[\mathbf{A} : \sigma] : \mathbf{C} : [\mathbf{A} : \sigma] + K\,[\sigma : \mathbf{A}\sigma]} & \text{if } \gamma > 0. \end{cases} \tag{2.5.9}$$

Observe that this expression gives *symmetric* elastoplastic moduli, agreeing with the associative character of the model under consideration.

Remarks 2.5.1.

1. It should be noted that one can use either α or q, as defined by $(2.5.4)_2$, to formulate the (isotropic) hardening law. The reasons for introducing q become apparent in the development that follows.
2. Extension of the model to account for other types of hardening is straightforward; in particular, we may formulate a kinematic hardening rule by introducing an additional internal variable $\bar{q}$ with an evolution equation of the form

$$\left.\begin{aligned} \bar{q} &:= -H\bar{\alpha}, \\ \dot{\bar{\alpha}} &= \dot{\varepsilon}^p. \end{aligned}\right\} \tag{2.5.10a}$$

   Then, in view of (2.5.7*b*) and (2.5.10*a*), the hardening law is formulated as follows. Set

$$\boldsymbol{\alpha} := \left\{ \begin{matrix} \alpha \\ \bar{\alpha} \end{matrix} \right\},$$

$$\boldsymbol{q} := \left\{ \begin{matrix} q \\ \bar{q} \end{matrix} \right\}, \tag{2.5.10b}$$

and

$$\mathbf{D} := \begin{bmatrix} K & \mathbf{0} \\ \mathbf{0}^T & H\mathbf{1} \end{bmatrix}.$$

Then the hardening law becomes

$$\dot{\boldsymbol{\alpha}} = \gamma \frac{\partial f(\boldsymbol{\sigma}, \boldsymbol{q})}{\partial \boldsymbol{q}}, \quad \boldsymbol{q} := -\mathbf{D}\boldsymbol{\alpha}, \tag{2.5.10c}$$

where $f(\boldsymbol{\sigma}, \boldsymbol{q})$ is the yield condition now defined as

$$f(\boldsymbol{\sigma}, \boldsymbol{q}) := \phi(\boldsymbol{\sigma} - \bar{\boldsymbol{q}}) + q - \sigma_Y \leq 0\,. \tag{2.5.10d}$$

All the results discussed below carry over to this more general constitutive model without essential modification.

2.6 The Principle of Maximum Plastic Dissipation

The principle of maximum plastic dissipation, often credited to von Mises (see Hill [1950], page 60), and subsequently considered by several authors, Mandel [1964] and Lubliner [1984,1986], plays a crucial role in the variational formulation of plasticity discussed in Chapter **5** which is the cornerstone of the finite-element approximation discussed subsequently. This principle is central in the modern mathematical formulation of plasticity; see e.g., Duvaut and Lions [1972], Johnson [1976,1978], Moreau [1976], and the recent account of Temam [1985]. Our presentation employing Lagrange multipliers and optimality conditions provides new insights into the fundamental role of this principle. For instance, loading/unloading conditions follow as part of the optimality conditions. To motivate our discussion, first we consider the case of perfect plasticity.

2.6.1 Classical Formulation. Perfect Plasticity

In its local form, the principle of maximum plastic dissipation states that, for *given* plastic strains $\boldsymbol{\varepsilon}^p$ among all possible stresses $\boldsymbol{\tau}$ satisfying the yield criterion, the plastic dissipation, which is now given for perfect plasticity (i.e., $\boldsymbol{q} = \mathbf{0}$) by

$$\boxed{\mathcal{D}^p[\boldsymbol{\tau}; \dot{\boldsymbol{\varepsilon}}^p] := \boldsymbol{\tau} : \dot{\boldsymbol{\varepsilon}}^p,} \tag{2.6.1}$$

attains its *maximum* for the *actual stress tensor* $\boldsymbol{\sigma}$, that is, let $\mathbb{E}_\sigma$ be the closure of the elastic range in stress space which we recall is defined as

$$\mathbb{E}_\sigma := \{\boldsymbol{\tau} \in \mathbb{S} \mid f(\boldsymbol{\tau}) \leq 0\}. \tag{2.6.2}$$

Then, the actual stress $\boldsymbol{\sigma} \in \mathbb{E}_\sigma$ is the argument of the maximum principle

$$\boxed{\mathcal{D}^p[\boldsymbol{\sigma}; \dot{\boldsymbol{\varepsilon}}^p] = \underset{\boldsymbol{\tau} \in \mathbb{E}_\sigma}{\mathsf{MAX}} \left\{\mathcal{D}^p[\boldsymbol{\tau}; \dot{\boldsymbol{\varepsilon}}^p]\right\}.} \tag{2.6.3}$$

The fundamental significance of this principle lies in the following classical result which completely defines the flow rule for a given yield condition.

Proposition 2.6.1. *Maximum plastic dissipation implies*
a. *associative flow rule in stress space; a condition which is often called normality in stress space,*
b. *loading/unloading conditions in Kuhn–Tucker complementarity form, and*
c. *convexity of the elastic range* $\mathbb{E}_{\sigma}$.

PROOF. To prove **(i)** and **(ii)**, we use the classical method of Lagrange multipliers, as follows. First, we transform the maximization principle into a *minimization* principle merely by changing the sign and considering as objective function $-\mathcal{D}^p[\boldsymbol{\tau}; \dot{\boldsymbol{\varepsilon}}^p]$. Next, we transform the constraint minimization problem into an *unconstrained* problem by introducing the cone of Lagrange multipliers

$$\mathbb{K}^p := \{\dot{\delta} \in \mathbb{L}^2(\mathcal{B}) \mid \dot{\delta} \geq 0\}, \tag{2.6.4}$$

and considering the Lagrangian functional $\mathcal{L}^p : \mathbb{S} \times \mathbb{K}^p \times \mathbb{S} \rightarrow \mathbb{R}$ defined as

$$\mathcal{L}^p(\boldsymbol{\tau}, \dot{\delta}; \dot{\boldsymbol{\varepsilon}}^p) := -\boldsymbol{\tau} : \dot{\boldsymbol{\varepsilon}}^p + \dot{\delta} f(\boldsymbol{\tau}), \tag{2.6.5}$$

where $\dot{\boldsymbol{\varepsilon}}^p \in \mathbb{S}$ is regarded as a *fixed but otherwise arbitrary function*. Then the solution to problem (2.6.3) is then given by the point $(\boldsymbol{\sigma}, \gamma) \in \mathbb{S} \times \mathbb{K}^p$ satisfying the classical Kuhn–Tucker optimality conditions , see, e.g., Luenberger [1984], page 314, Bertsekas[1982], or Strang [1986], page 724,

$$\boxed{\begin{aligned} &\frac{\partial \mathcal{L}^p(\boldsymbol{\sigma}, \gamma; \dot{\boldsymbol{\varepsilon}})}{\partial \boldsymbol{\tau}} \equiv -\dot{\boldsymbol{\varepsilon}}^p + \gamma \nabla f(\boldsymbol{\sigma}) = 0, \\ &\gamma \geq 0, \ f(\boldsymbol{\sigma}) \leq 0, \ \text{and } \gamma f(\boldsymbol{\sigma}) \equiv 0. \end{aligned}} \tag{2.6.6}$$

These conditions are precisely the statement of normality of the flow rule and loading/unloading conditions.

To see that the convexity condition **(iii)** on the elastic domain $\mathbb{E}_{\sigma}$ also follows from the principle of maximum plastic dissipation, it suffices to show that the function $f(\boldsymbol{\sigma})$ is *convex*, (in the sense defined below). To this end, we observe that, in view of (2.6.3), the extremum point $\boldsymbol{\sigma} \in \partial\mathbb{E}_{\sigma}$ satisfies the condition

$$\mathcal{D}^p[\boldsymbol{\sigma}; \dot{\boldsymbol{\varepsilon}}^p] \geq \mathcal{D}^p[\boldsymbol{\tau}; \dot{\boldsymbol{\varepsilon}}^p] \iff \boldsymbol{\sigma} : \dot{\boldsymbol{\varepsilon}}^p \geq \boldsymbol{\tau} : \dot{\boldsymbol{\varepsilon}}^p, \ \text{for all } \boldsymbol{\tau} \in \mathbb{E}_{\sigma}. \tag{2.6.7}$$

Accordingly, we have the inequality

$$[\boldsymbol{\tau} - \boldsymbol{\sigma}] : \dot{\boldsymbol{\varepsilon}}^p \leq 0, \ \text{for all } \boldsymbol{\tau} \in \mathbb{E}_{\sigma}, \tag{2.6.8}$$

which, in view of the flow rule $(2.6.6)_1$, reduces to

$$[\boldsymbol{\tau} - \boldsymbol{\sigma}] : \dot{\boldsymbol{\varepsilon}}^p = \gamma [\boldsymbol{\tau} - \boldsymbol{\sigma}] : \nabla f(\boldsymbol{\sigma}) \leq 0, \ \text{for all } \boldsymbol{\tau} \in \mathbb{E}_{\sigma}. \tag{2.6.9}$$

From the Kuhn–Tucker optimality condition, it follows that

$$\gamma [\boldsymbol{\tau} - \boldsymbol{\sigma}] : \nabla f(\boldsymbol{\sigma}) \leq \gamma f(\boldsymbol{\tau}), \ \text{for } \boldsymbol{\sigma} \in \partial\mathbb{E}_{\sigma} \text{ and any } \boldsymbol{\tau} \in \mathbb{E}_{\sigma}. \tag{2.6.10}$$

If $\gamma = 0$, the inequality is satisfied trivially. On the other hand, if $\gamma > 0$ and since $f(\boldsymbol{\sigma}) = 0$,

$$[\boldsymbol{\tau} - \boldsymbol{\sigma}]\colon \nabla f(\boldsymbol{\sigma}) \leq f(\boldsymbol{\tau}) - f(\boldsymbol{\sigma}) = f(\boldsymbol{\tau}) \leq 0, \text{ for any } \boldsymbol{\tau} \in \mathbb{E}_{\sigma}, \tag{2.6.11}$$

which coincides with (2.6.10); hence, convexity follows. □

The precise meaning of convexity is contained in the following:

Definition 2.6.1. A function $f\colon \mathbb{S} \to \mathbb{R}$ is said to be **convex** if

$$\boxed{f(\beta\boldsymbol{\sigma} + (1-\beta)\boldsymbol{\tau}) \leq \beta f(\boldsymbol{\sigma}) + (1-\beta) f(\boldsymbol{\tau}), \ \beta \in [0, 1].} \tag{2.6.12}$$

The geometric meaning of this definition is illustrated in Figure **2.6** for the one-dimensional case. If $f(\boldsymbol{\sigma})$ is smooth, definition (2.6.12) is equivalent to the following useful characterization of convexity employed in Proposition **6.1** above.

Lemma 2.6.1. *Assume that the function $f\colon \mathbb{S} \to \mathbb{R}$ is smooth. Then, $f(\boldsymbol{\sigma})$ is* **convex** *if and only if the following inequality holds*

$$\boxed{f(\boldsymbol{\tau}) - f(\boldsymbol{\sigma}) \geq (\boldsymbol{\tau} - \boldsymbol{\sigma})\colon \nabla f(\boldsymbol{\sigma}), \text{ for any } \boldsymbol{\tau}, \boldsymbol{\sigma} \in \mathbb{S}.} \tag{2.6.13}$$

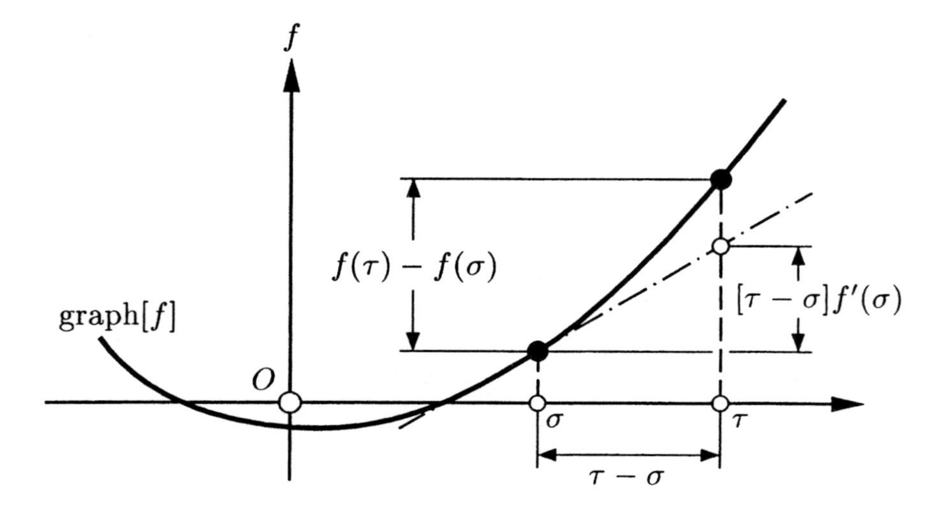

FIGURE 2-6. One dimensional example of smooth convex function $f\colon \mathbb{R} \to \mathbb{R}$, and a graphical illustration of the basic property $f(\tau) - f(\sigma) \geq (\tau - \sigma) f'(\sigma)$.

PROOF. (i) First assume that $f: \mathbb{S} \to \mathbb{R}$ is convex. Then, from definition (2.6.12),

$$f(\boldsymbol{\sigma}+\beta[\boldsymbol{\tau}-\boldsymbol{\sigma}]) \leq f(\boldsymbol{\sigma})+\beta[f(\boldsymbol{\tau})-f(\boldsymbol{\sigma})],\ \beta \in [0,1],\ \boldsymbol{\tau}, \boldsymbol{\sigma} \in \mathbb{S}, \quad (2.6.14))$$

which implies the relationship

$$\frac{f(\boldsymbol{\sigma}+\beta[\boldsymbol{\tau}-\boldsymbol{\sigma}])-f(\boldsymbol{\sigma})}{\beta} \leq f(\boldsymbol{\tau})-f(\boldsymbol{\sigma}),\ \beta \in [0,1]. \quad (2.6.15)$$

Taking the limit of (2.6.15) $\beta \to 0$ and noting that, by the chain rule,

$$\begin{aligned} \lim_{\beta\to 0} \frac{f(\boldsymbol{\sigma}+\beta[\boldsymbol{\tau}-\boldsymbol{\sigma}])-f(\boldsymbol{\sigma})}{\beta} &= \left.\frac{d}{d\beta}\right|_{\beta=0} f(\boldsymbol{\sigma}+\beta[\boldsymbol{\tau}-\boldsymbol{\sigma})] \\ &= \nabla f(\boldsymbol{\sigma}) : [\boldsymbol{\tau}-\boldsymbol{\sigma}], \end{aligned} \quad (2.6.16)$$

inequality (2.6.13) follows.

(ii) Conversely, assume that inequality (2.6.13) holds. For convenience, we set

$$\bar{\boldsymbol{\sigma}} := \beta\boldsymbol{\sigma}+(1-\beta)\boldsymbol{\tau},\ \beta \in [0,1]. \quad (2.6.17)$$

Applying (2.6.13) successively, we obtain

$$\left.\begin{aligned} f(\boldsymbol{\sigma})-f(\bar{\boldsymbol{\sigma}}) &\geq -(1-\beta)[\boldsymbol{\tau}-\boldsymbol{\sigma}] : \nabla f(\bar{\boldsymbol{\sigma}}), \\ f(\boldsymbol{\tau})-f(\bar{\boldsymbol{\sigma}}) &\geq \beta[\boldsymbol{\tau}-\boldsymbol{\sigma}] : \nabla f(\bar{\boldsymbol{\sigma}}). \end{aligned}\right\} \quad (2.6.18)$$

Multiplying the above equations by β and $(1-\beta)$, respectively, and adding the result we obtain

$$(1-\beta)f(\boldsymbol{\tau})+\beta f(\boldsymbol{\sigma})-f(\bar{\boldsymbol{\sigma}}) \geq 0, \quad (2.6.19)$$

which implies convexity of the function $f(\boldsymbol{\sigma})$. □

2.6.2 *General Associative Hardening Plasticity in Stress Space*

In a straightforward fashion we extend below the preceding arguments to the general plasticity model considered in Section **5**. Our main result is characterization of the general structure taken by *associative hardening laws* as a result of the principle of maximum plastic dissipation.

As in Section **6.1**, we let $\{\boldsymbol{\sigma}, \boldsymbol{q}\} \in \mathbb{E}_\sigma$ be the actual solution of the constitutive equations of classical plasticity summarized in BOX **2.1**, where $\mathbb{E}_\sigma$ is the closure of the elastic range in stress space defined by (2.2.5). Extending the definition (2.6.1),

$$\boxed{\mathcal{D}^p[\boldsymbol{\tau}, \boldsymbol{q}; \dot{\boldsymbol{\varepsilon}}^p, \dot{\boldsymbol{\alpha}}] := \boldsymbol{\tau} : \dot{\boldsymbol{\varepsilon}}^p + \boldsymbol{q} \cdot \dot{\boldsymbol{\alpha}},} \quad (2.6.20)$$

where the internal hardening variables $\boldsymbol{q}$ and $\boldsymbol{\alpha}$ in stress and strain space, respectively, are related through the equation $\boldsymbol{q} = -\nabla\mathcal{H}(\boldsymbol{\alpha})$, as in (2.2.44). In the present context, for fixed $\{\dot{\boldsymbol{\varepsilon}}^p, \dot{\boldsymbol{\alpha}}\}$, the principle of maximum plastic dissipation characterizes the *actual state* $\{\boldsymbol{\sigma}, \boldsymbol{q}\} \in \mathbb{E}_\sigma$ as the state among all *admissible states*

$\{\boldsymbol{\tau}, \boldsymbol{p}\} \in \mathbb{E}_\sigma$ for which plastic dissipation attains a maximum, i.e.,

$$\boxed{\mathcal{D}^p\big[\boldsymbol{\sigma}, \boldsymbol{q}; \dot{\boldsymbol{\varepsilon}}^p, \dot{\boldsymbol{\alpha}}\big] = \underset{(\boldsymbol{\tau}, \boldsymbol{p}) \in \mathbb{E}_\sigma}{\mathsf{MAX}} \big\{\mathcal{D}^p\big[\boldsymbol{\tau}, \boldsymbol{p}; \dot{\boldsymbol{\varepsilon}}^p, \dot{\boldsymbol{\alpha}}\big]\big\}.} \tag{2.6.21}$$

The extension of the result in Proposition **2.6.1** is now given by the following.

Proposition 2.6.2. *The principle of maximum plastic dissipation implies the following:*

a. *associativity of flow rule in stress space according to the relationship*

$$\boxed{\begin{aligned} \dot{\boldsymbol{\varepsilon}}^p &= \gamma \frac{\partial f(\boldsymbol{\sigma}, \boldsymbol{q})}{\partial \boldsymbol{\sigma}}, \\ \textit{where} \quad \boldsymbol{\sigma} &= \nabla W(\boldsymbol{\varepsilon} - \boldsymbol{\varepsilon}^p); \end{aligned}} \tag{2.6.22a}$$

b. *associativity of the hardening law in stress space in the sense that*

$$\boxed{\begin{aligned} \dot{\boldsymbol{\alpha}} &= \gamma \frac{\partial f(\boldsymbol{\sigma}, \boldsymbol{q})}{\partial \boldsymbol{q}}, \\ \textit{where} \quad \boldsymbol{q} &= -\nabla \mathcal{H}(\boldsymbol{\alpha}); \end{aligned}} \tag{2.6.22b}$$

c. *loading/unloading conditions in Kuhn–Tucker complementarity form, as*

$$\boxed{\begin{aligned} \gamma &\geq 0, \\ f(\boldsymbol{\sigma}, \boldsymbol{q}) &\leq 0, \\ \textit{and} \quad \gamma f(\boldsymbol{\sigma}, \boldsymbol{q}) &\equiv 0; \end{aligned}} \tag{2.6.22c}$$

d. *convexity of the elastic range, $\mathbb{E}_\sigma$.*

PROOF. The proof of **(i)**–**(iii)** above follows exactly the same lines as in Proposition **2.6.1**. Again one uses the method of Lagrange multipliers to remove the constraint that the admissible states be in $\mathbb{E}_\sigma$ by introducing the Lagrangian functional

$$\mathcal{L}^p(\boldsymbol{\tau}, \boldsymbol{p}, \dot{\delta}; \dot{\boldsymbol{\varepsilon}}^p, \dot{\boldsymbol{\alpha}}) := -\boldsymbol{\tau} : \dot{\boldsymbol{\varepsilon}}^p - \boldsymbol{p} \cdot \dot{\boldsymbol{\alpha}} + \dot{\delta} f(\boldsymbol{\tau}, \boldsymbol{p}), \tag{2.6.23}$$

for all admissible $\{\boldsymbol{\tau}, \boldsymbol{p}\} \in \mathbb{E}_\sigma$, with $\dot{\delta} \geq 0$ (i.e., $\dot{\delta} \in \mathbb{K}^p$, as defined by (2.6.4)). Then the classical Kuhn–Tucker optimality conditions for this Lagrangian yield (2.6.22*a*,*b*,*c*).

The proof of the convexity condition **(iv)** on $\mathbb{E}_\sigma$ also follows the same lines as in Proposition **2.1**. First, in view of (2.6.20), one observes that the optimal point $\{\boldsymbol{\sigma}, \boldsymbol{q}\} \in \partial\mathbb{E}_\sigma$ satisfies the condition

$$\mathcal{D}^p[\boldsymbol{\sigma}, \boldsymbol{q}; \dot{\boldsymbol{\varepsilon}}, \dot{\boldsymbol{\alpha}}] \geq \mathcal{D}^p[\boldsymbol{\tau}, \boldsymbol{p}; \dot{\boldsymbol{\varepsilon}}, \dot{\boldsymbol{\alpha}}] \iff [\boldsymbol{\tau} - \boldsymbol{\sigma}] : \dot{\boldsymbol{\varepsilon}}^p + [\boldsymbol{p} - \boldsymbol{q}] \cdot \dot{\boldsymbol{\alpha}} \leq 0, \tag{2.6.24}$$

for all $\{\boldsymbol{\tau}, \boldsymbol{p}\} \in \mathbb{E}_\sigma$. Substitution of the flow rule (2.6.22*a*) and the hardening law (2.6.22*b*) in (2.6.24) and using the fact that $\gamma \geq 0$ results in the inequality

$$[\boldsymbol{\tau} - \boldsymbol{\sigma}] : \partial_\sigma f(\boldsymbol{\sigma}, \boldsymbol{q}) + [\boldsymbol{p} - \boldsymbol{q}] \cdot \partial_q f(\boldsymbol{\sigma}, \boldsymbol{q}) \leq 0, \tag{2.6.25}$$

for $\{\sigma, q\} \in \partial\mathbb{E}_\sigma$ and all $\{\tau, p\} \in \mathbb{E}_\sigma$, which is the convexity condition on $\mathbb{E}_\sigma$. □

2.6.3 *Interpretation of Associative Plasticity as a Variational Inequality.*

Convex analysis is the natural arena for a rigorous formulation of classical plasticity. Ideas of convex analysis also play an increasingly important role in the numerical analysis of the algorithmic elastoplastic problem; see e.g., Moreau[1976,1977], Johnson [1976,1977,1978], Matthies[1978] and Glowinski and Le Tallec [1989]. Here we restrict ourselves to a derivation of a fundamental inequality in a simplified context and refer to the literature; e.g., Moreau [1976], or Temam [1985], for an in-depth treatment of the subject. As we shall see, this inequality is a direct consequence of the principle of maximum plastic dissipation.

Let $\chi : \mathbb{S} \to \mathbb{R}$ be the complementary stored-energy function, and let $\Theta : \mathbb{R}^m \to \mathbb{R}$ be the complementary hardening potential associated with W and $\mathcal{H}$, respectively. We define the elastic and hardening *tangent compliances* by the relationships

$$\boxed{\mathbf{C}^{-1} := \frac{\partial^2 \chi(\sigma)}{\partial \sigma^2}} \,,$$

and

$$\boxed{\mathbf{D}^{-1} := \frac{\partial^2 \Theta(q)}{\partial q^2}} \,. \tag{2.6.26}$$

Note that $\mathbf{C}^{-1}$ coincides with the inverse of the elastic tangent moduli; hence the notation employed above. We assume that $\mathbf{C}$ and $\mathbf{D}$ are *positive-definite* on $\mathbb{S}$ and $\mathbb{R}^m$, respectively. By using (2.2.21) and by time differentiating the inverse relationships (2.2.42) while using the (associative) flow rule and hardening law (2.6.22) along with (2.6.26), we arrive at

$$\left.\begin{aligned} \dot{\varepsilon}^p &\equiv \dot{\varepsilon} - \mathbf{C}^{-1}\dot{\sigma} = \gamma \frac{\partial f(\sigma, q)}{\partial \sigma} \,, \\ \dot{\alpha} &\equiv -\mathbf{D}^{-1}\dot{q} = \gamma \frac{\partial f(\sigma, q)}{\partial q} \,. \end{aligned}\right\} \tag{2.6.27}$$

Now let $\{\sigma, q\} \in \mathbb{E}_\sigma$ be the *actual state*, and let $\{\tau, p\}$ denote an arbitrary point in $\mathbb{E}_\sigma$. Contracting $(2.6.27)_1$ and $(2.6.27)_2$ with $(\sigma - \tau)$ and $(q - p)$, respectively, and adding the result, we obtain

$$\begin{aligned} &(\sigma - \tau) : \left[\dot{\varepsilon} - \mathbf{C}^{-1}\dot{\sigma}\right] - (q - p) \cdot \mathbf{D}^{-1}\dot{q} = \\ &\qquad \gamma\Big[(\sigma - \tau) : \frac{\partial f(\sigma, q)}{\partial \sigma} + (q - p) \cdot \frac{\partial f(\sigma, q)}{\partial q}\Big]. \end{aligned} \tag{2.6.28}$$

However, the right-hand side of (2.6.28) is precisely the difference between the dissipation $\mathcal{D}^p[\sigma, q; \dot{\varepsilon}^p, \dot{\alpha}]$ associated with *actual* state $\{\sigma, q\}$ and the dissipation

$\mathcal{D}^p[\boldsymbol{\tau}, \boldsymbol{p}; \dot{\boldsymbol{\varepsilon}}^p, \dot{\boldsymbol{\alpha}}]$ associated with an arbitrary admissible state $\{\boldsymbol{\tau}, \boldsymbol{p}\} \in \mathbb{E}_\sigma$ (see equations (2.6.24) and (2.6.25)), which is positive according to the principle of maximum plastic dissipation. Consequently, we arrive at the inequality

$$\boxed{(\boldsymbol{\sigma} - \boldsymbol{\tau}) : \left[\nabla \dot{\boldsymbol{u}} - \mathbf{C}^{-1}\dot{\boldsymbol{\sigma}}\right] - (\boldsymbol{q} - \boldsymbol{p}) \cdot \mathbf{D}^{-1}\dot{\boldsymbol{q}} \geq 0, \text{ for all } \{\boldsymbol{\tau}, \boldsymbol{p}\} \in \mathbb{E}_\sigma .} \tag{2.6.29}$$

By integrating (2.6.29) over $\mathcal{B}$ and using the kinematic relationship $\dot{\boldsymbol{\varepsilon}} = \nabla \dot{\boldsymbol{u}}$, we arrive at the *variational inequality*

$$\boxed{\int_{\mathcal{B}} \left\{(\boldsymbol{\sigma} - \boldsymbol{\tau}) : \left[\dot{\boldsymbol{\varepsilon}} - \mathbf{C}^{-1}\dot{\boldsymbol{\sigma}}\right] - (\boldsymbol{q} - \boldsymbol{p}) \cdot \mathbf{D}^{-1}\dot{\boldsymbol{q}}\right\} dV \geq 0, \text{ for all } \{\boldsymbol{\tau}, \boldsymbol{p}\} \in \mathbb{E}_\sigma .} \tag{2.6.30}$$

This variational equation and the weak formulation of the rate form of the momentum balance equation furnish a variational formulation of plasticity which has often been taken as the starting point for mixed formulations of elastoplasticity; see Johnson[1977], and Simo, Kennedy, and Taylor [1988]. We remark that (2.6.29) or (2.6.30) remain valid in situations for which the formulation developed so far no longer holds, such as in the case of elastic domains with a *nonsmooth boundary* $\partial\mathbb{E}_\sigma$.

We have shown that for associative plasticity, the flow rule and the hardening law follow from the principle of maximum plastic dissipation. Furthermore, the analysis in the last two sections underscores the intimate relationship between the Helmholtz free energy and the hardening law through the potential $\mathcal{H}(\boldsymbol{\alpha})$. Finally, for the model problem discussed in Section **5.2**, we have shown that our definition of internal energy exactly satisfies the first law of thermodynamics. This model problem encompasses practically all the cases of interest in metal plasticity. We summarize our conclusions in BOX **2.2**.

BOX 2.2. Thermodynamics of Associative Plasticity.

Given : $W = \frac{1}{2}(\boldsymbol{\varepsilon} - \boldsymbol{\varepsilon}^p) : \mathbf{C} : (\boldsymbol{\varepsilon} - \boldsymbol{\varepsilon}^p)$ and the function $\phi(\boldsymbol{\sigma})$,

i. select hardening potential $\mathcal{H}(\boldsymbol{\alpha})$;

ii. define the yield condition as $f(\boldsymbol{\sigma}, \boldsymbol{\alpha}) := \phi(\boldsymbol{\sigma} - \bar{\boldsymbol{q}}) - \frac{\partial \mathcal{H}}{\partial \alpha}(\boldsymbol{\alpha}) - \sigma_Y \leq 0$, with $\boldsymbol{q} = -\nabla\mathcal{H}(\boldsymbol{\alpha})$;

iii. define the free energy as $\psi := W + \mathcal{H}$, and postulate *maximum plastic dissipation*; and

iv. compute the flow rule and hardening law through the associative relationships (2.6.22)

2.7 Classical (Rate-Dependent) Viscoplasticity

In this section we extend the one-dimensional formulation of viscoplasticity, and outline a classical rate-dependent plasticity model of the Perzyna type which has been often considered in computational applications; see, e.g., Zienkiewicz and Cormeau [1974], Cormeau [1975], Hughes and Taylor [1978], Pinsky, Ortiz and Pister [1983], or Simo and Ortiz [1985]. For general treatments concerned with fundamental aspects of viscoplasticity we refer to Perzyna [1971] and Lubliner [1972].

2.7.1 Formulation of the Basic Governing Equations

As in rate-independent plasticity, in classical formulations of viscoplasticity, one also introduces an *elastic range* which is defined, in terms of a *loading function* $f(\boldsymbol{\sigma}, \boldsymbol{q})$ by the set

$$\mathbb{E}_{\sigma} := \{\boldsymbol{\sigma} \in \mathbb{S}\,,\ \boldsymbol{q} \in \mathbb{R}^m \mid f(\boldsymbol{\sigma}, \boldsymbol{q}) \le 0\}. \tag{2.7.1}$$

As noted in Chapter **1** a basic difference between viscoplasticity and rate-independent plasticity is that in the former model states $\{\boldsymbol{\sigma}, \boldsymbol{q}\}$, such that $f(\boldsymbol{\sigma}, \boldsymbol{q}) \geq 0$ that is, stress states outside the closure of the elastic range are permissible, whereas in the latter constitutive model such states are not allowed.

The equations of evolution for the internal viscoplastic variables $\{\boldsymbol{\varepsilon}^{vp}, \boldsymbol{q}\}$ are formulated in terms of a C^2 *monotonically increasing* function $g: \mathbb{R} \to \mathbb{R}_+$, such that $g(x) = 0$ iff $x = 0$, and are summarized for convenience in BOX **2.3.** In these equations $\eta \in (0, \infty)$ is a given material parameter, called *fluidity* of the model. For metals, typical choices for the function $g(x)$ are exponentials and power laws.

2.7.2 Interpretation as a Viscoplastic Regularization

Classical viscoplasticity admits an important alternative interpretation as the regularization of rate-independent plasticity. Explicitly, it will be shown that the viscoplastic constitutive model summarized in BOX **2.3** is viewed as the optimality conditions of a regularized *penalty functional*, with penalty parameter $1/\eta > 0$, of the maximum plastic dissipation function discussed in the preceding section. We recall that the optimality conditions of the maximum plastic dissipation function are to be the constitutive equations summarized in BOX **2.1.** This interpretation is particularly important when one is concerned with softening response in the rate-independent model which results in losing ellipticity of the incremental equations. In this context, the viscoplastic regularization can be viewed as a means of *regularizing* the rate-independent problem so that the governing equations for the dynamic problem remain hyperbolic. Such a regularization technique is closely related to the classical Yosida regularization in the theory of semigroup operators; see e.g., Pazy [1983, page 9].

According to the preceding interpretation of the viscoplastic regularization, if one considers *decreasing* values of the fluidity parameter $\eta \in (0, \infty)$ in the limit as $\eta \to 0$, one expects to recover the rate-independent formulation. Somewhat heuristically, one may argue that, as $\eta \to 0$, states outside of the loading surface are increasingly penalized and thus $f \to 0$ so that $\langle g(f)\rangle/\eta \to \gamma$ finite. An illustration of this fact is contained in the simple one-dimensional example discussed in Chapter **1**. Computationally, this property has been often exploited in the past as an algorithmic procedure for rate-independent plasticity; see e.g., Hughes and Taylor [1978] and Simo, Hjelmstad, and Taylor [1984].

2.7.3 Penalty Formulation of the Principle of Maximum Plastic Dissipation

We begin our analysis by considering the classical penalty formulation for the principle of maximum plastic dissipation. We recall that, in the context of constrained optimization theory, the basic idea underlying this technique is to transform a constrained minimization problem into a (sequence) of *unconstrained* problems by appending a *penalization function* of the constraints to the objective function. We explain this idea in detail in what follows. To simplify matters we restrict our attention to perfect viscoplasticity; i.e., we assume that $\boldsymbol{q} \equiv \boldsymbol{0}$. Furthermore, without loss of generality take $g(x) = x$, i.e., $g := identity$.

Once more consider the *constrained* minimization problem associated with maximum plastic dissipation, i.e.,

$$\boxed{\begin{aligned} &\underset{\boldsymbol{\tau} \in \mathbb{E}_\sigma}{\text{MIN}} \left\{-\mathcal{D}^{\text{vp}}[\boldsymbol{\tau}; \dot{\boldsymbol{\varepsilon}}^{\text{vp}}]\right\}, \\ &\mathcal{D}^{\text{vp}}[\boldsymbol{\tau}; \dot{\boldsymbol{\varepsilon}}^{\text{vp}}] := \boldsymbol{\tau} : \dot{\boldsymbol{\varepsilon}}^{\text{vp}}. \end{aligned}} \tag{2.7.2}$$

Let $\boldsymbol{\sigma} \in \mathbb{E}_\sigma$ be the solution to problem (2.7.2). Associated with this problem, we consider the following sequence of *unconstrained* minimization problems for

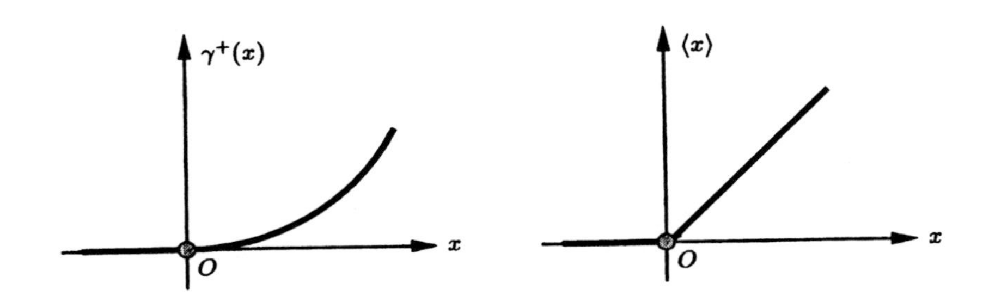

FIGURE 2-7. Graph of the penalty function $\gamma^+ : \mathbb{R} \to \mathbb{R}_+$, and its derivative $d\gamma^+/dx = \langle x\rangle$.

$\eta \in (0, \infty)$,

$$
\boxed{
\begin{aligned}
&\underset{\boldsymbol{\tau} \in \mathbb{S}}{\text{MIN}} \left\{ -\mathcal{D}^{\text{vp}}_{\eta}\left[\boldsymbol{\tau}; \dot{\boldsymbol{\varepsilon}}^{\text{vp}}\right] \right\}, \\
&\mathcal{D}^{\text{vp}}_{\eta}\left[\boldsymbol{\tau}; \dot{\boldsymbol{\varepsilon}}^{\text{vp}}\right] := \boldsymbol{\tau} : \dot{\boldsymbol{\varepsilon}}^{\text{vp}} + \frac{1}{\eta} \gamma^{+}[f(\boldsymbol{\tau})],
\end{aligned}
} \tag{2.7.3}
$$

Here, $\eta \in (0, \infty)$ is the so-called *penalty parameter*, whereas the function $\gamma^{+}: \mathbb{R} \rightarrow \mathbb{R}_{+}$ of the constraint $f(\boldsymbol{\tau}) \leq 0$ is called the *penalization* function and is subject to the following two requirements (see e.g., Luenberger [1984, page 366]):

1. γ^{+} is a C^{1} function;
2. $\gamma^{+}(x) \geq 0$, and $\gamma^{+}(x) = 0$ if and only if $x \leq 0$.

When the preceding conditions are satisfied, one refers to problem (2.7.3) as the *penalty regularization* of the constrained minimization problem (2.7.2). Let $\boldsymbol{\sigma}_{\eta} \in \mathbb{S}$ be the solution of problem (2.7.3). Under mild conditions on the smoothness of $\mathcal{D}^{\text{vp}}[\boldsymbol{\tau}; \dot{\boldsymbol{\varepsilon}}^{\text{vp}}]$, it can be shown that $\boldsymbol{\sigma}_{\eta} \rightarrow \boldsymbol{\sigma}$ as $\eta \rightarrow 0$. The proof of this result in the finite-dimensional case is rather straightforward; see, e.g., Luenberger [1984, Chapter **12**].

The computational advantage of problem (2.7.3) versus problem (2.7.2) should be clear. By augmenting the objective function $-\mathcal{D}^{\text{vp}}$ with the term $\frac{1}{\eta}\gamma^{+}[f(\boldsymbol{\tau})]$, we replace a constrained problem in which $\boldsymbol{\sigma} \in \mathbb{E}_{\sigma}$ with an unconstrained problem in which $\boldsymbol{\sigma} \in \mathbb{S}$. Note that, for $1/\eta$ large enough we achieve a *locally convex* version of the original problem which leads to weaker conditions on the existence of the minimizer $\boldsymbol{\sigma} \in \mathbb{S}$.

To relate the regularized version of the principle of maximum plastic dissipation given by (2.7.3) to the classical equations of viscoplasticity, we consider the following explicit expression for the function $\gamma^{+}: \mathbb{R} \rightarrow \mathbb{R}_{+}$. Let

$$
\boxed{
\gamma^{+}(x) := \begin{cases} \frac{1}{2}x^{2} & \Longleftrightarrow \quad x \geq 0 \\ 0 & \Longleftrightarrow \quad x \leq 0. \end{cases}
} \tag{2.7.4}
$$

Clearly, such a function satisfies conditions **1** and **2** above and, therefore, qualifies as a penalization function for problem (2.7.2). Furthermore, it is also clear that its derivative is given by (see Figure **2.7** for a graphical illustration)

$$
\boxed{
\frac{d}{dx}\gamma^{+}(x) = \langle x \rangle := \begin{cases} x & \Longleftrightarrow \quad x \geq 0 \\ 0 & \Longleftrightarrow \quad x \leq 0. \end{cases}
} \tag{2.7.5}
$$

In view of (2.7.5), it follows that the optimality condition for the unconstrained problem (2.7.3) yields the viscoplastic flow rule, since

$$
\boxed{
\frac{\partial \mathcal{D}^{\text{vp}}_{\eta}(\boldsymbol{\sigma}, \dot{\boldsymbol{\varepsilon}}^{\text{vp}})}{\partial \boldsymbol{\sigma}} = 0 \;\Rightarrow\; \dot{\boldsymbol{\varepsilon}}^{\text{vp}} = \frac{1}{\eta} \langle f(\boldsymbol{\sigma}) \rangle \frac{\partial f(\boldsymbol{\sigma})}{\partial \boldsymbol{\sigma}}.
} \tag{2.7.6}
$$

In conclusion, the viscoplastic constitutive model in BOX **2.3** is the penalty regularization of the rate-independent model in BOX **2.3**, and the solution of the viscoplastic problem converges to the solution of the rate-independent problem as the penalty (fluidity) parameter $\eta \rightarrow 0$.

Remark 2.7.1.

1. It is clear that the preceding arguments generalize immediately to the general case of rate-independent plasticity with internal hardening variable $\boldsymbol{q}$. In the general case, the penalty regularized functional associated with maximum plastic dissipation takes, the form

$$\boxed{\begin{aligned} &\underset{(\boldsymbol{\tau},\boldsymbol{p}) \in \mathbb{S} \times \mathbb{R}^m}{\text{MIN}} \left\{-\mathcal{D}_\eta^{\text{vp}}\left[\boldsymbol{\tau},\boldsymbol{p};\dot{\boldsymbol{\varepsilon}}^{\text{vp}},\dot{\boldsymbol{\alpha}}\right]\right\}, \\ &\mathcal{D}_\eta^{\text{vp}}\left[\boldsymbol{\tau},\boldsymbol{p};\dot{\boldsymbol{\varepsilon}}^{\text{vp}},\dot{\boldsymbol{\alpha}}\right] := \boldsymbol{\tau} : \dot{\boldsymbol{\varepsilon}}^{\text{vp}} + \boldsymbol{p}\cdot\dot{\boldsymbol{\alpha}} - \frac{1}{\eta}\gamma^{+}[f(\boldsymbol{\tau},\boldsymbol{p})], \end{aligned}} \tag{2.7.7}$$

where $\gamma^{+}: \mathbb{R} \rightarrow \mathbb{R}_{+}$ is defined by (2.7.4) with derivative given by (2.7.5). Then the optimality conditions for unconstrained minimization problems yield

$$\left.\begin{aligned} \frac{\partial \mathcal{L}}{\partial \boldsymbol{\sigma}} &= 0 \\ \frac{\partial \mathcal{L}}{\partial \boldsymbol{q}} &= 0 \end{aligned}\right\} \Rightarrow \boxed{\begin{aligned} \dot{\boldsymbol{\varepsilon}}^{\text{vp}} &= \frac{1}{\eta}\langle f(\boldsymbol{\sigma},\boldsymbol{q})\rangle \frac{\partial f(\boldsymbol{\sigma},\boldsymbol{q})}{\partial \boldsymbol{\sigma}} \\ \dot{\boldsymbol{\alpha}} &= \frac{1}{\eta}\langle f(\boldsymbol{\sigma},\boldsymbol{q})\rangle \frac{\partial f(\boldsymbol{\sigma},\boldsymbol{q})}{\partial \boldsymbol{q}} \end{aligned}} \tag{2.7.8}$$

which constitute the *associative version* of the flow rule and hardening law for classical viscoplasticity.

2. Although it appears that penalty regularization is an attractive procedure for obtaining the rate-independent limit, as $\eta \rightarrow 0$, one should keep in mind the well-known fact that, computationally, the unconstrained regularized problem becomes progressively ill-conditioned as the penalty parameter $1/\eta \rightarrow \infty$; see e.g., Bertsekas [1982] or Luenberger [1984]. From the standpoint of, numerical analysis such behavior is the source of serious difficulties that is overcome only when alternative formulations are employed, in particular, the *method of multipliers* or *method of augmented Lagrangians* of Hestenes [1969] and Powell [1969]. These techniques have received recently considerable attention in computational solid mechanics literature; see e.g., Fortin and Glowinski [1983] and Glowinski and Le Tallec [1989].
3. The difficulties alluded to above associated with the penalty regularization for enforcing the rate-independent limit, to a large extent, have motivated the active development of well-conditioned return-mapping algorithms which deal directly with the rate-independent limit. Remarkably, as illustrated in Chapter **1**, it is also possible to obtain the rate-dependent solution by a well-conditioned *closed-form* algorithm whose first step is the rate-independent solution, thus bypassing the characteristic ill-conditioning associated with penalty procedures. We discuss these ideas in detail in the following chapter.

BOX 2.3. Classical Associative Viscoplasticity.

i. Elastic stress strain relationships

$$\boldsymbol{\sigma} = \frac{\partial W(\boldsymbol{\varepsilon} - \boldsymbol{\varepsilon}^{\mathrm{vp}})}{\partial \boldsymbol{\varepsilon}} = \mathbf{C} : (\boldsymbol{\varepsilon} - \boldsymbol{\varepsilon}^{\mathrm{vp}})$$

$$\mathbf{C} := \frac{\partial^2 W(\boldsymbol{\varepsilon} - \boldsymbol{\varepsilon}^{\mathrm{vp}})}{\partial \boldsymbol{\varepsilon}^2} = \text{(constant) elastic moduli.}$$

ii. (Closure of) elastic domain in stress space

$$\mathbb{E}_\sigma = \left\{ (\boldsymbol{\sigma}, \boldsymbol{q}) \in \mathbb{S} \times \mathbb{R}^m \mid f(\boldsymbol{\sigma}, \boldsymbol{q}) \le 0 \right\}$$

iii.a. Flow rule and hardening law (*Perzyna* model)

$$\dot{\boldsymbol{\varepsilon}}^{\mathrm{vp}} = \gamma \frac{\partial f(\boldsymbol{\sigma}, \boldsymbol{q})}{\partial \boldsymbol{\sigma}}$$

$$\dot{\boldsymbol{q}} = -\gamma \mathbf{D} \frac{\partial f(\boldsymbol{\sigma}, \boldsymbol{q})}{\partial \boldsymbol{q}}$$

$$\gamma = \left\langle g\left(f(\boldsymbol{\sigma}, \boldsymbol{q}) \right) \right\rangle / \eta$$

where

$$g(x) \ \textit{monotone} \text{ with } g(x) = 0 \iff x \le 0$$

and

$$\langle x \rangle := \frac{x + |x|}{2} \quad \text{(ramp function).}$$

iii.b. Flow rule and hardening law (*generalized Duvaut–Lions model*)

$$\dot{\boldsymbol{\varepsilon}}^{\mathrm{vp}} = \mathbf{C}^{-1} : \left[\boldsymbol{\sigma} - \bar{\boldsymbol{\sigma}} \right] / \tau$$

$$\dot{\boldsymbol{q}} = -\left[\boldsymbol{q} - \bar{\boldsymbol{q}} \right] / \tau$$

where

$$(\bar{\boldsymbol{\sigma}}, \bar{\boldsymbol{q}}) = \begin{cases} (\boldsymbol{\sigma}, \boldsymbol{q}) & \text{if } (\boldsymbol{\sigma}, \boldsymbol{q}) \in \mathbb{E}_\sigma \\ \mathbf{P}[\boldsymbol{\sigma}, \boldsymbol{q}] & \text{if } (\boldsymbol{\sigma}, \boldsymbol{q}) \notin \mathbb{E}_\sigma . \end{cases}$$

Here, $\mathbf{P} : \mathrm{ext}(\mathbb{E}_\sigma) \to \partial \mathbb{E}_\sigma$ is the *closest point projection operator*, and

$$\mathrm{ext}(\mathbb{E}_\sigma) = \left\{ (\boldsymbol{\sigma}, \boldsymbol{q}) \in \mathbb{S} \times \mathbb{R}^m \mid f(\boldsymbol{\sigma}, \boldsymbol{q}) > 0 \right\}.$$

2.7.4 *The Generalized Duvaut-Lions Model*

As a follow to of our introductory discussion of one-dimensional viscoplasticity, in this section we examine an alternative formulation of rate-dependent plasticity closely related to a model originally proposed by Duvaut–Lions [1972]. To motivate the general structure of the three-dimensional model, first we consider the case of J_2 flow theory.

2.7.4.1 J_2 perfect viscoplasticity.

Assume that the loading function is given by the Mises condition as

$$f(\boldsymbol{\sigma}) := \|\text{dev}[\boldsymbol{\sigma}]\| - \sqrt{\tfrac{2}{3}}\sigma_Y \leq 0\,. \tag{2.7.9}$$

Further, consider the case of an associative viscoplastic flow rule with $g(x) \equiv x$, so that the model in BOX **2.3** reduces to

$$\dot{\boldsymbol{\varepsilon}}^{\text{vp}} = \frac{\langle f(\boldsymbol{\sigma})\rangle}{\eta}\boldsymbol{n}\,, \quad \boldsymbol{n} := \frac{\text{dev}[\boldsymbol{\sigma}]}{\|\text{dev}[\boldsymbol{\sigma}]\|}\,, \tag{2.7.10}$$

for $\eta \in (0, \infty)$. Assume that the elastic response is isotropic, so that the elasticity tensor takes the form

$$\mathbf{C} = \kappa \mathbf{1} \otimes \mathbf{1} + 2\mu[\mathbf{I} - \tfrac{1}{3}\mathbf{1} \otimes \mathbf{1}]\,, \tag{2.7.11}$$

where $\kappa = \lambda + \frac{2}{3}\mu > 0$ is the bulk modulus and $\mu > 0$ is the shear modulus. Now let

$$\boxed{\tau := \frac{\eta}{2\mu}} \tag{2.7.12}$$

be the relaxation time. Since $\boldsymbol{s} = \|\boldsymbol{s}\|\boldsymbol{n}$, now the viscoplastic flow rule (2.7.10) is written as

$$\dot{\boldsymbol{\varepsilon}}^{\text{vp}} = \begin{cases} [2\mu]^{-1}\dfrac{\boldsymbol{s} - \sqrt{\frac{2}{3}}\sigma_Y \boldsymbol{n}}{\tau} & \text{if } \|\boldsymbol{s}\| - \sqrt{\frac{2}{3}}\sigma_Y \geq 0 \\ \mathbf{0} & \text{otherwise,} \end{cases} \tag{2.7.13}$$

where we have set $\boldsymbol{s} := \text{dev}[\boldsymbol{\sigma}]$. Now for $f(\boldsymbol{\sigma}) := \|\boldsymbol{s}\| - \sqrt{\frac{2}{3}}\sigma_Y$, we interpret $\sqrt{\frac{2}{3}}\sigma_Y \boldsymbol{n}$ as the projection of $\boldsymbol{s} = \|\boldsymbol{s}\|\boldsymbol{n}$ on the boundary of the elastic domain $\partial\mathbb{E}_\sigma := \{\boldsymbol{s} \in \mathbb{S} \mid \|\boldsymbol{s}\| = \sqrt{\frac{2}{3}}\sigma_Y\}$. This interpretation suggests the following construction. Let

$$\text{ext}\,(\mathbb{E}_\sigma) := \left\{\boldsymbol{s} \in \mathbb{S} \mid f(\boldsymbol{\sigma}) := \|\boldsymbol{s}\| - \sqrt{\tfrac{2}{3}}\sigma_Y \geq 0\right\}\,. \tag{2.7.14}$$

Define the projection operator $\mathbf{P} : \mathbb{S} \to \mathbb{S}$ by the expression

$$\boxed{\bar{\boldsymbol{s}} = \mathbf{P}(\boldsymbol{s}) := \begin{cases} \boldsymbol{s} & \iff \boldsymbol{s} \in \text{int}\,(\mathbb{E}_\sigma), \\ \sqrt{\frac{2}{3}}\sigma_Y \dfrac{\boldsymbol{s}}{\|\boldsymbol{s}\|} & \iff \boldsymbol{s} \in \text{ext}\,(\mathbb{E}_\sigma). \end{cases}} \tag{2.7.15}$$

Clearly, $\mathbf{P} \circ \mathbf{P} = \mathbf{P}$ so that $\mathbf{P}$ is indeed a projection. Further, note that

$$\mathbf{C}^{-1} := \left[\frac{\kappa^{-1}}{9}\mathbf{1} \otimes \mathbf{1} + [2\mu]^{-1}(\mathbf{I} - \tfrac{1}{3}\mathbf{1} \otimes \mathbf{1})\right]\,. \tag{2.7.16}$$

Therefore, since $s = \text{dev}[\sigma]$ so that $\text{tr}\,[s] = 0$, from (2.7.16),

$$\mathbf{C}^{-1}\boldsymbol{n} = [2\mu]^{-1}\boldsymbol{n}\,. \tag{2.7.17}$$

Using (2.7.15) and (2.7.17), we reformulate (2.7.13) as

$$\boxed{\dot{\varepsilon}^{\text{vp}} = \mathbf{C}^{-1} : \frac{[s - \bar{s}]}{\tau}.} \tag{2.7.18}$$

The interpretation of equations (2.7.15) and (2.7.18) should be clear:

a. The projection $\mathbf{P}$ constrains $\bar{s} = \mathbf{P}(s)$ to lie within the closure of the elastic domain $\mathbb{E}_\sigma$ since $\mathbb{E}_\sigma$ is a circle, for $s \in \text{ext}\,(\mathbb{E}_\sigma)$, the point $\bar{s}$ is the closest point projection of s onto $\mathbb{E}_\sigma$.
b. For $s \in \text{ext}\,(\mathbb{E}_\sigma)$, the magnitude of the viscoplastic strain rate $\dot{\varepsilon}^{\text{vp}}$ given by (2.7.18) is proportional to the distance between the deviatoric state s and its projection onto $\partial\mathbb{E}_\sigma$; see Figure **2.8**.

2.7.4.2 Perfect viscoplasticity.

The conclusions derived above in the context of J_2 flow viscoplasticity carry over without modification to general viscoplasticity with no hardening.

Recall that, according to the well-known *projection theorem* (valid in a general Hilbert space, not necessarily finite-dimensional; see, e.g., Lang [1983]), if $\mathbb{E}_\sigma \subset \mathbb{S}$ is *convex*, given any $\sigma \in \text{ext}\,(\mathbb{E}_\sigma)$ there is a *unique point* $\bar{\sigma} \in \partial\mathbb{E}_\sigma$ which is closest to σ, that is, the problem

$$\boxed{\text{find} \quad \bar{\sigma} \in \mathbb{E}_\sigma, \text{ such that } \|\sigma - \bar{\sigma}\| = \underset{\tau \in \mathbb{E}_\sigma}{\text{MIN}} \|\sigma - \tau\|} \tag{2.7.19}$$

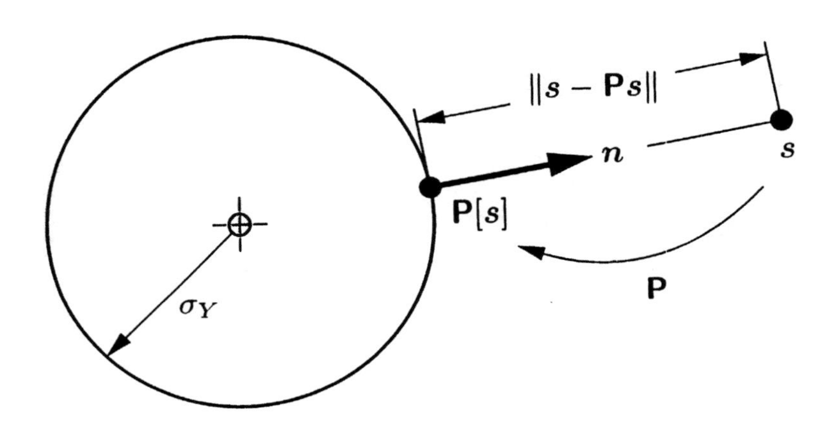

FIGURE 2-8. Geometric interpretation of the closest point projection $\mathbf{P}$ for J_2 flow theory.

has a unique solution $\bar{\sigma} \in \mathbb{E}_\sigma$. We write

$$\bar{\sigma} = \mathbf{P}(\sigma),$$

where

$$\mathbf{P} \circ \mathbf{P} = \mathbf{P} \tag{2.7.20}$$

is called the closest point projection.

Then the general viscoplasticity equation takes the form

$$\boxed{\dot{\varepsilon}^{vp} = \frac{\mathbf{C}^{-1}}{\tau} : [\sigma - \bar{\sigma}],} \tag{2.7.21}$$

which is identical in structure to (2.7.18).

2.7.4.3 General hardening viscoplasticity.

The extension of the preceding ideas to hardening response is accomplished by the following rate-dependent constitutive equations:

$$\boxed{\begin{aligned} \dot{\varepsilon}^{vp} &= \frac{\mathbf{C}^{-1}}{\tau} : [\sigma - \bar{\sigma}], \\ \dot{\alpha} &= \frac{\mathbf{D}^{-1}}{\tau} [q - \bar{q}], \end{aligned}} \tag{2.7.22}$$

where $\dot{q} = -\mathbf{D}\dot{\alpha}$ is a suitable set of internal hardening variables and $\mathbf{D}$ is the matrix of generalized hardening moduli. In addition, $\{\bar{\sigma}, \bar{q}\}$ is the solution of the *rate-independent problem*. For further details on the structure of this model, which is ideally suited to the case of multisurface plasticity, we refer to Simo, Kennedy, and Govindjee [1988].

3

Integration Algorithms for Plasticity and Viscoplasticity

As illustrated in Chapter **1**, the numerical solution of nonlinear boundary-value problems in solid mechanics is based on an *iterative* solution of a discretized version of the momentum balance equations. Typically, the following steps are involved:

1. The discretized momentum equations generate incremental motions which, in turn, are used to calculate the incremental strain history by kinematic relationships.
2. For a *given* incremental strain history, new values of the state variables $\{\boldsymbol{\sigma}, \boldsymbol{\varepsilon}^p, \boldsymbol{q}\}$ are obtained by integrating the *local* constitutive equations with given initial conditions.
3. The (discrete) momentum balance equation is tested for the computed stresses and, if violated, the iteration process is continued by returning to step **1**.

In most of the computational architectures currently in use, steps **1** and **3** are carried out at a global level by finite-element/finite-difference procedures. In this section we are concerned with step **2**, which is regarded as the *central problem of computational plasticity* as it corresponds to the main role played by constitutive equations in actual computations. The crucial aspect, illustrated in Chapter **1** and pointed out in Section **3.1** below, is the fact that from a computational standpoint this problem can always be regarded as strain-driven in the sense that the state variables are computed for a *given deformation history*.

The evolution equations of classical elastoplasticity, as summarized in BOX **2.1** of Chapter **2**, define a unilaterally constrained problem of evolution. By applying an implicit backward-Euler difference scheme, this problem is transformed into a *constrained-optimization* problem, governed by discrete Kuhn–Tucker conditions. The structure of this discrete problem, the fundamental role played by the discrete Kuhn–Tucker conditions, and the geometric interpretation of the solution as the closest point projection in the energy norm of trial elastic state onto the elastic domain, are considered in Section **3.2**. As an illustration of these ideas and to motivate the general algorithms developed in Section **3.6**, in Sections **3.3** and **3.4** we examine in detail the important case of J_2 flow theory with nonlinear kinematic and isotropic hardening rules. An interpretation of these algorithms as product formula algorithms emanating from an elastic-plastic operator split is given in Section **3.5**. This interpretation leads naturally to the notion of *elastic trial state*, first introduced in Section **3.2**, and provides the necessary framework for developing the general cutting-plane algorithm in Section **3.6**. We conclude the chapter by considering in Section **3.7** the generalization of this class of algorithms to classical viscoplasticity. For further reading see Simo and Hughes [1987].

The developments in this chapter generalize and unify a number of existing algorithmic schemes which, starting with the classical radial return algorithm of Wilkins [1964], have been largely restricted to J_2 flow theory. Representative algorithms include the extension of the radial return method in Krieg and Key [1976] to accommodate linear isotropic and kinematic hardening, the midpoint return map of Rice and Tracy [1973], and alternative formulations of elastic-predictor/plastic corrector methods, as summarized in Krieg and Krieg [1977]. To a large extent, it appears that return mapping (or "catching-up") algorithms have replaced classical treatments based on the elastoplastic tangent modulus, as in Marcal and King [1967] or Nayak and Zienkiewicz [1972]. We refer to Zienkiewicz [1977] for a review of this class of methods not considered here.

3.1 Basic Algorithmic Setup. Strain-Driven Problem

Let $[0, T] \subset \mathbb{R}$ be the time interval of interest. At time $t \in [0, T]$, we assume that the total and plastic strain fields and the internal variables are known, that is

$$\{\boldsymbol{\varepsilon}_n, \boldsymbol{\varepsilon}^p_n, \boldsymbol{q}_n\} \tag{3.1.1a}$$

are given data at t_n. Note that the *elastic strain* tensor and the *stress* tensor are regarded as *dependent* variables which are always obtained from the basic variables (3.1.1a) through the elastic stress-strain relationships

$$\boldsymbol{\varepsilon}^e_n := \boldsymbol{\varepsilon}_n - \boldsymbol{\varepsilon}^p_n\,, \qquad \boldsymbol{\sigma}_n = \nabla W(\boldsymbol{\varepsilon}^e_n). \tag{3.1.1b}$$

Let $\Delta\boldsymbol{u} : \mathcal{B} \to \mathbb{R}^{n_{dim}}$ be the incremental displacement field, which is *assumed to be given*. Here, $\mathcal{B} \subset \mathbb{R}^{n_{dim}}$ is the reference configuration of the body of interest. Without loss of generality we consider the case $n_{dim} = 3$ throughout. Then the basic problem is to update the fields (3.1.1) to $t_{n+1} \in [0, T]$ in a manner consistent with the elastoplastic constitutive equations developed in the previous chapter and summarized for convenience below:

$$\boxed{\begin{aligned} \dot{\boldsymbol{\varepsilon}} &= \nabla^s(\Delta\dot{\boldsymbol{u}}) \\ \dot{\boldsymbol{\varepsilon}}^p &= \gamma\boldsymbol{r}(\boldsymbol{\sigma}, \boldsymbol{q}) \\ \dot{\boldsymbol{q}} &= -\gamma\boldsymbol{h}(\boldsymbol{\sigma}, \boldsymbol{q})\,, \end{aligned}} \tag{3.1.2a}$$

subject to the unilateral Kuhn–Tucker complementarity conditions

$$\boxed{\begin{aligned} f(\boldsymbol{\sigma}, \boldsymbol{q}) &\le 0 \\ \gamma &\ge 0 \\ \gamma f(\boldsymbol{\sigma}, \boldsymbol{q}) &= 0. \end{aligned}} \tag{3.1.2b}$$

and with the initial conditions

$$\boxed{\{\boldsymbol{\varepsilon}, \boldsymbol{\varepsilon}^p, \boldsymbol{q}\}\big|_{t=t_n} = \{\boldsymbol{\varepsilon}_n, \boldsymbol{\varepsilon}_n^p, \boldsymbol{q}_n\}.} \tag{3.1.2c}$$

Here, $\nabla^s(\bullet)$ denotes the symmetric gradient. Observe that, once the plastic strain field is known, the stress field $\boldsymbol{\sigma}$ is computed from $(3.1.1b)_2$

3.1.1 Associative plasticity.

Recall that according to Proposition **2.6.2** the hardening law in associative plasticity is characterized by a potential function $\mathcal{H}\colon \mathbb{R}^m \to \mathbb{R}$ such that

$$\dot{\boldsymbol{q}} = -\nabla^2\mathcal{H}(\boldsymbol{\alpha})\dot{\boldsymbol{\alpha}},$$

where

$$\dot{\boldsymbol{\alpha}} = \gamma \frac{\partial f(\boldsymbol{\sigma}, \boldsymbol{q})}{\partial \boldsymbol{q}}. \tag{3.1.3}$$

Proceeding as in Section **2.6.3**, we introduce the Legendre transformation defined by (2.6.26) and set

$$\boxed{\mathbf{D} := \nabla^2\mathcal{H}(\boldsymbol{\alpha}) = [\nabla^2\Theta(\boldsymbol{q})]^{-1}.} \tag{3.1.4}$$

With this notation in hand, the associative version of the evolution equation (3.1.2*b*) is given by

$$\boxed{\begin{aligned} \dot{\boldsymbol{\varepsilon}} &= \nabla^s(\Delta\dot{\boldsymbol{u}}) \\ \dot{\boldsymbol{\varepsilon}}^p &= \gamma \frac{\partial f(\boldsymbol{\sigma}, \boldsymbol{q})}{\partial \boldsymbol{\sigma}} \\ \dot{\boldsymbol{q}} &= -\gamma \mathbf{D} \frac{\partial f(\boldsymbol{\sigma}, \boldsymbol{q})}{\partial \boldsymbol{q}}. \end{aligned}} \tag{3.1.2a)*$$

In what follows, we are concerned mainly with this associative plasticity model. Our developments, however, are general and apply without essential modification to the general nonassociative case covered by (3.1.2).

3.2 The Notion of Closest Point Projection

Equations (3.1.2) define a nonlinear problem of evolution, with initial conditions (3.1.2*c*), whose characteristic feature is the unilateral constraint condition

(3.1.2*b*). As discussed below, this continuum problem is transformed into a *discrete, constrained-optimization problem* by applying an *implicit backward-Euler* difference scheme. Thus, computationally, the solution of the constrained problem of evolution (3.1.2) collapses to the (iterative) solution of a convex mathematical programming problem. In fact, it is shown below that this problem reduces to the standard problem of finding the closest distance (in the energy norm) of a point (the trial state) to a convex set (the elastic domain). General solution strategies for this problem are deferred to Section **3.6**.

According to the preceding ideas, from (3.1.2*a*) by an implicit backward-Euler difference scheme and using initial conditions (3.1.2*c*), we obtain the *nonlinear* coupled system:

$$\boxed{\begin{aligned} \boldsymbol{\varepsilon}_{n+1} &= \boldsymbol{\varepsilon}_n + \nabla^s(\Delta \boldsymbol{u}) \quad \text{(trivial)} \\ \boldsymbol{\sigma}_{n+1} &= \nabla W(\boldsymbol{\varepsilon}_{n+1} - \boldsymbol{\varepsilon}^p_{n+1}) \\ \boldsymbol{\varepsilon}^p_{n+1} &= \boldsymbol{\varepsilon}^p_n + \Delta\gamma\, \partial_{\sigma} f(\boldsymbol{\sigma}_{n+1}, \boldsymbol{q}_{n+1}) \\ \boldsymbol{q}_{n+1} &= \boldsymbol{q}_n - \Delta\gamma\, \mathbf{D}\partial_{q} f(\boldsymbol{\sigma}_{n+1}, \boldsymbol{q}_{n+1})\,. \end{aligned}} \tag{3.2.1}$$

where $\Delta\gamma = \gamma_{n+1}\Delta t$. In addition, the discrete counterpart of the Kuhn–Tucker conditions (3.1.2b) become

$$\boxed{\begin{aligned} f(\boldsymbol{\sigma}_{n+1}, \boldsymbol{q}_{n+1}) &\le 0, \\ \Delta\gamma &\ge 0, \\ \Delta\gamma f(\boldsymbol{\sigma}_{n+1}, \boldsymbol{q}_{n+1}) &= 0\,. \end{aligned}} \tag{3.2.2}$$

As in the continuum case, Kuhn–Tucker conditions (3.2.2) define the appropriate notion of loading/unloading. These conditions are reformulated in a form directly amenable to computational implementation by introducing the following *trial elastic state*:

$$\left.\begin{aligned} \boldsymbol{\varepsilon}^{e\,\text{trial}}_{n+1} &:= \boldsymbol{\varepsilon}_{n+1} - \boldsymbol{\varepsilon}^p_n \\ \boldsymbol{\sigma}^{\text{trial}}_{n+1} &:= \nabla W(\boldsymbol{\varepsilon}^{e\,\text{trial}}_{n+1}) \\ \boldsymbol{q}^{\text{trial}}_{n+1} &:= \boldsymbol{q}_n \\ f^{\text{trial}}_{n+1} &:= f(\boldsymbol{\sigma}^{\text{trial}}_{n+1}, \boldsymbol{q}^{\text{trial}}_{n+1})\,. \end{aligned}\right\} \tag{3.2.3}$$

From a physical standpoint the trial elastic state is obtained by *freezing plastic flow* during the time step. Section **3.5** shows that this state arises naturally in the context of an *elastic-plastic operator split.* Observe that only *functional evaluations* are required in definition (3.2.3).

3.2.1 Plastic Loading. Discrete Kuhn–Tucker Conditions

From an algorithmic standpoint, a basic result is the fact that plastic loading or unloading is characterized exclusively in terms of the trial state, *provided the yield function is convex.* To appreciate the role of convexity and conditions (3.2.2) in this conclusion, for simplicity consider the following situation

1. assume constant generalized plastic moduli, i.e., $\mathbf{D} := \nabla^2 \mathcal{H}(\boldsymbol{\alpha}) \equiv$ constant; and
2. assume constant elasticities, i.e., $\mathbf{C} := \nabla^2 W(\boldsymbol{\varepsilon}^e) \equiv$ constant.

Then we have the following:

Lemma 3.1. *Let $f\colon \mathbb{S} \to \mathbb{R}$ be* **convex**, *and let f_{n+1}^{trial} be computed according to (3.2.3)$_4$. Then*

$$\boxed{f_{n+1}^{\text{trial}} \geq f_{n+1}.} \tag{3.2.4}$$

PROOF. By Lemma **2.6.1** the convexity assumption on f implies that

$$f_{n+1}^{\text{trial}} - f_{n+1} \geq \left[\boldsymbol{\sigma}_{n+1}^{\text{trial}} - \boldsymbol{\sigma}_{n+1}\right] : \partial_{\sigma} f_{n+1} + \left[\boldsymbol{q}_n - \boldsymbol{q}_{n+1}\right] : \partial_q f_{n+1}. \tag{3.2.5a}$$

However, from (3.2.1),

$$\left.\begin{aligned} \boldsymbol{\sigma}_{n+1} &= \boldsymbol{\sigma}_{n+1}^{\text{trial}} - \Delta\gamma \mathbf{C} : \partial_{\sigma} f_{n+1}, \\ \boldsymbol{q}_{n+1} &= \boldsymbol{q}_n - \Delta\gamma \mathbf{D} : \partial_q f_{n+1}. \end{aligned}\right\} \tag{3.2.5b}$$

Therefore, by substituting (3.2.5b) in (3.2.5a), we obtain

$$\begin{aligned} f_{n+1}^{\text{trial}} - f_{n+1} &\geq \Delta\gamma\left[\partial_{\sigma} f_{n+1} : \mathbf{C} : \partial_{\sigma} f_{n+1} + \partial_q f_{n+1} : \mathbf{D} : \partial_q f_{n+1}\right] \\ &=: \Delta\gamma\left[\|\partial_{\sigma} f_{n+1}\|_{\mathbf{C}}^2 + \|\partial_q f_{n+1}\|_{\mathbf{D}}^2\right], \end{aligned} \tag{3.2.6}$$

where $\|\bullet\|_{\mathbf{C}}$ denotes the norm induced by $\mathbf{C}$, and $\|\bullet\|_{\mathbf{D}}$ is the norm induced by $\mathbf{D}$. (We assume that $\mathbf{D}$ is positive-definite. Alternatively, the terms within brackets in (3.2.6) are positive by virtue of Assumption **2.1**.) Then the result follows by noting that $\Delta\gamma \geq 0$. □

Then conditions (3.2.2) and the preceding lemma imply the following computational statement of the loading/unloading conditions.

Proposition 3.1. *Loading/unloading is decided solely from f_{n+1}^{trial} according to the conditions*

$$\boxed{\begin{aligned} f_{n+1}^{\text{trial}} < 0 &\Rightarrow \text{elastic step} \Leftrightarrow \Delta\gamma = 0, \\ f_{n+1}^{\text{trial}} > 0 &\Rightarrow \text{plastic step} \Leftrightarrow \Delta\gamma > 0. \end{aligned}} \tag{3.2.7}$$

PROOF. (a) First, if $f_{n+1}^{\text{trial}} < 0$, then by Lemma **3.1** it follows that $f_{n+1} < 0$. Then the discrete Kuhn–Tucker condition $\Delta\gamma f_{n+1} = 0$ implies $\Delta\gamma = 0$. Thus, $\boldsymbol{\varepsilon}_{n+1}^p \equiv \boldsymbol{\varepsilon}_n^p$, and the process is elastic.

(b) On the other hand, if $f_{n+1}^{\text{trial}} > 0$, then $\boldsymbol{\varepsilon}^{e\,\text{trial}}_{n+1}$ cannot be feasible, that is, $\boldsymbol{\varepsilon}^{e\,\text{trial}}_{n+1} \neq \boldsymbol{\varepsilon}_{n+1}$. Thus, we require that $\Delta\gamma \neq 0$. Since $\Delta\gamma$ cannot be negative it follows that $\Delta\gamma > 0$. Then the discrete Kuhn–Tucker condition $\Delta\gamma f_{n+1} = 0$ implies that $f_{n+1} = 0$, and the step is plastic. □

The solution to equations (3.2.1) is amenable to a compelling geometric interpretation which can be exploited numerically.

3.2.2 Geometric Interpretation

We consider three different cases and examine the geometric interpretation associated with the discrete problem (3.2.1) assuming that the elasticity tensor $\mathbf{C}$ is constant.

i. For *perfect plasticity* (i.e., $\boldsymbol{q} \equiv \mathbf{0}$), the solution $\varepsilon^e_{n+1} := \varepsilon_{n+1} - \varepsilon^p_{n+1}$ in *strain space* is the *closest point projection* in the energy norm of the trial state $\varepsilon^{e\,\text{trial}}_{n+1}$ onto the yield surface, that is

$$\boxed{\varepsilon^e_{n+1} = \text{ARG}\left[\underset{f[\nabla W(\varepsilon^e_{n+1})] \leq 0}{\text{MIN}} \left\{\tfrac{1}{2}\,\|\varepsilon^{e\,\text{trial}}_{n+1} - \varepsilon^e_{n+1}\|^2_{\mathbf{C}}\right\}\right]} \tag{3.2.8}$$

This geometric interpretation, illustrated in Figure **3.1**, follows at once by noting that the Lagrangian function associated with this constrained problem is expressed as

$$\mathcal{L}(\varepsilon^e_{n+1}, \Delta\gamma) := \tfrac{1}{2}\,\|\varepsilon^{e\,\text{trial}}_{n+1} - \varepsilon^e_{n+1}\|^2_{\mathbf{C}} + \Delta\gamma f[\nabla W(\varepsilon^e_{n+1})], \tag{3.2.9}$$

and the corresponding Kuhn–Tucker optimality conditions are

$$\left.\begin{aligned} \frac{\partial \mathcal{L}}{\partial \varepsilon^e_{n+1}} &= \mathbf{C} : \left[-\varepsilon^{e\,\text{trial}}_{n+1} + \varepsilon^e_{n+1} + \Delta\gamma \left.\frac{\partial f}{\partial \sigma}\right|_{n+1}\right] = \mathbf{0} \\ \Delta\gamma \geq 0&, \quad \Delta\gamma f[\nabla W(\varepsilon^e_{n+1})] = 0 \end{aligned}\right\} \tag{3.2.10}$$

which coincide with equations (3.2.1) for $\boldsymbol{h} \equiv \mathbf{0}$.

ii. For *perfect plasticity* a similar interpretation holds in *stress space*. From $(3.2.1)_2$, since $\mathbf{C} = \text{constant}$,

$$\boldsymbol{\sigma}_{n+1} = \boldsymbol{\sigma}^{\text{trial}}_{n+1} - \Delta\gamma \mathbf{C} : \nabla f(\boldsymbol{\sigma}_{n+1}), \tag{3.2.11}$$

where $\boldsymbol{\sigma}^{\text{trial}}_{n+1} = \mathbf{C} : \left[\varepsilon_{n+1} - \varepsilon^p_n\right]$. It follows that, for an associative flow rule, $\boldsymbol{\sigma}_{n+1}$ is the *closest point projection* onto the yield surface of the trial elastic stress $\boldsymbol{\sigma}^{\text{trial}}_{n+1}$ in the inner product induced by the compliance tensor $\mathbf{C}^{-1}$, that is,

$$\boldsymbol{\sigma}_{n+1} = \text{ARG}\left[\underset{\boldsymbol{\sigma} \in \mathbb{E}_\sigma}{\text{MIN}} \left\{\tfrac{1}{2}\,\|\boldsymbol{\sigma}^{\text{trial}}_{n+1} - \boldsymbol{\sigma}\|^2_{\mathbf{C}^{-1}}\right\}\right], \tag{3.2.12}$$

where $\|\boldsymbol{\sigma}\|_{\mathbf{C}^{-1}} := \sqrt{\boldsymbol{\sigma} : \mathbf{C}^{-1} : \boldsymbol{\sigma}}$ is the energy norm and $\mathbb{E}_\sigma$ is the closure of the elastic domain defined by (2.2.5). In conclusion, $\boldsymbol{\sigma}_{n+1}$ is the *closest point projection of $\boldsymbol{\sigma}^{\text{trial}}_{n+1}$ onto the yield surface in the energy norm.*

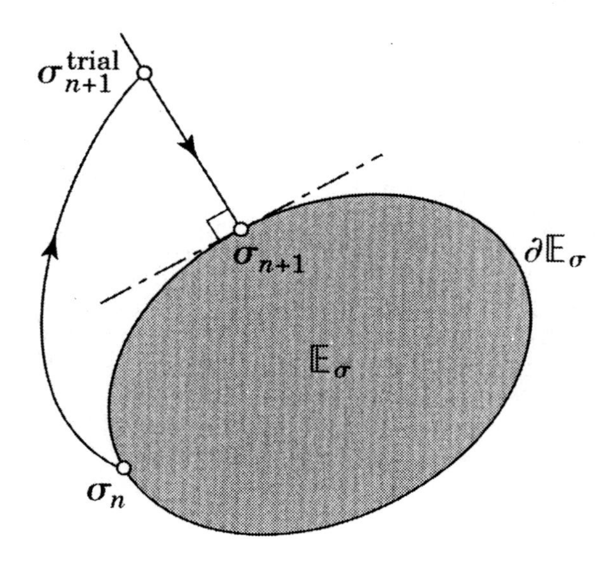

FIGURE 3-1. Geometric illustration of the concept of closest point projection.

iii. For *associative hardening plasticity* with hardening law of the form recorded in (3.1.2a), the interpretation given above admits the following generalization. Assume that the generalized plastic moduli $\mathbf{D}$ are *constant* and *positive-definite*. Let $\mathbf{G}$ be the *block-diagonal, positive-definite* matrix defined as

$$\mathbf{G} := \begin{bmatrix} \mathbf{C}^{-1} & \mathbf{O} \\ \mathbf{O} & \mathbf{D}^{-1} \end{bmatrix}. \tag{3.2.13}$$

Then the actual solution $\{\boldsymbol{\sigma}_{n+1}, \boldsymbol{q}_{n+1}\}$ is the *closest point projection* of the trial state $\{\boldsymbol{\sigma}_{n+1}^{\text{trial}}, \boldsymbol{q}_n\}$ onto the boundary $\partial\mathbb{E}_\sigma$ of the elastic range in the norm induced by the metric $\mathbf{G}$. Accordingly, $\{\boldsymbol{\sigma}_{n+1}, \boldsymbol{q}_{n+1}\}$ is the argument of the following minimum principle:

$$\boxed{\{\boldsymbol{\sigma}_{n+1}, \boldsymbol{q}_{n+1}\} = \text{ARG}\left[\underset{(\boldsymbol{\sigma}, \boldsymbol{q}) \in \mathbb{E}_\sigma}{\text{MIN}} \left\{ \tfrac{1}{2} \|\boldsymbol{\sigma}_{n+1}^{\text{trial}} - \boldsymbol{\sigma}\|_{\mathbf{C}^{-1}}^2 + \tfrac{1}{2} \|\boldsymbol{q}_n - \boldsymbol{q}\|_{\mathbf{D}^{-1}}^2 \right\}\right],} \tag{3.2.14}$$

where $\|\boldsymbol{q}\|_{\mathbf{D}^{-1}} := \sqrt{\boldsymbol{q} : \mathbf{D}^{-1} : \boldsymbol{q}}$ is the norm induced by $\mathbf{D}^{-1}$. This geometric interpretation, again follows by noting that the Lagrangian functional associated with this constrained problem is expressed as

$$\mathcal{L}(\boldsymbol{\sigma}, \boldsymbol{q}, \gamma) := \tfrac{1}{2} \|\boldsymbol{\sigma}_{n+1}^{\text{trial}} - \boldsymbol{\sigma}\|_{\mathbf{C}^{-1}}^2 + \tfrac{1}{2} \|\boldsymbol{q}_n - \boldsymbol{q}\|_{\mathbf{D}^{-1}}^2 + \Delta\gamma f(\boldsymbol{\sigma}, \boldsymbol{q}). \tag{3.2.15}$$

The associated Kuhn–Tucker optimality conditions are given by

$$\left.\begin{aligned} \left.\frac{\partial \mathcal{L}}{\partial \boldsymbol{\sigma}}\right|_{n+1} &= \mathbf{C}^{-1}:\left[-\boldsymbol{\sigma}_{n+1}^{\text{trial}} + \boldsymbol{\sigma}_{n+1}\right] + \Delta\gamma \left.\frac{\partial f}{\partial \boldsymbol{\sigma}}\right|_{n+1} = \mathbf{0}, \\ \left.\frac{\partial \mathcal{L}}{\partial \boldsymbol{q}}\right|_{n+1} &= \mathbf{D}^{-1}:\left[-\boldsymbol{q}_{n} + \boldsymbol{q}_{n+1}\right] + \Delta\gamma \left.\frac{\partial f}{\partial \boldsymbol{q}}\right|_{n+1} = \mathbf{0}, \\ f(\boldsymbol{\sigma}_{n+1}, \boldsymbol{q}_{n+1}) \le 0, \quad \Delta\gamma &\ge 0, \quad \Delta\gamma f(\boldsymbol{\sigma}_{n+1}, \boldsymbol{q}_{n+1}) = 0, \end{aligned}\right\} \tag{3.2.16}$$

which coincide with equations (3.2.1).

Before addressing the formulation of general solution techniques for the discrete optimization problem (3.2.1), in the following two sections we consider the particular and important case of J_2 flow theory. Historically, the first numerical algorithms for rate-independent plasticity, notably the radial return method of Wilkins [1964], were formulated in the context of J_2 flow theory. Only recently a general methodology has emerged for solving the general case. This is the subject of Section **3.6**.

3.3 Example 3.1. J_2 Plasticity. Nonlinear Isotropic/Kinematic Hardening

The algorithm developed herein is a particular instance of equations (3.1.2) and applies to both the three-dimensional and *plane-strain* cases. The simplicity of the von Mises yield condition — a hypersphere in stress deviator space — enables one to obtain essentially a closed-form solution of equations (3.1.2) resulting in the so-called radial return method, originally proposed by Wilkins [1964]. This simple return-mapping strategy, however, does not hold for the plane-stress situation, as discussed in Example **3.2**.

3.3.1 Radial Return Mapping

We recall from Section **2.3** that the plastic internal variables are $\boldsymbol{q} := \{\alpha, \boldsymbol{\beta}\}$. Further, recall that the *relative stress* is defined as $\boldsymbol{\xi} := \text{dev}[\boldsymbol{\sigma}] - \boldsymbol{\beta}$. Now let $\boldsymbol{n}_{n+1} := \frac{\boldsymbol{\xi}_{n+1}}{\|\boldsymbol{\xi}_{n+1}\|}$ be the unit vector field normal to the Mises yield surface at the *end* of a typical time step $[t_n, t_{n+1}]$. In view of $(2.3.4)_2$, the general equations (3.2.1) take the form

$$\left.\begin{aligned} \boldsymbol{\varepsilon}_{n+1}^{p} &= \boldsymbol{\varepsilon}_{n}^{p} + \Delta\gamma\boldsymbol{n}_{n+1}, \\ \alpha_{n+1} &= \alpha_n + \sqrt{\tfrac{2}{3}}\Delta\gamma, \\ \boldsymbol{\beta}_{n+1} &= \boldsymbol{\beta}_n + \sqrt{\tfrac{2}{3}}\Delta H_{n+1}\boldsymbol{n}_{n+1}, \end{aligned}\right\} \tag{3.3.1}$$

where $\Delta H_{n+1} := H(\alpha_{n+1}) - H(\alpha_n)$. In addition, the trial state becomes

$$\left.\begin{aligned} \boldsymbol{s}_{n+1}^{\text{trial}} &:= \boldsymbol{s}_n + 2\mu\Delta\boldsymbol{e}_{n+1}, \\ \boldsymbol{\xi}_{n+1}^{\text{trial}} &:= \boldsymbol{s}_{n+1}^{\text{trial}} - \boldsymbol{\beta}_n, \end{aligned}\right\} \tag{3.3.2}$$

where $\boldsymbol{e} := \text{dev}[\boldsymbol{\varepsilon}]$ is the strain deviator $\boldsymbol{s} := \text{dev}[\boldsymbol{\sigma}]$ is the stress deviator, and μ is the shear modulus. Next we show that the solution of (3.3.1) reduces to the solution of a scalar equation for the consistency parameter $\Delta\gamma$. In view of of the relationship $\boldsymbol{s}_{n+1} = \boldsymbol{s}_{n+1}^{\text{trial}} - \Delta\gamma 2\mu \boldsymbol{n}_{n+1}$, we see that $\boldsymbol{\xi}_{n+1}$ is expressed in terms of $\boldsymbol{\xi}_{n+1}^{\text{trial}}$ according to the expression

$$\begin{aligned} \boldsymbol{\xi}_{n+1} &:= \boldsymbol{s}_{n+1} - \boldsymbol{\beta}_{n+1} \\ &= \boldsymbol{\xi}_{n+1}^{\text{trial}} - \left[2\mu\Delta\gamma + \sqrt{\tfrac{2}{3}}\Delta H_{n+1}\right]\boldsymbol{n}_{n+1}. \end{aligned} \tag{3.3.3}$$

Next, to arrive at the algorithmic counterpart of the consistency condition (2.3.8), we note that by definition, $\boldsymbol{\xi}_{n+1} = \|\boldsymbol{\xi}_{n+1}\|\boldsymbol{n}_{n+1}$. Hence, from (3.3.3) the unit normal $\boldsymbol{n}_{n+1}$ is determined exclusively in terms of the trial elastic stress $\boldsymbol{\xi}_{n+1}^{\text{trial}}$ as

$$\boxed{\boldsymbol{n}_{n+1} \equiv \boldsymbol{\xi}_{n+1}^{\text{trial}} / \|\boldsymbol{\xi}_{n+1}^{\text{trial}}\|.} \tag{3.3.4}$$

By taking the dot product of (3.3.3) with $\boldsymbol{n}_{n+1}$ and noting that $\|\boldsymbol{\xi}_{n+1}\| - \sqrt{\tfrac{2}{3}}K(\alpha_{n+1}) = 0$, we obtain the following *scalar* (generally nonlinear) equation that determines the consistency parameter $\Delta\gamma$

$$\boxed{\begin{aligned} g(\Delta\gamma) &:= -\sqrt{\tfrac{2}{3}}K(\alpha_{n+1}) + \|\boldsymbol{\xi}_{n+1}^{\text{trial}}\| \\ &\qquad - \left\{2\mu\Delta\gamma + \sqrt{\tfrac{2}{3}}\left[H(\alpha_{n+1}) - H(\alpha_n)\right]\right\} = 0, \\ \alpha_{n+1} &= \alpha_n + \sqrt{\tfrac{2}{3}}\Delta\gamma. \end{aligned}} \tag{3.3.5}$$

Equation (3.3.5) is effectively solved by a *local* Newton iterative procedure since $g(\Delta\gamma)$ is a *convex* function, and then convergence of the Newton procedure is guaranteed. Details of the local Newton procedure are summarized for convenience in BOX **3.1**.

Remark 3.3.1. If the kinematic/isotropic hardening law is *linear* of the form (3.3.6), equation (3.3.5) is amenable to closed-form solution that results in the generalizing the radial return algorithm in Krieg and Key [1976]. Set

$$\boxed{f_{n+1}^{\text{trial}} := \|\boldsymbol{\xi}_{n+1}^{\text{trial}}\| - \sqrt{\tfrac{2}{3}}[\sigma_Y + \beta\bar{H}'\alpha_n],} \tag{3.3.6}$$

where $\beta \in [0, 1]$, $\sigma_Y > 0$ is the flow stress in pure tension, and $\bar{H}' > 0$ is a *given* material parameter that characterizes the hardening response. Substituting (3.3.6) in (3.3.5),

$$\boxed{2\mu\Delta\gamma = \frac{f_{n+1}^{\text{trial}}}{1 + \frac{\bar{H}'}{3\mu}}\ .} \tag{3.3.7}$$

Since $\Delta H = \sqrt{\tfrac{2}{3}}(1-\beta)\bar{H}'\Delta\gamma$, the update procedure is completed by substituting (3.3.7) in formulas (3.3.1).

BOX 3.1. Consistency Condition. Determination of $\Delta\gamma$.

1. Initialize.

$$\Delta\gamma^{(0)} = 0$$

$$\alpha_{n+1}^{(0)} = \alpha_n$$

2. Iterate.

DO UNTIL: $|g(\Delta\gamma^{(k)})| < \text{TOL}$,

$k \leftarrow k + 1$

2.1. Compute iterate $\Delta\gamma^{(k+1)}$:

$$g(\Delta\gamma^{(k)}) := -\sqrt{\tfrac{2}{3}}\, K(\alpha_{n+1}^{(k)}) + \|\boldsymbol{\xi}_{n+1}^{\text{trial}}\| - \left\{2\mu\Delta\gamma^{(k)} + \sqrt{\tfrac{2}{3}}\left[H(\alpha_{n+1}^{(k)}) - H(\alpha_n)\right]\right\}$$

$$Dg(\Delta\gamma^{(k)}) := -2\mu\left\{1 + \frac{H'[\alpha_{n+1}^{(k)}] + K'[\alpha_{n+1}^{(k)}]}{3\mu}\right\}$$

$$\Delta\gamma^{(k+1)} = \Delta\gamma^{(k)} - \frac{g[\Delta\gamma^{(k)}]}{Dg[\Delta\gamma^{(k)}]}$$

2.2. Update equivalent plastic strain.

$$\alpha_{n+1}^{(k+1)} = \alpha_n + \sqrt{\tfrac{2}{3}}\,\Delta\gamma^{(k+1)}$$

For convenience, a step-by-step description of the algorithm discussed above is summarized in BOX **3.2** below. The geometric interpretation of the algorithm is contained in Figure **3.2**. Next, we obtain *consistent* elastoplastic tangent moduli by linearizing the two-step, return-mapping algorithm. These moduli relate incremental strains and incremental stresses and play a crucial role in the overall solution strategy of a boundary-value problem. Their significance becomes apparent in Chapter **5**, where the variational structure of plasticity and its numerical implementation are discussed in detail.

3.3.2 Exact Linearization of the Algorithm

By differentiating the *algorithmic* expression $\boldsymbol{\sigma}_{n+1} = \kappa(\text{tr}\,[\boldsymbol{\varepsilon}_{n+1}])\mathbf{1} + 2\mu(\boldsymbol{e}_{n+1} - \Delta\gamma\boldsymbol{n}_{n+1})$ for the stress tensor, one obtains

$$\begin{aligned} d\boldsymbol{\sigma}_{n+1} &= \mathbf{C} : d\boldsymbol{\varepsilon}_{n+1} - 2\mu[d\Delta\gamma\boldsymbol{n}_{n+1} + \Delta\gamma d\boldsymbol{n}_{n+1}] \\ &= \left[\mathbf{C} - 2\mu\boldsymbol{n}_{n+1} \otimes \frac{\partial\Delta\gamma}{\partial\boldsymbol{\varepsilon}_{n+1}} - 2\mu\Delta\gamma\frac{\partial\boldsymbol{n}_{n+1}}{\partial\boldsymbol{\varepsilon}_{n+1}}\right] : d\boldsymbol{\varepsilon}_{n+1}\,, \end{aligned} \qquad (3.3.8)$$

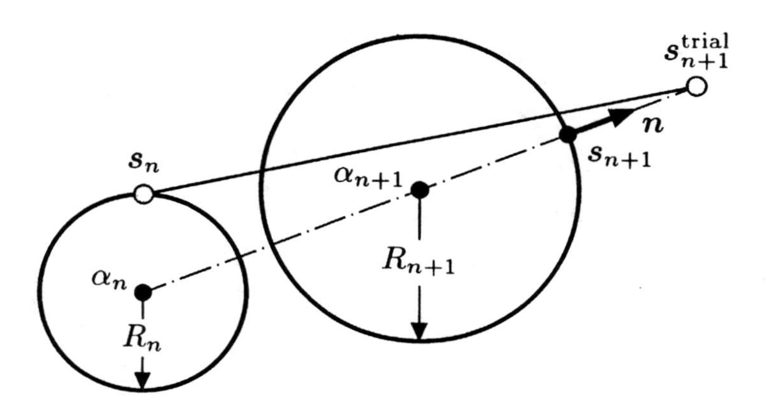

FIGURE 3-2. Geometric interpretation of the return-mapping algorithm for the von Mises yield condition and isotropic/kinematic hardening.

where $\mathbf{C} := \kappa \mathbf{1} \otimes \mathbf{1} + 2\mu(\mathbf{I} - \frac{1}{3}\mathbf{1} \otimes \mathbf{1})$ is the elasticity tensor. To carry out the computation further, the following result is used.

Lemma 3.2. *The derivative of the unit normal field* $\boldsymbol{n}(\boldsymbol{\xi}) := \frac{\boldsymbol{\xi}}{\|\boldsymbol{\xi}\|}$ *is given by the formula*

$$\boxed{\frac{\partial \boldsymbol{n}}{\partial \boldsymbol{\xi}} = \frac{1}{\|\boldsymbol{\xi}\|}[\mathbf{I} - \boldsymbol{n} \otimes \boldsymbol{n}].} \tag{3.3.9}$$

PROOF. The result easily follows with the aid of the directional derivative. First we note that, for an arbitrary vector $\boldsymbol{h} \in \mathbb{R}^6$,

$$\left.\frac{d}{d\zeta}\right|_{\zeta=0} \|\boldsymbol{\xi} + \zeta \boldsymbol{h}\| = \frac{\boldsymbol{\xi} : \boldsymbol{h}}{\|\boldsymbol{\xi}\|} \equiv \boldsymbol{n} : \boldsymbol{h}. \tag{3.3.10}$$

Then by the chain rule it follows that

$$\left.\frac{d}{d\zeta}\boldsymbol{n}(\boldsymbol{\xi} + \zeta \boldsymbol{h})\right|_{\zeta=0} = \frac{\boldsymbol{h} - (\boldsymbol{n} : \boldsymbol{h})\boldsymbol{n}}{\|\boldsymbol{\xi}\|} \equiv \left[\frac{\mathbf{I} - \boldsymbol{n} \otimes \boldsymbol{n}}{\|\boldsymbol{\xi}\|}\right] : \boldsymbol{h}, \tag{3.3.11}$$

so that (3.3.9) holds. □

As in the general case, the term $\partial \Delta\gamma / \partial \boldsymbol{\varepsilon}_{n+1}$ in (3.3.8) is obtained by differentiating the *scalar* consistency condition $(3.3.5)_1$. Accordingly,

$$\frac{\partial \Delta\gamma}{\partial \boldsymbol{\varepsilon}_{n+1}} = \left[1 + \frac{K'(\alpha_{n+1}) + H'(\alpha_{n+1})}{3\mu}\right]^{-1} \boldsymbol{n}_{n+1}. \tag{3.3.12}$$

Substituting (3.3.9) and (3.3.12) in (3.3.8), after some manipulation, produces the expression summarized in BOX **3.2**.

BOX 3.2. Radial Return Algorithm.
Nonlinear Isotropic/Kinematic Hardening.

1. Compute *trial elastic* stress.

$$\boldsymbol{e}_{n+1} = \boldsymbol{\varepsilon}_{n+1} - \tfrac{1}{3}(\mathrm{tr}[\boldsymbol{\varepsilon}_{n+1}])\mathbf{1}$$

$$\boldsymbol{s}^{\mathrm{trial}}_{n+1} = 2\mu(\boldsymbol{e}_{n+1} - \boldsymbol{e}^{p}_{n})$$

$$\boldsymbol{\xi}^{\mathrm{trial}}_{n+1} = \boldsymbol{s}^{\mathrm{trial}}_{n+1} - \boldsymbol{\beta}_{n}$$

2. Check yield condition

$$f^{\mathrm{trial}}_{n+1} := \|\boldsymbol{\xi}^{\mathrm{trial}}_{n+1}\| - \sqrt{\tfrac{2}{3}}\,K(\alpha_n)$$

IF $f^{\mathrm{trial}}_{n+1} \leq 0$ THEN:

Set $(\bullet)_{n+1} = (\bullet)^{\mathrm{trial}}_{n+1}$ & *EXIT.*

ENDIF.

3. Compute $\boldsymbol{n}_{n+1}$ and find $\Delta\gamma$ from BOX **3.1**. Set

$$\boldsymbol{n}_{n+1} := \frac{\boldsymbol{\xi}^{\mathrm{trial}}_{n+1}}{\|\boldsymbol{\xi}^{\mathrm{trial}}_{n+1}\|},$$

$$\alpha_{n+1} := \alpha_n + \sqrt{\tfrac{2}{3}}\,\Delta\gamma$$

4. Update back stress, plastic strain and stress

$$\boldsymbol{\beta}_{n+1} = \boldsymbol{\beta}_n + \sqrt{\tfrac{2}{3}}\,\left[H(\alpha_{n+1}) - H(\alpha_n)\right]\boldsymbol{n}_{n+1}$$

$$\boldsymbol{e}^{p}_{n+1} = \boldsymbol{e}^{p}_{n} + \Delta\gamma\,\boldsymbol{n}_{n+1}$$

$$\boldsymbol{\sigma}_{n+1} = \kappa\,\mathrm{tr}[\boldsymbol{\varepsilon}_{n+1}]\mathbf{1} + \boldsymbol{s}^{\mathrm{trial}}_{n+1} - 2\mu\Delta\gamma\,\boldsymbol{n}_{n+1}$$

5. Compute *consistent elastoplastic tangent moduli*

$$\mathbf{C}_{n+1} = \kappa\mathbf{1}\otimes\mathbf{1} + 2\mu\theta_{n+1}\left[\mathbf{I} - \tfrac{1}{3}\mathbf{1}\otimes\mathbf{1}\right] - 2\mu\bar{\theta}_{n+1}\boldsymbol{n}_{n+1}\otimes\boldsymbol{n}_{n+1}$$

$$\theta_{n+1} := 1 - \frac{2\mu\Delta\gamma}{\|\boldsymbol{\xi}^{\mathrm{trial}}_{n+1}\|}$$

$$\bar{\theta}_{n+1} := \frac{1}{1 + \frac{[K' + H']_{n+1}}{3\mu}} - (1 - \theta_{n+1})$$

Remark 3.3.2.

1. For the case of perfect plasticity, $\boldsymbol{\beta} \equiv \mathbf{0}$, $\sqrt{\tfrac{2}{3}}\sigma_Y = R \equiv$ constant, and the algorithm summarized in BOX **3.2** reduces to the classical radial return of Wilkins

[1964]. The case of linear isotropic/kinematic hardening rules corresponds to the extension of the radial return algorithm proposed by Krieg and Key [1976]. The extension to nonlinear hardening rules and the development of the consistent tangent moduli were considered in Simo and Taylor [1985]. The origin of the notion of consistent tangent moduli is found in Hughes and Taylor [1978] Nagtegaal [1982].

2. The backward Euler method can be replaced by the generalized midpoint rule in the derivation of the discrete equations (3.2.1), as in Ortiz and Popov [1985] or Simo and Taylor [1986]. For J_2 flow theory this results in the return map proposed in Rice and Tracey [1973] which is second-order accurate. However, the overall superiority of the radial return method relative to other return schemes is conclusively established in Krieg and Krieg [1977]; Schreyer, Kulak, and Kramer [1979]; and Yoder and Whirley [1984]. The same conclusions hold for other plasticity models, see Loret and Prevost [1986] and Ortiz and Popov [1985].
3. Note that, according to the algorithm in BOX **3.1**, the values $(\bullet)^{i}_{n+1}$ are calculated *based solely on the converged values* $(\bullet)_n$ at the beginning of the time step $t = t_n$. The (nonconverged) values $(\bullet)^{(k)}_{n+1}$ at the previous iteration play no explicit role in the stress update. In fact, if the elastic trial stress $\mathbf{s}^{\text{trial}^{(k+1)}}_{n+1}$ at the $(k+1)th$ iteration were computed from the *nonconverged* stresses $\mathbf{s}^{(k)}_{n+1}$ at the previous iteration (rather than from *converged* stress $\mathbf{s}_n$ as in BOX **3.2**), then the "continuum" elastoplastic tangent moduli (2.3.0) is the *consistent tangent* for this particular algorithm. However, use of an iterative scheme based on intermediate nonconverged values is questionable for a problem which is physically path-dependent. In addition, if "unloading" within the iterative process occurs, a new iteration is necessary starting from the converged stresses $\mathbf{s}_n$.
4. The expression for the consistent tangent moduli in BOX **3.2** should be compared with equation (2.3.0) for the "continuum" elastoplastic tangent. As a result of the radial return algorithm, the shear modulus μ enters into the "consistent" tangent moduli scaled by the factor θ_{n+1}. Observe that $\theta_{n+1} \leq 1$ and that, for large time steps, $\mathbf{s}^{\text{trial}}_{n+1}$ may be far outside the yield surface so that θ_{n+1} is significantly less than unity. In addition, since $\bar{\theta}_{n+1} = \Delta\gamma + \theta_{n+1} - 1$, the bound $\Delta\gamma - 1 < \bar{\theta}_{n+1} \leq \Delta\gamma$. Therefore, for large time steps, the consistent tangent moduli may differ significantly from the "continuum" elastoplastic tangent (2.3.0). Note, however, that as $\Delta t \to 0$, $\Delta\gamma \to 0$, and the consistent and continuum tangent moduli coincide. This result is a manifestation of the *consistency* between the algorithm and the continuum problem.

3.4 Example 3.2. Plane-Stress J_2 Plasticity. Kinematic/Isotropic Hardening

For the plane-stress case, a simple radial return violates the plane-stress condition and thus is no longer applicable. The basic idea in the algorithm discussed below, proposed in Simo and Taylor [1986], is to perform the return mapping directly in the constrained plane-stress subspace by using equations (2.4.10). Thus, by construction, the plane stress condition is identically satisfied. From a computational standpoint, the proposed procedure is a particular instance of the general closest point projection iteration, discussed in Section **3.6**, for solving the discrete equations (3.2.1). It is seen that for plane stress, the J_2 flow theory solution of (3.2.1) reduces to the iterative solution of a simple nonlinear equation of the form (3.3.5). In addition, the exact linearization of the algorithm is obtained, leading to a closed-form expression for the consistent tangent moduli.

3.4.1 Return-Mapping Algorithm

For the case at hand, starting from equations (2.4.10), a backward-Euler difference scheme yields the following approximation to the plastic return mapping:

$$\left.\begin{aligned} \boldsymbol{\varepsilon}^p_{n+1} &= \boldsymbol{\varepsilon}^p_n + \Delta\gamma \mathbf{P}\boldsymbol{\xi}_{n+1}, \\ \alpha_{n+1} &= \alpha_n + \Delta\gamma\sqrt{\tfrac{2}{3}}\bar{f}_{n+1}, \\ \tilde{\boldsymbol{\beta}}_{n+1} &= \tilde{\boldsymbol{\beta}}_n + \tfrac{2}{3}\Delta\gamma H'\boldsymbol{\xi}_{n+1}, \end{aligned}\right\} \tag{3.4.1}$$

where $\boldsymbol{\xi}_{n+1}$ and $\bar{f}_{n+1}$ are defined as

$$\left.\begin{aligned} \boldsymbol{\xi}_{n+1} &:= \boldsymbol{\sigma}_{n+1} - \tilde{\boldsymbol{\beta}}_{n+1}, \\ \bar{f}_{n+1} &:= \sqrt{\boldsymbol{\xi}^T_{n+1}\mathbf{P}\boldsymbol{\xi}_{n+1}}\,. \end{aligned}\right\} \tag{3.4.2}$$

Using the elastic stress-strain relationships, $\boldsymbol{\sigma}_{n+1} = \mathbf{C}[\boldsymbol{\varepsilon}_{n+1} - \boldsymbol{\varepsilon}^p_{n+1}] \equiv \boldsymbol{\sigma}^{\text{trial}}_{n+1} - \mathbf{C}\Delta\boldsymbol{\varepsilon}^p_{n+1}$ yields the following sequential update procedure:

$$\left.\begin{aligned} \boldsymbol{\varepsilon}_{n+1} &= \boldsymbol{\varepsilon}_n + \nabla^s \boldsymbol{u}, \\ \boldsymbol{\sigma}^{\text{trial}}_{n+1} &= \mathbf{C}[\boldsymbol{\varepsilon}_{n+1} - \boldsymbol{\varepsilon}^p_n], \\ \boldsymbol{\xi}^{\text{trial}}_{n+1} &= \boldsymbol{\sigma}^{\text{trial}}_{n+1} - \tilde{\boldsymbol{\beta}}_n, \\ \boldsymbol{\xi}_{n+1} &= \frac{1}{1 + \tfrac{2}{3}\Delta\gamma H'}\,\boldsymbol{\Xi}(\Delta\gamma)\mathbf{C}^{-1}\boldsymbol{\xi}^{\text{trial}}_{n+1}, \\ \tilde{\boldsymbol{\beta}}_{n+1} &= \tilde{\boldsymbol{\beta}}_n + \Delta\gamma\,\tfrac{2}{3}\,H'\boldsymbol{\xi}_{n+1}, \\ \boldsymbol{\sigma}_{n+1} &= \boldsymbol{\xi}_{n+1} + \tilde{\boldsymbol{\beta}}_{n+1}, \\ \alpha_{n+1} &= \alpha_n + \sqrt{\tfrac{2}{3}}\Delta\gamma\,\bar{f}_{n+1}. \end{aligned}\right\} \tag{3.4.3}$$

Here, $\boldsymbol{\Xi}(\Delta\gamma)$ plays the role of a modified (*algorithmic*) elastic tangent matrix and is defined as

$$\boxed{\boldsymbol{\Xi}(\Delta\gamma) := \left[\mathbf{C}^{-1} + \frac{\Delta\gamma}{1 + \tfrac{2}{3}H'\Delta\gamma}\mathbf{P}\right]^{-1}}\,. \tag{3.4.4}$$

The update formulas (3.4.3) depend parametrically on the plastic Lagrange multiplier $\Delta\gamma$ which is determined by enforcing the consistency condition *at time* t_{n+1}, i.e.,

$$\boxed{f^2(\Delta\gamma) := \tfrac{1}{2}\bar{f}_{n+1}^2 - \tfrac{1}{3}\left[K(\alpha_n + \sqrt{\tfrac{2}{3}}\Delta\gamma\bar{f}_{n+1})\right]^2 = 0\,,} \tag{3.4.5}$$

where $\bar{f}_{n+1}$ is defined by $(3.4.2)_2$ with $\boldsymbol{\xi} = \boldsymbol{\xi}_{n+1}$. Note that $\boldsymbol{\xi}_{n+1}$ is a nonlinear function of $\Delta\gamma$ as defined by $(3.4.3)_4$ and (3.4.4). Therefore, the discrete consistency condition (3.4.5) furnishes a nonlinear scalar equation which is to be solved for $\Delta\gamma$. For the case of isotropic elasticity, condition (3.4.5) has a particularly simple form because of the structure of matrices $\mathbf{P}$ and $\mathbf{C}$, and is easily solved by elementary methods, as shown below.

3.4.2 Consistent Elastoplastic Tangent Moduli

Tangent moduli consistent with the integration algorithm are developed by linearizing the algorithm (3.4.3). Although in the limit, as the step size $h \to 0$, one recovers the classical elastoplastic moduli defined by (2.4.16), for finite values of h, use of the consistent tangent moduli is essential to preserve the quadratic rate of asymptotic convergence that characterizes Newton's method. The methodology parallels that followed in the continuum problem. By differentiating the algorithm, we obtain the formulas

$$\left.\begin{aligned} d\boldsymbol{\sigma}_{n+1} &= \mathbf{C}\left[d\boldsymbol{\varepsilon}_{n+1} - d\Delta\gamma\mathbf{P}\boldsymbol{\xi}_{n+1} - \Delta\gamma\mathbf{P}(d\boldsymbol{\sigma}_{n+1} - d\bar{\boldsymbol{\beta}}_{n+1})\right], \\ d\alpha_{n+1} &= \sqrt{\tfrac{2}{3}}\left[\bar{f}_{n+1}d\Delta\gamma + \Delta\gamma d\bar{f}_{n+1}\right], \\ d\bar{\boldsymbol{\beta}}_{n+1} &= \frac{\tfrac{2}{3}H'}{1 + \tfrac{2}{3}\Delta\gamma H'}\left(d\Delta\gamma\boldsymbol{\xi}_{n+1} + \Delta\gamma d\boldsymbol{\sigma}_{n+1}\right), \end{aligned}\right\} \tag{3.4.6}$$

where $\bar{f}_{n+1}$ is defined by $(3.4.2)_2$ and $d\bar{f}_{n+1}$ is computed by differentiating this expression. From $(3.4.6)_1$ and $(3.4.6)_3$, it follows that

$$\left.\begin{aligned} d\boldsymbol{\sigma}_{n+1} &= \boldsymbol{\Xi}(\Delta\gamma)\left[d\boldsymbol{\varepsilon}_{n+1} - \frac{d\Delta\gamma}{1 + \tfrac{2}{3}\Delta\gamma H'}\mathbf{P}\boldsymbol{\xi}_{n+1}\right], \\ d\boldsymbol{\xi} &= \frac{1}{1 + \tfrac{2}{3}H'\Delta\gamma}\left[d\boldsymbol{\sigma}_{n+1} - \tfrac{2}{3}H'\boldsymbol{\xi}_{n+1}d\Delta\gamma\right]. \end{aligned}\right\} \tag{3.4.7}$$

Differentiating the consistency condition (3.4.5) at t_{n+1} by using $(3.4.2)_2$ yields the consistency equation

$$df_{n+1} = 0 \Rightarrow \left(1 - \tfrac{2}{3}K'_{n+1}\Delta\gamma\right)\boldsymbol{\xi}_{n+1}^T\mathbf{P}d\boldsymbol{\xi}_{n+1} - \tfrac{2}{3}K'_{n+1}\bar{f}_{n+1}^2 d\Delta\gamma = 0. \tag{3.4.8}$$

By using $(3.4.6)_3$ and (3.4.7), we solve (3.4.8) for $d\Delta\gamma$ to obtain the following expression:

$$\boxed{d\Delta\gamma = \frac{\theta_1}{(1 + \beta_{n+1})}\frac{\boldsymbol{\xi}_{n+1}^T\mathbf{P}\boldsymbol{\Xi}d\boldsymbol{\varepsilon}_{n+1}}{\boldsymbol{\xi}_{n+1}^T\mathbf{P}\boldsymbol{\Xi}\mathbf{P}\boldsymbol{\xi}_{n+1}}\,,} \tag{3.4.9}$$

where

$$\left.\begin{aligned} \theta_1 &:= 1 + \tfrac{2}{3} H' \Delta\gamma\,, \\ \theta_2 &:= 1 - \tfrac{2}{3} K'_{n+1} \Delta\gamma, \\ \beta_{n+1} &:= \tfrac{2}{3} \frac{\theta_1}{\theta_2} \bar{f}^2_{n+1} \frac{\left[K'_{n+1}\theta_1 + H'\theta_2\right]}{\boldsymbol{\xi}^T_{n+1} \mathbf{P} \boldsymbol{\Xi} \mathbf{P} \boldsymbol{\xi}_{n+1}}\,. \end{aligned}\right\} \tag{3.4.10}$$

Finally, from (3.4.9), $(3.4.10)_1$, and $(3.4.7)_1$, we obtain the expression for the consistent elastoplastic tangent matrix as

$$\boxed{\left.\frac{d\boldsymbol{\sigma}}{d\boldsymbol{\varepsilon}}\right|_{n+1} = \boldsymbol{\Xi} - \frac{\boldsymbol{N}_{n+1} \otimes \boldsymbol{N}_{n+1}}{1 + \beta_{n+1}}\,,} \tag{3.4.11}$$

where we have set

$$\boxed{\boldsymbol{N}_{n+1} := \frac{\boldsymbol{\Xi} \mathbf{P} \boldsymbol{\xi}_{n+1}}{\sqrt{\boldsymbol{\xi}^T_{n+1} \mathbf{P} \boldsymbol{\Xi} \mathbf{P} \boldsymbol{\xi}_{n+1}}}.} \tag{3.4.12}$$

Remark 3.4.1. Observe that, as the time step $h \to 0$, $\Delta\gamma \to 0$. From expressions $(3.4.9)_2$ and $(3.4.10)_2$, it follows that $\theta_1 \to 1$ and $\theta_2 \to 1$, as $h \to 0$. Hence

$$h \to 0 \;\Rightarrow\; \boldsymbol{\Xi}(\Delta\gamma) \to \mathbf{C} \text{ and } \beta_{n+1} \to \beta, \tag{3.4.13}$$

where β is given by (2.4.15). Therefore, the "consistent" elastoplastic moduli (3.4.11) reduce to the classical elastoplastic moduli given by (2.4.16), as $h \to 0$. This shows that algorithm (3.4.3) is *consistent* with problem (2.4.10).

3.4.3 Implementation

For isotropic elastic response, implementation of the algorithm discussed above takes a remarkably simple form. Employing the same notation as in Section **2.4**, we define

$$\boldsymbol{\eta} := \boldsymbol{Q}^T \boldsymbol{\xi} \equiv \left[\frac{\xi_{11} + \xi_{22}}{\sqrt{2}} \quad \frac{-\xi_{11} + \xi_{22}}{\sqrt{2}} \quad \xi_{12}\right]^T, \tag{3.4.14}$$

where $\boldsymbol{Q}$ is given by $(2.4.12\text{b})_1$. In addition, we define an *elastic trial* state given by $\boldsymbol{\sigma}^{\text{trial}}_{n+1}$, $\boldsymbol{\xi}^{\text{trial}}_{n+1}$ and $\boldsymbol{\eta}^{\text{trial}}_{n+1}$, by setting

$$\left.\begin{aligned} \boldsymbol{\sigma}^{\text{trial}}_{n+1} &:= \mathbf{C}[\boldsymbol{\varepsilon}_{n+1} - \boldsymbol{\varepsilon}^p_n], \\ \boldsymbol{\xi}^{\text{trial}}_{n+1} &:= \boldsymbol{\sigma}^{\text{trial}}_{n+1} - \tilde{\boldsymbol{\beta}}_n, \\ \boldsymbol{\eta}^{\text{trial}}_{n+1} &:= \boldsymbol{Q}^T \boldsymbol{\xi}^{\text{trial}}_{n+1}. \end{aligned}\right\} \tag{3.4.15}$$

Using relationships (2.4.12b,c) the basic update formula $(3.4.3)_2$ takes the form

$$\boldsymbol{\xi}_{n+1} = \left[(1 + \tfrac{2}{3}\Delta\gamma H')\mathbf{I} + \Delta\gamma \boldsymbol{\Lambda}_{\mathbf{P}} \boldsymbol{\Lambda}_{\mathbf{C}}\right]^{-1} \boldsymbol{\xi}^{\text{trial}}_{n+1} =: \boldsymbol{\Gamma}(\Delta\gamma) \boldsymbol{\xi}^{\text{trial}}_{n+1}, \tag{3.4.16}$$

where $\boldsymbol{\Gamma}(\Delta\gamma)$ is a diagonal matrix given by

$$\boldsymbol{\Gamma}(\Delta\gamma) := \text{DIAG}\left[\frac{1}{1+\left(\frac{E}{3(1-\nu)}+\frac{2}{3}H'\right)\Delta\gamma}, \frac{1}{1+\left(2\mu+\frac{2}{3}H'\right)\Delta\gamma}, \frac{1}{1+\left(2\mu+\frac{2}{3}H'\right)\Delta\gamma}\right] \tag{3.4.17}$$

In terms of the η variables, the consistency condition (3.4.5) takes a simple form. For convenience, we set

$$\begin{aligned} \bar{f}^2(\Delta\gamma) &:= \frac{\frac{1}{3}(\eta_{11}^{\text{trial}})^2}{\left[1+\left(\frac{E}{3(1-\nu)}+\frac{2}{3}H'\right)\Delta\gamma\right]^2} + \frac{(\eta_{22}^{\text{trial}})^2+2(\eta_{12}^{\text{trial}})^2}{\left[1+\left(2\mu+\frac{2}{3}H'\right)\Delta\gamma\right]^2}, \\ R^2(\Delta\gamma) &:= \frac{1}{3}K^2\left[\alpha_n+\sqrt{\tfrac{2}{3}}\Delta\gamma\bar{f}(\Delta\gamma)\right], \end{aligned} \tag{3.4.18}$$

where $\sqrt{2}R(\Delta\gamma)$ is the radius of the yield surface defined in terms of the hardening rule $(2.4.10)_5$. With this notation at hand, now equation (3.4.5) reads

$$f^2(\Delta\gamma) = \frac{1}{2}\bar{f}^2(\Delta\gamma) - R^2(\Delta\gamma), \quad \Delta\gamma \geq 0. \tag{3.4.19}$$

It is readily shown that the function $\bar{f}^2(\Delta\gamma)$ *monotonically* decreases for $\Delta\gamma \in [0, \infty)$ and further that

$$\lim_{\Delta\gamma\to\infty} \bar{f}^2(\Delta\gamma) \equiv \lim_{\Delta\gamma\to\infty} \frac{d}{d\Delta\gamma}\bar{f}^2(\Delta\gamma) = 0. \tag{3.4.20}$$

Thus, for the physically meaningful case of a monotonically increasing hardening law, (3.4.19) has a *unique* solution $\Delta\gamma \geq 0$. In particular, linear and saturation laws of the exponential type are often used, i.e.,

$$K(\alpha) = \sigma_Y + \bar{K}\alpha + (\bar{K}_\infty - \bar{K}_0)[1 - \exp(-\delta\alpha)]. \tag{3.4.21}$$

Here, $\sigma_Y > 0$, $\bar{K} > 0$, $\bar{K}_\infty > \bar{K}_0$, and $\delta > 0$ are material constants.

Remark 3.4.2. Equation (3.4.19) is ideally suited for a local iterative solution procedure employing Newton's method. Note that in most realistic applications for which the hardening law is nonlinear, such as (3.4.21), a local iterative solution is always necessary. As shown in Example **3.1**, this is the case even for plane strain with the von Mises yield condition; see BOX **3.1**. Thus, the additional effort required to solve (3.4.19) because of the presence of $\bar{f}^2(\Delta\gamma)$ is negligible.

A step-by-step implementation of the algorithm discussed above is summarized for convenience in BOX **3.3**.

BOX 3.3. Return Mapping Algorithm for Plane Stress.

1. Update strain tensor. Compute trial elastic stresses

$$\boldsymbol{\varepsilon}_{n+1} = \boldsymbol{\varepsilon}_n + \nabla^s \boldsymbol{u}$$

$$\boldsymbol{\sigma}^{\text{trial}} = \mathbf{C}[\boldsymbol{\varepsilon}_{n+1} - \boldsymbol{\varepsilon}_n^p]$$

$$\boldsymbol{\xi}^{\text{trial}} = \boldsymbol{\sigma}^{\text{trial}} - \tilde{\boldsymbol{\beta}}_n$$

2. IF $f_{n+1}^{\text{trial}} < 0$ THEN: *EXIT.*

ELSE: Solve $f(\Delta\gamma) = 0$ for $\Delta\gamma$ (Consistency)

$$f^2(\Delta\gamma) := \tfrac{1}{2}\bar{f}^2(\Delta\gamma) - R^2(\Delta\gamma) \equiv 0$$

$$\bar{f}^2(\Delta\gamma) := \frac{1}{2}\frac{\frac{1}{3}(\xi_{11}^{\text{trial}} + \xi_{22}^{\text{trial}})^2}{\left\{1 + \left(\frac{E}{3(1-\nu)} + \frac{2}{3}H'\right)\Delta\gamma\right\}^2} + \frac{\frac{1}{2}(\xi_{11}^{\text{trial}} - \xi_{22}^{\text{trial}})^2 + 2(\eta_{12}^{\text{trial}})^2}{\left[1 + \left(2\mu + \frac{2}{3}H'\right)\Delta\gamma\right]^2}$$

$$R^2(\Delta\gamma) := \tfrac{1}{3}K^2\left[\alpha_n + \sqrt{\tfrac{2}{3}}\Delta\gamma\bar{f}(\Delta\gamma)\right]$$

3. Compute modified (*algorithmic*) elastic tangent moduli

$$\boldsymbol{\Xi} := \left[\mathbf{C}^{-1} + \frac{\Delta\gamma}{1 + \frac{2}{3}\Delta\gamma H'}\mathbf{P}\right]^{-1}$$

4. Update stress, plastic strain, back-stress and equivalent strain

$$\boldsymbol{\xi}_{n+1} = \frac{1}{1 + \frac{2}{3}\Delta\gamma H'}\boldsymbol{\Xi}(\Delta\gamma)\mathbf{C}^{-1}\boldsymbol{\xi}^{\text{trial}}$$

$$\tilde{\boldsymbol{\beta}}_{n+1} = \tilde{\boldsymbol{\beta}}_n + \Delta\gamma\tfrac{2}{3}H'\boldsymbol{\xi}_{n+1}$$

$$\boldsymbol{\sigma}_{n+1} = \boldsymbol{\xi}_{n+1} + \tilde{\boldsymbol{\beta}}_{n+1}$$

$$\alpha_{n+1} = \alpha_n + \sqrt{\tfrac{2}{3}}\Delta\gamma\bar{f}(\Delta\gamma)$$

$$\boldsymbol{\varepsilon}_{n+1}^p = \boldsymbol{\varepsilon}_n^p + \Delta\gamma\mathbf{P}\boldsymbol{\xi}_{n+1}$$

5. Compute *consistent* elastoplastic tangent moduli

$$\left.\frac{d\boldsymbol{\sigma}}{d\boldsymbol{\varepsilon}}\right|_{n+1} = \boldsymbol{\Xi} - \frac{[\boldsymbol{\Xi}\mathbf{P}\boldsymbol{\xi}_{n+1}][\boldsymbol{\Xi}\mathbf{P}\boldsymbol{\xi}_{n+1}]^T}{\boldsymbol{\xi}_{n+1}^T\mathbf{P}\boldsymbol{\Xi}\mathbf{P}\boldsymbol{\xi}_{n+1} + \bar{\beta}_{n+1}}$$

$$\theta_1 := 1 + \tfrac{2}{3}H'\Delta\gamma \quad \theta_2 := 1 - \tfrac{2}{3}K'_{n+1}\Delta\gamma$$

$$\bar{\beta}_{n+1} := \frac{2}{3}\frac{\theta_1}{\theta_2}(K'_{n+1}\theta_1 + H'\theta_2)\boldsymbol{\xi}_{n+1}^T\mathbf{P}\boldsymbol{\xi}_{n+1}$$

6. Update ε_{33} strain

$$\varepsilon_{33_{n+1}} = \frac{-\nu}{E}(\sigma_{11_{n+1}} + \sigma_{22_{n+1}}) - (\varepsilon_{11_{n+1}}^p + \varepsilon_{22_{n+1}}^p)$$

ENDIF

3.4.4 Accuracy Assessment. Isoerror Maps

Next, attention is focused on assessing the accuracy of the proposed algorithm by numerical testing. For this purpose, isoerror maps are often developed based on a strain-controlled homogeneous problem. The procedure is employed by a number of authors, e.g., Krieg and Krieg [1977]; Schreyer, Kulak, and Kramer [1979]; Iwan and Yoder [1983]; Ortiz and Popov [1985]; Ortiz and Simo [1986]; and Simo and Taylor [1986]. Although this technique usefully assesses the overall accuracy of the algorithm it should not be regarded as a replacement of a rigorous accuracy and stability analysis. In the present context, we restrict our discussion to an outline of constructing isoerror maps. (See also Schreyer, Kulak and Kramer [1979].)

Three points on the yield surface are selected which represent a wide range of possible states of stress. These points, labeled A, B, and C in Figure 3.3, correspond to uniaxial, biaxial, and pure shear stress, respectively. To construct the isoerror maps, for each selected point on the yield surface we consider a sequence of specified normalized strain increments. Then the stresses, corresponding to the (homogeneous) states of strain prescribed in this manner, are computed by applying the algorithm. At each point the normalization parameters are chosen as the elastic strains associated with initial yielding. Without loss of generality, the calculation is performed in terms of principal values of the strain and stress tensors, i.e., it is assumed that $\varepsilon_{12} = 0$. Results are reported as the relative root mean square of the error between the exact and computed solution, which is obtained according to the

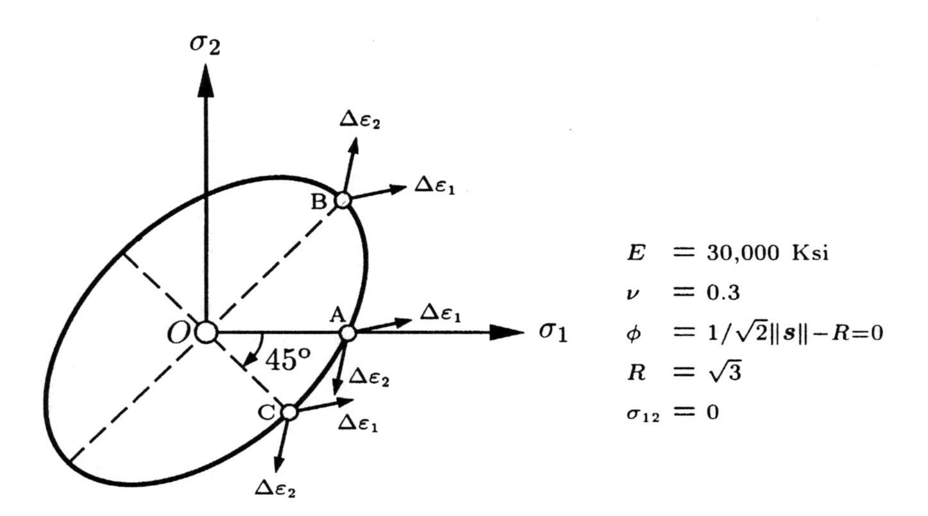

FIGURE 3-3. Plane-stress yield surface. Points for isoerror maps.

expression

$$\boxed{\delta := \frac{\sqrt{(\boldsymbol{\sigma} - \boldsymbol{\sigma}^*) : (\boldsymbol{\sigma} - \boldsymbol{\sigma}^*)}}{\sqrt{\boldsymbol{\sigma}^* : \boldsymbol{\sigma}^*}} \times 100 .} \tag{3.4.22}$$

Here, $\boldsymbol{\sigma}$ is the result obtained by applying the algorithm, whereas $\boldsymbol{\sigma}^*$ is the exact solution corresponding to the specified strain increment. The exact solution for any given strain increment is obtained by repeatedly applying the algorithm with increasing numbers of subincrements. The value for which further sub-incrementing produces no change in the numerical result is taken as the exact solution.

The isoerror maps corresponding to points A, B, and C are shown in Figures **3.4** through **3.6**. The values reported here were obtained for a von Mises yield condition with no hardening and a Poisson's ratio of 0.3. Observe that Figures **3.5** and **3.6** exhibit a symmetry which may be expected from the location of points B and C on the yield surface. From these results, it may be concluded that the level of error observed is roughly equivalent to that previously reported in the literature for other return-mapping algorithms. As a rule, good accuracy (within 5 percent) is obtained for moderate strain increments of the order of the characteristic yield

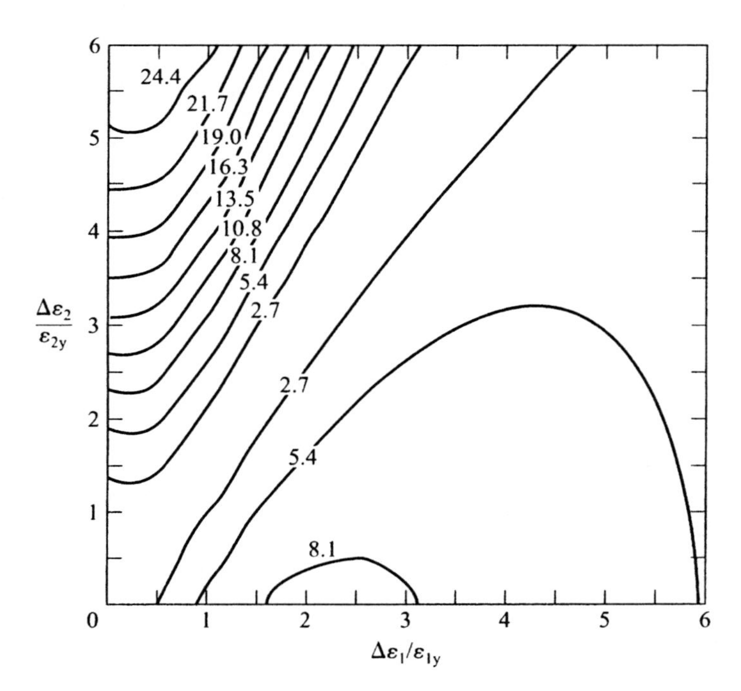

FIGURE 3-4. Isoerror map corresponding to point A on the yield surface.

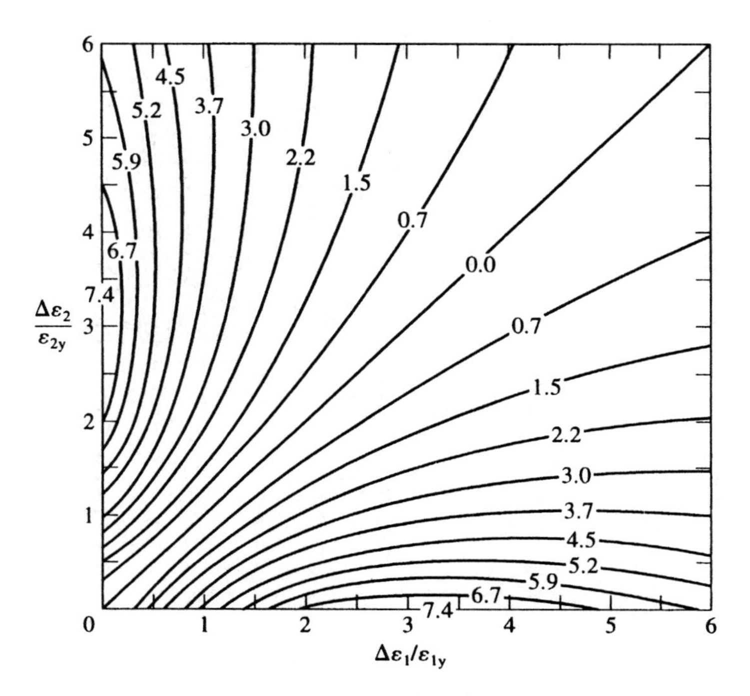

FIGURE 3-5. Isoerror map corresponding to point B on the yield surface.

strains. It is also noted that exact results for any strain increment are obtained for radial loading along both symmetry axes, as expected.

3.4.5 Closed-Form Exact Solution of the Consistency Equation

Simo and Govindjee [1988] noted that for *linear* kinematic hardening and *certain* forms of isotropic hardening the discrete consistency equation (3.4.19) reduces to a *quartic equation* that can be solved in closed form. Because of the practical importance of the resulting algorithm in large scale computations we discuss below details pertaining to this exact solution. In particular, we show the following:

a. There is *one and only one* positive root of the discrete consistency equation. As shown below, this follows at once by inspecting an appropriate graphical interpretation.
b. The *unique positive root* is determined *directly* by a modified version of a classical solution procedure for quartic equations. The resulting closed-form algorithm *involves only real arithmetic*.

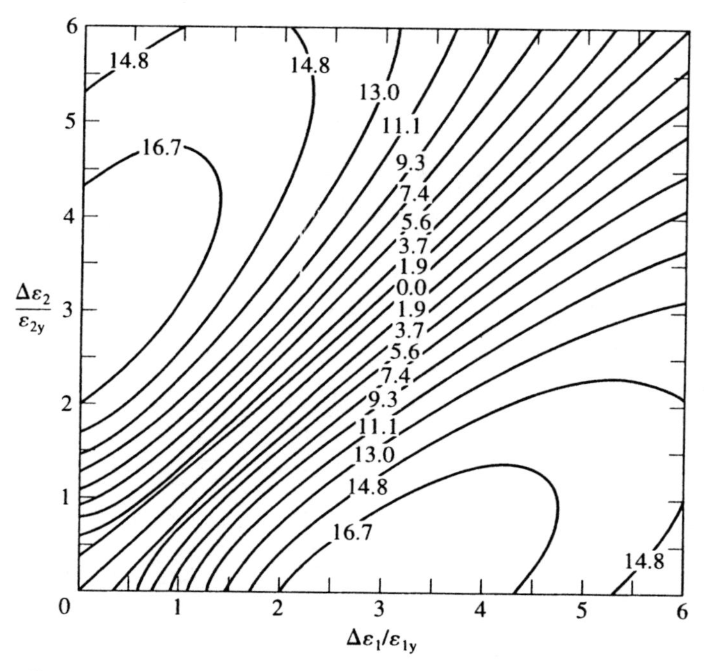

FIGURE 3-6. Iso error map corresponding to point C on the yield surface.

3.4.5.1 Pure kinematic hardening.

For the case of *pure kinematic hardening* the shape of the yield surface remains unchanged, i.e., $R(\Delta\gamma) = constant$. By defining

$$\left.\begin{aligned} A^2 &= \frac{(\eta_{11}^{\text{trial}} + \eta_{22}^{\text{trial}})^2}{6R^2} \\ B^2 &= \frac{\frac{1}{2}(\eta_{11}^{\text{trial}} - \eta_{22}^{\text{trial}})^2 + 2(\eta_{12}^{\text{trial}})^2}{R^2} \\ C &= \frac{1}{3(1-\nu)} + \frac{2H'}{3E} \end{aligned}\right\} \tag{3.4.23}$$

and

$$D = \frac{2\mu}{E} + \frac{2H'}{3E} ,$$

problem (3.4.18)–(3.4.19) is reduced to finding the positive zeros of the following very special quartic equation

$$\boxed{\begin{aligned}\bar{\gamma}^4 + \left[\frac{2}{C} + \frac{2}{D}\right]\bar{\gamma}^3 + \left[\frac{4}{CD} + \frac{1 - B^2}{D^2} + \frac{1 - A^2}{C^2}\right]\bar{\gamma}^2 \\ + 2\left[\frac{1 - B^2}{CD^2} + \frac{1 - A^2}{DC^2}\right]\bar{\gamma} + \frac{1 - A^2 - B^2}{C^2D^2} = 0.\end{aligned}} \tag{3.4.24}$$

In this equation $\bar{\gamma}$ is a *nondimensional* variable defined as $\bar{\gamma} := E\Delta\gamma$, where E is Young's modulus. Further insight into the nature of equation (3.4.24) is obtained by letting

$$x = 1 + C\bar{\gamma},$$

and (3.4.25)

$$y = 1 + D\bar{\gamma},$$

so that (3.4.24) (or (3.4.19)) reduces to finding the intersection of a quartic and a straight line, i.e.,

$$\boxed{\begin{aligned}&\frac{A^2}{x^2} + \frac{B^2}{y^2} = 1 \\ &y = \frac{D}{C}x + (1 - \frac{D}{C}).\end{aligned}} \tag{3.4.26}$$

Note that $(3.4.26)_1$ has orthogonal asymptotes $x = \pm|A|$ and $y = \pm|B|$. Further, observe that by introducing the change of variables

$$x = Ar\cos(\theta)$$

and (3.4.27)

$$y = Br\sin(\theta),$$

equation $(3.4.26)_1$ takes the following simple form:

$$r^2\sin^2(2\theta) = 4. \tag{3.4.28}$$

This is illustrated graphically in Figure **3.7.** Since $(3.4.25)_2$ always has positive slope and passes through the point (1,1), direct inspection of Figure **3.7** reveals that problem (3.4.26) (or, equivalently, (3.4.24)) has *one negative root*, *one positive root* (the one of interest), and a *pair of complex conjugate roots*. Keeping the above observation in mind, one has the following algorithm, summarized in BOX **3.4**, which derives from Galois theory (see, for example, Hungerford [1974] or Herstein[1964]).

3.4.5.2 Isotropic hardening.

The standard formulation of *linear* isotropic hardening, on the other hand, leads to a reduced plane-stress problem with *no closed-form* solution. To elaborate, recall

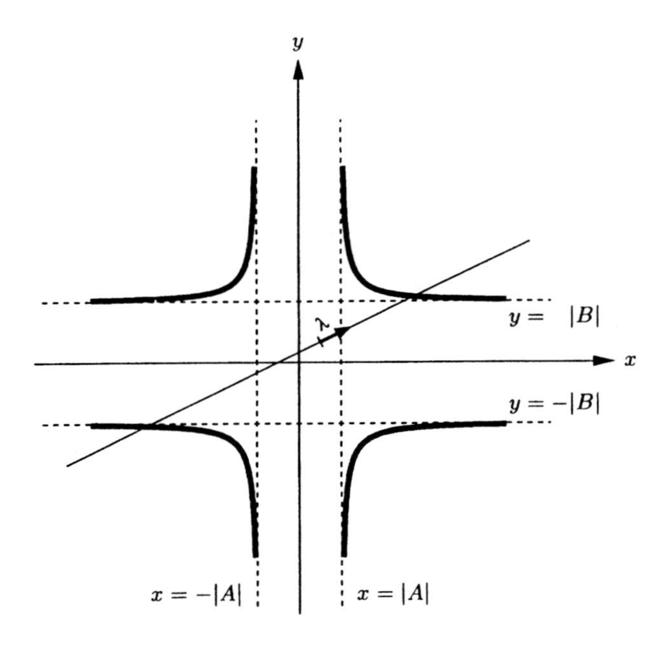

FIGURE 3-7. Geometric view of the consistency problem in xy phase plane.

that the linear isotropic hardening model is given as

$$R(\alpha) = \sqrt{\tfrac{2}{3}}\,[\sigma_Y + K\alpha] \ , \tag{3.4.29}$$

where α is the *equivalent plastic strain*, with equation of evolution

$$\dot{\alpha} = \gamma\sqrt{\tfrac{2}{3}\,\boldsymbol{\xi}^T\mathbf{P}\boldsymbol{\xi}} \ . \tag{3.4.30}$$

The discrete version of this hardening law yields

$$R(\Delta\gamma) := \sqrt{\tfrac{2}{3}}\,(\sigma_Y + K(\alpha_n + \sqrt{\tfrac{2}{3}}\,\Delta\gamma\,\bar{f}(\Delta\gamma))) \ . \tag{3.4.31}$$

Substituting (3.4.31) in (3.4.19) leads to a transcendental equation which does not admit a closed-form solution.

Nevertheless, forms of the isotropic hardening law that lead to a closed form solution are possible. As an example, consider the internal hardening variable q governed by the equation of evolution

$$\dot{q} = \tfrac{2}{3}\gamma\boldsymbol{\xi}^T\mathbf{P}\boldsymbol{\xi} \ , \tag{3.4.32}$$

and define

$$R^2(q) := \tfrac{2}{3}\left(\sigma_Y^2 + \tilde{K}^2 q\right) . \tag{3.4.33}$$

It easily follows that

$$R^2(\Delta\gamma) := \left(\tfrac{2}{3}\sigma_Y^2 + \tfrac{4}{9}\tilde{K}^2 q_n\right) + \tfrac{4}{9}\tilde{K}^2 \Delta\gamma \bar{f}^2(\Delta\gamma)\,. \tag{3.4.34}$$

Substituting (3.4.34) into (3.4.19) again yields a quartic equation with the following form:

$$\begin{aligned}
&\bar{\gamma}^4 + \left[\frac{2}{C} + \frac{F(A^2D^2 + B^2C^2)}{C^2D^2J} + \frac{2}{D}\right]\bar{\gamma}^3 \\
&+ \left[\frac{1}{D^2} + \frac{1}{C^2} + \frac{4}{CD} + \frac{2F(A^2D + B^2C)}{C^2D^2J} - \frac{A^2}{C^2J} - \frac{B^2}{D^2J}\right]\bar{\gamma}^2 \\
&+ \left[\frac{2(C + D)}{C^2D^2} - \frac{A^2(2D - F) + B^2(2C - F)}{C^2D^2J}\right]\bar{\gamma} \\
&+ \frac{J - A^2 - B^2}{C^2D^2J} = 0,
\end{aligned} \tag{3.4.35}$$

where $J = \frac{2}{3}\sigma_Y^2 + \frac{4}{9}\tilde{K}^2 q_n$ and $F = \frac{4\tilde{K}^2}{9E}$. Equation (3.4.35) possesses the same properties as (3.4.24) and, therefore, it is also solvable in closed form by the algorithm in BOX **3.4**.

Remarks 3.4.1.

1. A word on computational effort is in order. Our implementation of the algorithm in BOX **3.4** is approximately equivalent to four to six Newton iterations without line search. The comparisons were made on a CONVEX C-1 superminicomputer running under CONVEX UNIX 6.1. No attempt was made to optimize our implementations (written in C and executed in scalar, and not vectorized mode). We note, however, that the algorithm in BOX **3.4** is amenable to vectorization, a fact which constitutes the main advantage of this closed-form procedure. By contrast, local Newton iterative procedures are typically not amenable to vectorization, and vectorized implementations typically rely on stipulating *a fixed number of iterations*, as in the DYNA codes, Hallquist [1988]. Although, in the present situation, Newton's method is guaranteed to converge regardless of the initial trial state, the required number of iterations depends crucially on the location of the trial stress.
2. The procedure discussed above is also applicable to more general plasticity models with a quadratic strain-energy function and yield conditions.

BOX 3.4. Solution Algorithm for the Quartic Equation.

Solve :

$$\bar{\gamma}^4 + a\bar{\gamma}^3 + b\bar{\gamma}^2 + c\bar{\gamma} + d = 0^*$$

for its *only positive root.*

1. Let

$$p := \left(\frac{b}{3}\right)^2 - \frac{1}{3}(ac - 4d)$$

$$q := \left(\frac{b}{3}\right)^3 - \frac{b(ac - 4d)}{6} + \frac{a^2 d - 4bd + c^2}{2}$$

2. If $q^2 - p^3 \geq 0$ then let

$$y := \sqrt[3]{q + \sqrt{q^2 - p^3}} + \sqrt[3]{q - \sqrt{q^2 - p^3}} + \frac{b}{3}$$

Else let

$$y := 2\sqrt{p}\cos\left[\frac{1}{3}\arccos\left(\frac{q}{p\sqrt{p}}\right)\right] + \frac{b}{3}$$

3. Let

$$Q := \sqrt{\frac{a^2}{4} - b + y}$$

$$S := \frac{3a^2}{4} - 2b - Q^2$$

4. If $Q^2 > \text{TOL}^{**}$ then let

$$T := \frac{4ab - 8c - a^3}{4Q}$$

Else let

$$T := 2\sqrt{y^2 - 4d}$$

5. If $T + S > 0$ then let the desired root

$$\bar{\gamma} = -\frac{a}{4} + \frac{Q}{2} + \frac{\sqrt{S + T}}{2}$$

Else let

$$\bar{\gamma} = -\frac{a}{4} - \frac{Q}{2} + \frac{\sqrt{S - T}}{2}$$

* For convenience, we define $a := (\frac{2}{C} + \frac{2}{D})$, $b := (\frac{4}{CD} + \frac{1-B^2}{D^2} + \frac{1-A^2}{C^2})$, $c := 2(\frac{1-B^2}{CD^2} + \frac{1-A^2}{DC^2})$, and $d := \frac{1-A^2-B^2}{C^2D^2}$.

** TOL is chosen to approximate machine zero for the word size employed.

3.5 Interpretation. Operator Splits and Product Formulas

The examples developed in the preceding two sections constitute specific illustrations of the general equations (3.2.1). These discrete equations may be viewed as a *two-step-algorithm*:

1. an elastic *trial* predictor defined by formulas (3.2.3), followed by
2. a plastic corrector that performs the closest point projection of the trial state onto the yield surface.

In this section we show that the above two-step algorithm may be interpreted as a *product formula algorithm* emanating from an *elastic-plastic* operator split of the elastoplastic constitutive equations. This interpretation is particularly useful in analyzing and developing algorithms. In Section **3.6**, for instance, the general notion of return mapping is exploited to develop the cutting-plane algorithm. To motivate the basic methodology we consider the following elementary example. For a detailed account of product formulas and operator split methods we refer to Chorin et al. [1978] and references therein.

3.5.1 Example 3.3. Lie's Formula

Consider the following linear initial-value problem governing the evolution of $\boldsymbol{x}(t) \in \mathbb{R}^N$,

$$\left.\begin{aligned} \dot{\boldsymbol{x}}(t) &= \boldsymbol{A}\boldsymbol{x}(t) \equiv \left[\boldsymbol{A}^1 + \boldsymbol{A}^2\right]\boldsymbol{x}(t) \\ \boldsymbol{x}(t_n) &= \boldsymbol{x}_n \,, \end{aligned}\right\} \tag{3.5.1}$$

where the (linear) operator $\boldsymbol{A} : \mathbb{R}^N \to \mathbb{R}^N$ admits the *additive decomposition* $\boldsymbol{A} = \boldsymbol{A}^1 + \boldsymbol{A}^2$. Of course, the exact solution of (3.5.1) at time $t_{n+1} = t_n + h$, $h > 0$, is given by $\boldsymbol{x}(t_{n+1}) = \exp[(\boldsymbol{A}^1 + \boldsymbol{A}^2)h]\boldsymbol{x}_n$. To approximate this solution, we proceed as follows. Consider the *split* problems:

<u>Problem 1</u>   <u>Problem 2</u>

$$\left.\begin{aligned} \dot{\bar{\boldsymbol{x}}}(t) &= \boldsymbol{A}^1\bar{\boldsymbol{x}}(t) \\ \bar{\boldsymbol{x}}(t_n) &= \boldsymbol{x}_n \end{aligned}\right\} \qquad \left.\begin{aligned} \dot{\boldsymbol{x}}(t) &= \boldsymbol{A}^2\boldsymbol{x}(t) \\ \boldsymbol{x}(t_n) &= \bar{\boldsymbol{x}}(t_{n+1}). \end{aligned}\right\} \tag{3.5.2}$$

Note that the solution of problem 1 is taken as the initial condition for problem 2 and that both problems do indeed add up to the original problem (3.5.1). This sequential solution scheme defines a *product-formula* algorithm of the form

$$\boxed{\boldsymbol{x}_{n+1} = \exp[\boldsymbol{A}^1h]\exp[\boldsymbol{A}^2h]\boldsymbol{x}_n \,,} \tag{3.5.3}$$

where $\exp[\boldsymbol{A}^kh]$, $(k = 1, 2)$, are the (exact) solutions of problems 1 and 2, respectively. Of course, the product-formula algorithm (3.5.3) does not furnish the exact solution to the initial problem (3.5.1) (unless $\boldsymbol{A}^1$ and $\boldsymbol{A}^2$ commute). However, it is easily shown that (3.5.3) defines a *first-order accurate* algorithm:

$$\|\exp[\boldsymbol{A}^1h]\exp[\boldsymbol{A}^2h] - \exp[(\boldsymbol{A}^1 + \boldsymbol{A}^2)h]\| = Ch^2 + \mathcal{O}(h^3), \tag{3.5.4}$$

where $C > 0$ is a constant (in fact, the commutator or Lie bracket of $\boldsymbol{A}^1$ and $\boldsymbol{A}^2$). Furthermore, repeated application of formula (3.5.3) with a decreasing time step yields the exact solution:

$$\lim_{n\to\infty}\left[\exp(\boldsymbol{A}^1 h/n)\exp(\boldsymbol{A}^2 h/n)\right]^n = \exp\left[(\boldsymbol{A}^1+\boldsymbol{A}^2)h\right]. \tag{3.5.5}$$

This is the so-called Lie's formula, see Abraham, Marsden, and Ratiu [1984]. One can show that the above results essentially carry over for the case in which $\boldsymbol{A}$ is a nonlinear operator and the "exact algorithms" that solve problems 1 and 2 are replaced by (first-order accurate) consistent algorithms.

3.5.2 Elastic-Plastic Operator Split

Now we apply the basic idea illustrated in the example above to the elastoplastic problem of evolution (3.1.2). To this end, we introduce the following additive split:

$$\underset{\text{Total}}{\left.\begin{aligned}\dot{\boldsymbol{\varepsilon}} &= \nabla^s(\Delta\dot{\boldsymbol{u}})\\ \dot{\boldsymbol{\varepsilon}}^p &= \gamma\partial_\sigma f(\boldsymbol{\sigma},\boldsymbol{q})\\ \dot{\boldsymbol{q}} &= -\gamma\boldsymbol{h}(\boldsymbol{\sigma},\boldsymbol{q})\end{aligned}\right\}} = \underset{\text{Elastic predictor}}{\left.\begin{aligned}\dot{\boldsymbol{\varepsilon}} &= \nabla^s(\Delta\dot{\boldsymbol{u}})\\ \dot{\boldsymbol{\varepsilon}}^p &= \boldsymbol{0}\\ \dot{\boldsymbol{q}} &= \boldsymbol{0}\end{aligned}\right\}} + \underset{\text{Plastic Corrector}}{\left.\begin{aligned}\dot{\boldsymbol{\varepsilon}} &= \boldsymbol{0}\\ \dot{\boldsymbol{\varepsilon}}^p &= \gamma\partial_\sigma f(\boldsymbol{\sigma},\boldsymbol{q})\\ \dot{\boldsymbol{q}} &= -\gamma\boldsymbol{h}(\boldsymbol{\sigma},\boldsymbol{q})\end{aligned}\right\}} \tag{3.5.6}$$

Conceptually, a product-formula algorithm is constructed as follows. The elastic predictor problem is solved with the initial conditions (3.1.2b) which are the *converged* values of the previous time step. This produces a *trial elastic state* which, if outside of the yield surface, is taken as the initial conditions for the solution of the plastic corrector problem. The objective of this second step is to restore consistency by "returning" the trial stress to the yield surface. This is pictorially indicated in Figure **3.8**.

3.5.3 Elastic Predictor. Trial Elastic State

First we observe that the elastic predictor problem admits an *exact solution* which merely reduces to a *geometric* update

$$\boxed{\begin{aligned}\boldsymbol{\varepsilon}_{n+1} &= \boldsymbol{\varepsilon}_n + \nabla^s(\Delta\boldsymbol{u})\\ \boldsymbol{\varepsilon}^{p^{\text{trial}}}_{n+1} &= \boldsymbol{\varepsilon}^p_n\\ \boldsymbol{q}^{\text{trial}}_{n+1} &= \boldsymbol{q}_n\,,\end{aligned}} \tag{3.5.7}$$

where $\Delta\boldsymbol{u}$ is the specified displacement increment over the time step $[t_n, t_{n+1}]$. In addition, the stress tensor associated with this trial elastic state is computed by *functional evaluation* simply by using the elastic stress-strain relationships, i.e.,

$$\boxed{\boldsymbol{\sigma}^{\text{trial}}_{n+1} = \nabla W(\boldsymbol{\varepsilon}_{n+1} - \boldsymbol{\varepsilon}^p_n).} \tag{3.5.8}$$

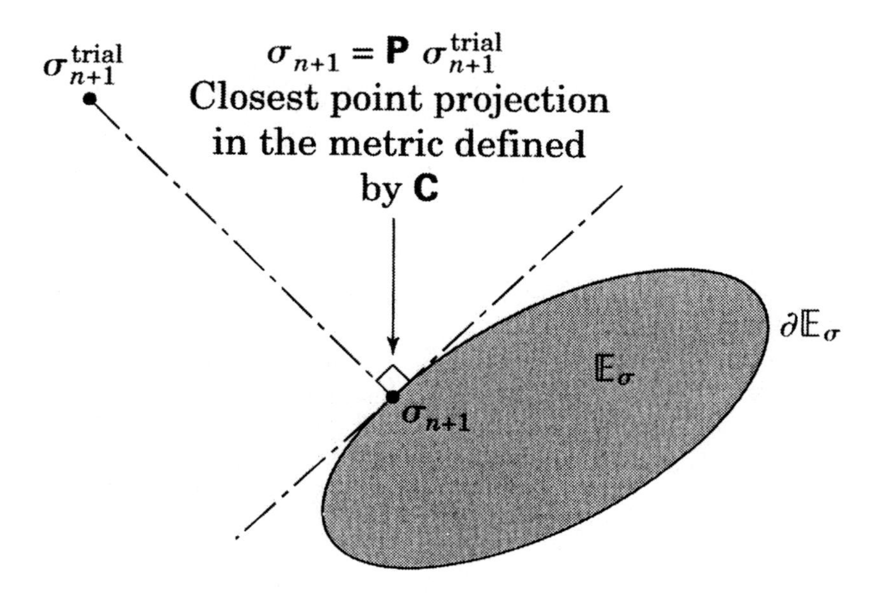

FIGURE 3-8. Conceptual representation of the elastic predictor–plastic return-mapping algorithm for perfect plasticity (no hardening).

In fact, these are the equations defining the trial state in Section **3.2**. There, it was shown that, assuming $f(\bullet, \bullet)$ is convex, if $f(\sigma_{n+1}^{\text{trial}}, \boldsymbol{q}_n) \leq 0$, the process is elastic and the trial state is the final state. On the other hand, if $f(\sigma_{n+1}^{\text{trial}}, \boldsymbol{q}_n) > 0$, the Kuhn–Tucker loading/unloading conditions are violated by the trial state which now lies outside the yield surface. Then consistency is restored by the return-mapping algorithm.

3.5.4 Plastic Corrector. Return Mapping

Dividing by $\Delta\gamma$, the plastic corrector problem is rephrased as

$$\left.\begin{aligned} \frac{d\boldsymbol{\varepsilon}^p(\Delta\gamma)}{d\Delta\gamma} &= \partial_\sigma f\left\{\nabla W\left[\boldsymbol{\varepsilon}_{n+1} - \boldsymbol{\varepsilon}^p(\Delta\gamma)\right], \boldsymbol{q}(\Delta\gamma)\right\} \\ \frac{d\boldsymbol{q}(\Delta\gamma)}{d\Delta\gamma} &= -\boldsymbol{h}\left\{\nabla W\left[\boldsymbol{\varepsilon}_{n+1} - \boldsymbol{\varepsilon}^p(\Delta\gamma)\right], \boldsymbol{q}(\Delta\gamma)\right\} \end{aligned}\right\} \tag{3.5.9}$$

subject to

$$\{\boldsymbol{\varepsilon}^p(\Delta\gamma), \boldsymbol{q}(\Delta\gamma)\}\big|_{\Delta\gamma=0} = \{\boldsymbol{\varepsilon}_n^p, \boldsymbol{q}_n\}. \tag{3.5.10}$$

For associative plasticity, by using the elastic stress-strain relationships and the hardening relationships, problem (3.5.9) is formulated in terms of the stress and the internal variables as

$$\boxed{\begin{aligned}\frac{d\sigma(\Delta\gamma)}{d\Delta\gamma} &= -\mathbf{C}(\Delta\gamma) : \partial_\sigma f\{\sigma(\Delta\gamma), \boldsymbol{q}(\Delta\gamma)\} \\ \frac{d\boldsymbol{q}(\Delta\gamma)}{d\Delta\gamma} &= -\mathbf{D}(\Delta\gamma)\partial_{\boldsymbol{q}} f\{\sigma(\Delta\gamma), \boldsymbol{q}(\Delta\gamma)\}\end{aligned}} \tag{3.5.11}$$

subject to

$$\boxed{\{\sigma(\Delta\gamma), \boldsymbol{q}(\Delta\gamma)\}\big|_{\Delta\gamma=0} = \{\sigma_{n+1}^{\text{trial}}, \boldsymbol{q}_n\},} \tag{3.5.12}$$

where $\mathbf{C}(\Delta\gamma) := \nabla^2 W[\varepsilon - \varepsilon^p(\Delta\gamma)]$ is the elasticity tensor and $\mathbf{D}(\Delta\gamma) := \left[\nabla^2\Theta(\boldsymbol{q}(\Delta\gamma))\right]^{-1}$ is the tensor of generalized plastic moduli. Form (3.5.11) is preferred when the elasticity tensor and the plastic moduli are constant (the usual case). The solution of (3.5.11)–(3.5.12) is a curve $\Delta\gamma \in \mathbb{R}_+ \rightarrow [\sigma(\Delta\gamma), \boldsymbol{q}(\Delta\gamma)]$ which starts at the trial state. Consistency is enforced by determining the intersection of this curve with the boundary $\partial\mathbb{E}_\sigma$ of the elastic domain, equivalently, by solving the following problem:

$$\boxed{\begin{aligned}&\text{Find } \Delta\gamma \in \mathbb{R}_+ \text{ such that} \\ &\bar{f}(\Delta\gamma) := f\{\sigma(\Delta\gamma), \boldsymbol{q}(\Delta\gamma)\} = 0.\end{aligned}} \tag{3.5.13}$$

Remarks 3.5.1. To gain further insight into the nature of problem (3.5.11) and nonlinear equation (3.5.13), we examine the return map $\Delta\gamma \mapsto \{\sigma(\Delta\gamma), \boldsymbol{q}(\Delta\gamma)\}$ under the assumption of associative plasticity. Multiplying $(3.5.11)_1$ by $\partial_\sigma f$ and $(3.5.11)_2$ by $\partial_{\boldsymbol{q}} f$, adding the result, and using the chain rule, we obtain

$$\begin{aligned}\frac{d}{d\Delta\gamma} f\{\sigma(\Delta\gamma), \boldsymbol{q}(\Delta\gamma)\} &= -\partial_\sigma f : \mathbf{C} : \partial_\sigma f - \partial_{\boldsymbol{q}} f : \mathbf{D} : \partial_{\boldsymbol{q}} f \\ &= -\left\|\partial_\sigma f\right\|_{\mathbf{C}}^2 - \left\|\partial_{\boldsymbol{q}} f\right\|_{\mathbf{D}}^2 < 0,\end{aligned} \tag{3.5.14}$$

where $\gamma > 0$. Here, $\|\bullet\|_{\mathbf{C}}$ and $\|\bullet\|_{\mathbf{D}}$ denote the norms induced by the (Riemanian) metrics $\mathbf{C}(\Delta\gamma)$ and $\mathbf{D}(\Delta\gamma)$, respectively. Equivalently, (3.5.14) is the norm induced by the block-diagonal metric $\mathbf{G}$ defined by (3.2.13). Since the function $f(\sigma, \boldsymbol{q})$ is *convex*, it follows that the system (3.5.11) is dissipative. In addition, the function $\hat{f}(\Delta\gamma) := f\{\sigma(\Delta\gamma), \boldsymbol{q}(\Delta\gamma)\}$ is monotonically decreasing with the shape indicated in Figure **3.9**. Note that $\bar{f}(\Delta\gamma) = 0$ is ideally suited to a solution by Newton's method which becomes a globally convergent algorithm for the present case.

Conceptually, once $\Delta\gamma > 0$ is determined, the stresses and internal variables are obtained by setting $\sigma_{n+1} = \sigma(\Delta\gamma)$, and so on. The algorithms developed in Section **3.3** and Section **3.4** may, in fact, be regarded as particular examples of numerical schemes that approximate the flow $\Delta\gamma \rightarrow \{\sigma(\Delta\gamma), \boldsymbol{q}(\Delta\gamma)\}$ associated

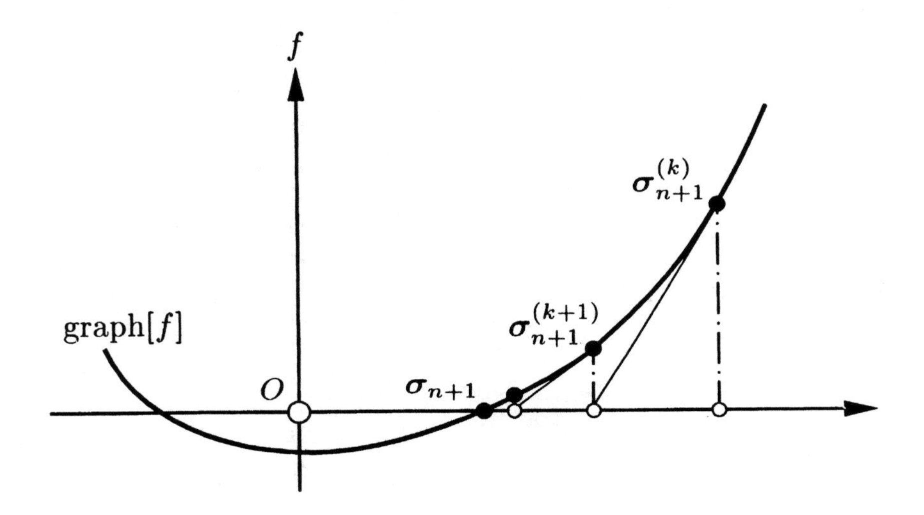

FIGURE 3-9. Shape of the function $\bar{f}(\Delta\gamma) := f[\sigma(\Delta\gamma)]$.

with problem (3.5.11) and starting at $(\sigma_{n+1}^{\text{trial}}, \boldsymbol{q}_n)$ for $\Delta\gamma = 0$. The general situation is considered next.

3.6 General Return-Mapping Algorithms

In what follows, motivated by the examples discussed in Sections **3.3** and **3.4** and with the background provided by Section **3.5**, we present two general algorithms for numerically solving the plastic corrector problem (3.3.2). We emphasize that these two algorithms apply to the case of a general yield condition, flow rule, and hardening law.

3.6.1 General Closest Point Projection

This algorithm is the extension of the procedure discussed in Example **3.2** to the general case governed by equations (3.2.1). Conceptually, the underlying idea is rather simple and is explained below in the simpler context of perfect plasticity. The general case is summarized in BOX **3.5**.

1. Assume *plastic loading*, that is, $f_{n+1}^{\text{trial}} > 0$ so that, by Lemma **3.1**, $\Delta\gamma > 0$.

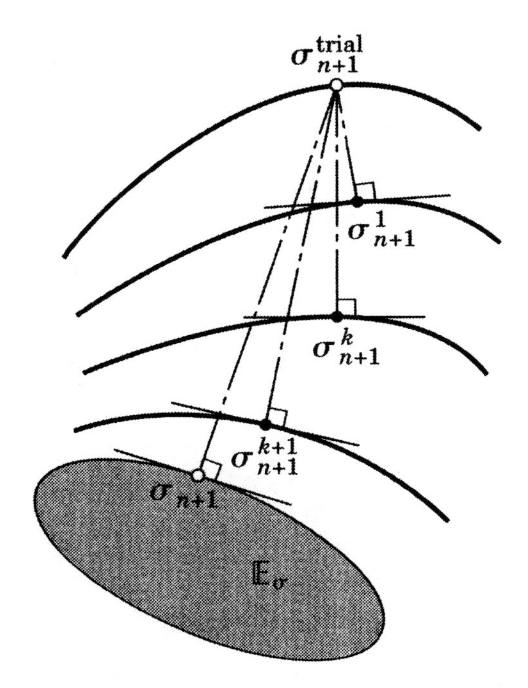

FIGURE 3-10. A geometric interpretation of the closest point projection algorithm in stress space. At each iterate $(\bullet)^{(k)}$, the constraint is linearized to find the intersection (cut) with $f = 0$. The next iterate $(\bullet)^{(k+1)}$, located on level set $f_{n+1}^{(k+1)} > 0$, is the closest point of that level set to the previous iterate $(\bullet)^{(k)}$ in the metric defined by the elasticities $\mathbf{C}$.

Define the plastic flow residual $\boldsymbol{R}_{n+1}$ and yield condition:

$$\left.\begin{aligned} \boldsymbol{R}_{n+1} &:= -\varepsilon_{n+1}^{p} + \varepsilon_{n}^{p} + \Delta\gamma\,\partial_\sigma f_{n+1} \\ f_{n+1} &:= f(\sigma_{n+1}) \end{aligned}\right\} \tag{3.6.1}$$

where $\sigma_{n+1} = \mathbf{C} : [\varepsilon_{n+1} - \varepsilon_{n+1}^{p}]$.

2. Linearize the above equations. Since ε_{n+1} is *fixed during the return-mapping stage*, it follows that $\Delta\varepsilon_{n+1}^{p^{(k)}} = -\mathbf{C}^{-1} : \Delta\sigma_{n+1}^{(k)}$, and one is led to the linearized problem

$$\left.\begin{aligned} \boldsymbol{R}_{n+1}^{(k)} + [\boldsymbol{\Xi}_{n+1}^{(k)}]^{-1} : \Delta\sigma_{n+1}^{(k)} + \Delta^{2}\gamma_{n+1}^{(k)}\partial_\sigma f_{n+1}^{(k)} &= \mathbf{0} \\ f_{n+1}^{(k)} + \partial_\sigma f_{n+1}^{(k)} : \Delta\sigma_{n+1}^{(k)} &= 0 \end{aligned}\right\} \tag{3.6.2}$$

where $\boldsymbol{\Xi} := \left[\mathbf{C}^{-1} + \Delta\gamma\,\partial_{\sigma\sigma}^{2} f\right]^{-1}$ is the exact Hessian matrix.

3a. Solve the linearized problem to obtain $\Delta^2\gamma_{n+1}^{(k)}$ and $\Delta\varepsilon_{n+1}^{p(k)}$:

$$\left.\begin{aligned}
\Delta^2\gamma_{n+1}^{(k)} &:= \frac{f_{n+1}^{(k)} - \boldsymbol{R}_{n+1}^{(k)} : \boldsymbol{\Xi}_{n+1}^{(k)} : \partial_\sigma f_{n+1}^{(k)}}{\partial_\sigma f_{n+1}^{(k)} : \boldsymbol{\Xi}_{n+1}^{(k)} : \partial_\sigma f_{n+1}^{(k)}} \\
\Delta\boldsymbol{\sigma}_{n+1}^{(k)} &:= \boldsymbol{\Xi}_{n+1}^{(k)} : \left[-\boldsymbol{R}_{n+1}^{(k)} - \Delta^2\gamma_{n+1}^{(k)}\partial_\sigma f_{n+1}^{(k)}\right]
\end{aligned}\right\} \tag{3.6.3}$$

and

$$\Delta\boldsymbol{\varepsilon}_{n+1}^{p^{(k)}} := -\mathbf{C}^{-1} : \Delta\boldsymbol{\sigma}_{n+1}^{(k)}$$

3b. Update the plastic strain $\boldsymbol{\varepsilon}_{n+1}^{(k)}$ and consistency parameter $\Delta\gamma_{n+1}^{(k)}$:

$$\boldsymbol{\varepsilon}_{n+1}^{p^{(k+1)}} = \boldsymbol{\varepsilon}_{n+1}^{p^{(k)}} + \Delta\boldsymbol{\varepsilon}_{n+1}^{p^{(k)}}$$

and

$$\Delta\gamma_{n+1}^{(k+1)} = \Delta\gamma_{n+1}^{(k)} + \Delta^2\gamma_{n+1}^{(k)}. \tag{3.6.4}$$

The procedure summarized above is simply a systematic application of Newton's method to the system of equations (3.6.1) that results in the computation of the *closest point projection* from the trial state onto the yield surface. A geometric interpretation of the iteration scheme is contained in Figure **3.10**, and the general case is summarized for convenience in BOX **3.5**. From a physical viewpoint, the algorithm is a systematic procedure for finding the *intermediate configuration*, which is defined by ε_{n+1}^{p} and $\boldsymbol{q}_{n+1}$ for a given strain ε_{n+1}.

For comparison with the the cutting-plane algorithm developed below, we summarize the basic characteristics of the closest point projection algorithm.

4. It is an **implicit** procedure that involves solving a local 6×6 system of equations.
5. Normality is enforced at the final (**unknown**) iterate.

3.6.2 Consistent Elastoplastic Moduli. Perfect Plasticity

An important advantage of the algorithm summarized in BOX **3.5** lies in the fact that it can be *exactly linearized in closed form*. This leads to the notion of *consistent* — as opposed to *continuum* — elastoplastic tangent moduli. The former are obtained essentially by enforcing the consistency condition on the discrete *algorithmic* problem, whereas the later notion results from the classical consistency condition on the *continuum* problem. In what follows, we illustrate the derivation of these algorithmic tangent moduli. For simplicity, attention is restricted to perfect plasticity. An identical but more elaborated computation applies to the general case summarized in BOX **3.5**. (The concept of *consistent linearization* was introduced in Hughes and Pister [1978], and *consistent alogrithmic moduli* were first derived in Hughes and Taylor [1978] for viscoplasticity.)

BOX 3.5. General Closest Point Projection Iteration.

1. Initialize: $k = 0, \boldsymbol{\varepsilon}_{n+1}^{p^{(0)}} = \boldsymbol{\varepsilon}_n^p, \boldsymbol{\alpha}_{n+1}^{(0)} = \boldsymbol{\alpha}_n, \Delta\gamma_{n+1}^{(0)} = 0.$

2. Check yield condition and evaluate flow rule/hardening law residuals

$$
\begin{aligned}
\boldsymbol{\sigma}_{n+1}^{(k)} &:= \nabla W\big(\boldsymbol{\varepsilon}_{n+1} - \boldsymbol{\varepsilon}_{n+1}^{p^{(k)}}\big)\\
\boldsymbol{q}_{n+1}^{(k)} &:= -\nabla\mathcal{H}(\boldsymbol{\alpha}_{n+1}^{(k)})\\
f_{n+1}^{(k)} &:= f(\boldsymbol{\sigma}_{n+1}^{(k)}, \boldsymbol{q}_{n+1}^{(k)})\\
\boldsymbol{R}_{n+1}^{(k)} &:= \left\{\begin{array}{c} -\boldsymbol{\varepsilon}_{n+1}^{p^{(k)}} + \boldsymbol{\varepsilon}_n^p \\ -\boldsymbol{\alpha}_{n+1}^{(k)} + \boldsymbol{\alpha}_n \end{array}\right\} + \Delta\gamma_{n+1}^{(k)} \left\{\begin{array}{c} \partial_{\sigma} f_{n+1} \\ \partial_{q} f_{n+1} \end{array}\right\}^{(k)}
\end{aligned}
$$

IF: $f_{n+1}^{(k)} < \text{TOL}_1$ and $\|\boldsymbol{R}_{n+1}^{(k)}\| < \text{TOL}_2$ THEN: *EXIT.*

3. Compute elastic moduli and consistent tangent moduli

$$
\begin{aligned}
\mathbf{C}_{n+1}^{(k)} &:= \nabla^2 W(\boldsymbol{\varepsilon}_{n+1} - \boldsymbol{\varepsilon}_{n+1}^{p^{(k)}})\\
\mathbf{D}_{n+1}^{(k)} &:= -\nabla^2\mathcal{H}(\boldsymbol{\alpha}_{n+1}^{(k)})
\end{aligned}
$$

$$
\big[\mathbf{A}_{n+1}^{(k)}\big]^{-1} := \begin{bmatrix} \big[\mathbf{C}_{n+1}^{-1} + \Delta\gamma_{n+1}\partial_{\sigma\sigma}^2 f_{n+1}\big] & \Delta\gamma_{n+1}\partial_{\sigma q}^2 f_{n+1} \\ \Delta\gamma_{n+1}\partial_{q\sigma}^2 f_{n+1} & \big[\mathbf{D}_{n+1}^{-1} + \Delta\gamma_{n+1}\partial_{qq} f_{n+1}\big] \end{bmatrix}^{(k)}
$$

4. Obtain increment to consistency parameter

$$
\Delta^2\gamma_{n+1}^{(k)} := \frac{f_{n+1}^{(k)} - \big[\partial_{\sigma} f_{n+1}^{(k)} \partial_{q} f_{n+1}^{(k)}\big]^T \mathbf{A}_{n+1}^{(k)} \boldsymbol{R}_{n+1}^{(k)}}{\big[\partial_{\sigma} f_{n+1}^{(k)} \partial_{q} f_{n+1}^{(k)}\big]^T \mathbf{A}_{n+1}^{(k)} \left\{\begin{array}{c} \partial_{\sigma} f_{n+1} \\ \partial_{q} f_{n+1} \end{array}\right\}^{(k)}}
$$

5. Obtain incremental plastic strains and internal variables

$$
\left\{\begin{array}{c} \Delta\boldsymbol{\varepsilon}_{n+1}^{p^{(k)}} \\ \Delta\boldsymbol{\alpha}_{n+1}^{(k)} \end{array}\right\} = \begin{bmatrix} \mathbf{C}_{n+1}^{-1} & \mathbf{0} \\ \mathbf{0} & \mathbf{D}_{n+1}^{-1} \end{bmatrix}^{(k)} \mathbf{A}_{n+1}^{(k)} \left[\boldsymbol{R}_{n+1}^{(k)} + \Delta^2\gamma_{n+1}^{(k)} \left\{\begin{array}{c} \partial_{\sigma} f_{n+1} \\ \partial_{q} f_{n+1} \end{array}\right\}^{(k)}\right]
$$

6. Update state variables and consistency parameter

$$
\begin{aligned}
\boldsymbol{\varepsilon}_{n+1}^{p^{(k+1)}} &= \boldsymbol{\varepsilon}_{n+1}^{p^{(k)}} + \Delta\boldsymbol{\varepsilon}_{n+1}^{p^{(k)}}\\
\boldsymbol{\alpha}_{n+1}^{(k+1)} &= \boldsymbol{\alpha}_{n+1}^{(k)} + \Delta\boldsymbol{\alpha}_{n+1}^{(k)}\\
\Delta\gamma_{n+1}^{(k+1)} &= \Delta\gamma_{n+1}^{(k)} + \Delta^2\gamma_{n+1}^{(k)}
\end{aligned}
$$

Set $k \leftarrow k + 1$ and GO TO **2.**

By differentiating the elastic stress-strain relationships and the discrete (algorithmic) flow rule $(3.6.1)_1$ (with attention restricted to the case of perfect plasticity), we obtain

$$\left.\begin{aligned} d\boldsymbol{\sigma}_{n+1} &= \mathbf{C}_{n+1} : (d\boldsymbol{\varepsilon}_{n+1} - d\boldsymbol{\varepsilon}^p_{n+1}) \\ \text{and} \quad d\boldsymbol{\varepsilon}^p_{n+1} &= \Delta\gamma_{n+1}\partial^2_{\sigma\sigma} f(\boldsymbol{\sigma}_{n+1}) : d\boldsymbol{\sigma}_{n+1} + d\Delta\gamma_{n+1}\partial_\sigma f(\boldsymbol{\sigma}_{n+1}) \,. \end{aligned}\right\} \tag{3.6.5}$$

Thus, one obtains the algorithmic relationship

$$d\boldsymbol{\sigma}_{n+1} = \boldsymbol{\Xi}_{n+1} : \left[d\boldsymbol{\varepsilon}_{n+1} - d\Delta\gamma_{n+1}\partial_\sigma f(\boldsymbol{\sigma}_{n+1})\right], \tag{3.6.6}$$

where $\boldsymbol{\Xi}_{n+1}$ are *algorithmic* moduli defined as

$$\boldsymbol{\Xi}_{n+1} := \left[\mathbf{C}^{-1}_{n+1} + \Delta\gamma_{n+1}\partial^2_{\sigma\sigma} f(\boldsymbol{\sigma}_{n+1})\right]^{-1}. \tag{3.6.7}$$

On the other hand, differentiating the discrete consistency condition $f(\boldsymbol{\sigma}) = 0$ yields

$$\partial_\sigma f(\boldsymbol{\sigma}_{n+1}) : d\boldsymbol{\sigma}_{n+1} = 0\,. \tag{3.6.8}$$

Thus, from (3.6.6) and (3.6.8),

$$d\Delta\gamma_{n+1} = \frac{\partial_\sigma f : \boldsymbol{\Xi}_{n+1} : d\boldsymbol{\varepsilon}_{n+1}}{\partial_\sigma f_{n+1} : \boldsymbol{\Xi}_{n+1} : \partial_\sigma f_{n+1}}\,. \tag{3.6.9}$$

Finally, substituting (3.57) into (3.6.6) yields the expression for the algorithmic elastoplastic tangent moduli

$$\boxed{\begin{aligned} \left.\frac{d\boldsymbol{\sigma}}{d\boldsymbol{\varepsilon}}\right|_{n+1} &= \boldsymbol{\Xi}_{n+1} - \boldsymbol{N}_{n+1} \otimes \boldsymbol{N}_{n+1} \\ \boldsymbol{N}_{n+1} &:= \frac{\boldsymbol{\Xi}_{n+1} : \partial_\sigma f(\boldsymbol{\sigma}_{n+1})}{\sqrt{\partial_\sigma f(\boldsymbol{\sigma}_{n+1}) : \boldsymbol{\Xi}_{n+1} : \partial_\sigma f(\boldsymbol{\sigma}_{n+1})}}\,. \end{aligned}} \tag{3.6.10}$$

Note that the structure of (3.6.10) is analogous to expression (3.4.11) derived in Example **3.2** (see subsection **3.4.2**). The preceding derivation shows that all that is needed to obtain the *algorithmic* tangent moduli is to replace the elastic moduli $\mathbf{C}_{n+1}$ in the expression for the *continuum* elastoplastic moduli by the *algorithmic* moduli $\boldsymbol{\Xi}_{n+1}$ defined by (3.6.10).

Remarks 3.6.1.

1. It should be noted that symmetry of the generalized moduli $\mathbf{A}$ depends crucially on the choice of variables employed and the potential relationships (3.1.3) connecting $\boldsymbol{q}$ and $\boldsymbol{\alpha}$.
2. The main drawback associated with the closest point iterative procedure summarized in BOX **3.5** is the need for computing the gradients of the flow rule and hardening laws, that is,

$$\begin{bmatrix} \partial^2_{\sigma\sigma} f(\boldsymbol{\sigma}, \boldsymbol{q}) & \partial^2_{q\sigma} f(\boldsymbol{\sigma}, \boldsymbol{q}) \\ \partial^2_{\sigma q} f(\boldsymbol{\sigma}, \boldsymbol{q}) & \partial^2_{qq} f(\boldsymbol{\sigma}, \boldsymbol{q}) \end{bmatrix}. \tag{3.6.11}$$

This task may prove exceedingly laborious for complicated plasticity models.

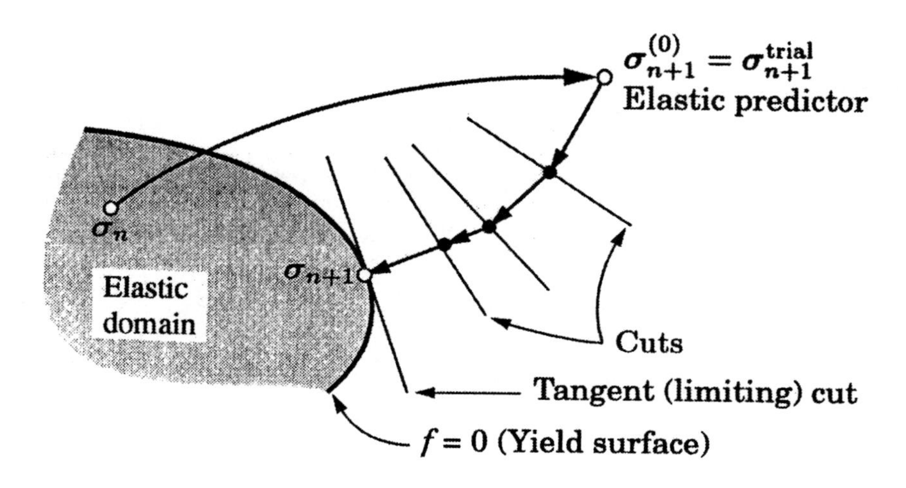

FIGURE 3-11. Geometric interpretation of the cutting-plane algorithm in stress space. At each iterate $(\bullet)^{(k)}$ the constraint is linearized about $(\bullet)^{(k)}$. The intersection of the plane normal to $f^{(k)} = 0$ with the level set $f^{(k+1)}$ determines the next iterate $(\bullet)^{(k+1)}$.

3.6.3 Cutting-Plane Algorithm

The main goal of this algorithm, proposed in Simo and Ortiz [1985] and further analyzed in Ortiz and Simo [1986], is to bypass the need for computing the gradients (3.6.11). The algorithm falls within the class of *convex cutting-plane methods* of constrained optimization; see Luenberger [1984, Sections **13.6** and **13.7**]. The basic idea relies crucially on the notion of operator splitting, as discussed in Section **3.5**, and involves the following steps:

1. Assume *plastic loading* so that $f_{n+1}^{\text{trial}} > 0$ so that, by Lemma **3.1**, $\Delta\gamma_{n+1} > 0$. Then integrate **explicitly** the return-mapping equations (3.5.11) over an interval of length $\Delta^2\gamma$ as yet undetermined.
2. **Linearize** the constraint equation (3.5.13), and solve for the length $\Delta^2\gamma$.
3. Update $\Delta\gamma$ and the state variables, and check for satisfaction of the consistency condition (3.5.13). Return to step **1** if the constraint is violated.

The procedure is summarized in BOX **3.6**. It should be noted that convergence of the algorithm toward the final value of the state variables is obtained at a *quadratic rate*. A geometric interpretation of the algorithm is contained in Figure **3.11.** The return path $\Delta\gamma \mapsto \{\boldsymbol{\sigma}(\Delta\gamma), \boldsymbol{q}(\Delta\gamma)\}$ is approximated by a sequence of straight segments on the space $\mathbb{S} \times \mathbb{R}^m$ of stresses and internal variables $(\boldsymbol{\sigma}, \boldsymbol{q})$.

To some extent the basic characteristics of the algorithm are opposite those of the closest point projection algorithm and are summarized below:

4. It is an **explicit** procedure that involves only functional evaluations.
5. Normality is enforced at the initial (**known**) iterate.

BOX 3.6. General Convex Cutting-Plane Algorithm.

1. Initialize: $k = 0$, $\boldsymbol{\varepsilon}_{n+1}^{p^{(0)}} = \boldsymbol{\varepsilon}_n^p$, $\boldsymbol{q}_{n+1}^{(0)} = \boldsymbol{q}_n$, $\Delta\gamma_{n+1}^{(0)} = 0$.
2. Compute stresses, hardening moduli, and yield function

$$\boldsymbol{\sigma}_{n+1}^{(k)} = \nabla W\big[\boldsymbol{\varepsilon}_{n+1} - \boldsymbol{\varepsilon}_{n+1}^{p^{(k)}}\big]$$

$$\boldsymbol{h}_{n+1}^{(k)} = \boldsymbol{h}\big[\boldsymbol{\sigma}_{n+1}^{(k)}, \boldsymbol{q}_{n+1}^{(k)}\big]$$

$$f_{n+1}^{(k)} = f\big[\boldsymbol{\sigma}_{n+1}^{(k)}, \boldsymbol{q}_{n+1}^{(k)}\big]$$

IF $f_{n+1}^{(k)} \leq$ TOL THEN: *EXIT*.

ELSE:

3. Compute increment to plastic consistency parameter

$$\Delta^2\gamma_{n+1}^{(k)} := \frac{f_{n+1}^{(k)}}{\partial_{\sigma} f_{n+1}^{(k)} : \mathbf{C}_{n+1}^{(k)} : \partial_{\sigma} f_{n+1}^{(k)} + \partial_{\sigma} f_{n+1}^{(k)} \cdot \boldsymbol{h}_{n+1}^{(k)}}$$

4. Update state variables and consistency parameter

$$\boldsymbol{\varepsilon}_{n+1}^{p^{(k+1)}} = \boldsymbol{\varepsilon}_{n+1}^{p^{(k)}} + \Delta^2\gamma_{n+1}^{(k)} \partial_{\sigma} f_{n+1}^{(k)}$$

$$\boldsymbol{q}_{n+1}^{(k+1)} = \boldsymbol{q}_{n+1}^{(k)} - \Delta^2\gamma_{n+1}^{(k)} \boldsymbol{h}_{n+1}^{(k)}$$

$$\Delta\gamma_{n+1}^{(k+1)} = \Delta\gamma_{n+1}^{(k)} + \Delta^2\gamma_{n+1}^{(k)}$$

Set $k \leftarrow k + 1$ and GO TO **2.**

ENDIF

Remark 3.6.2. Although the simplicity of the algorithm in BOX **3.6** leads to a very attractive computational scheme for large scale calculations, it appears that exact linearization of the algorithm cannot be obtained in closed form. Thus global solution strategies involving quasi-Newton methods are required.

For further reading on return-mapping alogrithms, see Ortiz and Martin [1989].

3.7 Extension of General Algorithms to Viscoplasticity

The general algorithms developed in the precedings section are readily modified to accommodate classical viscoplasticity. As motivation, we start with the important case of classical J_2 viscoplasticity.

3.7.1 Motivation. J_2 Viscoplasticity

To develop the extension of the algorithm presented in Section **3.3.1**, we recall that classical viscoplasticity is obtained from rate-independent plasticity by replacing the consistency parameter $\gamma > 0$ with the constitutive equation

$$\boxed{\gamma = \frac{\langle f(\boldsymbol{\sigma}, \boldsymbol{q})\rangle}{\eta}\,, \qquad \eta \in (0, \infty)\,.} \tag{3.7.1}$$

For J_2 viscoplasticity the flow potential in BOX **3.7** has the expression $f = \|\boldsymbol{\xi}\| - \sqrt{\tfrac{2}{3}}\left[\sigma_Y + \beta \bar{H}'\alpha\right]$ where, for simplicity, we have assumed *linear* isotropic/kinematic hardening. As in the rate-independent theory, $\beta \in [0, 1]$ is a material parameter. Assuming that $f^{\text{trial}}_{n+1} > 0$ so that viscoplastic loading takes place, then an implicit backward-Euler difference scheme yields the counterpart of the algorithmic equations (3.3.1):

$$\left.\begin{aligned}
\boldsymbol{\epsilon}^{\text{vp}}_{n+1} &= \boldsymbol{\varepsilon}^{\text{vp}}_{n} + \frac{f_{n+1}}{\eta}\,\Delta t\,\boldsymbol{n}_{n+1} \\
\alpha_{n+1} &= \alpha_n + \sqrt{\tfrac{2}{3}}\,\frac{f_{n+1}}{\eta}\,\Delta t \\
\boldsymbol{\beta}_{n+1} &= \boldsymbol{\beta}_n + \tfrac{2}{3}(1-\beta)\bar{H}'\,\frac{f_{n+1}}{\eta}\,\Delta t\,\boldsymbol{n}_{n+1}\,,
\end{aligned}\right\} \tag{3.7.2}$$

where $\boldsymbol{n}_{n+1} := \boldsymbol{\xi}_{n+1}/\|\boldsymbol{\xi}_{n+1}\|$. An argument identical to that presented in Section **3.3.1** leads to the result

$$\boxed{\boldsymbol{n}_{n+1} = \boldsymbol{\xi}^{\text{trial}}_{n+1}/\|\boldsymbol{\xi}^{\text{trial}}_{n+1}\|,} \tag{3.7.3}$$

along with the condition

$$\boxed{\|\boldsymbol{\xi}_{n+1}\| = \|\boldsymbol{\xi}^{\text{trial}}_{n+1}\| - \frac{2\mu\Delta t}{\eta}\langle f_{n+1}\rangle\left[1 + (1-\beta)\frac{\bar{H}'}{3\mu}\right],} \tag{3.7.4}$$

where $\boldsymbol{\xi}_{n+1} := \boldsymbol{\sigma}_{n+1} - \boldsymbol{\beta}_{n+1}$, and $\boldsymbol{\xi}^{\text{trial}}_{n+1}$ is defined by (3.3.2). From an algorithmic standpoint the only difference with the rate-independent case concerns the enforcement of the counterpart of the consistency condition. Now it follows from (3.7.4) that

$$\boxed{\Delta\gamma_{n+1} := \langle f_{n+1}\rangle\frac{\Delta t}{\eta} = \frac{\langle f^{\text{trial}}_{n+1}\rangle/2\mu}{\frac{\tau}{\Delta t} + \left[1 + \frac{\bar{H}'}{3\mu}\right]}\,, \qquad \tau := \frac{\eta}{2\mu}\,,} \tag{3.7.5}$$

where f^{trial}_{n+1} is defined by (3.2.3) and τ is the *relaxation time* (see (2.7.12)). The preceding analysis easily extends to the case of nonlinear kinematic/isotropic hardening by considering a local iterative procedure analogous to that summarized in BOX **3.1.**

3.7.1.1 Linearization.

By differentiating the algorithm along the lines discussed in Section **3.3.2**, one obtains the *algorithmic consistent viscoplastic tangent moduli*. In particular, for $\Delta\gamma_{n+1}$,

$$\frac{\partial f^{\text{trial}}_{n+1}}{\partial \boldsymbol{\varepsilon}_{n+1}} = 2\mu \boldsymbol{n}_{n+1} \Rightarrow \frac{\partial \Delta\gamma_{n+1}}{\partial \boldsymbol{\varepsilon}_{n+1}} = \frac{\boldsymbol{n}_{n+1}}{\frac{\eta}{2\mu\Delta t} + \left[1 + \frac{\bar{H}'}{3\mu}\right]} . \tag{3.7.6}$$

Moreover, since $\boldsymbol{\sigma}_{n+1} = \kappa \ \text{tr}[\boldsymbol{\varepsilon}_{n+1}]\mathbf{1} + 2\mu(\boldsymbol{e}^p_{n+1} - \boldsymbol{e}^p_n - \Delta\gamma_{n+1}\boldsymbol{n}_{n+1})$, Lemma **3.2** and (3.7.6) yield the expression recorded in BOX **3.7**, which also includes a step-by-step summary of the algorithm.

Remark 3.7.1. From expression (3.7.5),

$$\Delta\gamma_{n+1} := f_{n+1}\frac{\Delta t}{\eta} \quad \rightarrow \quad \frac{f^{\text{trial}}_{n+1}/2\mu}{1 + \frac{\bar{H}'}{3\mu}} , \qquad \text{as } \tau/\Delta t \ \rightarrow \ 0 , \tag{3.7.7}$$

which coincides with expression (3.3.7) for γ_{n+1} in the rate-independent case. This illustrates the fact that, as the ratio of the relaxation time over the time step goes to zero, i.e., $\tau/\Delta t \rightarrow 0$, one recovers the rate-independent limit; in agreement with the conclusions obtained in Section **1.9.1** in analyzing the continuum problem.

3.7.2 Closest Point Projection

The iterative procedure is analogous to that for the rate-independent case. One simply needs to observe the following:

1. For perfect viscoplasticity, $\Delta\gamma_{n+1} = \Delta t f_{n+1}/\eta$, where $f_{n+1} = f(\boldsymbol{\sigma}_{n+1})$.
2. Since $\boldsymbol{R}_{n+1} = -\boldsymbol{\varepsilon}^{\text{vp}}_{n+1} + \boldsymbol{\varepsilon}^{\text{vp}}_n + \Delta\gamma_{n+1}\nabla f_{n+1}$, the linearization yields

$$\frac{\partial \boldsymbol{R}_{n+1}}{\partial \boldsymbol{\sigma}_{n+1}} = \left[\mathbf{C}^{-1} + \Delta\gamma_{n+1}\nabla^2 f_{n+1}\right] + \frac{\Delta t}{\eta}\left[\nabla f_{n+1} \otimes \nabla f_{n+1}\right] \tag{3.7.8}$$

where, for simplicity, we have restricted our attention to perfect viscoplasticity. The general case is handled along similar lines. With these observations in mind, for convenience the iterative scheme is summarized in BOX **3.8**.

3.7.3 A Note on Notational Conventions

In order to minimize confusion, we wish to review our notational conventions. The symbol Δ is used to denote an incremental quantity, such as the increment over a time step, or an increment between successive iterations. We typically do not use separate notations to distinguish between these two cases, relying on context to make the intent clear. On the other hand, we often encounter situations in which the rate-of-slip, γ, appears in both incremental forms within an alogrithm, and even a single equation; see, e.g., Box **3.5**. In these cases, we adhere to the following conventions: $\Delta\gamma = \gamma\,\Delta t$ denotes the increment of γ over a time step, and $\Delta^2\gamma$ denotes

BOX 3.7. J_2-Viscoplasticity.
Linear Isotropic/Kinematic Hardening.

1. Compute *trial elastic* stress:

$$\boldsymbol{e}_{n+1} = \boldsymbol{\varepsilon}_{n+1} - \tfrac{1}{3}\,(\mathrm{tr}[\boldsymbol{\varepsilon}_{n+1}])\mathbf{1}$$

$$\boldsymbol{s}_{n+1}^{\mathrm{trial}} = 2\mu\left(\boldsymbol{e}_{n+1} - \boldsymbol{e}_n^p\right)$$

$$\boldsymbol{\xi}_{n+1}^{\mathrm{trial}} = \boldsymbol{s}_{n+1}^{\mathrm{trial}} - \boldsymbol{\beta}_n$$

2. Check viscoplastic flow potential

$$f_{n+1}^{\mathrm{trial}} := \left\|\boldsymbol{\xi}_{n+1}^{\mathrm{trial}}\right\| - \sqrt{\tfrac{2}{3}}\,(\sigma_Y + \beta\bar{H}'\alpha_n)$$

IF: $f_{n+1}^{\mathrm{trial}} \leq 0$

Set $(\bullet)_{n+1} = (\bullet)_{n+1}^{\mathrm{trial}}$ & *EXIT*

ENDIF

3. Compute $\boldsymbol{n}_{n+1}$ and $\Delta\gamma_{n+1} := f_{n+1}\Delta t/\eta$

$$\boldsymbol{n}_{n+1} := \boldsymbol{\xi}_{n+1}^{\mathrm{trial}} / \left\|\boldsymbol{\xi}_{n+1}^{\mathrm{trial}}\right\|$$

$$\Delta\gamma_{n+1} := \frac{f_{n+1}^{\mathrm{trial}}/2\mu}{\frac{\eta}{2\mu\Delta t} + \left[1 + \frac{\bar{H}'}{3\mu}\right]}$$

4. Update back stress, viscoplastic strain, and stress

$$\boldsymbol{\beta}_{n+1} = \boldsymbol{\beta}_n + \tfrac{2}{3}\,(1-\beta)\bar{H}'\Delta\gamma_{n+1}\boldsymbol{n}_{n+1}$$

$$\alpha_{n+1} = \alpha_n + \sqrt{\tfrac{2}{3}}\,\Delta\gamma_{n+1}$$

$$\boldsymbol{\varepsilon}_{n+1}^{\mathrm{vp}} = \boldsymbol{\varepsilon}_n^{\mathrm{vp}} + \Delta\gamma_{n+1}\boldsymbol{n}_{n+1}$$

$$\boldsymbol{\sigma}_{n+1} = \kappa\ \mathrm{tr}[\boldsymbol{\varepsilon}_{n+1}]\mathbf{1} + \boldsymbol{s}_{n+1}^{\mathrm{trial}} - 2\mu\Delta\gamma_{n+1}\boldsymbol{n}_{n+1}$$

5. Compute *consistent viscoplastic tangent moduli*

$$\mathbf{C}_{n+1} = \kappa\mathbf{1}\otimes\mathbf{1} + 2\mu\theta_{n+1}\left[\mathbf{I} - \tfrac{1}{3}\mathbf{1}\otimes\mathbf{1}\right] - 2\mu\bar{\theta}_{n+1}\boldsymbol{n}_{n+1}\otimes\boldsymbol{n}_{n+1}$$

$$\theta_{n+1} := \left[1 - \frac{2\mu\Delta\gamma_{n+1}}{\left\|\boldsymbol{\xi}_{n+1}^{\mathrm{trial}}\right\|}\right]$$

$$\bar{\theta}_{n+1} := \left[\frac{1}{\frac{\eta}{2\mu\Delta t} + \left(1 + \frac{\bar{H}'}{3\mu}\right)} - \frac{2\mu\Delta\gamma_{n+1}}{\left\|\boldsymbol{\xi}_{n+1}^{\mathrm{trial}}\right\|}\right]$$

the increment of $\Delta\gamma$ between iterations. For example, we write

$$\Delta\gamma^{(k+1)} = \Delta\gamma^{(k)} + \Delta^2\gamma^{(k)}, \tag{3.7.9}$$

where k is the iteration number.

BOX 3.8. Perfect Viscoplasticity.
Closest Point Projection Iteration.

1. Initialize: $k = 0$, $\varepsilon_{n+1}^{\mathrm{vp}(0)} = \varepsilon_n^{\mathrm{vp}}$, $\boldsymbol{\sigma}_{n+1}^{(0)} := \nabla W(\nabla^s \boldsymbol{u}_{n+1} - \varepsilon_n^{\mathrm{vp}})$.

   IF: $f(\boldsymbol{\sigma}_{n+1}^{(0)}) \leq 0$

   Set $\boldsymbol{\varepsilon}_{n+1}^{\mathrm{vp}} = \boldsymbol{\varepsilon}_n^{\mathrm{vp}}$ & *EXIT*

   ELSE:

2. Return mapping iterative algorithm

   2.a. Compute residuals

$$\boldsymbol{\sigma}_{n+1}^{(k)} := \nabla W\left(\nabla^s \boldsymbol{u}_{n+1} - \boldsymbol{\varepsilon}_{n+1}^{\mathrm{vp}^{(k)}}\right)$$

$$f_{n+1}^{(k)} := f\left(\boldsymbol{\sigma}_{n+1}^{(k)}\right)$$

$$\Delta\gamma_{n+1}^{(k)} := \Delta t f_{n+1}^{(k)}/\eta$$

$$\boldsymbol{R}_{n+1}^{(k)} := -\boldsymbol{\varepsilon}_{n+1}^{\mathrm{vp}(k)} + \boldsymbol{\varepsilon}_n^{\mathrm{vp}} + \Delta\gamma_{n+1}^{(k)} \nabla f\left(\boldsymbol{\sigma}_{n+1}^{(k)}\right)$$

   2.b. Check convergence

   IF: $\left\|\boldsymbol{R}_{n+1}^{(k)}\right\| < \mathrm{TOL}_1$ THEN:

   Set $\boldsymbol{\varepsilon}_{n+1}^{\mathrm{vp}} = \boldsymbol{\varepsilon}_{n+1}^{\mathrm{vp}^{(k)}}$ & *EXIT*

   ELSE:

   2.b. Compute consistent (algorithmic) tangent moduli

$$\mathbf{C}_{n+1}^{(k)} := \nabla^2 W\left(\nabla^s \boldsymbol{u}_{n+1} - \boldsymbol{\varepsilon}_{n+1}^{\mathrm{vp}^{(k)}}\right)$$

$$\boldsymbol{\Xi}_{n+1}^{(k)^{-1}} := \left[\mathbf{C}_{n+1}^{-1} + \Delta\gamma_{n+1}^{(k)} \nabla^2 f\left(\boldsymbol{\sigma}_{n+1}^{(k)}\right)\right]$$

$$\boldsymbol{\Xi}_{n+1}^{\mathrm{vp}^{(k)}} := \left[\boldsymbol{\Xi}_{n+1}^{(k)^{-1}} + \frac{\Delta t}{\eta} \nabla f_{n+1}^{(k)} \otimes \nabla f_{n+1}^{(k)}\right]^{-1}$$

   2.c. Compute kth increments

$$\Delta\boldsymbol{\varepsilon}_{n+1}^{\mathrm{vp}^{(k)}} = \mathbf{C}_{n+1}^{-1} : \boldsymbol{\Xi}_{n+1}^{\mathrm{vp}^{(k)}} : \boldsymbol{R}_{n+1}^{(k)}$$

   2.d. Update viscoplastic strain

$$\boldsymbol{\varepsilon}_{n+1}^{\mathrm{vp}^{(k+1)}} = \boldsymbol{\varepsilon}_{n+1}^{\mathrm{vp}^{(k)}} + \Delta\boldsymbol{\varepsilon}_{n+1}^{\mathrm{vp}^{(k)}}$$

   Set $k \leftarrow k + 1$ & GO TO **2.a**.

4

Discrete Variational Formulation and Finite-Element Implementation

In this chapter we address in detail the variational formulation and numerical implementation of classical plasticity and viscoplasticity in the context of the finite-element method. As noted in Chapter **2**, the variational setting of classical plasticity leads naturally to a *variational inequality* typically formulated in stress space. This is the framework adopted by several authors, notably Johnson [1976a,b, 1978]. On the other hand, our formulation transforms this inequality into a variational equality by introducing a Lagrange multiplier at the outset which is interpreted as the *consistency parameter*. Furthermore, the yield condition is formulated in strain space. We show that these steps are in fact crucial to obtain a variational framework suitable for the implementing the *strain-driven*, return-mapping algorithms examined in detail in Chapter **3**.

By now it is well established that displacement-based, finite-element methods may lead to grossly inaccurate numerical solutions in the presence of constraints, such as incompressibility or nearly incompressible response; see e.g., Hughes [1987, Chapter **4**] for a review and an illustration of the difficulties involved in the context of linear incompressible elasticity. As first noted in Nagtegaal, Parks, and Rice [1974], the classical assumption of incompressible plastic flow in metal plasticity is the source of similar numerical difficulties. Finite-element approximations based on mixed variational formulations have provided a useful framework in the context of which constrained problems can be successfully tackled. A large body of literature exists on the subject, which has its point of departure in the pioneering work of Herrmann [1965], Taylor, Pister, and Herrmann [1968], Key [1969], and Nagtegaal, Park, and Rice [1974]. Review accounts of several aspects of this exponentially growing area are in several textbooks, e.g., Ciarlet [1978, Chapter **7**], Oden and Carey [1983, Chapter **4**], Carey and Oden [1984, Chapter **3**], Girault and Raviart [1986, Chapter **III**], Hughes [1987, Chapter **4**], Johnson [1987, Chapter **11**], Zienkiewicz and Taylor [1989, Chapter **12**], and others. See also Taylor et al. [1986].

To retain the simplicity and computational convenience afforded by *strain-driven*, return-mapping algorithms and, at the same time, properly account for nearly incompressible response, a class of methods, called *assumed-strain methods*, has gained considerable popularity in recent years. Direct precedents of this methodology are in the work of Nagtegaal, Parks, and Rice [1974] and the reduced and selective-reduced integration techniques, introduced by Zienkiewicz, Taylor, and Too [1971] and Doherty, Wilson, and Taylor [1969], which are equivalent to certain mixed methods as shown in Malkus and Hughes [1978]. For the nearly incompressible problem, the structure of assumed-strain methods widely used nowadays was originally proposed in Hughes [1980] and is commonly referred to as the *B-bar method.* For linearized and finite strain elasticity and plasticity, Simo, Taylor, and Pister [1985] showed that B-bar methods result from finite-element approximations constructed on the basis of a *three-field variational formulation.* The fact that general assumed strain methods can be made consistent with a three-field variational formulation of the Hu-Washizu type was first pointed out in Simo and Hughes [1986].

The preceding remarks motivate our development in this chapter of a variational formulation of plasticity suitable for constructing three-field, finite-element approximations of the elastoplastic boundary-value problem. In this development, the variational counterpart of the notion of plastic dissipation introduced in Chapter **2** plays a central role. In particular, we show in Section **4.2** that a suitable time discretization of the dissipation function leads to a discrete Lagrangian whose Euler–Lagrange equations produce the weak forms of the momentum balance equation and the strain-displacement relationships. In addition, the weak form of the *closest point projection* algorithm is obtained as an Euler–Lagrange equation that constitutes the discrete counterpart of the the plastic flow rule and the hardening law. Finally, the discrete Kuhn–Tucker loading/unloading conditions also appear as Euler–Lagrange equations.

In Section **4.3** we show that one recovers the computational architecture of assumed strain methods within this variational framework by assuming that the flow rule and loading/unloading conditions hold strongly (pointwise). As already pointed out, the implementation of plasticity models in a finite-element method becomes particularly simple in the context of an assumed-strain method. Essentially, the procedure reduces to *testing independently* at each quadrature point of the element whether the elastic trial state violates the yield condition. If this is the case at a particular quadrature point, one simply applies a local return-mapping algorithm at the quadrature point level to restore consistency. The validity of this simple scheme relies crucially on the statement of the yield criterion in strain space. For a given yield condition in stress space, a strain-space formulation is obtained merely by using the *pointwise* stress-strain relationships. (In way of contrast, see Hinton and Owen [1980] for earlier approaches to integrating constitutive equations of plasticity.)

4.1 Review of Some Basic Notation

In this section we introduce some of the notation necessary for our subsequent developments. Our presentation is informal and technical details are omitted. We illustrate the basic definitions needed with a few elementary examples and refer the reader to standard textbooks for further details and the proper functional analysis setting. See, e.g., Vainberg [1964], Mikhlin [1970, Part **II**], Gelfand and Fomin [1963], Luenberger [1972], Oden and Reddy [1976], and Troutman [1983]. Readers familiar with elementary calculus of variations may proceed directly to Section **4.2.**

4.1.1 Gateaux Variation

Let $\mathbb{V}$ denote an appropriate function space, typically a Banach space with dual $\mathbb{V}^*$, and duality pairing $\langle \bullet, \bullet \rangle_{\mathbb{V}} : \mathbb{V}^* \times \mathbb{V} \to \mathbb{R}$. Given a functional $\Pi : \mathbb{V} \to \mathbb{R}$ and a point $\boldsymbol{u}_0 \in \mathbb{V}$, one defines the *Gateaux variation* at $\boldsymbol{u}_0 \in \mathbb{V}$ in the direction $\boldsymbol{\eta} \in \mathbb{V}$ as the following limit (whenever it exists)

$$\boxed{\delta \Pi(\boldsymbol{u}_0, \boldsymbol{\eta}) := \lim_{\alpha \to 0} \frac{\Pi(\boldsymbol{u}_0 + \alpha\boldsymbol{\eta}) - \Pi(\boldsymbol{u}_0)}{\alpha} = \frac{d}{d\alpha} \Pi(\boldsymbol{u}_0 + \alpha\boldsymbol{\eta})\Big|_{\alpha=0} .} \tag{4.1.1}$$

This definition generalizes the notion of the directional derivative of functions in Euclidean space. However, it does not imply the stronger notion of (Fréchet) differentiability. We recall that a functional $\Pi : \mathbb{V} \to \mathbb{R}$ is *Fréchet* differentiable at $\boldsymbol{u}_0 \in \mathbb{V}$ if there exists a linear functional, denoted by $D\Pi(\boldsymbol{u}_0)$ and called Fréchet derivative, such that

$$\frac{\Pi(\boldsymbol{u}_0 + \boldsymbol{\eta}) - \Pi(\boldsymbol{u}_0) - D\Pi(\boldsymbol{u}_0) \cdot \boldsymbol{\eta}}{\|\boldsymbol{\eta}\|_{\mathbb{V}}} \to 0 , \text{ as } \|\boldsymbol{\eta}\|_{\mathbb{V}} \to 0, \tag{4.1.2}$$

where $\| \bullet \|_{\mathbb{V}}$ denotes the norm in $\mathbb{V}$. One refers to $D\Pi(\boldsymbol{u}_0) \cdot \boldsymbol{\eta}$ as the (Fréchet) derivative at $\boldsymbol{u}_0$ in the direction $\boldsymbol{\eta}$. The Gateaux derivative $\delta\Pi(\boldsymbol{u}_0; \boldsymbol{\eta})$ coincides with $D\Pi(\boldsymbol{u}_0) \cdot \boldsymbol{\eta}$ if the following two technical conditions hold (see Troutman [1983]):

i. $\delta\Pi(\boldsymbol{u}_0, \boldsymbol{\eta})$ is *linear and continuous* in $\boldsymbol{\eta} \in \mathbb{V}$.
ii. $|\delta\Pi(\boldsymbol{u}, \boldsymbol{\eta}) - \delta\Pi(\boldsymbol{u}_0, \boldsymbol{\eta})| \to 0$ as $\boldsymbol{u} \to \boldsymbol{u}_0$, uniformly for $\boldsymbol{u}$ in the unit ball about $\boldsymbol{u}_0$ in $\mathbb{V}$.

Geometrically, $D\Pi(\boldsymbol{u}_0)$ determines the *best linear approximation* to Π at $\boldsymbol{u}_0$, as the following example illustrates.

EXAMPLE: 4.1.1. Let $\Pi : \mathcal{B} \subset \mathbb{R}^2 \to \mathbb{R}$ be a real function of two variables $\boldsymbol{x} = (x_1, x_2) \in \mathcal{B}$, assumed to be continuous and differentiable. Then, the *best linear approximation* to $\Pi(\boldsymbol{x})$ at $\boldsymbol{x}_0$ is the tangent plane, $L_{\boldsymbol{x}_0}\Pi(\boldsymbol{x})$ at $\boldsymbol{x}_0$, defined as

$$L_{\boldsymbol{x}_0}\Pi(\boldsymbol{x}) := \Pi(\boldsymbol{x}_0) + D\Pi(\boldsymbol{x}_0) \cdot (\boldsymbol{x} - \boldsymbol{x}_0) , \tag{4.1.3a}$$

where $D\Pi(\boldsymbol{x}_0)$ is the vector with components

$$D\Pi(\boldsymbol{x}_0) := \left\{ \frac{\partial \Pi(\boldsymbol{x}_0)}{\partial x_1} , \frac{\partial \Pi(\boldsymbol{x}_0)}{\partial x_2} \right\} . \tag{4.1.3b}$$

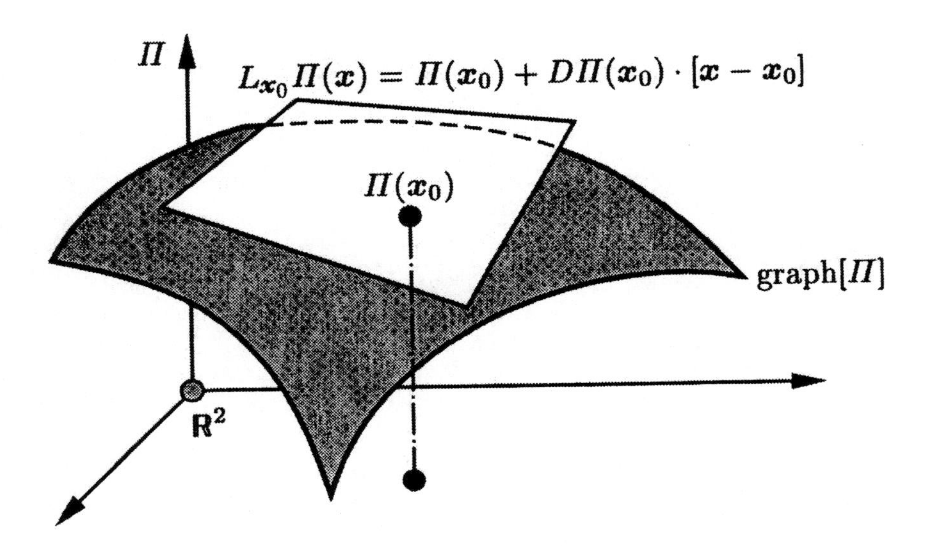

FIGURE 4.1. Illustration of the Fréchet derivative as the best linear approximation for a real function of two real variables (i.e., a surface).

See Figure **4.1** for an illustration.

Formally, we shall always assume the stronger condition of Fréchet differentiability. However, in practical calculations, the Gateaux derivative formula furnishes the most convenient tool for calculating the Fréchet derivative.

EXAMPLE: 4.1.2. Consider curves $u : [a, b] \to \mathbb{R}$ which are continuous and differentiable with a continuous derivative in $[a, b] \subset \mathbb{R}$, and have *fixed ends* such that $u(a) = u(b) = 0$. Following standard notation, we write $u \in C_0^1([a, b], \mathbb{R})$.*

The *arc-length* of such curves is a functional $\Pi : C_0^1([a, b], \mathbb{R}) \to \mathbb{R}$ defined by the familiar expression

$$\Pi(u) = \int_a^b \sqrt{1 + [u'(x)]^2}dx. \qquad (4.1.4a)$$

*This is a Banach space with the norm $\|u\|_{C^1} := \underset{x\in[a,b]}{\text{SUP}} [|u(x)| + |u'(x)|]$.

Then the Gateaux variation at curve $u_0 \in C_0^1([a, b], \mathbb{R})$ in the direction $\eta(x) \in C_0^1([a, b], \mathbb{R})$ is given by

$$\begin{aligned}
\delta \Pi(u, \eta) &= \frac{d}{d\alpha} \int_a^b \sqrt{1 + [u'(x) + \alpha\eta'(x)]^2} dx \Bigg|_{\alpha=0} \\
&= \int_a^b \frac{d}{d\alpha} \sqrt{1 + [u'(x) + \alpha\eta'(x)]^2} dx \Bigg|_{\alpha=0} \\
&= \int_a^b \frac{u'(x)}{\sqrt{1 + [u'(x)]^2}} \eta'(x) dx.
\end{aligned} \tag{4.1.4b}$$

In the classical literature on the calculus of variations (see, e.g., Gelfand and Fomin [1963]), the function $u_0(x) + \alpha\eta(x)$, for $\alpha > 0$, is called a *variation* of u_0. See Figure **4.2** for a graphical illustration.

4.1.2 The Functional Derivative

Given a functional $\Pi : \mathbb{V} \to \mathbb{R}$, the *functional derivative* of Π at $\boldsymbol{u} \in \mathbb{V}$ is an element of $\mathbb{V}^*$, denoted by $\frac{\delta \Pi}{\delta \boldsymbol{u}}(\boldsymbol{u}) \in \mathbb{V}^*$, which satisfies the relationship

$$\boxed{\delta \Pi(\boldsymbol{u}, \eta) =: \left\langle \frac{\delta \Pi}{\delta \boldsymbol{u}}(\boldsymbol{u}), \eta \right\rangle_{\mathbb{V}}.} \tag{4.1.5}$$

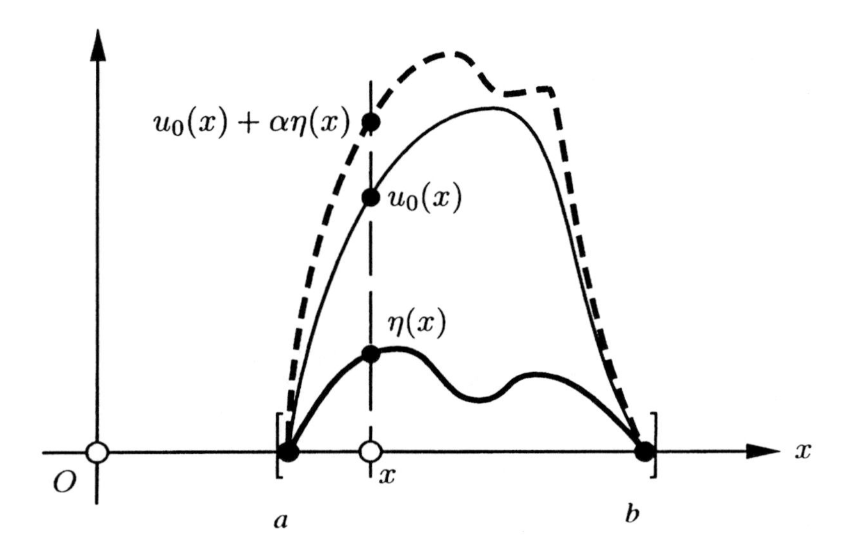

FIGURE 4.2. A geometric illustration of the variation $u_0(x) + \alpha\eta(x)$ of a function $u_0 \in C_0^1([a, b], \mathbb{R})$ for $\eta \in C_0^1([a, b], \mathbb{R})$.

In applications, (4.1.5) boils down to integration by parts of the expression for the Gateaux derivative, as the following example illustrates.

EXAMPLE: 4.1.3. Consider the same setting as in Example **4.1.2.** Integration of (4.1.4*b*) by parts yields

$$\delta\Pi(u,\eta) = -\int_a^b \left[\frac{u'(x)}{\sqrt{1+[u'(x)]^2}}\right]' \eta(x)dx + \left.\frac{u'(x)\eta(x)}{\sqrt{1+[u'(x)]^2}}\right|_a^b . \tag{4.1.6}$$

We note that the boundary terms vanish since, by assumption, $\eta(a) = \eta(b) = 0$. In the present situation, the duality pairing $\langle \bullet, \bullet \rangle_{\mathbb{V}}$ reduces to integration over the interval $[a, b]$ ($\mathbb{L}_2$-pairing). Thus, from (4.1.6) and (4.1.5),

$$\begin{aligned} \left\langle \frac{\delta\Pi}{\delta u}(u), \eta \right\rangle_{\mathbb{V}} &= -\int_a^b \left[\frac{u'(x)}{\sqrt{1+[u'(x)]^2}}\right]' \eta(x)\,dx \\ \Rightarrow \frac{\delta\Pi(u)}{\delta u} &= -\left[\frac{u'(x)}{\sqrt{1+[u'(x)]^2}}\right]' . \end{aligned} \tag{4.1.7}$$

The notion of functional derivative of a given functional provides the abstract version of the classical Euler–Lagrange equations. These equations constitute necessary conditions for a function to be an extremal, as shown below.

4.1.3 Euler–Lagrange Equations

In applications, the interest in the functional derivative is the direct consequence of the following classical result which represents the extension of the standard calculus test for extremal points of a function to the calculus of variations.

Proposition 4.1. *The necessary condition for a functional $\Pi : \mathbb{V} \to \mathbb{R}$ to have a local extremum (maximum, minimum, or saddle) at a point $\boldsymbol{u}_0 \in \mathbb{V}$ is that*

$$\boxed{\frac{\delta\Pi}{\delta \boldsymbol{u}}(\boldsymbol{u}_0) = 0 .} \tag{4.1.8}$$

These are the so-called Euler–Lagrange equations.

PROOF. Consider the real-valued function $\phi : \mathbb{V} \to \mathbb{R}$ defined for *fixed* but otherwise arbitrary $\boldsymbol{\eta} \in \mathbb{V}$ by

$$\phi(\epsilon) := \Pi(\boldsymbol{u}_0 + \epsilon\boldsymbol{\eta}) .$$

By hypothesis, $\phi(\epsilon)$ has an extremal point for $\epsilon = 0$. From elementary calculus,

$$\phi'(0) = \left.\frac{d}{d\epsilon}\Pi(\boldsymbol{u}_0 + \epsilon\boldsymbol{\eta})\right|_{\epsilon=0} = 0 .$$

By definition (4.1.1) of the variational derivative and definition (4.1.8) of the functional derivative,

$$\phi'(0) = \delta\Pi(\boldsymbol{u}_0, \boldsymbol{\eta}) = \left\langle \frac{\delta\Pi}{\delta\boldsymbol{u}}(\boldsymbol{u}_0), \boldsymbol{\eta} \right\rangle_{\mathbb{V}} = 0 \,.$$

Since $\boldsymbol{\eta} \in \mathbb{V}$ is *arbitrary*, the result follows by the properties of duality pairing. □

In the classical literature of the calculus of variations, the fact that $\left\langle \frac{\delta\Pi}{\delta\boldsymbol{u}}(\boldsymbol{u}_0), \boldsymbol{\eta} \right\rangle_{\mathbb{V}}$ implies that $\frac{\delta\Pi}{\delta\boldsymbol{u}}(\boldsymbol{u}) = \boldsymbol{0}$ for arbitrary functions in $C^0[\mathcal{B}, \mathbb{R}^3]$ is called the Dubois–Raymond–Lagrange lemma; see Gelfand and Fomin [1963], or Troutman [1983].

EXAMPLE: 4.1.4. From (4.1.7), the Euler–Lagrange equation for the problem considered in Examples **4.2** and **4.3** is expressed as

$$\left[\frac{u'(x)}{\sqrt{1 + [u'(x)]^2}}\right]' = 0 \;\Rightarrow\; u'(x) = constant. \tag{4.1.9}$$

Thus, $u(x) = Ax + B$, where A, B are constants. Since $u(a) = u(b) = 0$, it follows that $u(x) = 0$ for $x \in [a, b]$, that is, the solution of the Euler–Lagrange equation is a straight segment connecting points $(a, 0)$, and $(b, 0)$ (the shortest distance between a and b).

As a final illustration, consider the following important example.

EXAMPLE: 4.1.5. Let the functional $\Pi : \mathbb{V} \to \mathbb{R}$ be defined by the expression

$$\boxed{\Pi(\boldsymbol{u}) = \int_{\mathcal{B}} \{W[\boldsymbol{x}, \nabla\boldsymbol{u}(\boldsymbol{x})] - \boldsymbol{b}(\boldsymbol{x}) \cdot \boldsymbol{u}(\boldsymbol{x})\}\, dx \,,} \tag{4.1.10a}$$

where $\mathcal{B} \subset \mathbb{R}^3$, and $\boldsymbol{u}(\boldsymbol{x}) = \boldsymbol{0}$ for $\boldsymbol{x} \in \partial\mathcal{B}$ (the boundary of $\mathcal{B}$). Application of the Gateaux derivative formula yields

$$\begin{aligned} \delta\Pi(\boldsymbol{u}, \boldsymbol{\eta}) &:= \int_{\mathcal{B}} \frac{d}{d\alpha} \{W[\boldsymbol{x}, \nabla(\boldsymbol{u} + \alpha\boldsymbol{\eta})] - \boldsymbol{b} \cdot (\boldsymbol{u} + \alpha\boldsymbol{\eta})\} \Big|_{\alpha=0} dx \\ &= \int_{\mathcal{B}} [\nabla W(\boldsymbol{x}, \nabla\boldsymbol{u}) : \nabla\boldsymbol{\eta} - \boldsymbol{b} \cdot \boldsymbol{\eta}] dx \,, \end{aligned} \tag{4.1.10b}$$

so that application of Green's formula results in

$$\begin{aligned} \delta\Pi(\boldsymbol{u}, \boldsymbol{\eta}) := &- \int_{\mathcal{B}} \Big\{ \operatorname{div}[\nabla W(\boldsymbol{x}, \nabla\boldsymbol{u})] + \boldsymbol{b} \Big\} \cdot \boldsymbol{\eta} dx \\ &+ \int_{\partial\mathcal{B}} \boldsymbol{\eta} \cdot [\nabla^T W(\boldsymbol{x}, \nabla\boldsymbol{u})\boldsymbol{n}] d\Gamma, \end{aligned} \tag{4.1.10c}$$

where $\boldsymbol{n}(\boldsymbol{x})$ denotes the unit normal field to $\partial\mathcal{B}$. The boundary term *vanishes* because of the homogeneous boundary condition $\boldsymbol{\eta}|_{\partial\mathcal{B}} = \boldsymbol{0}$. Thus, the functional derivative takes the form

$$\boxed{\frac{\delta \Pi}{\delta \boldsymbol{u}}(\boldsymbol{u}) = -\text{div}\,\left\{\nabla W\big[\boldsymbol{x}, \nabla \boldsymbol{u}(\boldsymbol{x})\big]\right\} - \boldsymbol{b}(\boldsymbol{x})\,.} \tag{4.1.10d}$$

If $W(\boldsymbol{x}, \nabla \boldsymbol{u})$ is the strain energy and $\boldsymbol{b}(\boldsymbol{x})$ is the body force, then the Euler–Lagrange equations $\delta \Pi / \delta \boldsymbol{u} = \mathbf{0}$ are simply the equilibrium equations of elastostatics; see Gurtin [1972] for a detailed discussion.

4.2 General Variational Framework for Elastoplasticity

In this section we examine the variational formulation of classical elastoplasticity. Our objective is to define a *discrete functional* whose associated Euler–Lagrange equations yield the discrete equations of elastoplasticity, as summarized in Section **3.2** of Chapter **3**. The construction of this discrete functional relies crucially on the notion of plastic dissipation discussed in detail in Section **2.6** of Chapter **2**, and proceeds as follows.

i. We discretize the time interval $[0, T] \in \mathbb{R}_+$ of interest in nonoverlapping intervals $[0, T] = \bigcup_{n=0}^{N} [t_n, t_{n+1}]$. In a typical interval $[t_n, t_{n+1}]$, we assume that the initial data at t_n are given.

ii. We express the *total energy available* at time t_n in terms of the state variables at t_{n+1} as the sum of the potential energy at time t_{n+1} and the incremental dissipation in the interval $[t_n, t_{n+1}]$ computed by a backward Euler difference scheme. This leads to a *discrete functional* (Lagrangian) in terms of the unknown state variables at t_{n+1}.

iii. We show that the Euler–Lagrange equations associated with this discrete Lagrangian produce the equilibrium equations, the discrete versions of the flow rule and hardening law, and the discrete form of the loading conditions in Kuhn–Tucker form. We recall that the discrete flow rule and hardening law are the mathematical expression of the closest point projection algorithms discussed in detail in Chapter **3**.

To carry out the program outlined above, we start by recalling some further notation. We let $\mathbb{V}$ denote the space of kinematically admissible variations (or virtual displacements), defined as

$$\mathbb{V} := \left\{\boldsymbol{\eta} : \mathcal{B} \to \mathbb{R}^{n_{\text{dim}}} \mid \boldsymbol{\eta} \in [\mathbb{H}^1(\mathcal{B})]^{n_{\text{dim}}}\,;\ \boldsymbol{\eta}|_{\partial_u \mathcal{B}} = \mathbf{0}\right\}, \tag{4.2.1a}$$

where $n_{\text{dim}} \leq 3$ is the spatial dimension and $\mathbb{H}^1(\mathcal{B})$ denotes the Sobolev space of functions with square-integrable first-order derivatives.†. Further, we recall from Chapter **2** that $\partial_u \mathcal{B} \subset \partial \mathcal{B}$ and $\partial_\sigma \mathcal{B} \subset \partial \mathcal{B}$ are the parts of the boundary $\partial \mathcal{B}$ where

†Note that $\mathbb{V} = \mathbb{V}^*$. The duality pairing for the computation of functional derivatives is simply the $\mathbb{L}_2$ inner product, i.e., $\langle \boldsymbol{\eta}_1, \boldsymbol{\eta}_2 \rangle := \int_{\mathcal{B}} \boldsymbol{\eta}_1 \cdot \boldsymbol{\eta}_2 dV$.

displacement and stresses are respectively prescribed according to

$$u|_{\partial_u \mathcal{B}} = \bar{u},$$

and (4.2.1b)

$$\sigma n|_{\partial_\sigma \mathcal{B}} = \bar{t}.$$

Finally, we let $[0, T] \subset \mathbb{R}_+$ be the time interval of interest and denote the *current time* by $t \in [0, T]$. In the exposition that follows, we shall designate the value of the variable at time $\xi \in [0, t]$ by a *subscript* ξ. In addition, to simplify the notation, explicit indication of the spatial argument $x \in \mathcal{B}$ is often omitted.

4.2.1 Variational Characterization of Plastic Response

With the preceding notation in hand, we proceed to define the following functionals.

4.2.1.1 Free energy functional

We introduce the notion of *total free energy* available at current time t, by the following three-field functional of the *Hu–Washizu* type:

$$\boxed{\begin{aligned} \mathcal{P}_t(u_t, \varepsilon_t - \varepsilon_t^p, q, \sigma_t) := \int_{\mathcal{B}} \{\Psi(\varepsilon_t - \varepsilon_t^p, q) + \sigma_t : (\nabla^s u_t - \varepsilon_t)\}\, dV \\ + \mathcal{P}_{\text{ext}}(u_t). \end{aligned}} \tag{4.2.2}$$

Here, Ψ is the free energy defined in terms of the stored-energy function W and the contribution of the hardening variable $\mathcal{H}$ by (2.6.22*b*). For simplicity, we assume that $\mathcal{H}(\alpha)$ is quadratic so that, by the Legendre transformation, we trivially express $\mathcal{H}$ as a function of q. Therefore we set

$$\boxed{\begin{aligned} \mathcal{H} &= \tfrac{1}{2} q \cdot \mathbf{D}^{-1} q, \\ \Psi(\varepsilon - \varepsilon^p, q) &:= W(\varepsilon - \varepsilon^p) + \tfrac{1}{2} q \cdot \mathbf{D}^{-1} q. \end{aligned}} \tag{4.2.3}$$

Identical developments hold for a general form of $\mathcal{H}$. In addition, $\mathcal{P}_{\text{ext}}(u_t)$ denotes the potential energy of the external loading at current time, which, under the assumption of dead loading, is given by

$$\mathcal{P}_{\text{ext}}(u_t) := -\int_{\mathcal{B}} \rho b \cdot u_t dV - \int_{\partial_\sigma \mathcal{B}} \bar{t} \cdot u_t d\Gamma. \tag{4.2.4}$$

In (4.2.2) we have assumed that $\{u_t, \varepsilon_t, \sigma_t, q_t\}$ are independent variables. Then in this context, σ_t may be regarded as a Lagrange multiplier that enforces (weakly) the constraint $\nabla^s u_t - \varepsilon_t = \mathbf{0}$. The generality afforded by the three-field functional (4.2.2) is warranted by the class of mixed, finite-element methods considered in Section **4.3.**

4.2.1.2 Plastic dissipation functional

In addition to the free energy functional (4.2.2), we introduce the Lagrangian functional associated with the *plastic dissipation over the entire body* up to current time $t \in \mathbb{R}_+$. Recall from Section **2.6** that plastic dissipation is given by

$$\mathcal{D}^p_\xi := \dot{\boldsymbol{\varepsilon}}^p_\xi : \nabla W(\boldsymbol{\varepsilon}_\xi - \boldsymbol{\varepsilon}^p_\xi) - \dot{\boldsymbol{q}}_\xi \cdot \mathbf{D}^{-1}\boldsymbol{q}_\xi \geq 0, \tag{4.2.5a}$$

where the variables $\{\boldsymbol{\varepsilon}_\xi, \boldsymbol{\varepsilon}^p_\xi, \boldsymbol{q}_\xi\}$ are constrained to lie in the closure of the elastic domain. This constraint can be removed through the standard method of Lagrange multipliers, by considering the following functional associated with the the total dissipation up to time t

$$\boxed{\mathcal{L}^p_t := \int_0^t \int_{\mathcal{B}} \left\{\mathcal{D}^p_\xi - \gamma_\xi f\left[\nabla W(\boldsymbol{\varepsilon}_\xi - \boldsymbol{\varepsilon}^p_\xi), \boldsymbol{q}_\xi\right]\right\} dV d\xi.} \tag{4.2.5b}$$

Here, $\xi \mapsto \gamma_\xi \in \mathbb{K}^p$ is the Lagrange multiplier to be interpreted as the plastic consistency parameter. $\mathbb{K}^p$ is the set (a positive cone) defined as

$$\boxed{\mathbb{K}^p := \left\{\gamma \mid \gamma \in \mathbb{L}_2(\mathcal{B}),\ \gamma \geq 0\right\},} \tag{4.2.6}$$

and $f[\nabla W(\boldsymbol{\varepsilon}_\xi - \boldsymbol{\varepsilon}^p_\xi), \boldsymbol{q}_\xi] = 0$ defines the plastic yield surface at time $\xi \in [0, t]$. Although the yield function is given in stress space, it should be noted that $\mathcal{L}^p_t$ is formulated in strain space by replacing the stress tensor with the expression $\nabla W(\boldsymbol{\varepsilon}_\xi - \boldsymbol{\varepsilon}^p_\xi)$. Note further that we use $\nabla W(\boldsymbol{\varepsilon}_\xi - \boldsymbol{\varepsilon}^p_\xi)$, *not* $\boldsymbol{\sigma}_\xi$, as one would expect, in this transformation to strain space. This observation is crucial for the developments that follow.

4.2.2 *Discrete Lagrangian for elastoplasticity*

Next, we proceed to discretize functionals introduced above and focus our attention on a typical time interval $[t_n, t_{n+1}]$.

4.2.2.1 Discrete plastic dissipation functional

We consider a *discrete plastic dissipation functional* obtained from its continuum counterpart, defined by (4.2.5), by means of a backward Euler difference scheme. Let $t = t_{n+1}$ be the current time, and let $t_{n+1} = t_n + \Delta t$, where $\Delta t > 0$ is the time step. Then from (4.2.5)

$$\begin{aligned}\mathcal{L}^p_{n+1} = \mathcal{L}^p_n &+ \int_{t_n}^{t_{n+1}} \int_{\mathcal{B}} \left[\dot{\boldsymbol{\varepsilon}}^p_\xi : \nabla W(\boldsymbol{\varepsilon}_\xi - \boldsymbol{\varepsilon}^p_\xi) - \dot{\boldsymbol{q}}_\xi \cdot \mathbf{D}^{-1}\boldsymbol{q}_\xi\right] dV d\xi \\ &- \int_{t_n}^{t_{n+1}} \int_{\mathcal{B}} \gamma_\xi f\left[\nabla W(\boldsymbol{\varepsilon}_\xi - \boldsymbol{\varepsilon}^p_\xi), \boldsymbol{q}_\xi\right] dV d\xi.\end{aligned} \tag{4.2.7}$$

For convenience, in what follows, we use the following notation

$$\left.\begin{aligned} \chi_{n+1} &:= \{\boldsymbol{u}_{n+1}, \boldsymbol{\varepsilon}_{n+1}, \boldsymbol{\sigma}_{n+1}, \boldsymbol{\varepsilon}^p_{n+1}, \boldsymbol{q}_{n+1}, \Delta\gamma\}, \\ W_{n+1} &:= W(\boldsymbol{\varepsilon}_{n+1} - \boldsymbol{\varepsilon}^p_{n+1}), \\ f_{n+1} &:= f(\nabla W(\boldsymbol{\varepsilon}_{n+1} - \boldsymbol{\varepsilon}^p_{n+1}), \boldsymbol{q}_{n+1}). \end{aligned}\right\} \tag{4.2.8}$$

In addition, we assume that the history of the state variables up to time t_n, which is collectively denoted by χ_ξ for $\xi \in [0, t_n]$, *is given* and remains fixed through the developments that follow. Then, with this observation in mind, evaluation of (4.2.7) by a backward Euler algorithm results in the following expression:

$$\boxed{\begin{aligned} \mathcal{L}^p(\chi_{n+1}) := \mathcal{L}^p_n + \int_B \{[\boldsymbol{\varepsilon}^p_{n+1} - \boldsymbol{\varepsilon}^p_n] : \nabla W_{n+1} - \Delta\gamma f_{n+1} \\ - \boldsymbol{q}_{n+1} \cdot \mathbf{D}^{-1}(\boldsymbol{q}_{n+1} - \boldsymbol{q}_n)\} dV, \end{aligned}} \tag{4.2.9}$$

where interchange of spatial and time integration is allowed by the assumed smoothness of the integrand, and we have set $\Delta\gamma := \gamma\Delta t$. The functional $\mathcal{L}^p(\chi_{n+1})$ furnishes the algorithmic approximation to the plastic dissipation $\mathcal{L}^p_t$ up to time $t = t_{n+1}$.

4.2.2.2 The discrete Lagrangian

Now a discrete variational formulation of classical elastoplasticity is obtained by a functional denoted by $\widehat{\mathcal{P}}_n(\chi_{n+1})$, and defined as the *the total free energy available at time t_n expressed in terms of the state χ_{n+1} at t_{n+1}*. Accordingly, the discrete functional $\widehat{\mathcal{P}}_n(\chi_{n+1})$ is obtained as the sum of the energy $\mathcal{P}_{n+1}(\chi_{n+1})$ at t_{n+1} and the incremental dissipation in $[t_n, t_{n+1}]$:

$$\boxed{\text{Total free energy}\Big|_n = \text{Total free energy}\Big|_{n+1} + \text{Dissipation}\Big|_n^{n+1}.} \tag{4.2.10}$$

In view of this definition, it follows that $\widehat{\mathcal{P}}_n(\chi_{n+1})$ is given by the relationship

$$\widehat{\mathcal{P}}_n(\chi_{n+1}) := \mathcal{P}_{n+1}(\chi_{n+1}) + [\mathcal{L}^p(\chi_{n+1}) - \mathcal{L}^p_n]. \tag{4.2.11}$$

Then substituting of (4.2.9) into (4.2.11) yields the explicit expression

$$\boxed{\widehat{\mathcal{P}}_n(\chi_{n+1}) := \int_B \widehat{\mathcal{L}}_n(\chi_{n+1}) dV + \mathcal{P}_{\text{ext}}(\boldsymbol{u}_{n+1}),} \tag{4.2.12a}$$

where $\widehat{\mathcal{L}}_n(\chi_{n+1})$ is the *discrete Lagrangian* associated with $\widehat{\mathcal{P}}_n(\chi_{n+1}) - \mathcal{P}_{\text{ext}}(\boldsymbol{u}_{n+1})$, which is given by

$$\boxed{\begin{aligned} \widehat{\mathcal{L}}_n(\chi_{n+1}) : &= W_{n+1} + \tfrac{1}{2}\boldsymbol{q}_{n+1} \cdot \mathbf{D}^{-1}\boldsymbol{q}_{n+1} + \boldsymbol{\sigma}_{n+1} : [\nabla^s \boldsymbol{u}_{n+1} - \boldsymbol{\varepsilon}_{n+1}] \\ &- \Delta\gamma f_{n+1} + [\boldsymbol{\varepsilon}^p_{n+1} - \boldsymbol{\varepsilon}^p_n] : \nabla W_{n+1} - \boldsymbol{q}_{n+1} \cdot \mathbf{D}^{-1}(\boldsymbol{q}_{n+1} - \boldsymbol{q}_n). \end{aligned}} \tag{4.2.12b}$$

Below we show that the Euler–Lagrange equations associated with this functional furnish the discrete equations of elastoplasticity. To carry out the standard variational argument, we shall formally assume that $\boldsymbol{q} \in \mathbb{V}_q, \boldsymbol{\sigma} \in \mathbb{V}_\sigma$, and $\boldsymbol{\varepsilon} \in \mathbb{V}_\varepsilon$,

$$\left.\begin{aligned} \mathbb{V}_\sigma &:= \{\boldsymbol{\sigma}: \mathcal{B} \to \mathbb{S} \quad | \quad \sigma_{ij} \in \mathbb{L}_2(\mathcal{B})\}, \\ \mathbb{V}_\varepsilon &:= \{\boldsymbol{\varepsilon}: \mathcal{B} \to \mathbb{S} \quad | \quad \epsilon_{ij} \in \mathbb{L}_2(\mathcal{B})\}, \\ \mathbb{V}_q &:= \{\boldsymbol{q}: \mathcal{B} \to \mathbb{R}^m \quad | \quad q_i \in \mathbb{L}_2(\mathcal{B})\}. \end{aligned}\right\} \tag{4.2.13}$$

Since our calculations are formal we shall not elaborate on the significance of this functional setting and refer to Temam [1985] for an in-depth discussion of these and related topics.

4.2.3 Variational Form of the Governing Equations

Formally, the Euler–Lagrange equations associated with the functional (4.2.12) yield (i) the equilibrium equations, (ii) the strain-displacement relationships, (iii) the elastic constitutive equations, (iv) the closest point projection algorithm, and (v) the discrete Kuhn–Tucker conditions. The explicit result is given in the following.

Proposition 4.1. *The stationarity conditions of the time-discretized functional (4.2.12) result in the following weak forms of the governing equations:*

$$\begin{aligned} \delta\widehat{\mathcal{P}}_n(\chi_{n+1}, \boldsymbol{\eta}) &= \int_{\mathcal{B}} \left[\boldsymbol{\sigma}_{n+1} : \nabla^s\boldsymbol{\eta} - \rho\boldsymbol{b} \cdot \boldsymbol{\eta}\right] dV - \int_{\partial_\sigma\mathcal{B}} \bar{\boldsymbol{t}} \cdot \boldsymbol{\eta}\, d\Gamma \\ &= 0, \end{aligned} \tag{4.2.14a}$$

$$\delta\widehat{\mathcal{P}}_n(\chi_{n+1}, \boldsymbol{\tau}) = \int_{\mathcal{B}} \boldsymbol{\tau} : \left[\nabla^s\boldsymbol{u}_{n+1} - \boldsymbol{\varepsilon}_{n+1}\right] dV = 0, \tag{4.2.14b}$$

$$\begin{aligned} \delta\widehat{\mathcal{P}}_n(\chi_{n+1}, \boldsymbol{\xi}) &= \int_{\mathcal{B}} \boldsymbol{\xi} : \{\left[-\boldsymbol{\sigma}_{n+1} + \nabla W_{n+1}\right] \\ &\qquad + \mathbf{C}_{n+1} : \left[\boldsymbol{\varepsilon}^p_{n+1} - \boldsymbol{\varepsilon}^p_n - \Delta\gamma\,\partial_\sigma f_{n+1}\right]\} dV \\ &= 0, \end{aligned} \tag{4.2.14c}$$

$$\begin{aligned} \delta\widehat{\mathcal{P}}_n(\chi_{n+1}, \boldsymbol{\xi}^p) &= -\int_{\mathcal{B}} \boldsymbol{\xi}^p : \mathbf{C}_{n+1} : \left[\boldsymbol{\varepsilon}^p_{n+1} - \boldsymbol{\varepsilon}^p_n - \Delta\gamma\,\partial_\sigma f_{n+1}\right] dV \\ &= 0, \end{aligned} \tag{4.2.14d}$$

$$\begin{aligned} \delta\widehat{\mathcal{P}}_n(\chi_{n+1}, \boldsymbol{p}) &= -\int_{\mathcal{B}} \boldsymbol{p} \cdot \left[\mathbf{D}^{-1}(\boldsymbol{q}_{n+1} - \boldsymbol{q}_n) + \Delta\gamma\,\partial_q f_{n+1}\right] dV \\ &= 0, \end{aligned} \tag{4.2.14e}$$

$$\delta\widehat{\mathcal{P}}_n(\chi_{n+1}, \lambda) = \int_{\mathcal{B}} \lambda f_{n+1}\, dV = 0, \tag{4.2.14f}$$

for arbitrary displacement variation $\boldsymbol{\eta} \in \mathbb{V}$, stress variations $\boldsymbol{\tau} \in \mathbb{V}_\sigma$, strain $\boldsymbol{\xi} \in \mathbb{V}_\varepsilon$, $\boldsymbol{\xi}^p \in \mathbb{V}_\varepsilon$, and variations $\boldsymbol{p} \in \mathbb{V}_q$, $\lambda \in \mathbb{K}^p$. In these equations, $\mathbf{C}_{n+1} := \nabla^2 W_{n+1}$ denotes the (generally nonconstant) tensor of elastic moduli.

PROOF. The first two equations follow by a straightforward argument analogous to that employed in Example **4.1.5**. The proof of the remaining equations is based on the following relationships obtained by applying the chain rule:

$$\left.\begin{aligned}
\frac{d}{d\alpha}\ \nabla W(\varepsilon_{n+1}+\alpha\xi-\varepsilon^p_{n+1})\big|_{\alpha=0} &= \mathbf{C}_{n+1}:\xi,\\
\frac{d}{d\alpha}\ \nabla W(\varepsilon_{n+1}-\varepsilon^p_{n+1}-\alpha\xi^p)\big|_{\alpha=0} &= -\mathbf{C}_{n+1}:\xi^p,\\
\frac{d}{d\alpha}\ f\big[\nabla W(\varepsilon_{n+1}+\alpha\xi-\varepsilon^p_{n+1}),q_{n+1}\big]\big|_{\alpha=0} &= \partial_\sigma f_{n+1}:\mathbf{C}_{n+1}:\xi,\\
\frac{d}{d\alpha}\ f\big[\nabla W(\varepsilon_{n+1}-\varepsilon^p_{n+1}-\alpha\xi^p),q_{n+1}\big]\big|_{\alpha=0} &= -\partial_\sigma f_{n+1}:\mathbf{C}_{n+1}:\xi^p,
\end{aligned}\right\} \tag{4.2.15}$$

Using relationships $(4.2.15)_{2,4}$, from (4.2.14b),

$$\begin{aligned}
\delta\widehat{\mathcal{P}}_n(\chi_{n+1},\xi^p) = -\int_{\mathcal{B}}\{&\xi^p:[-\nabla W_{n+1}+\nabla W_{n+1}]\\
&+\xi^p:\mathbf{C}_{n+1}:\big[\varepsilon^p_{n+1}-\varepsilon^p_n-\Delta\gamma\,\partial_\sigma f_{n+1}\big]\}dV\\
&=0,
\end{aligned} \tag{4.2.16}$$

which reduces to (4.2.14*d*) after cancelling like terms. A similar calculation yields the remaining equations. □

Corollary 4.1. *The Euler–Lagrange equations associated with the discrete Lagrangian* (4.2.12) *take the form*

$$\frac{\delta\widehat{\mathcal{P}}_n(\chi_{n+1})}{\delta u_{n+1}} = \begin{cases} -\operatorname{div}\sigma_{n+1}-\rho b=\mathbf{0}\,, & \text{in } \mathcal{B}\\ \sigma n-\bar{t}=\mathbf{0}\,, & \text{on } \partial_\sigma\mathcal{B}\,, \end{cases} \tag{4.2.17a}$$

$$\frac{\delta\widehat{\mathcal{P}}_n(\chi_{n+1})}{\delta \varepsilon_{n+1}} = \nabla^s u_{n+1}-\varepsilon_{n+1}=\mathbf{0}\,, \tag{4.2.17b}$$

$$\frac{\delta\widehat{\mathcal{P}}_n(\chi_{n+1})}{\delta \sigma_{n+1}} = -\sigma_{n+1}+\nabla W_{n+1}=\mathbf{0}\,, \tag{4.2.17c}$$

$$\frac{\delta\widehat{\mathcal{P}}_n(\chi_{n+1})}{\delta \varepsilon^p_{n+1}} = -\varepsilon^p_{n+1}+\varepsilon^p_n+\Delta\gamma\,\partial_\sigma f_{n+1}=0\,, \tag{4.2.17d}$$

$$\frac{\delta\widehat{\mathcal{P}}_n(\chi_{n+1})}{\delta q_{n+1}} = -\mathbf{D}^{-1}(q_{n+1}-q_n)-\Delta\gamma\,\partial_q f_{n+1}=0\,, \tag{4.2.17e}$$

$$\left.\begin{aligned}
\frac{\delta\widehat{\mathcal{P}}_n(\chi_{n+1})}{\delta \Delta\gamma} &= f_{n+1}\le 0,\\
\Delta\gamma &\ge 0,\\
\Delta\gamma f_{n+1} &= 0.
\end{aligned}\right\} \tag{4.2.17f}$$

Proof. This follows immediately from variational equations (4.2.14*a*–*f*) by integrating by parts and standard arguments. □

Remark 4.2.1. Assume that $\mathbf{C} := \nabla^2 W$ is *constant*. Combining (4.2.17c) and (4.2.17d),

$$\left.\begin{aligned} \sigma_{n+1}^{trial} &:= \mathbf{C} : [\varepsilon_{n+1} - \varepsilon_n^p], \\ \sigma_{n+1} &= \sigma_{n+1}^{trial} - \Delta\gamma \mathbf{C} : \partial_\sigma f_{n+1}. \end{aligned}\right\} \tag{4.2.18}$$

which coincides with the expression of the *closest point projection* algorithm, as discussed in Chapter **3**.

4.2.4 Extension to Viscoplasticity

The variational formulation of viscoplasticity follows along the same lines as the rate-independent case discussed above, and is based on the viscoplastic regularization of maximum plastic dissipation considered in Section **2.7**. First, since $\Delta\gamma \geq 0$ is no longer an independent variable, we set

$$\chi_{n+1} := \{\boldsymbol{u}_{n+1}, \varepsilon_{n+1}, \sigma_{n+1}, \varepsilon_{n+1}^{\mathrm{vp}}, \boldsymbol{q}_{n+1}\}. \tag{4.2.19}$$

Next, we recall from Chapter **2** that the viscoplastic regularization of the dissipation function takes the form

$$\boxed{\begin{aligned} \mathcal{D}_{\eta\,\xi}^{\mathrm{vp}} := \dot{\varepsilon}_\xi^{\mathrm{vp}} &: \nabla W(\varepsilon_\xi - \varepsilon_\xi^{\mathrm{vp}}) - \dot{\boldsymbol{q}}_\xi \cdot \mathbf{D}^{-1}\boldsymbol{q}_\xi \\ &- \frac{1}{\eta}\gamma_\xi^+\Big\{f\big(\nabla W(\varepsilon_\xi - \varepsilon_\xi^{\mathrm{vp}})\big)\Big\}, \end{aligned}} \tag{4.2.20}$$

where $\eta > 0$ is the viscosity coefficient and $\gamma_\xi^+ : \mathbb{R} \to \mathbb{R}_+$ is defined by (2.7.4). Therefore, the viscoplastic counterpart of the time-discretized functional (4.2.9) takes the following form

$$\begin{aligned} \mathcal{L}^{\mathrm{vp}}(\chi_{n+1}) &:= \mathcal{L}_n^{\mathrm{vp}} + \int_{t_n}^{t_{n+1}} \int_{\mathcal{B}} \mathcal{D}_{\eta\,\xi}^{\mathrm{vp}} dV d\xi \\ &= \int_{\mathcal{B}} \{[\varepsilon_{n+1}^{\mathrm{vp}} - \varepsilon_n^{\mathrm{vp}}] : \nabla W_{n+1} - \frac{\Delta t}{\eta}\gamma_{n+1}^+(f_{n+1}) \\ &\quad - \boldsymbol{q}_{n+1} \cdot \mathbf{D}^{-1}(\boldsymbol{q}_{n+1} - \boldsymbol{q}_n)\} dV. \end{aligned} \tag{4.2.21}$$

Then for viscoplasticity, the counterpart of the discrete functional (4.2.11) becomes

$$\widehat{\mathcal{P}}_n(\chi_{n+1}) := \mathcal{P}_{n+1}(\chi_{n+1}) + [\mathcal{L}^{\mathrm{vp}}(\chi_{n+1}) - \mathcal{L}_n^{\mathrm{vp}}], \tag{4.2.22}$$

which, in view of (4.2.21), can be written as follows

$$\begin{aligned} \widehat{\mathcal{P}}_n(\chi_{n+1}) := \int_{\mathcal{B}} \Big\{ & W_{n+1} + \tfrac{1}{2}\boldsymbol{q}_{n+1} \cdot \mathbf{D}^{-1}\boldsymbol{q}_{n+1} + \sigma_{n+1} : [\nabla^s \boldsymbol{u}_{n+1} - \varepsilon_{n+1}] \\ & + [\varepsilon_{n+1}^{\mathrm{vp}} - \varepsilon_n^{\mathrm{vp}}] : \nabla W_{n+1} - \frac{\Delta t}{\eta}\gamma_{n+1}^+(f_{n+1}) \\ & - \boldsymbol{q}_{n+1} \cdot \mathbf{D}^{-1}(\boldsymbol{q}_{n+1} - \boldsymbol{q}_n)\Big\} dV + \mathcal{P}_{\mathrm{ext}}. \end{aligned} \tag{4.2.23}$$

Finally, the first three Euler–Lagrange equations remain unchanged and are given by (4.2.17a–c) (or by (4.2.14a–c) in weak form). Now the next two Euler–Lagrange equations, which are given in the rate-independent case by (4.2.17d,e), are replaced by

$$\left.\begin{aligned} \frac{\delta \widehat{\mathcal{P}}_n(\chi_{n+1})}{\delta \varepsilon^{\mathrm{vp}}_{n+1}} &= -\varepsilon^{\mathrm{vp}}_{n+1} + \varepsilon^{\mathrm{vp}}_{n} + \frac{\Delta t \langle f_{n+1}\rangle}{\eta}\,\partial_\sigma f_{n+1} = 0,\\ \frac{\delta \widehat{\mathcal{P}}_n(\chi_{n+1})}{\delta \boldsymbol{q}_{n+1}} &= -\mathbf{D}^{-1}(\boldsymbol{q}_{n+1} - \boldsymbol{q}_n) - \frac{\Delta t \langle f_{n+1}\rangle}{\eta}\,\partial_q f_{n+1} = 0, \end{aligned}\right\} \tag{4.2.24}$$

where $f_{n+1} := f\big[\nabla W(\varepsilon_{n+1} - \varepsilon^{\mathrm{vp}}_{n+1}), \boldsymbol{q}_{n+1}\big]$.

4.3 Finite-Element Formulation. Assumed-Strain Method

To develop the class of assumed-strain methods considered in this section, we shall assume in what follows that the flow rule, the hardening law, and the loading/unloading conditions hold pointwise. Accordingly, we assume that (4.2.17d–f) hold, so that the variables $\{\varepsilon^p_{n+1}, \Delta\gamma, \boldsymbol{q}_{n+1}\}$ are determined from the strain tensor ε_{n+1} by the closest point projection algorithm. The manner in which this is accomplished in the finite-element method is explained below.

4.3.1 Matrix and Vector Notation

Following common usage in the finite-element literature, in the exposition that follows we abandon tensor notation in favor of matrix and vector notation. The following standard conventions are used:

1. Second-order tensors in three-dimensional Euclidean space are mapped into column vectors according to a convention which distinguishes $\mathbb{V}_\sigma$ (contravariant) from $\mathbb{V}_\varepsilon$ (covariant):

$$\varepsilon = \begin{Bmatrix} \epsilon_{11} \\ \epsilon_{22} \\ \epsilon_{33} \\ 2\epsilon_{12} \\ 2\epsilon_{13} \\ 2\epsilon_{23} \end{Bmatrix},$$

$$\varepsilon^p = \begin{Bmatrix} \epsilon^p_{11} \\ \epsilon^p_{22} \\ \epsilon^p_{33} \\ 2\epsilon^p_{12} \\ 2\epsilon^p_{13} \\ 2\epsilon^p_{23} \end{Bmatrix},$$

$$\nabla^s \boldsymbol{u} = \begin{Bmatrix} u_{1,1} \\ u_{2,2} \\ u_{3,3} \\ u_{1,2} + u_{2,1} \\ u_{1,3} + u_{3,1} \\ u_{2,3} + u_{3,2} \end{Bmatrix}, \tag{4.3.1a}$$

and

$$\boldsymbol{\sigma} = \begin{Bmatrix} \sigma_{11} \\ \sigma_{22} \\ \sigma_{33} \\ \sigma_{12} \\ \sigma_{13} \\ \sigma_{23} \end{Bmatrix}.$$

2. Fourth-order tensors are mapped onto matrices. Of particular importance is the elasticity tensor $\mathbf{C} = \nabla^2 W$ which in matrix notation takes the explicit form

$$\mathbf{C} = \begin{bmatrix} C_{1111} & C_{1122} & C_{1133} & C_{1112} & C_{1113} & C_{1123} \\ & C_{2222} & C_{2233} & C_{2212} & C_{2213} & C_{2223} \\ & & C_{3333} & C_{3312} & C_{3313} & C_{3323} \\ & & & C_{1212} & C_{1213} & C_{1223} \\ & & & & C_{1313} & C_{1323} \\ & & & & & C_{2323} \end{bmatrix}. \tag{4.3.1b}$$

3. According to the preceding conventions, the *fourth-order unit tensor* **I** and the *second order unit tensor* **1** become

$$\mathbf{I} = \begin{bmatrix} 1 & 0 & 0 & 0 & 0 & 0 \\ 0 & 1 & 0 & 0 & 0 & 0 \\ 0 & 0 & 1 & 0 & 0 & 0 \\ 0 & 0 & 0 & 1 & 0 & 0 \\ 0 & 0 & 0 & 0 & 1 & 0 \\ 0 & 0 & 0 & 0 & 0 & 1 \end{bmatrix},$$

and (4.3.1c)

$$\mathbf{1} = \begin{Bmatrix} 1 \\ 1 \\ 1 \\ 0 \\ 0 \\ 0 \end{Bmatrix}.$$

4. Contraction between second-order tensors is replaced by a dot product, and application of a fourth-order tensor to a second-order tensor reduces to a matrix transformation; for instance,

$$\left.\begin{aligned} \boldsymbol{\sigma} : \boldsymbol{\varepsilon} = \sigma_{ij}\varepsilon_{ij} \quad &\Longleftrightarrow \quad \boldsymbol{\sigma} \cdot \boldsymbol{\varepsilon} = \boldsymbol{\sigma}^T \boldsymbol{\varepsilon}, \\ \boldsymbol{\sigma} = \mathbf{C} : \boldsymbol{\varepsilon}^e \quad &\Longleftrightarrow \quad \boldsymbol{\sigma} = \mathbf{C}\boldsymbol{\varepsilon}^e. \end{aligned}\right\} \tag{4.3.1d}$$

Similar conventions apply to plane-strain, plane-stress and axisymmetric problems.

4.3.2 Summary of Governing Equations

According to the preceding ideas, the relevant set of variational equations, repeated below for convenience, is the following:

$$\delta \mathcal{P}(\chi_{n+1}, \boldsymbol{\eta}) = \int_{\mathcal{B}} (\boldsymbol{\sigma}_{n+1} \cdot \nabla^s \boldsymbol{\eta} - \rho \boldsymbol{b} \cdot \boldsymbol{\eta}) dV - \int_{\partial_\sigma \mathcal{B}} \bar{\boldsymbol{t}} \cdot \boldsymbol{\eta} d\Gamma = 0, \quad (4.3.2a)$$

$$\delta \mathcal{P}(\chi_{n+1}, \boldsymbol{\tau}) = \int_{\mathcal{B}} \boldsymbol{\tau} \cdot (\nabla^s \boldsymbol{u}_{n+1} - \boldsymbol{\varepsilon}_{n+1}) dV = 0, \quad (4.3.2b)$$

$$\delta \mathcal{P}(\chi_{n+1}, \boldsymbol{\xi}) = \int_{\mathcal{B}} \boldsymbol{\xi} \cdot \big[- \boldsymbol{\sigma}_{n+1} + \nabla W(\boldsymbol{\varepsilon}_{n+1} - \boldsymbol{\varepsilon}^p_{n+1}) \big] dV = 0\,, \quad (4.3.2c)$$

where vector notation is used. These equations are supplemented by the local Euler–Lagrange equations (4.2.14d–f) that define $\boldsymbol{\varepsilon}^p_{n+1}$, $\Delta\gamma$, and $\boldsymbol{q}_{n+1}$ in terms of $\boldsymbol{\varepsilon}_{n+1}$. In what follows, for simplicity we restrict our attention to perfect plasticity. Accordingly,

$$\boldsymbol{\varepsilon}^p_{n+1} = \boldsymbol{\varepsilon}^p_n + \Delta\gamma \partial_\sigma f \big[\nabla W(\boldsymbol{\varepsilon}_{n+1} - \boldsymbol{\varepsilon}^p_{n+1})\big], \quad (4.3.2d)$$

$$\left. \begin{aligned} \Delta\gamma \geq 0, \quad f\big[\nabla W(\boldsymbol{\varepsilon}_{n+1} - \boldsymbol{\varepsilon}^p_{n+1})\big] \leq 0\,, \\ \Delta\gamma f\big[\nabla W(\boldsymbol{\varepsilon}_{n+1} - \boldsymbol{\varepsilon}^p_{n+1})\big] = 0\,. \end{aligned} \right\} \quad (4.3.2e)$$

The treatment of viscoplasticity follows the same lines as rate-independent plasticity, and therefore we shall omit further details.

4.3.3 Discontinuous Strain and Stress Interpolations

The finite-element formulation discussed below is based on *discontinuous interpolations* of stress and strain over a typical element $\mathcal{B}^e \subset \mathbb{R}^{n_{\text{dim}}}$ of a discretization $\mathcal{B} \approx \bigcup_{e=1}^{n_{\text{en}}} \mathcal{B}^e$. Our goal is to recover a displacement-like, finite-element architecture. We start by introducing a finite-dimensional, approximating subspace for stresses $\mathbb{V}^h_\sigma$, defined as

$$\mathbb{V}^h_\sigma := \big\{\boldsymbol{\sigma}^h \in \mathbb{V}_\sigma \mid \boldsymbol{\sigma}^h|_{\mathcal{B}^e} = \boldsymbol{S}(\boldsymbol{x})\boldsymbol{c}_e\,,\ \boldsymbol{c}_e \in \mathbb{R}^m\big\}\,. \quad (4.3.3)$$

Here, $\boldsymbol{S}(\boldsymbol{x})$ is a $6 \times m$ matrix of prescribed functions, generally given in terms of natural coordinates. Explicit examples are given below. Similarly, one introduces a strain, finite-element approximating subspace $\mathbb{V}^h_\varepsilon$ defined as

$$\mathbb{V}^h_\varepsilon := \big\{\boldsymbol{\varepsilon}^h \in \mathbb{V}_\varepsilon \mid \boldsymbol{\varepsilon}^h|_{\mathcal{B}^e} = \boldsymbol{E}(\boldsymbol{x})\boldsymbol{a}_e\,,\ \boldsymbol{a}_e \in \mathbb{R}^m\big\}\,, \quad (4.3.4)$$

where $\boldsymbol{E}(\boldsymbol{x})$ is a $6 \times m$ matrix of prescribed functions. In general, $\boldsymbol{S}(\boldsymbol{x}) \neq \boldsymbol{E}(\boldsymbol{x})$. In what follows, for notational simplicity, the superscript h is omitted. Since the approximations (4.3.3) and (4.3.4) are discontinuous over the elements, the variational equations (4.3.2*b*,*c*) hold for *each element* $\mathcal{B}^e$. Substituting (4.3.3) and (4.3.4) in (4.3.2*b*) and solving for the element parameters $\boldsymbol{a}_e \in \mathbb{R}^m$ yields the representation

$$\boxed{\boldsymbol{\varepsilon}_{n+1}\big|_{\mathcal{B}^e} = \bar{\nabla}^s \boldsymbol{u} := \boldsymbol{E}(\boldsymbol{x})\boldsymbol{H}^{-1}\int_{\mathcal{B}^e} \boldsymbol{S}^T \nabla^s \boldsymbol{u}\, dV,} \tag{4.3.5}$$

where

$$\boldsymbol{H} := \int_{\mathcal{B}^e} \boldsymbol{S}^T \boldsymbol{E}\, dV\,. \tag{4.3.6}$$

Similarly, by substituting (4.3.3) and (4.3.4) in (4.3.2*c*), we find the following representation for the stress tensor

$$\boxed{\boldsymbol{\sigma}_{n+1}\big|_{\mathcal{B}^e} = \boldsymbol{S}(\boldsymbol{x})\boldsymbol{H}^{-T}\int_{\mathcal{B}^e} \boldsymbol{E}^T \nabla W(\boldsymbol{\varepsilon}_{n+1} - \boldsymbol{\varepsilon}^p_{n+1})\, dV\,.} \tag{4.3.7}$$

Note that by virtue of interpolations (4.3.3) and (4.3.4), we obtain a *discrete approximation* to the symmetric gradient operator $\nabla^s(\bullet)$, which we denote by $\bar{\nabla}^s(\bullet)$ in subsequent developments. This discrete gradient is crucial in the formulation described below.

4.3.4 Reduced Residual. Generalized Displacement Model

The structure of the preceding strain and stress interpolation has the remarkable property that only the discrete gradient operator enters in the resulting expression of the discretized weak form (4.3.2*a*). Moreover, the stress representation (4.3.7) does not appear explicitly in the formulation and is relevant only in the stress recovery phase. The explicit result is contained in the following

Proposition 4.2. *The momentum balance equation in the assumed-strain approach is identical in form to a displacement model, provided the gradient operator* $\nabla^s(\bullet)$ *is replaced by the discrete gradient operator* $\bar{\nabla}^s(\bullet)$ *defined by* (4.3.5)*:*

$$\boxed{\delta\mathcal{P}^h_e(\chi_{n+1}, \boldsymbol{\eta}) := \int_{\mathcal{B}^e} \nabla W\big(\bar{\nabla}^s \boldsymbol{u}_{n+1} - \boldsymbol{\varepsilon}^p_{n+1}\big) \cdot \bar{\nabla}^s \boldsymbol{\eta}\, dV - G_{\text{ext}}\big|_{\mathcal{B}^e},} \tag{4.3.8}$$

where $G_{\text{ext}}\big|_{\mathcal{B}^e}$ *is the restriction to a typical element* $\mathcal{B}^e$ *of external virtual work:*

$$G_{\text{ext}}(\boldsymbol{\eta}) := -\delta\mathcal{P}_{\text{ext}}(\boldsymbol{u}_{n+1}, \boldsymbol{\eta}) = \int_{\mathcal{B}} \rho \boldsymbol{b} \cdot \boldsymbol{\eta}\, dV + \int_{\partial_\sigma \mathcal{B}} \bar{\boldsymbol{t}} \cdot \boldsymbol{\eta}\, d\Gamma. \tag{4.3.9}$$

PROOF. Using of (4.3.5), (4.3.6) and (4.3.7), equation (4.3.2a) is rewritten as follows:

$$\delta\mathcal{P}^h_e(\chi_{n+1}, \boldsymbol{\eta}) := \int_{\mathcal{B}^e} \nabla^s \boldsymbol{\eta} \cdot \boldsymbol{\sigma}_{n+1}\, dV - G_{\text{ext}}\big|_{\mathcal{B}^e}$$

$$= \int_{\mathcal{B}^e} \nabla^s \boldsymbol{\eta} \cdot \left[\boldsymbol{S} \boldsymbol{H}^{-T} \int_{\mathcal{B}^e} \boldsymbol{E}^T \nabla W(\boldsymbol{\varepsilon}_{n+1} - \boldsymbol{\varepsilon}^p_{n+1})\, dV \right] dV - G_{\text{ext}}\big|_{\mathcal{B}^e}$$

$$= \int_{\mathcal{B}^e} \nabla W(\boldsymbol{\varepsilon}_{n+1} - \boldsymbol{\varepsilon}^p_{n+1}) \cdot \left[\boldsymbol{E} \boldsymbol{H}^{-1} \int_{\mathcal{B}^e} \boldsymbol{S}^T \nabla^s \boldsymbol{\eta}\, dV \right] dV - G_{\text{ext}}\big|_{\mathcal{B}^e}$$

$$= \int_{\mathcal{B}^e} \nabla W(\bar{\nabla}^s \boldsymbol{u}_{n+1} - \boldsymbol{\varepsilon}^p_{n+1}) \cdot \bar{\nabla}^s \boldsymbol{\eta}\, dV - G_{\text{ext}}\big|_{\mathcal{B}^e}. \tag{4.3.10}$$

□

4.3.5 Closest Point Projection Algorithm

Expression (4.3.8) becomes completely determined once $\boldsymbol{\varepsilon}^p_{n+1}$ is defined in terms of $\boldsymbol{\varepsilon}^p_n$ and the *discrete strain* $\boldsymbol{\varepsilon}_{n+1} := \bar{\nabla}^s \boldsymbol{u}_{n+1}$. This is accomplished at each Gauss point by solving the local equations (4.3.2*d*,*e*) that define the closest point projection algorithm in strain space. For this purpose the general closest point projection iteration summarized in BOX **3.5** is employed. Notice that this algorithm is identical to the return-mapping algorithms developed in Chapter **3** except that now the strain tensor is evaluated by the discrete gradient operator as $\boldsymbol{\varepsilon}_{n+1} = \bar{\nabla}^s \boldsymbol{u}_{n+1}$. Consequently, the implementation of return-mapping algorithms in an assumed-strain method is identical to the implementation of a displacement model.

For completeness, a step-by-step summary of the computational procedure for the case of ideal plasticity is contained in BOX **4.1**. A derivation of the expression for the consistent, discrete, tangent stiffness matrix is given in the next subsection.

Remarks 4.3.3.

1. The process summarized in BOX **4.1** determines the plastic strains $\boldsymbol{\varepsilon}^p_{n+1}$ and plastic consistency parameter $\Delta\gamma$ for *given* strains $\boldsymbol{\varepsilon}_{n+1} = \bar{\nabla}^s \boldsymbol{u}$. From a physical viewpoint, this is equivalent to determining the *unloaded configuration* defined by $\boldsymbol{\varepsilon}^p_{n+1}$. However, observe that the intermediate configuration is determined *only up to an infinitesimal rotation* $\boldsymbol{w}^p_{n+1}(\boldsymbol{x})$, since only the symmetric part $\boldsymbol{\varepsilon}^p_{n+1}$ of the local plastic displacement gradient is defined. This situation is analogous to that found in finite-strain plasticity.
2. In particular, as shown in Chapter **9**, the interpretation in the closest point projection algorithm in strain space of finding the unloaded configuration for a given strain history, has its counterpart in the finite deformation case within the context of a multiplicative decomposition of the deformation gradient.
3. The algorithm summarized in BOX **4.1** is performed at each Gauss point. Hence, consistency and loading/unloading conditions are established *independently* at each Gauss point of the element. Recall that, for a von Mises yield condition (not subject to the plane-stress constraint), convergence is attained in one iteration, and the algorithm reduces to the classical radial return method of Wilkins [1964].

BOX 4.1. Closest Point Projection Iteration *at Each Gauss Point.*

1. Compute elastic predictor $\bar{\boldsymbol{\sigma}}_{n+1}^{(0)} := \nabla W(\bar{\nabla}^s \boldsymbol{u}_{n+1} - \boldsymbol{\varepsilon}_n^p)$

IF $f(\bar{\boldsymbol{\sigma}}_{n+1}^{(0)}) \leq 0$ Elastic Gauss point

Set $\boldsymbol{\varepsilon}_{n+1}^p = \boldsymbol{\varepsilon}_n^p$ and *EXIT*.

ELSE $[f(\bar{\boldsymbol{\sigma}}_{n+1}^{(0)}) > 0]$ Plastic Gauss point

Set $\Delta\gamma^{(0)} = 0$ GO TO **2**.

ENDIF

2. Return-mapping iterative algorithm (closest point projection)

2a. Compute residuals

$$\bar{\boldsymbol{\sigma}}_{n+1}^{(k)} := \nabla W\left[\bar{\nabla}^s \boldsymbol{u}_{n+1} - \boldsymbol{\varepsilon}_{n+1}^{p^{(k)}}\right]$$

$$f_{n+1}^{(k)} := f\left[\bar{\boldsymbol{\sigma}}_{n+1}^{(k)}\right]$$

$$\boldsymbol{r}_{n+1}^{(k)} := \boldsymbol{\varepsilon}_{n+1}^{p^{(k)}} - \boldsymbol{\varepsilon}_n^p - \Delta\gamma^{(k)} \nabla f\left[\bar{\boldsymbol{\sigma}}_{n+1}^{(k)}\right]$$

2b. Check convergence

IF $\left\|\boldsymbol{r}_{n+1}^{(k)}\right\| > \mathrm{TOL}_1$ or $f\left[\bar{\boldsymbol{\sigma}}_{n+1}^{(k)}\right] > \mathrm{TOL}_2$ THEN

2c. Compute consistent (algorithmic) tangent moduli

$$\mathbf{C}_{n+1}^{(k)} := \nabla^2 W\left[\bar{\nabla}^s \boldsymbol{u}_{n+1} - \boldsymbol{\varepsilon}_{n+1}^{p^{(k)}}\right]$$

$$\boldsymbol{\Xi}_{n+1}^{(k)} := \left[\mathbf{C}_{n+1}^{-1} + \Delta\gamma^{(k)} \nabla^2 f(\bar{\boldsymbol{\sigma}}_{n+1}^{(k)})\right]^{-1}$$

2d. Compute kth increments

$$\Delta^2\gamma^{(k)} := \frac{f_{n+1}^{(k)} + \left[\nabla f_{n+1}^{(k)}\right]^T \boldsymbol{\Xi}_{n+1}^{(k)} \boldsymbol{r}_{n+1}^{(k)}}{\left[\nabla f_{n+1}^{(k)}\right]^T \boldsymbol{\Xi}_{n+1}^{(k)} \nabla f_{n+1}^{(k)}}$$

$$\Delta\boldsymbol{\varepsilon}_{n+1}^{p^{(k)}} = \mathbf{C}_{n+1}^{(k)^{-1}} \boldsymbol{\Xi}_{n+1}^{(k)} \left[\Delta^2\gamma^{(k)} \nabla f_{n+1}^{(k)} - \boldsymbol{r}_{n+1}^{(k)}\right]$$

2e. Update plastic strains and consistency parameter

$$\Delta\gamma^{(k+1)} = \Delta\gamma^{(k)} + \Delta^2\gamma^{(k)}$$

$$\boldsymbol{\varepsilon}_{n+1}^{p^{(k+1)}} = \boldsymbol{\varepsilon}_{n+1}^{p^{(k)}} + \Delta\boldsymbol{\varepsilon}_{n+1}^{p^{(k)}}$$

Set $k \leftarrow k + 1$ and GO TO **2a**.

ELSE

Set $\boldsymbol{\varepsilon}_{n+1}^p = \boldsymbol{\varepsilon}_{n+1}^{p^{(k)}}$

ENDIF

EXIT.

4.3.6 Linearization. Consistent Tangent Operator

The expression for the tangent stiffness operator associated with the residual (4.3.8) remains to be determined. Here, the notion of *consistent tangent moduli* introduced in Chapter **3** and obtained by exact linearization of the closest point projection algorithm, plays an essential role. The derivation is based on the following observations:

1. The discrete gradient operator $\bar{\nabla}^s \boldsymbol{u}_{n+1}$ defined by (4.3.5) is a *linear function* of $\boldsymbol{u}_{n+1}$; we write

$$\bar{\nabla}^s \boldsymbol{u}_{n+1} = \bar{\varepsilon}(\boldsymbol{u}_{n+1}) \,. \tag{4.3.11a}$$

2. The plastic strain ε^p_{n+1} and the consistency parameter $\Delta\gamma$ are *nonlinear functions* of ε_{n+1} and the plastic strain ε^p_n defined by the algorithm in BOX **4.1**. Since ε^p_n is regarded as a *fixed given* history variable, the only remaining independent variable is $\boldsymbol{u}_{n+1}$. Therefore we write

$$\varepsilon^p_{n+1} = \widehat{\varepsilon}^p(\boldsymbol{u}_{n+1}) \,, \; \Delta\gamma = \Delta\widehat{\gamma}(\boldsymbol{u}_{n+1}) \,. \tag{4.3.11b}$$

3. In view of expression (4.3.8), from the two preceding observations, the residual becomes a function of $\boldsymbol{u}_{n+1}$, and we write

$$\delta\mathcal{P}^h_e(\chi_{n+1}, \boldsymbol{\eta}) := G^h_e(\boldsymbol{u}_{n+1}, \boldsymbol{\eta}) \,, \; \boldsymbol{\eta} \in \mathbb{V} \,. \tag{4.3.11c}$$

Now the objective is to obtain the linearization of (4.3.11c) about an intermediate state $\boldsymbol{u}^{(i)}_{n+1}$, for a *given incremental* displacement field $\Delta\boldsymbol{u}^{(i+1)}_{n+1} \in \mathbb{V}$. Here, the superscript refers to the global iterative scheme, typically a Newton or quasi-Newton method, and not to the local constitutive iterations in BOX **4.1**. To simplify the notation, however, superscripts are omitted in what follows.

4.3.6.1 Linearization of the discrete-gradient operator

It is clear from expression (4.3.5) that $\bar{\varepsilon}(\boldsymbol{u}_{n+1}) = \bar{\nabla}^s \boldsymbol{u}_{n+1}$ is *linear* in $\boldsymbol{u}_{n+1}$. Therefore,

$$\boxed{\delta\bar{\varepsilon}(\boldsymbol{u}_{n+1}, \Delta\boldsymbol{u}) = \bar{\nabla}^s(\Delta\boldsymbol{u}) \,.} \tag{4.3.12}$$

However, as noted in Simo, Taylor, and Pister [1985], this simple result no longer holds in the finite-strain case since $\bar{\nabla}^s(\bullet)$ becomes a nonlinear function of the configuration.

4.3.6.2 Linearization of the closest point projection algorithm.

To simplify the notation, we set

$$\delta\varepsilon^p_{n+1} := \delta\widehat{\varepsilon}^p(\boldsymbol{u}_{n+1}, \Delta\boldsymbol{u}) \,,$$

and (4.3.13)

$$\delta\Delta\gamma := \delta\Delta\widehat{\gamma}(\boldsymbol{u}_{n+1}, \Delta\boldsymbol{u}) \,.$$

Now the derivation follows the same steps already discussed in Chapter **3**. First, we take the derivative of (4.3.2*d*) to obtain

$$\delta\varepsilon^{p}_{n+1} = \Delta\gamma\nabla^{2} f_{n+1}\mathbf{C}_{n+1}\left[\bar{\nabla}^{s}(\Delta\boldsymbol{u}_{n+1}) - \delta\varepsilon^{p}_{n+1}\right] + \delta\Delta\gamma\nabla f_{n+1}. \tag{4.3.14}$$

Next, we introduce the *algorithmic* tangent moduli defined as

$$\boxed{\boldsymbol{\Xi}_{n+1} := \left[\mathbf{C}^{-1}_{n+1} + \Delta\gamma\nabla^{2} f_{n+1}\right]^{-1},} \tag{4.3.15}$$

and rephrase equation (4.3.14) as

$$\mathbf{C}_{n+1}\left[\bar{\nabla}^{s}(\Delta\boldsymbol{u}_{n+1}) - \delta\varepsilon^{p}_{n+1}\right] = \boldsymbol{\Xi}_{n+1}\left[\bar{\nabla}^{s}(\Delta\boldsymbol{u}_{n+1}) - \delta\Delta\gamma\nabla f_{n+1}\right]. \tag{4.3.16}$$

On the other hand, by taking the directional derivative of the constraint condition $(4.3.2e)_1$, we obtain the differentiated version of the consistency condition as

$$[\nabla f_{n+1}]^{T}\mathbf{C}_{n+1}\left[\bar{\nabla}^{s}(\Delta\boldsymbol{u}_{n+1}) - \delta\varepsilon^{p}_{n+1}\right] = 0\,. \tag{4.3.17}$$

Now the expression for $\delta\Delta\gamma$ is obtained by combining (4.3.16) and (4.3.17). A straightforward manipulation yields

$$\boxed{\delta\Delta\gamma = \frac{[\nabla f_{n+1}]^{T}\boldsymbol{\Xi}_{n+1}\bar{\nabla}^{s}(\Delta\boldsymbol{u}_{n+1})}{[\nabla f_{n+1}]^{T}\boldsymbol{\Xi}_{n+1}\nabla f_{n+1}}\,.} \tag{4.3.18}$$

Substitution of (4.3.18) in (4.3.16) leads to the expression

$$\boxed{\begin{aligned}\mathbf{C}_{n+1}\left[\bar{\nabla}^{s}\boldsymbol{u}_{n+1} - \delta\varepsilon^{p}_{n+1}\right] &= \left[\boldsymbol{\Xi}_{n+1} - \boldsymbol{N}_{n+1}\boldsymbol{N}^{T}_{n+1}\right]\bar{\nabla}^{s}(\Delta\boldsymbol{u}_{n+1})\\ \boldsymbol{N}_{n+1} &:= \frac{\boldsymbol{\Xi}_{n+1}\nabla f_{n+1}}{\sqrt{[\nabla f_{n+1}]^{T}\boldsymbol{\Xi}_{n+1}\nabla f_{n+1}}}\end{aligned}} \tag{4.3.19}$$

4.3.6.3 Consistent linearization of the residual. Tangent moduli.

Finally, we obtain the linearization of the reduced equilibrium equation (4.3.11*c*) in the direction of the incremental displacement $\Delta\boldsymbol{u} \in \mathbb{V}$, as follows. By applying the chain rule, from (4.3.2*a*) we obtain

$$\delta G^{h}_{e}(\boldsymbol{u}_{n+1}, \Delta\boldsymbol{u}) = \int_{\mathcal{B}^{e}} [\bar{\nabla}\eta]^{T}\mathbf{C}_{n+1}\left[\delta\bar{\varepsilon}(\boldsymbol{u}_{n+1}, \Delta\boldsymbol{u}) - \delta\widehat{\varepsilon}^{p}(\boldsymbol{u}_{n+1}, \Delta\boldsymbol{u})\right]dV\,. \tag{4.3.20}$$

Then substituting (4.3.12) and (4.3.19) in (4.3.20) produces the desired result:

$$\boxed{\delta G^{h}_{e}(\boldsymbol{u}_{n+1}, \Delta\boldsymbol{u}) = \int_{\mathcal{B}^{e}} [\bar{\nabla}^{s}\eta]^{T}\left[\boldsymbol{\Xi}_{n+1} - \boldsymbol{N}_{n+1}\boldsymbol{N}^{T}_{n+1}\right]\bar{\nabla}^{s}(\Delta\boldsymbol{u})dV\,.} \tag{4.3.21}$$

Expression (4.3.21) has the following noteworthy features.

1. Recall that for perfect plasticity the elastoplastic tangent moduli are given by $\mathbf{C} - \boldsymbol{n}\boldsymbol{n}^{T}$, where $\boldsymbol{n} := \mathbf{C}\nabla f\big/\sqrt{[\nabla f]^{T}\mathbf{C}\nabla f}$. Expression (4.3.21) shows that, in an algorithmic context, the consistent (algorithmic) elastoplastic tangent moduli are obtained from the continuum moduli by replacing the elasticity tensor $\mathbf{C}$ with the *algorithmic* tangent elasticities $\boldsymbol{\Xi}_{n+1}$ defined by (4.3.15).

2. Expression (4.3.21) is identical in structure to a displacement model with the gradient operator $\nabla^s(\bullet)$ replaced by the discrete gradient operator $\bar{\nabla}^s(\bullet)$.

4.3.7 Matrix Expressions

To illustrate expressions (4.3.8) and (4.3.21) for the residual and tangent operator in a familiar context, let $N_a(\boldsymbol{x})$, $a = 1, \ldots, n_{\text{en}}$, denote the shape functions of a typical element $\mathcal{B}^e \subset \mathbb{R}^{n_{\text{dim}}}$ with n_{en} nodes.

$$\boldsymbol{u}^h\big|_{\mathcal{B}^e} = \sum_{a=1}^{n_{\text{en}}} N_a(\boldsymbol{x})\boldsymbol{d}_a \;\Rightarrow\; \nabla^s\boldsymbol{u}^h\big|_{\mathcal{B}^e} = \sum_{a=1}^{n_{\text{en}}} \boldsymbol{B}_a\boldsymbol{d}_a\,, \tag{4.3.22}$$

where $\boldsymbol{B}_a$ is the matrix expression of the symmetric gradient of the shape functions N_a. By (4.3.5),

$$\left.\begin{aligned} \bar{\nabla}^s\boldsymbol{u}^h\big|_{\mathcal{B}^e} &= \sum_{a=1}^{n_{\text{en}}} \bar{\boldsymbol{B}}_a\boldsymbol{d}_a \\ \bar{\boldsymbol{B}}_a &:= \boldsymbol{E}(\boldsymbol{x})\boldsymbol{H}^{-1}\int_{\mathcal{B}^e} \boldsymbol{S}^T\boldsymbol{B}_a dV \\ \boldsymbol{H} &:= \int_{\mathcal{B}^e} \boldsymbol{S}^T\boldsymbol{E}dV. \end{aligned}\right\} \tag{4.3.23}$$

Then the contribution of node a in element $\mathcal{B}^e$ to the momentum equation (4.3.10) is given by

$$\boxed{\boldsymbol{r}_a^e := \int_{\mathcal{B}^e} \bar{\boldsymbol{B}}_a^T\left\{\nabla W(\bar{\nabla}^s\boldsymbol{u}_{n+1} - \boldsymbol{\varepsilon}_{n+1}^p)\right\} dV\,,} \tag{4.3.24a}$$

where $\boldsymbol{\varepsilon}_{n+1}^p$ is computed from the algorithm in BOX **4.1**. Finally, the contribution to the tangent stiffness matrix associated with nodes a and b of element $\mathcal{B}^e$ is obtained from (4.3.21) as

$$\boxed{\boldsymbol{k}_{ab}^e := \int_{\mathcal{B}^e} \bar{\boldsymbol{B}}_a^T[\boldsymbol{\Xi}_{n+1} - \boldsymbol{N}_{n+1}\boldsymbol{N}_{n+1}^T]\bar{\boldsymbol{B}}_b dV\,,} \tag{4.3.24b}$$

where $\boldsymbol{\Xi}_{n+1}$ is given by (4.3.15).

4.3.8 Variational Consistency of Assumed-Strain Methods

The preceding arguments show that a *variational formulation* based on the Hu–Washizu principle is equivalent to a *generalized displacement* method in which the standard discrete gradient operator $\boldsymbol{B}_a := \nabla^s N_a$ is replaced by an assumed $\bar{\boldsymbol{B}}_a$ given by (4.3.23). Here, we shall be concerned with the converse problem and consider the conditions for which an assumed-strain method with $\bar{\boldsymbol{B}}_a$ *given a priori*, and not necessarily by (4.3.23), is variationally consistent.

The question of variational consistency is relevant because of the following two reasons.

i. The convergence analysis of assumed-strain methods is brought into correspondence with the analysis of mixed methods for which a large body of literature exists. In particular, the convergence of certain assumed-strain methods, such as the $\bar{\boldsymbol{B}}$-method for quasi-incompressibility discussed in the next section, is automatically settled once the variational equivalence is established.
ii. As illustrated in the next section, by exploiting the variational consistency of the method, one can develop expressions for the stiffness matrix which are better suited to computation.

To simplify our presentation, attention focuses on linear elasticity. However, our results carry over to the nonlinear situation by straightforward linearization. Consider an assumed-strain method in which strains and stresses are computed according to the following expressions

$$\boldsymbol{\varepsilon}^h|_{\mathcal{B}^e} = \bar{\boldsymbol{B}}\boldsymbol{d}^e\,,$$

and (4.3.25)

$$\boldsymbol{\sigma}^h|_{\mathcal{B}^e} = \mathbf{C}\bar{\boldsymbol{B}}\boldsymbol{d}^e\,,$$

where $\bar{\boldsymbol{B}}$ is *given a priori,* and we have employed the following matrix notation:

$$\begin{aligned}\bar{\boldsymbol{B}} &= \left[\bar{\boldsymbol{B}}_1\ \bar{\boldsymbol{B}}_2\ \ldots\ \bar{\boldsymbol{B}}_{n_{\text{en}}}\right]\\ (\boldsymbol{d}^e)^T &= \left[(\boldsymbol{d}_1^e)^T\ (\boldsymbol{d}_2^e)^T\ \ldots\ (\boldsymbol{d}_{n_{\text{en}}}^e)^T\right].\end{aligned} \tag{4.3.26}$$

Assumption (4.3.25) leads to admissible variations $\boldsymbol{\xi}^h \in \mathbb{V}_\varepsilon^h$ and $\boldsymbol{\tau}^h \in \mathbb{V}_\sigma^h$ given by

$$\boldsymbol{\xi}^h\big|_{\mathcal{B}^e} = \bar{\boldsymbol{B}}\boldsymbol{w}^e\,,\quad \boldsymbol{\tau}^h\big|_{\mathcal{B}^e} = \mathbf{C}\bar{\boldsymbol{B}}\boldsymbol{w}^e\,, \tag{4.3.27}$$

for arbitrary $\boldsymbol{w}^e \in \mathbb{R}^{n_{\text{en}}\times n_{\text{dim}}}$, where $n_{\text{dim}} \leq 3$ is the spatial dimension of the problem. Then substituting (4.3.25) and (4.3.27) in the Hu–Washizu variational equations, (4.3.2) yields

$$\left.\begin{aligned}G_e^h :&= \boldsymbol{w}^e\cdot\int_{\mathcal{B}^e}\boldsymbol{B}^T\mathbf{C}\bar{\boldsymbol{B}}\,dV\boldsymbol{d}^e - G_{\text{ext}}^h|_{\mathcal{B}^e}\\ 0 &= \boldsymbol{w}^e\cdot\int_{\mathcal{B}^e}\bar{\boldsymbol{B}}^T\mathbf{C}\left[\boldsymbol{B}-\bar{\boldsymbol{B}}\right]dV\boldsymbol{d}^e\\ 0 &= \boldsymbol{w}^e\cdot\int_{\mathcal{B}^e}\bar{\boldsymbol{B}}^T\left[-\mathbf{C}\bar{\boldsymbol{B}}\boldsymbol{d}^e+\mathbf{C}\bar{\boldsymbol{B}}\boldsymbol{d}^e\right]dV.\end{aligned}\right\} \tag{4.3.28}$$

The third equation is satisfied identically. The second equation, on the other hand, is satified if the following condition holds:

$$\boxed{\int_{\mathcal{B}^e}\bar{\boldsymbol{B}}^T\mathbf{C}\bar{\boldsymbol{B}}\,dV = \int_{\mathcal{B}^e}\bar{\boldsymbol{B}}^T\mathbf{C}\boldsymbol{B}\,dV.} \tag{4.3.29}$$

This orthogonality condition, first derived in Simo and Hughes [1986], furnishes the requirement for an assumed-strain method to be variationally consistent. Note that the right-hand side of (4.3.29) yields an equivalent expression for the stiffness matrix which is better suited to computation than the standard expression for assumed-strain methods furnished by the left-hand side of (4.3.29) since $\bar{\boldsymbol{B}}$ is usually a fully populated matrix, whereas $\boldsymbol{B}$ is sparse. Further details on implementation are discussed below.

4.4 Application. B-Bar Method for Incompressibility

As an important application of the ideas developed above, we examine the so-called B-bar method, proposed by Hughes [1980] for extending the methodology in Nagtegaal, Parks, and Rice [1974] and reformulated in the context of a Hu–Washizu type of variational principle in Simo, Taylor, and Pister [1985]. This class of finite-element procedures is the most widely used presently in large scale inelastic computations, for instance, the Livermore codes, see Hallquist [1984]. Because of its practical relevance, we present an outline of this methodology and discuss two possible implementations. The basic idea is to construct an assumed-strain method in which *only the dilatational part* of the displacement gradient is the independent variable. The main motivation is the development of a finite element scheme that properly accounts for the incompressibility constraint emanating from the volume-preserving nature of plastic flow. The situation is analogous to that found in incompressible elasticity.

4.4.1 Assumed-Strain and Stress Fields

According to the preceding ideas, one introduces a scalar *volume-like* variable $\Theta \in \mathbb{L}_2(\mathcal{B})$ and then considers the following assumed-strain field:

$$\boxed{\boldsymbol{\varepsilon} := \operatorname{dev}[\nabla^s \boldsymbol{u}] + \tfrac{1}{3}\Theta\mathbf{1}\,,} \tag{4.4.1}$$

where, as usual, $\operatorname{dev}[\bullet] := (\bullet) - \frac{1}{3}\operatorname{tr}[\bullet]\mathbf{1}$ denotes the deviator of the indicated argument. Similarly, one introduces a *mean-stress* variable $p \in \mathbb{L}_2(\mathcal{B})$, so that the assumed-stress field takes the form

$$\boxed{\boldsymbol{\sigma} := \operatorname{dev}\big[\nabla W(\boldsymbol{\varepsilon} - \boldsymbol{\varepsilon}^p)\big] + p\mathbf{1}.} \tag{4.4.2}$$

For subsequent developments, it proves convenient to rephrase (4.4.1) and (4.4.2) in an alternative form in terms of *projection operators* as follows. Define the following fourth-order tensors

$$\mathsf{P}_{\text{dev}} := \mathbf{I} - \tfrac{1}{3}\mathbf{1}\otimes\mathbf{1}$$

and (4.4.3)

$$\mathsf{P}_{\text{vol}} := \tfrac{1}{3}\mathbf{1}\otimes\mathbf{1}\,.$$

In matrix notation, (4.4.3) takes the form

$$\boxed{\mathsf{P}_{\text{dev}} = \mathbf{I} - \tfrac{1}{3}\mathbf{1}\mathbf{1}^T}$$

and (4.4.4)

$$\boxed{\mathbf{P}_{\text{vol}} = \tfrac{1}{3}\,\mathbf{1}\mathbf{1}^T}\,.$$

We note the following properties:

$$\left.\begin{aligned} \mathbf{P}_{\text{dev}}\mathbf{P}_{\text{vol}} &= \mathbf{P}_{\text{vol}}\mathbf{P}_{\text{dev}} = \mathbf{0}\,, \\ \mathbf{P}_{\text{dev}} + \mathbf{P}_{\text{vol}} &= \mathbf{I}\,, \\ \mathbf{P}_{\text{dev}}^2 &= \mathbf{P}_{\text{dev}}\,, \\ \mathbf{P}_{\text{vol}}^2 &= \mathbf{P}_{\text{vol}}\,, \end{aligned}\right\} \tag{4.4.5}$$

which are easily checked by direct calculation and make $\mathbf{P}_{\text{dev}}$ and $\mathbf{P}_{\text{vol}}$ *orthogonal projections*. Their physical meaning should be clear. $\mathbf{P}_{\text{dev}}$ and $\mathbf{P}_{\text{vol}}$ are orthogonal projections that map a second-order tensor into its deviatoric and volumetric parts, respectively. Then the assumed-strain and stress fields (4.4.1) and (4.4.2) read

$$\boxed{\begin{aligned} \boldsymbol{\varepsilon} &= \mathbf{P}_{\text{dev}}\big[\nabla^s \boldsymbol{u}\big] + \tfrac{1}{3}\,\Theta\mathbf{1} \\ \boldsymbol{\sigma} &= \mathbf{P}_{\text{dev}}\big[\nabla W(\boldsymbol{\varepsilon} - \boldsymbol{\varepsilon}^p)\big] + p\mathbf{1}\,. \end{aligned}} \tag{4.4.6}$$

4.4.2 Weak Forms

In the present context, the variational structure of the assumed-strain method takes the following form. Let

$$\chi_{n+1} := \{\boldsymbol{u}_{n+1}, \Theta_{n+1}, p_{n+1}\} \tag{4.4.7}$$

be the set of independent variables. As before, assume that the plastic variables $\{\boldsymbol{\varepsilon}^p_{n+1}, \Delta\gamma\}$ are defined locally in terms of χ_{n+1} and the history of plastic strain $\boldsymbol{\varepsilon}^p_n$, up to time t_n, by the local equations (4.3.2*d,e*). We have the following result.

Proposition 4.3. *For the assumed-strain and stress fields defined by* (4.4.6) *the weak forms* (4.3.2*a,b,c*) *reduce to*

$$\begin{aligned} G(\chi_{n+1}, \boldsymbol{\eta}) := &\int_{\mathcal{B}} \big[\nabla^s\boldsymbol{\eta}\cdot \operatorname{dev}\big[\nabla W(\boldsymbol{\varepsilon}_{n+1} - \boldsymbol{\varepsilon}^p_{n+1})\big] + p\,\operatorname{div}\boldsymbol{\eta}\big]\,dV \\ &- \int_{\mathcal{B}} \rho\boldsymbol{b}\cdot\boldsymbol{\eta}\,dV - \int_{\partial_\sigma\mathcal{B}} \bar{\boldsymbol{t}}\cdot\boldsymbol{\eta}\,d\Gamma = 0, \end{aligned} \tag{4.4.8a}$$

$$\Gamma(\chi_{n+1}, q) := \int_{\mathcal{B}} q[\operatorname{div}\boldsymbol{u}_{n+1} - \Theta_{n+1}]\,dV = 0, \tag{4.4.8b}$$

and

$$H(\chi_{n+1}, \phi) := \int_{\mathcal{B}} \phi\left(-p_{n+1} + \tfrac{1}{3}\operatorname{tr}\big[\nabla W(\boldsymbol{\varepsilon}_{n+1} - \boldsymbol{\varepsilon}^p_{n+1})\big]\right) dV = 0\,, \tag{4.4.8c}$$

for any $\boldsymbol{\eta} \in \mathbb{V}$, $q \in \mathbb{L}_2(\mathcal{B})$, *and* $\phi \in \mathbb{L}_2(\mathcal{B})$.

PROOF. The proof follows by application of properties (4.4.5). In particular, from expressions (4.4.3) and (4.4.6), we observe that

$$\left.\begin{aligned} \mathbf{P}_{\text{dev}}\mathbf{1} &= \mathbf{0}, \\ \mathbf{P}_{\text{dev}}[\nabla^s \boldsymbol{u}_{n+1} - \boldsymbol{\varepsilon}_{n+1}] &= \mathbf{0}\,. \end{aligned}\right\} \tag{4.4.9}$$

By using these relationships along with properties (4.4.5),

$$\begin{aligned} \boldsymbol{\tau} \cdot [\nabla^s \boldsymbol{u}_{n+1} - \boldsymbol{\varepsilon}_{n+1}] =& (\mathbf{P}_{\text{dev}}\boldsymbol{\tau}) \cdot (\nabla^s \boldsymbol{u}_{n+1} - \boldsymbol{\varepsilon}_{n+1}) \\ &+ (\mathbf{P}_{\text{vol}}\boldsymbol{\tau}) \cdot (\nabla^s \boldsymbol{u}_{n+1} - \boldsymbol{\varepsilon}_{n+1}) \\ =& \boldsymbol{\tau} \cdot \big(\mathbf{P}_{\text{dev}}(\nabla^s \boldsymbol{u}_{n+1} - \boldsymbol{\varepsilon}_{n+1})\big) \\ &+ \tfrac{1}{3}\operatorname{tr}[\boldsymbol{\tau}]\mathbf{1} \cdot (\nabla^s \boldsymbol{u}_{n+1} - \boldsymbol{\varepsilon}_{n+1}) \\ =& \tfrac{1}{3}\operatorname{tr}[\boldsymbol{\tau}]\operatorname{tr}[\nabla^s \boldsymbol{u}_{n+1} - \boldsymbol{\varepsilon}_{n+1}] \\ =& q(\operatorname{div} \boldsymbol{u}_{n+1} - \Theta_{n+1})\,, \end{aligned} \tag{4.4.10}$$

where $q := \frac{1}{3}\operatorname{tr}[\boldsymbol{\tau}]$. This proves variational equation (4.4.8*b*). A similar argument holds for (4.4.8*a*) and (4.4.8*c*). □

Remark 4.4.4. Assuming that the discrete plastic flow rule and discrete Kuhn–Tucker conditions (4.3.2*d*,*e*) hold pointwise, one can easily show that (4.4.8*a*,*b*,*c*) are the Euler–Lagrange equations of the three-field variational principle

$$\boxed{\begin{aligned} \mathcal{P}(\chi_{n+1}) :=& \int_{\mathcal{B}} \big[W(\boldsymbol{\varepsilon}_{n+1} - \boldsymbol{\varepsilon}^p_{n+1}) + p\,(\operatorname{div} \boldsymbol{u}_{n+1} - \Theta_{n+1})\big]\,dV \\ &+ \mathcal{P}_{\text{ext}}(\boldsymbol{u}_{n+1}), \end{aligned}} \tag{4.4.11}$$

where $\boldsymbol{\varepsilon}_{n+1} := \mathbf{P}_{\text{dev}}[\nabla^s \boldsymbol{u}_{n+1}] + \frac{1}{3}\,\Theta_{n+1}\mathbf{1}$.

4.4.3 Discontinuous Volume/Mean-Stress Interpolations

As in Section **4.3** we consider discontinuous interpolations of the assumed-stress and strain fields, and introduce the following approximating subspace for both the volume and the mean-stress fields

$$\mathbb{V}^h_{\text{vol}} := \left\{\Theta^h \in \mathbb{L}_2(\mathcal{B}) \mid \Theta^h|_{\mathcal{B}^e} = \boldsymbol{\Gamma}^T(\boldsymbol{x})\boldsymbol{\Theta}^e\,;\ \boldsymbol{\Theta}^e \in \mathbb{R}^m\right\}, \tag{4.4.12}$$

where $\boldsymbol{\Gamma}^T(\boldsymbol{x}) := [\Gamma_1(\boldsymbol{x}), \ \ldots\ , \Gamma_m(\boldsymbol{x})]$ is a vector of m-prescribed local functions. Then substituting in (4.4.8*b*,*c*) and omitting the superscript h yields

$$\left.\begin{aligned} \Theta_{n+1} &= \overline{\operatorname{div}}\, \boldsymbol{u}_{n+1} := \boldsymbol{\Gamma}^T(\boldsymbol{x})\boldsymbol{H}^{-1}\int_{\mathcal{B}^e} \boldsymbol{\Gamma}\operatorname{div} \boldsymbol{u}_{n+1}\,dV, \\ p_{n+1} &= \boldsymbol{\Gamma}^T(\boldsymbol{x})\boldsymbol{H}^{-1}\int_{\mathcal{B}^e} \boldsymbol{\Gamma}\,\tfrac{1}{3}\operatorname{tr}\big[\nabla W(\boldsymbol{\varepsilon}_{n+1} - \boldsymbol{\varepsilon}^p_{n+1})\big]dV, \\ \boldsymbol{H} &:= \int_{\mathcal{B}^e} \boldsymbol{\Gamma}\boldsymbol{\Gamma}^T\,dV. \end{aligned}\right\} \tag{4.4.13}$$

On the basis of these interpolations, two alternative implementations of the method are possible, as discussed next.

4.4.4 Implementation 1. B-Bar Approach

As in Section **3**, the first implementational procedure within the present variational framework hinges on the following result.

Proposition 4.4. *The mean-stress term in the weak form (4.4.8a) can be expressed in the alternative form as*

$$\left.\begin{aligned} \int_{\mathcal{B}^e} p \operatorname{div} \boldsymbol{\eta} dV &= \int_{\mathcal{B}^e} \tfrac{1}{3}\operatorname{tr}\left[\nabla W(\boldsymbol{\varepsilon}_{n+1} - \boldsymbol{\varepsilon}^p_{n+1})\right]\overline{\operatorname{div}}\, \boldsymbol{\eta} dV, \\ \boldsymbol{\varepsilon}_{n+1} &= \operatorname{dev}\left[\nabla^s \boldsymbol{u}_{n+1}\right] + \tfrac{1}{3}(\overline{\operatorname{div}}\, \boldsymbol{u}_{n+1})\mathbf{1}, \\ \overline{\operatorname{div}}\, \boldsymbol{u}_{n+1} &:= \boldsymbol{\Gamma}^T(\boldsymbol{x})\boldsymbol{H}^{-1}\int_{\mathcal{B}^e} \boldsymbol{\Gamma} \operatorname{div} \boldsymbol{u}_{n+1} dV. \end{aligned}\right\} \tag{4.4.14}$$

PROOF. This is analogous to Proposition **4.2**. □

To complete the implementation, we use matrix notation and set

$$\left.\begin{aligned} \operatorname{div} \boldsymbol{u}_{n+1} &= \sum_{a=1}^{n_{en}} \boldsymbol{b}_a^T \boldsymbol{d}_{a,n+1}, \\ \overline{\operatorname{div}}\, \boldsymbol{u}_{n+1} &= \sum_{a=1}^{n_{en}} \bar{\boldsymbol{b}}_a^T \boldsymbol{d}_{a,n+1}. \end{aligned}\right\} \tag{4.4.15}$$

From (4.4.13a),

$$\boxed{\begin{aligned} \bar{\boldsymbol{b}}_a^T &:= \boldsymbol{\Gamma}^T \boldsymbol{H}^{-1} \int_{\mathcal{B}^e} \boldsymbol{\Gamma} \boldsymbol{b}_a^T dV, \\ \boldsymbol{B}_a^{\text{vol}} &:= \tfrac{1}{3}\mathbf{1}\boldsymbol{b}_a^T, \\ \bar{\boldsymbol{B}}_a^{\text{vol}} &:= \tfrac{1}{3}\mathbf{1}\bar{\boldsymbol{b}}_a^T, \\ \bar{\boldsymbol{B}}_a &:= [\boldsymbol{B}_a - \boldsymbol{B}_a^{\text{vol}}] + \bar{\boldsymbol{B}}_a^{\text{vol}}, \end{aligned}} \tag{4.4.16}$$

where $\boldsymbol{B}_a := \nabla^s N_a(\boldsymbol{x})$ is the standard, discrete, symmetric gradient matrix. Furthermore, by orthogonality

$$\bar{\boldsymbol{B}}_a\left\{\operatorname{dev}\left[\nabla W(\boldsymbol{\varepsilon}_{n+1} - \boldsymbol{\varepsilon}^p_{n+1})\right]\right\} = \boldsymbol{B}_a\left\{\operatorname{dev}\left[\nabla W(\boldsymbol{\varepsilon}_{n+1} - \boldsymbol{\varepsilon}^p_{n+1})\right]\right\}. \tag{4.4.17}$$

By using (4.4.14), the contribution of node a of element $\mathcal{B}^e$ to the momentum equation (4.4.8a) takes the form

$$\boxed{\boldsymbol{r}_a^e := \int_{\mathcal{B}^e} \bar{\boldsymbol{B}}_a^T\left\{\nabla W(\bar{\nabla}^s \boldsymbol{u}_{n+1} - \boldsymbol{\varepsilon}^p_{n+1})\right\} dV,} \tag{4.4.18}$$

which is the same result obtained in Section **4.3**, equation (4.3.24a). Then the tangent stiffness matrix is given by the same expression (4.3.24b).

4.4.5 Implementation 2. Mixed Approach

An alternative implementation of the present assumed-strain method can be developed directly from the mixed formulation (4.4.8a,b,c). As shown below, the main advantage over the previous implementation is that computations are performed with the standard discrete gradient matrix $\boldsymbol{B}_a$, which is sparse, instead of the $\bar{\boldsymbol{B}}_a$-matrix, which is fully populated.

The main steps involved in the present implementation are the following.

1. In addition to computing $\overline{\operatorname{div}\,\boldsymbol{u}_{n+1}}$, one also computes p_{n+1} by using (4.4.13).
2. Now the contribution to the momentum balance equation of node a of a typical element $\mathcal{B}^e$ is computed by setting

$$\boxed{\boldsymbol{r}_a^e := \int_{\mathcal{B}^e} \boldsymbol{B}_a^T \left\{ p_{n+1}\mathbf{1} + \mathbf{P}_{\text{dev}}\big[W(\bar{\nabla}^s \boldsymbol{u}_{n+1} - \boldsymbol{\varepsilon}_{n+1}^p)\big]\right\} dV\,.} \tag{4.4.19}$$

3. Setting $\boldsymbol{\Xi}_{\text{ep}} := \boldsymbol{\Xi}_{n+1} - \boldsymbol{N}_{n+1}\boldsymbol{N}_{n+1}^T$, the contribution to the tangent stiffness matrix of nodes a, b of a typical element $\mathcal{B}^e$ is computed according to the expression

$$\boxed{\begin{aligned} \boldsymbol{k}_{ab}^e = &\int_{\mathcal{B}^e} \boldsymbol{B}_a^T[\mathbf{P}_{\text{dev}}\boldsymbol{\Xi}_{\text{ep}}\mathbf{P}_{\text{dev}}]\boldsymbol{B}_b dV \\ &+ \int_{\mathcal{B}^e} \boldsymbol{B}_a^T[\mathbf{P}_{\text{dev}}\boldsymbol{\Xi}_{\text{ep}}\mathbf{1}]\tfrac{1}{3}\bar{\boldsymbol{b}}_b^T dV \\ &+ \int_{\mathcal{B}^e} \tfrac{1}{3}\bar{\boldsymbol{b}}_a[\mathbf{1}^T\boldsymbol{\Xi}_{\text{ep}}\mathbf{P}_{\text{dev}}]\boldsymbol{B}_b dV \\ &+ \int_{\mathcal{B}^e} [\bar{\boldsymbol{b}}_a\bar{\boldsymbol{b}}_b^T]\tfrac{1}{9}(\mathbf{1}^T\boldsymbol{\Xi}_{\text{ep}}\mathbf{1})dV \end{aligned}} \tag{4.4.20}$$

This expression follows by noting that

$$\boxed{\bar{\boldsymbol{B}}_a = \mathbf{P}_{\text{dev}}\boldsymbol{B}_a + \tfrac{1}{3}\mathbf{1}\bar{\boldsymbol{b}}_a^T} \tag{4.4.21}$$

and using properties (4.4.5).

Remark 4.4.5. The advantages of this second implementation are apparent in the common case encountered in most applications for which elasticity is linear, with uncoupled mean-stress/deviatoric response, and plastic flow is isochoric. The reasons for this observation are as follows.

1. Evaluation of p_{n+1} is trivially accomplished once Θ_{n+1} is computed from $(4.4.13)_1$, since

$$p_{n+1} = \kappa\Theta_{n+1}\,, \quad \kappa \equiv \text{bulk modulus.} \tag{4.4.22}$$

Note that *no additional computations* are involved with respect to implementation **1** since only the evaluation of $\bar{\boldsymbol{b}}_a$ is needed, which is also required in implementation **1**.

2. No coupling terms appear in the calculation of the stiffness matrix. This follows by noting that plastic flow is uncoupled from volumetric response. Consequently,

$$\boldsymbol{\Xi}_{\text{ep}}\mathbf{1} = \kappa\mathbf{1} \;\Rightarrow\; \mathbf{P}_{\text{dev}}\boldsymbol{\Xi}_{\text{ep}}\mathbf{1} = \mathbf{0}\,. \tag{4.4.23}$$

Therefore, (4.4.20) reduces to

$$\boxed{\boldsymbol{k}^e_{ab} = \int_{\mathcal{B}^e} \boldsymbol{B}^T_a[\boldsymbol{\Xi}_{\text{ep}} - \kappa\mathbf{1}\mathbf{1}^T]\boldsymbol{B}_b dV + \kappa\int_{\mathcal{B}^e}[\bar{\boldsymbol{b}}_a\bar{\boldsymbol{b}}^T_b]dV} \tag{4.4.24}$$

3. Because of the linearity of the mean-stress/volume elastic response, the calculation of the second term in (4.4.24) reduces to a mere *rank-one* update. This follows from $(4.4.16)_1$ and definition $(4.4.13)_3$ of $\boldsymbol{H}$, since

$$\boxed{\kappa\int_{\mathcal{B}^e}[\bar{\boldsymbol{b}}_a\bar{\boldsymbol{b}}^T_b]dV = \kappa\left[\int_{\mathcal{B}^e}\boldsymbol{\Gamma}(\boldsymbol{x})\boldsymbol{b}^T_a dV\right]^T \boldsymbol{H}^{-1}\left[\int_{\mathcal{B}^e}\boldsymbol{\Gamma}(\boldsymbol{x})\boldsymbol{b}^T_b dV\right].} \tag{4.4.25}$$

4.4.6 Examples and Remarks on Convergence

Typical examples of the formulation discussed in this section include the following:

i. A four-node quadrilateral element with bilinear isoparametric interpolation functions N_a for displacements and $\boldsymbol{\Gamma} = [1]$ constant over $\mathcal{B}^e$. Essentially, this is the *mean dilatation* formulation advocated by Nagtegaal, Parks, and Rice [1974]. This is a widely used element known not to satisfy the LBB condition. However, application of a mean-stress filtering procedure renders the discrete mean-stress field convergent; see Pitkaranta and Stenberg [1984].

ii. A nine-node element with biquadratic isoparametric interpolative functions N_a for displacements and $\boldsymbol{\Gamma} = [1\ x\ y]^T$, where (x, y) are defined in terms of natural coordinates by the standard isoparametric mapping. This is an optimal element known to satisfy the LBB condition, see Oden and Carey [1983].

We emphasize that the convergence properties of this class of assumed-strain methods is the direct consequence of their variational consistency and follows at once from the convergent characteristics of the corresponding mixed, finite-element formulations.

4.5 Numerical Simulations

A number of numerical simulations are presented that illustrate the performance of the return-mapping algorithms and the practical importance of consistent tangent operators in a Newton solution procedure. These simulations exhibit the significant loss in rate of convergence that occurs when the elastoplastic "continuum" tangent is used in place of the tangent consistently derived from the integration algorithm. The overall robustness of the algorithm is significantly enhanced by combining the classical Newton procedure with a line search algorithm. This strategy is suggested by a number of authors, e.g., see Dennis and Schnabel [1983] or Luenberger [1984]. The specific algorithm used is a linear line search which is invoked whenever a computed energy norm is more than 0.6 of a previous value in the load step (see Matthies and Strang [1979]). Computations are performed with an enhanced version of the general purpose, nonlinear, finite-element computer program FEAP, developed by R.L. Taylor and described in Chapter **24** of Zienkiewicz [1977]. Convergence is measured in terms of the (discrete) *energy norm*, which is computed from the residual vector $\boldsymbol{R}(\boldsymbol{d}_{n+1})$ and the incremental nodal displacement vector $\Delta \boldsymbol{d}_{n+1}$ as

$$\Delta E\big(\boldsymbol{d}_{n+1}^{(i)}\big) := \big[\Delta \boldsymbol{d}_{n+1}^{(i+1)}\big]^T \boldsymbol{R}\big(\boldsymbol{d}_{n+1}^{(i)}\big) . \tag{4.5.1}$$

Alternative discrete norms may be used in place of (4.5.1), in particular the Euclidean norm of the residual force vector. In terms of the energy norm (4.5.1), our termination criteria for the Newton solution strategy takes the following form

$$\Delta E\big(\boldsymbol{d}_{n+1}^{(i)}\big) \leq 10^{-9} \Delta E\big(\boldsymbol{d}_{n+1}^{(1)}\big) . \tag{4.5.2}$$

Although it would appears that this convergence criterion provides an exceedingly severe condition difficult to satisfy, through the numerical examples it is shown that criterion (4.5.2) is easily satisfied with a rather small number of iterations, when the consistent tangent operator is used.

4.5.1 Plane-Strain J_2 Flow Theory

Two simulations are considered that employ the return-mapping algorithms in BOX **3.1** and BOX **3.2**, along with a four-node bilinear isoparametric quadrilateral element with constant mean-stress and volume fields, as described above.

EXAMPLE: 4.5.1. Thick-Walled Cylinder Subject to Internal Pressure. An infinitely long thick-walled cylinder with a 5-m inner radius and a 15-m outer radius is subject to internal pressure. The properties of the material are $E = 70$ MPa, $\nu = 0.2$. The isotropic and kinematic hardening rules are of the exponential type, defined according to the expressions

$$\left.\begin{aligned} h(\alpha) &:= \bar{K}_\infty - [\bar{K}_\infty - \bar{K}_0]\exp[-\delta\alpha] + \bar{H}'\alpha \\ K(\alpha) &:= \beta h(\alpha) \\ H'(\alpha) &:= (1-\beta)h(\alpha) , \ \ \beta \in [0, 1] . \end{aligned}\right\} \tag{4.5.3}$$

Note that $\beta = 0$ and $\beta = 1$ correspond to the limiting cases of pure kinematic and pure isotropic hardening rules, respectively. The values for the parameters in

TABLE 4.1. Example **4.5.1.** Iterations for Each Step.

Step	1	2	3	4	5
State	el	el-pl	el-pl	pl	pl
Continuum	2	6	9	10	6
Consistent	2	5	7	5	3

(4.5.3) are

$$\left.\begin{aligned} \bar{K}_0 &= 0.2437 \text{ MPa}, \\ \bar{K}_\infty &= 0.343 \text{ MPa}, \\ \delta &= 0.1, \\ \bar{H}' &= 0.15 \text{ MPa}, \\ \text{and} \qquad & \\ \beta &= 0.1\,. \end{aligned}\right\} \tag{4.5.4}$$

The internal pressure is increased linearly in time until the entire cylinder yields. The finite-element mesh employed in the calculation is shown in Figure **4.3**. The size of the time step is selected to achieve yielding of the entire section in two time steps involving plastic deformation. The position of the elastic-plastic interface in these two time steps is depicted in Figure **4.4**.

The calculation is performed with both the "continuum" and the "consistent" elastoplastic tangent, and the results are shown in Table **4.1**. In spite of the better performance exhibited by the "consistent" tangent, no substantial reduction in the required number of iterations for convergence is obtained except in the fully plastic situation because of the extreme simplicity of this boundary-value problem. The next example confirms this observation.

EXAMPLE: 4.5.2. Perforated Strip Subject to Uniaxial Extension. We consider the *plane-strain* problem of an infinitely long rectangular strip with a circular hole.

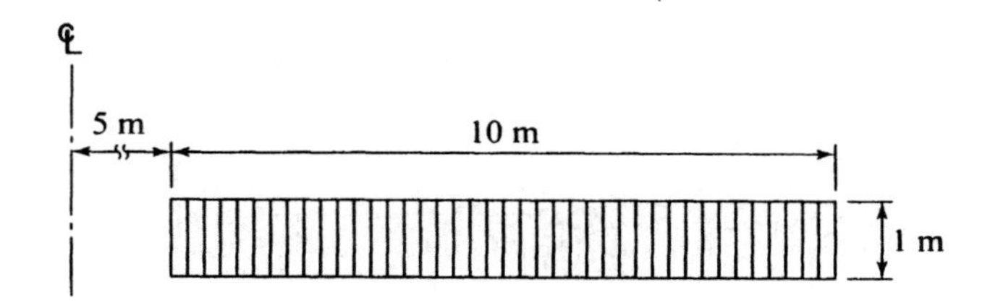

FIGURE 4.3. Thick-walled cylinder. Finite-element mesh.

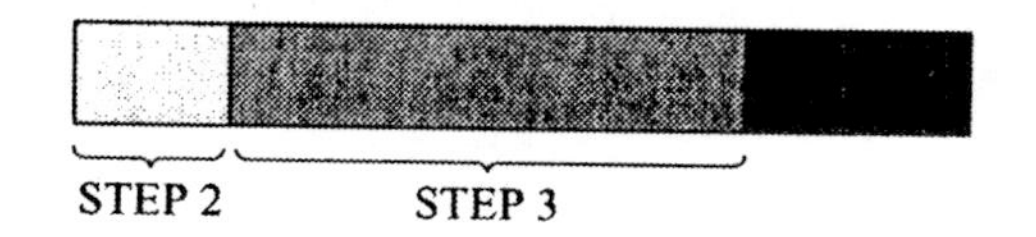

FIGURE 4.4. Thick-walled cylinder. Elastic-plastic interface.

subjected to increasing extension. The elastic properties of the material are taken as $E = 70$ MPa, $\nu = 0.2$, and the parameters in the saturation type of hardening rule (4.5.3) are $\bar{K}_0 = \bar{K}_\infty = 0.243$ MPa, $\bar{H}' = 0$, and $\beta = 1$ (perfectly plastic behavior). Loading is performed by *controlling* the vertical *displacement* of the top and bottom boundaries of the rectangular strip. The finite-element mesh employed is shown in Figure **4.5**. For obvious symmetry reasons, only 1/4 of the strip is considered.

The evolution of the elastic-plastic interface with increasing straining of the strip is shown in Figure **4.6**. To plot these results, the stresses computed at the Gauss points of a typical element are projected onto the nodal points by bilinear interpolation functions. Related "smoothing" procedures are discussed in Zienkiewicz [1977] (Section **11.5**, and references therein), and are often used for filtering spurious pressure modes (e.g., Lee, Gresho, and Sani [1979] and Hughes [1987]).

The calculation is performed with both the "continuum" and the "consistent" tangent operators. The number of iterations for convergence is summarized in Table **4.2**. The numerical values of the energy norm in a typical iteration are displayed in Table **4.3** and the values of the Euclidean norm of the residual for the same iteration in Table **4.4**.

The superior performance of the "consistent" tangent is apparent from these results. Note that the Euclidean norm of the residual lags behind the energy norm in the iterative process.

This example also provides a severe test for the global performance of the Newton solution strategy. Although the calculation is completed successfully with a time step of $\Delta t = 0.0125$, the iterative procedure diverges for twice this value. However, when the Newton solution procedure is combined with a line search, as

TABLE 4.2. Example **4.5.2**. Iterations for Each Step.

Step	1	2	3	4	5
State	el	el-pl	el-pl	el-pl	el-pl
Continuum	2	13	23	23	22
Consistent	2	5	5	4	5

TABLE 4.3. Example **4.5.2**. Energy Norm Values for Step 4.

Iteration	1	2	3	4	5	6
Continuum	0.14 E+2	0.80 E−2	0.61 E−3	0.18 E−3	0.89 E−4	0.47 E−4
Consistent	0.14 E+2	0.11 E−1	0.77 E−4	0.10 E−9	–	–
Iteration	7	8	9	10	11	12
Continuum	0.27 E−4	0.16 E−4	0.97 E−5	0.59 E−5	0.36 E−5	0.22 E−5
Consistent	–	–	–	–	–	–
Iteration	13	14	15	16	17	18
Continuum	0.13 E−5	0.85 E−6	0.52 E−6	0.32 E−6	0.20 E−6	0.12 E−6
Consistent	–	–	–	–	–	–
Iteration	19	20	21	22	23	
Continuum	0.77 E−7	0.47 E−7	0.29 E−7	0.18 E−7	0.11 E−7	
Consistent	–	–	–	–	–	

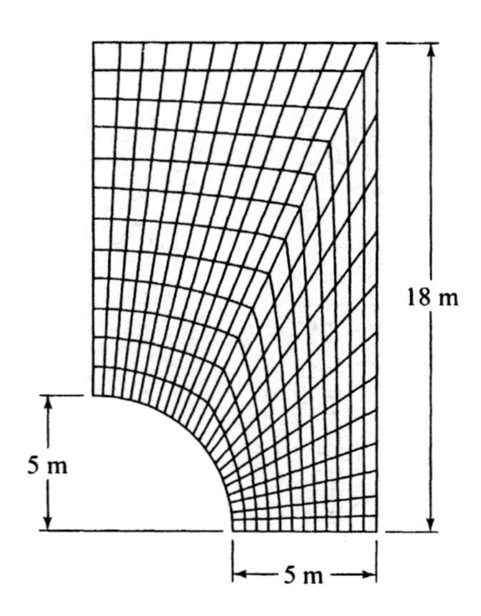

FIGURE 4.5. Plane-strain strip with a circular hole. Finite-element mesh.

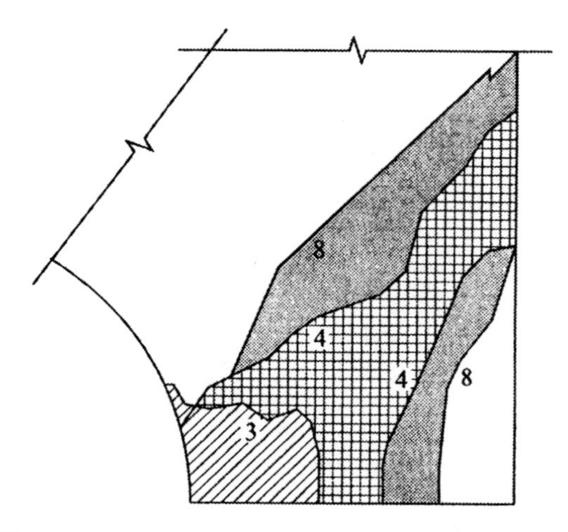

FIGURE 4.6. Plane-strain strip with a circular hole. Elastic-plastic boundary.

described in Matthies and Strang [1979], global convergence is attained for a step size $\Delta t = 0.1$.

4.5.2 Plane-Stress J_2 Flow Theory

Two numerical simulations are considered that employ the return-mapping algorithm summarized in BOX **3.3**.

TABLE 4.4. Example **4.5.2**. Residual Norm Values for Step 4.

Iteration	1	2	3	4	5	6
Continuum	0.25 E+3	0.74 E+1	0.22 E+1	0.11 E+1	0.75 E+0	0.55 E+0
Consistent	0.25 E+3	0.74 E+1	0.84 E+0	0.66 E−3	0.35 E−8	–
Iteration	7	8	9	10	11	12
Continuum	0.41 E+0	0.32 E+0	0.25 E+0	0.20 E+0	0.15 E+0	0.12 E+0
Consistent	–	–	–	–	–	–
Iteration	13	14	15	16	17	18
Continuum	0.98 E−1	0.78 E−1	0.61 E−1	0.48 E−1	0.38 E−1	0.30 E−1
Consistent	–	–	–	–	–	–
Iteration	19	20	21	22	23	
Continuum	0.23 E−1	0.18 E−1	0.14 E−1	0.11 E−1	0.91 E−2	
Consistent	–	–	–	–	–	

TABLE 4.5. Example **4.5.3**. Error Norms for Time-Step 0.01.

Energy Norm

Iteration	**Step**			
#	1	2	3	4
1	0.112 E+01	0.709 E−01	0.716 E−01	0.725 E−01
2	0.277 E−03	0.597 E−04	0.248 E−03	0.250 E−04
3	0.218 E−04	0.100 E−04	0.262 E−04	0.336 E−05
4	0.504 E−06	0.494 E−07	0.139 E−07	0.944 E−06
5	0.142 E−08	0.153 E−12	0.245 E−10	0.302 E−08
6	0.441 E−14	0.396 E−22	0.201 E−19	0.650 E−13
7	0.690 E−25	0.183 E−33	0.180 E−33	0.334 E−22
8	–	–	–	0.392 E−33

Residual Norm

Iteration	**Step**			
#	1	2	3	4
1	0.114 E+02	0.285 E+01	0.285 E+01	0.285 E+01
2	0.388 E−01	0.263 E−01	0.450 E−01	0.230 E−01
3	0.387 E−01	0.319 E−01	0.439 E−01	0.963 E−02
4	0.624 E−02	0.165 E−02	0.942 E−03	0.422 E−02
5	0.215 E−03	0.237 E−05	0.360 E−04	0.279 E−03
6	0.411 E−06	0.361 E−10	0.135 E−08	0.120 E−05
7	0.155 E−11	0.175 E−15	0.164 E−15	0.276 E−10
8	–	–	–	0.189 E−15

EXAMPLE: 4.5.3. Extension of a Strip with a Circular Hole. The geometry and finite-element mesh for the problem considered are shown in Figure **4.7**. A unit thickness is assumed and the calculation is performed by imposing uniform displacement control on the upper boundary. For obvious symmetry considerations, only one-quarter of the specimen is analyzed. A total of 164 four-node isoparametric quadrilaterals with bilinear interpolation of the displacement field is employed in the calculation. It should be noted that no special treatment of the incompressibility constraint is needed for plane-stress problems. A von Mises yield condition with linear isotropic hardening is considered. The elastic constants and nonzero parameters in hardening law (4.5.3) are $E = 70$, $\nu = 0.2$, $\bar{K}_0 = \bar{K}_\infty = 0.243$, $\bar{H}' = 2.24$, and $\beta = 1$.

The problem is first solved using prescribed increments of vertical displacement on the upper boundary of 0.04 followed by three subsequent equal increments of 0.01. The resulting spread of the plastic zone is shown in Figure **4.8**. Note that the spread of the plastic zone across the entire cross-section is achieved in the third load increment. The values of the $\mathbb{H}^1$-energy norm and the Euclidean norm of the residual for the entire calculation are shown in Table **4.5**. These results exhibit an asymptotic rate of approximately quadratic convergence. No line search is required for this time step.

To demonstrate the robustness of the solution procedure, the problem described above was also solved using two equal increments $\Delta t = 0.07$.

The values of the energy and residual norms for the entire iterative process are shown in Table **4.6**. An approximately quadratic rate of asymptotic convergence is again exhibited. Finally, to demonstrate the possible range of application of the proposed procedure, the problem was solved using two increments of $\Delta t = 0.5$. Nonconverged solutions corresponding to the first two iterates are shown in Figure **4.9**, and the converged solutions for the two time steps, labeled 1 and 2, are shown in Figure **4.10**. These results demonstrate that, even with the entire specimen plastified in the first two iterations, the solution procedure still produces a meaningful converged solution.

The values of the energy and residual norms for the entire iterative process along with the number of line searches performed in each iteration are shown in Table **4.7**. For this very large loading step an approximately quadratic rate of asymptotic convergence is still exhibited.

EXAMPLE: 4.5.4. Bending of a Strip with a Circular Notch. The problem considered is pure bending of a strip of finite width with two symmetric circular notches, as shown in Figure **4.11**. By noting symmetry and asymmetry, only one quarter of the region need be modeled. The finite-element mesh, also shown in Figure **4.11**, consists of 252 four-node isoparametric elements with bilinear interpolation functions. Loading is applied by prescribing the boundary condition as a

TABLE 4.6. Example **4.5.3**. Error Norms for Time-Step 0.07.

Iteration	**Energy Norm**		**Residual Norm**	
#	Step 1	Step 2	Step 1	Step 2
1	0.344 E+01	0.355 E+01	0.199 E+02	0.199 E+02
2	0.149 E−01	0.139 E−02	0.123 E+00	0.102 E+00
3	0.997 E−01	0.287 E−03	0.880 E+00	0.612 E−01
4	0.162 E−01	0.103 E−04	0.718 E+00	0.109 E−01
5	0.386 E−02	0.300 E−07	0.385 E+00	0.646 E−03
6	0.716 E−05	0.707 E−12	0.150 E−01	0.317 E−05
7	0.608 E−06	0.497 E−21	0.407 E−02	0.693 E−10
8	0.973 E−08	—	0.547 E−03	—
9	0.480 E−13	—	0.163 E−05	—
10	0.325 E−23	—	0.147 E−10	—

TABLE 4.7. Example **4.5.3**. Error Norms for Time-Step 0.5.

Energy Norm				Residual Norm	
Step (Line Search)				Step	
1		2		1	2
0.175E +03	(0)	0.182E +03	(2)	0.142E +03	0.142E +03
0.438E +00		0.434E −01		0.546E +00	0.207E +00
0.502E +01	(7)	0.344E −02		0.133E +01	0.141E +00
0.877E −01		0.182E −03		0.115E +01	0.393E −01
0.207E +01	(8)	0.172E −05		0.155E +01	0.576E −02
0.244E −01		0.154E −09		0.113E +01	0.433E −04
0.113E −01		0.186E −17		0.732E +00	0.505E −08
0.228E −01	(2)	0.103E −30		0.545E +00	0.347E −14
0.172E −02				0.284E +00	
0.358E −03				0.176E +00	
0.122E −04				0.196E −01	
0.119E −04				0.943E −02	
0.279E −06				0.191E −02	
0.135E −09				0.525E −04	
0.512E −16				0.302E −07	
0.717E −29				0.136E −13	

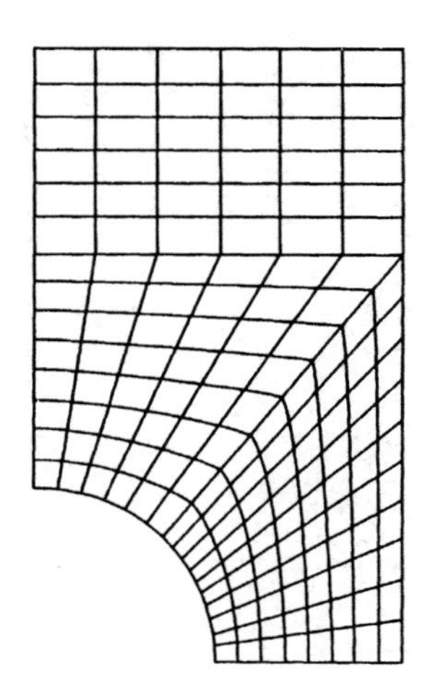

FIGURE 4.7. Plane-stress strip with a circular hole. Finite-element mesh.

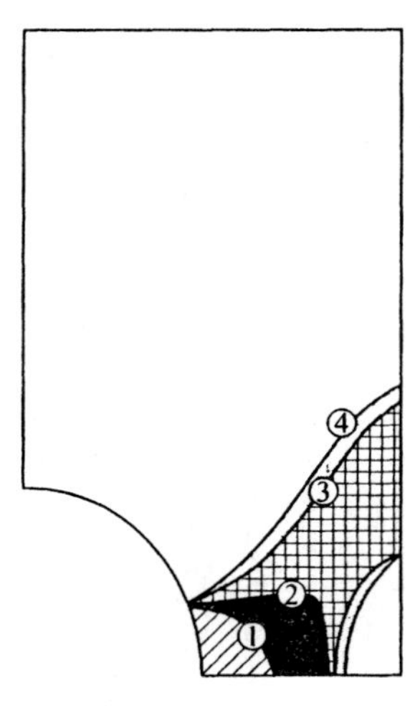

FIGURE 4.8. Plane stress strip with a circular hole. Elastic-plastic interface.

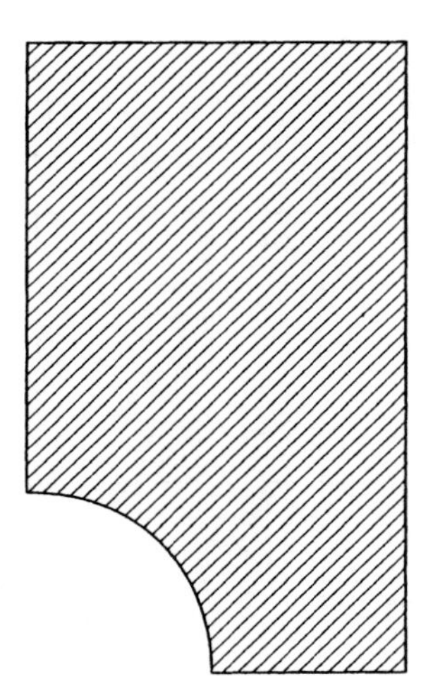

FIGURE 4.9. Plane-stress strip with a circular hole. Non-converged solution for time step $\Delta t = 0.5$ corresponding to the first two iterations.

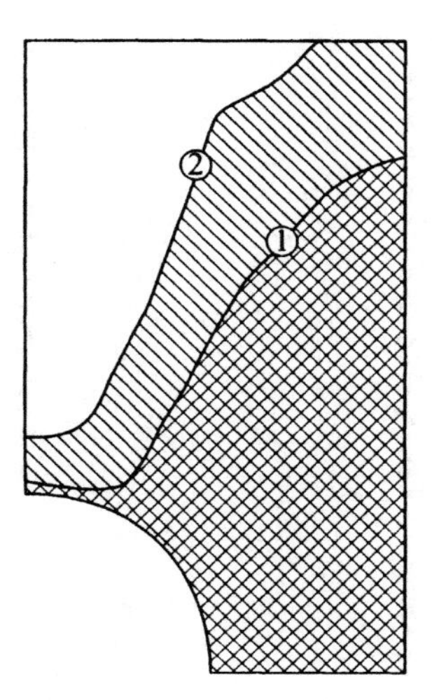

FIGURE 4.10. Plane-stress strip with a circular hole. Converged solutions for the first two time steps.

linear-varying, vertical displacement along the upper boundary, that is,

$$v(x, y, t)\big|_{y=9} = \frac{x}{b}t, \quad 0 \le x \le b, \tag{4.5.5}$$

where b is a constant given by $b = 10 - 2.5\sqrt{2}$. Four loading increments of equal size, corresponding to $\Delta t = 0.1$, are considered. For the geometry described above, two sets of material parameters were analyzed.

Case (a): *Linear isotropic hardening.* The nonzero material parameters are chosen as

$$E = 100, \quad \nu = 0.3, \quad \bar{K}_0 = \bar{K}_\infty = 1.0, \quad \bar{H}' = 5.0, \quad \beta = 1. \tag{4.5.6}$$

The results of the numerical simulation are shown in Figure **4.12**. To provide an idea of the computational effort involved in the calculation, the error in the $\mathbb{H}^1$-energy norm and the Euclidean norm of the residual for each iteration are given in Table **4.8**.

Case (b): *Combined linear isotropic-kinematic hardening.* The nonzero material parameters are chosen as

$$E = 100, \quad \nu = 0.3, \quad \bar{K}_0 = \bar{K}_\infty = 1.0, \quad \bar{H}' = 5.0, \quad \beta = 1/5. \tag{4.5.7}$$

The results of the numerical simulation are shown in Figure **4.13**, and the corresponding values of the energy and residual norms for each iteration are summarized in Table **4.9**.

Again the quadratic rate of asymptotic convergence of the Newton iterative scheme is exhibited by these results. Note that although included in the solution scheme, no line searches are required during the iterative process.

TABLE 4.8. Example **4.5.4**. Error Norms for Pure Isotropic Hardening.

Energy Norm

Iteration	**Step**			
#	1	2	3	4
1	0.403 E+01	0.409 E+01	0.417 E+01	0.296 E+01
2	0.148 E−02	0.127 E−01	0.228 E−02	0.109 E−02
3	0.155 E−03	0.658 E−02	0.695 E−04	0.359 E−04
4	0.178 E−06	0.213 E−04	0.616 E−07	0.429 E−07
5	0.467 E−12	0.134 E−06	0.415 E−09	0.755 E−13
6	0.152 E−22	0.647 E−12	0.143 E−17	0.517 E−24
7	–	0.793 E−22	0.180 E−31	–

Residual Norm

Iteration	**Step**			
#	1	2	3	4
1	0.205 E+02	0.205 E+02	0.205 E+02	0.170 E+02
2	0.977 E−01	0.234 E+00	0.186 E+00	0.112 E+00
3	0.153 E+00	0.535 E+00	0.729 E−01	0.382 E−01
4	0.358 E−02	0.557 E−01	0.185 E−02	0.186 E−02
5	0.614 E−05	0.345 E−02	0.208 E−03	0.273 E−05
6	0.354 E−10	0.883 E−05	0.131 E−07	0.716 E−11
7	–	0.967 E−10	0.179 E−14	–

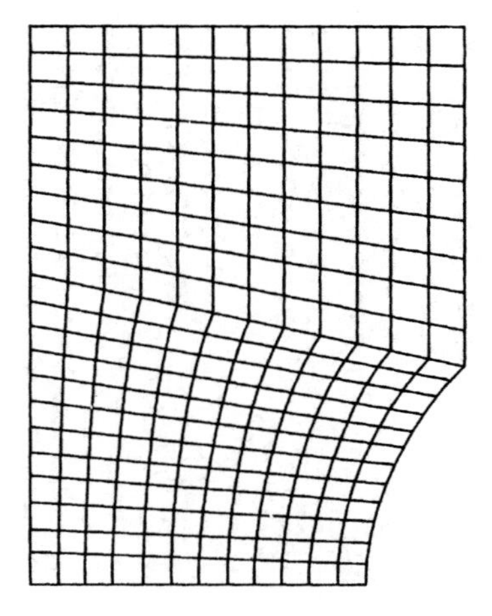

FIGURE 4.11. Bending of a plane-stress strip with a circular notch. Finite-element mesh.

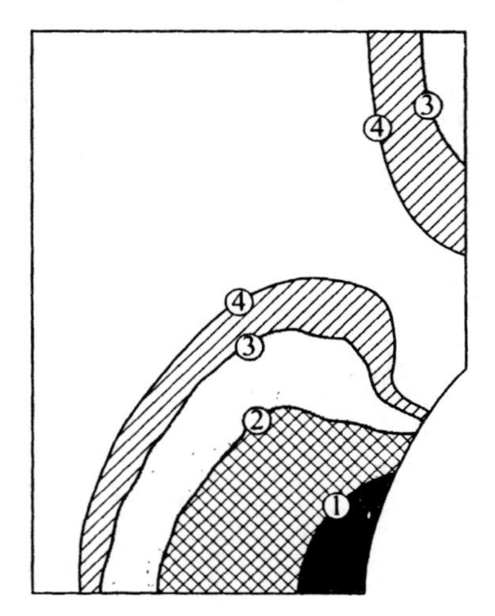

FIGURE 4.12. Bending of a plane-stress strip with a circular notch. Elastic-plastic interface for pure isotropic hardening.

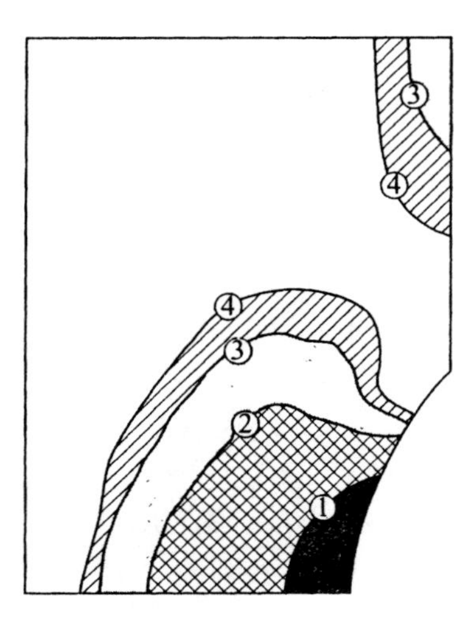

FIGURE 4.13. Bending of a plane-stress strip with a circular notch. Elastic-plastic interface for combined isotropic-kinematic hardening.

TABLE 4.9. Example **4.5.4**. Error Norms for Combined Isotropic-Kinematic Hardening.

Energy Norm

Iteration	**Step**			
#	1	2	3	4
1	0.403E +01	0.409E +01	0.417E +01	0.296E +01
2	0.148E −02	0.128E −01	0.214E −02	0.719E −03
3	0.155E −03	0.659E −02	0.716E −04	0.136E −04
4	0.178E −06	0.213E −04	0.221E −06	0.110E −07
5	0.467E −12	0.134E −06	0.388E −09	0.202E −14
6	0.152E −22	0.642E −12	0.153E −17	0.201E −27
7	–	0.779E −22	0.166E −31	–

Residual Norm

Iteration	**Step**			
#	1	2	3	4
1	0.205E +02	0.205E +02	0.205E +02	0.170E +02
2	0.977E −01	0.233E +00	0.191E +00	0.120E +00
3	0.153E +00	0.536E +00	0.761E −01	0.261E −01
4	0.358E −02	0.557E −01	0.420E −02	0.941E −03
5	0.614E −05	0.345E −02	0.202E −03	0.449E −06
6	0.354E −10	0.880E −05	0.133E −07	0.143E −12
7	–	0.958E −10	0.172E −14	–

5

Nonsmooth Multisurface Plasticity and Viscoplasticity

In this chapter we address the formulation and numerical implementation of the classical plasticity and viscoplasticity models discussed in Chapter **2** extended to accommodate elastic domains whose boundaries $\partial\mathbb{E}_\sigma$ are nonsmooth. In particular, we are concerned with the case in which $\partial\mathbb{E}_\sigma$ is composed of several smooth yield surfaces which intersect nonsmoothly, leading to the presence of singular points or "*corners*" in the boundary of the elastic domain.

Plasticity models possessing nonsmooth yield surfaces are used in many engineering applications, especially in soil and rock mechanics, see e.g., Nemat-Nasser [1983], Desai and Siriwardane [1984] and Chen [1984]. The *Cam–Clay* and *Critical State* models, the *Cap* model, and *Mohr–Coulomb* models are familiar examples that arise in soil mechanics. Representative algorithmic treatments of these and related models are in Aravas [1986]; Loret and Prevost [1986]; DiMaggio and Sandler [1971]; Sandler, DiMaggio, and Baladi [1976]; Sandler and Rubin [1979]; Resende and Martin [1986]; and Simo, Ju, Pister, and Taylor [1987]. In metal plasticity, the classical *Tresca* yield criterion (see Hill [1950]) furnishes another example of a model in which the the boundary of the elastic domain is nonsmooth. In structural mechanics, yield conditions formulated in terms of stress resultants often exhibit corners. Typical examples include *Illushin's* yield condition for shells and lower bounds to the Mises yield condition for beams of the type considered by Neal [1961] and Simo, Hjelmstad, and Taylor [1984]. Finally, J_2 corner theories of the type considered by Christoffersen and Hutchinson [1979] play an important role in some recent numerical studies of localization of deformation, as in Tvergaard, Needleman, and Lo [1981], Triantafyllidis, Needleman, and Tvergaard [1982], and Needleman and Tvergaard [1984]. Alternative formulations of J_2 flow theory that exhibit a corner-like effect and are suited for large-scale computation have been proposed in Hughes and Shakib [1986] and Simo [1987b]. These models agree with certain experimental results on tangential plastic loading reported by Budiansky et al. [1951].

The extension of classical plasticity models to accommodate nonsmooth yield surfaces goes back to the fundamental work of Koiter [1960], and Mandel [1964,1965]. Modern formulations of plasticity employing convex analysis, as in Moreau [1976], Suquet [1981] or Temam [1985], encompass these classical treatments as a particular case. (See also, Matthies [1979].) Here, we follow the classical approach and formulate multisurface plasticity models as direct extensions of the models discussed in Chapter **2**. Since all the notions introduced in that chapter extend straightforwardly to multisurface plasticity, our developments here serve largely as a review in a slightly more general context of ideas already discussed.

The algorithmic treatment of multisurface plasticity also follows the same lines already discussed in Chapter **3**. The only crucial difference lies in the algorithmic characterization of plastic loading by the trial elastic state. In the present context, the classical Kuhn–Tucker complementarity conditions, which are closely related to loading/unloading conditions discussed by Koiter [1960], again provide the only useful characterization of plastic loading. However, in contrast with the situation encountered in single-surface plasticity, the actual implementation of these conditions is not straightforward and an iterative procedure must be adopted. Our treatment here follows the approach advocated in Simo, Kennedy, and Govindjee [1988] which is inspired by classical ideas of convex mathematical programming.

In contrast with rate-independent plasticity, the extension of Perzyna-type models considered in Chapter **2** is, in general, not meaningful for nonsmooth multiple loading surfaces. However, the notion of closest point projection underlying models of the Duvaut–Lions type is immediately generalizable to the case in which the boundary of the elastic domain is nonsmooth. Moreover, we show that the algorithmic treatment of this class of models is remarkably simple. In fact, a *closed-form*, unconditionally stable algorithm can be constructed from the trial state and the rate-independent solution. In a sense, this methodology opposes the view taken in standard computational treatments of viscoplasticity where the rate-independent solution is obtained from the viscoplastic solution as the inviscid limit; see e.g., Zienkiewicz and Cormeau [1974]; Cormeau [1975]; Hughes and Taylor [1978]; or Pinsky, Ortiz, and Pister [1983].

5.1 Rate-Independent Multisurface Plasticity. Continuum Formulation

As in Chapter **2**, $\mathcal{B} \subset \mathbb{R}^3$ denotes the reference configuration with smooth boundary $\partial\mathcal{B}$, and $\boldsymbol{u} : \mathcal{B} \to \mathbb{R}^3$ is the displacement field of particles $\boldsymbol{x} \in \mathcal{B}$. Further, $\boldsymbol{\varepsilon} = \nabla^s \boldsymbol{u}$ denotes the strain tensor, $\{\boldsymbol{\varepsilon}^p, \boldsymbol{q}\}$ the plastic-strain tensor and the hardening variables, respectively, and $\boldsymbol{\sigma} \in \mathbb{S}$ denotes the stress tensor.

5.1.1 Summary of Governing Equations

As usual, the *elastic response* is characterized by a strain-energy function $W(\boldsymbol{\varepsilon} - \boldsymbol{\varepsilon}^p)$ leading to stress-strain relationships of the form

$$\boldsymbol{\sigma} = \nabla W(\boldsymbol{\varepsilon} - \boldsymbol{\varepsilon}^p),$$

where (5.1.1)

$$\mathbf{C} := \nabla^2 W(\varepsilon - \varepsilon^p),$$

is the tensor of elastic moduli, typically assumed constant. The essential feature of multisurface plasticity is the characterization of the elastic domain $\mathbb{E}_\sigma$. We assume that $\mathbb{E}_\sigma$ is a convex subset of $\mathbb{S} \times \mathbb{R}^q$ defined as

$$\boxed{\mathbb{E}_\sigma := \left\{(\boldsymbol{\sigma}, \boldsymbol{q}) \in \mathbb{S} \times \mathbb{R}^q \mid f_\alpha(\boldsymbol{\sigma}, \boldsymbol{q}) \leq 0\,, \text{ for all } \alpha \in [1, 2, \ldots, m]\right\},} \tag{5.1.2}$$

where $f_\alpha(\boldsymbol{\sigma}, \boldsymbol{q})$ are $m \geq 1$ functions intersecting possibly nonsmoothly; see Figure **5.1**. Accordingly, the boundary $\partial\mathbb{E}_\sigma$ of $\mathbb{E}_\sigma$ is generally nonsmooth and is given by

$$\boxed{\partial\mathbb{E}_\sigma := \left\{(\boldsymbol{\sigma}, \boldsymbol{q}) \in \mathbb{S} \times \mathbb{R}^q \mid f_\alpha(\boldsymbol{\sigma}, \boldsymbol{q}) = 0\,, \text{ for some } \alpha \in [1, 2, \ldots, m]\right\}.} \tag{5.1.3}$$

In what follows, attention is restricted to the case in which the $m \geq 1$ functions $f_\alpha(\boldsymbol{\sigma}, \boldsymbol{q})$ are *smooth* and define *independent* (*nonredundant*) constraints at any $(\boldsymbol{\sigma}, \boldsymbol{q}) \in \partial\mathbb{E}_\sigma$.* With this definition of the closure of the elastic range, the evolution of plastic strain is governed by the following flow rule, often referred to as Koiter's rule (see Koiter[1953], Mandel [1965] and the review articles of Koiter [1960] and Naghdi[1960]):

$$\boxed{\dot{\varepsilon}^p = \sum_{\alpha=1}^{m} \gamma^\alpha \partial_\sigma f_\alpha(\boldsymbol{\sigma}, \boldsymbol{q}),} \tag{5.1.4}$$

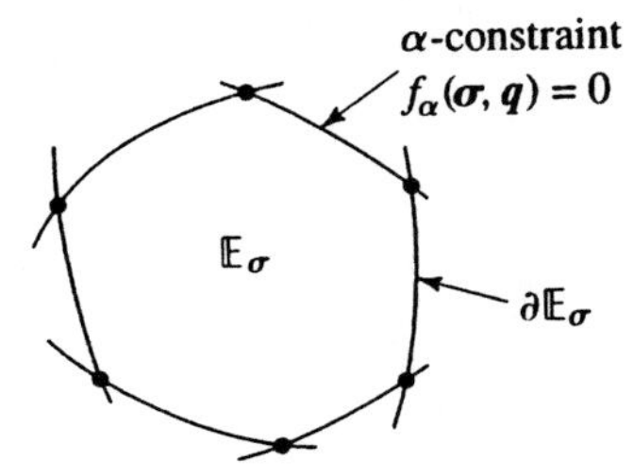

FIGURE 5.1. Illustration of the elastic domain in multisurface plasticity. Note the possible presence of "corner points" on the boundary of the elastic domain defined by the yield surface.

---

*The fact that $\dim \mathbb{E}_\sigma = 6 + q$ is finite, limits the number of independent surfaces which can intersect at one point $(\boldsymbol{\sigma}, \boldsymbol{q}) \in \partial\mathbb{E}_\sigma$ so that the vectors $\{\partial_\sigma f_\alpha(\boldsymbol{\sigma}, \boldsymbol{q})\}$ (and $\{\partial_q f_\alpha(\boldsymbol{\sigma}, \boldsymbol{q})\}$) remain *linearly independent*. For example, if $\boldsymbol{q} = \mathbf{0}$ and $\dim \mathbb{E}_\sigma = 6$ then at most six independent surfaces can intersect at one point.

where, for simplicity, attention is restricted to the *associative* case. Here γ^α are $m \geq 1$ *plastic consistency parameters*, which satisfy the following *Kuhn–Tucker complementary conditions* for $\alpha = 1, 2, \ldots, m$,*

$$\boxed{\gamma^\alpha \geq 0, \quad f_\alpha(\boldsymbol{\sigma}, \boldsymbol{q}) \leq 0,}$$

and (5.1.5)

$$\boxed{\gamma^\alpha f_\alpha(\boldsymbol{\sigma}, \boldsymbol{q}) \equiv 0\,,}$$

along with the consistency requirement

$$\boxed{\gamma^\alpha \dot{f}_\alpha(\boldsymbol{\sigma}, \boldsymbol{q}) \equiv 0\ .} \tag{5.1.6}$$

Conditions (5.1.5) and (5.1.6) are essentially the multisurface plasticity counterpart of those in Koiter[1960, equation (2.19)], and have been employed by several authors, notably, Maier[1970] and Maier and Griegson[1979].

Similarly, the generalization of the general evolution equations to multisurface plasticity for the hardening variables takes the form

$$\dot{\boldsymbol{q}} = -\sum_{\alpha=1}^{m} \gamma^\alpha \boldsymbol{h}_\alpha(\boldsymbol{\sigma}, \boldsymbol{q}), \tag{5.1.7}$$

whereas now the *associative* or potential form of this hardening law is written as

$$\boxed{\dot{\boldsymbol{\alpha}} := -\mathbf{D}^{-1}\dot{\boldsymbol{q}} = \sum_{\beta=1}^{m} \gamma^\beta \partial_{\boldsymbol{q}} f_\beta(\boldsymbol{\sigma}, \boldsymbol{q}).} \tag{5.1.8}$$

As in Chapter **2**, we refer to $\mathbf{D}$ as the matrix of generalized hardening moduli. For simplicity throughout our exposition, we assume that $\mathbf{D} \in \mathbb{R}^q \times \mathbb{R}^q$ is a *constant, symmetric, positive-definite* matrix. The same arguments discussed in detail in Chapter **2** show that the evolutionary equation (5.1.8) is the result of the principle of maximum plastic dissipation.

5.1.2 Loading/Unloading Conditions

The crucial aspect of multisurface plasticity concerns the statement of the loading/unloading conditions in a form suitable for algorithmic implementation. To this end, let $m_{\text{adm}} \leq m$ be the number of constraints active at a given point $(\boldsymbol{\sigma}, \boldsymbol{q}) \in \partial\mathbb{E}_\sigma$, and let $\mathbb{J}_{\text{adm}}$ be the set of m_{adm} indices associated with these constraints. By definition,

$$\boxed{\mathbb{J}_{\text{adm}} := \{\beta \in \{1, 2, \ldots, m\} \mid f_\beta(\boldsymbol{\sigma}, \boldsymbol{q}) = 0\}.} \tag{5.1.9}$$

---

*The summation convention on repeated indices is *not* enforced in this chapter.

To proceed further, we make an additional explicit assumption which is the counterpart in multisurface plasticity of Assumption **2.1** in Chapter **2**.

Assumption 5.1. It is assumed that the flow rule (5.1.4) and the hardening law (5.1.8) obey the following inequality at any $(\boldsymbol{\sigma}, \boldsymbol{q}) \in \partial\mathbb{E}_\sigma$:

$$\boxed{\begin{aligned} & g_{\alpha\beta}(\boldsymbol{\sigma}, \boldsymbol{q}) := \left[\partial_\sigma f_\alpha : \mathbf{C} : \partial_\sigma f_\beta + \partial_q f_\alpha \cdot \mathbf{D}\partial_q f_\beta\right] \\ & \sum_{\alpha \in \mathbb{J}_{\text{adm}}} \sum_{\beta \in \mathbb{J}_{\text{adm}}} \xi^\alpha g_{\alpha\beta}(\boldsymbol{\sigma}, \boldsymbol{q})\xi^\beta > 0, \text{ for } \xi^\alpha \in \mathbb{R}. \end{aligned}} \tag{5.1.10}$$

For perfect plasticity recall that this assumption follows from the standard requirement that $\boldsymbol{\xi} : \mathbf{C} : \boldsymbol{\xi} > 0$ for all $\boldsymbol{\xi}^T = \boldsymbol{\xi}$. In the present situation, this assumption amounts to requiring that the matrix $[g_{\alpha\beta}]$ be *positive-definite.* Now let $\alpha \in \mathbb{J}_{\text{adm}}$. By the chain rule along with (5.1.4) and (5.1.8), the value of $\dot{f}_\alpha$ is given as

$$\begin{aligned} \dot{f}_\alpha &= \partial_\sigma f_\alpha : \mathbf{C} : \dot{\boldsymbol{\varepsilon}} - \sum_{\beta \in \mathbb{J}_{\text{adm}}} \left[\partial_\sigma f_\alpha : \mathbf{C} : \partial_\sigma f_\beta + \partial_q f_\alpha \cdot \mathbf{D}\partial_q f_\beta\right]\gamma^\beta \\ &= \partial_\sigma f_\alpha : \mathbf{C} : \dot{\boldsymbol{\varepsilon}} - \sum_{\beta \in \mathbb{J}_{\text{adm}}} g_{\alpha\beta}(\boldsymbol{\sigma}, \boldsymbol{q})\gamma^\beta , \end{aligned} \tag{5.1.11}$$

where $g_{\alpha\beta}(\boldsymbol{\sigma}, \boldsymbol{q})$ are defined in (5.1.10). Now the same argument employed in Section **2.2.2** (see equation (2.2.16)) shows that

$$\boxed{\text{If } (\boldsymbol{\sigma}, \boldsymbol{q}) \in \partial\mathbb{E}_\sigma \text{ and } \alpha \in \mathbb{J}_{\text{adm}}, \text{ then } \dot{f}_\alpha(\boldsymbol{\sigma}, \boldsymbol{q}) \le 0.} \tag{5.1.12}$$

With this background, we formulate the loading/unloading conditions as follows.

Proposition 5.1. *Let $\dot{\boldsymbol{\varepsilon}}$ be given. The Kuhn–Tucker conditions (5.1.5), the consistency requirement (5.1.6), and Assumption* **5.1** *imply the following (strain-space) loading conditions*

$$\boxed{\begin{aligned} & \textit{If } \mathbb{J}_{\text{adm}} \equiv \emptyset, \textit{ then } \dot{\boldsymbol{\varepsilon}}^p = \mathbf{0} \textit{ and } \dot{\boldsymbol{q}} = \mathbf{0}. \\ & \textit{If } \mathbb{J}_{\text{adm}} \neq \emptyset, \textit{ then} \\ & \quad \textbf{i. } \textit{if } \partial_\sigma f_\alpha : \mathbf{C} : \dot{\boldsymbol{\varepsilon}} \le 0, \textit{ for all } \alpha \in \mathbb{J}_{\text{adm}} \Rightarrow \dot{\boldsymbol{\varepsilon}}^p = 0, \textit{ and } \dot{\boldsymbol{q}} = \mathbf{0}; \\ & \quad \textbf{ii. } \textit{if } \partial_\sigma f_\alpha : \mathbf{C} : \dot{\boldsymbol{\varepsilon}} > 0 \textit{ for at least one } \alpha \in \mathbb{J}_{\text{adm}} \Rightarrow \dot{\boldsymbol{\varepsilon}}^p \neq 0, \textit{ and } \dot{\boldsymbol{q}} \neq \mathbf{0}. \end{aligned}} \tag{5.1.13}$$

PROOF. First, if $\mathbb{J}_{\text{adm}} = \emptyset$, then $f_\alpha < 0$ for all $\alpha = 1, 2, \ldots, m$, so that, by $(5.1.5)_3$, $\gamma^\alpha \equiv 0$. Thus, $(5.1.13)_1$ holds. Next, consider the case $\mathbb{J}_{\text{adm}} \neq \emptyset$.

i. Let $\alpha \in \mathbb{J}_{\text{adm}} \neq \emptyset$. Since $\gamma^\alpha \dot{f}_\alpha = 0$ and $\gamma^\alpha \ge 0$, from (5.1.11),

$$\sum_{\alpha \in \mathbb{J}_{\text{adm}}} \sum_{\beta \in \mathbb{J}_{\text{adm}}} \gamma^\alpha g_{\alpha\beta}\gamma^\beta = \sum_{\alpha \in \mathbb{J}_{\text{adm}}} \gamma^\alpha \partial_\sigma f_\alpha : \mathbf{C} : \dot{\boldsymbol{\varepsilon}} \le 0 , \tag{5.1.14}$$

since, by hypothesis, $\partial_{\sigma} f_{\alpha} : \mathbf{C} : \dot{\varepsilon} \leq 0$ for all $\alpha \in \mathbb{J}_{\text{adm}}$. However, by assumption, $g_{\alpha\beta}$ is positive-definite. Thus $\gamma^{\alpha} = 0$ for all $\alpha \in J_{\text{adm}}$, and **i** holds.

ii. Let $\alpha \in \mathbb{J}_{\text{adm}} \neq \emptyset$ be such that $\partial_{\sigma} f_{\alpha} : \mathbf{C} : \dot{\varepsilon} > 0$. Suppose it were possible that $\dot{\varepsilon}^{p} = \mathbf{0}$ and $\dot{\boldsymbol{q}} = \mathbf{0}$. Then, (5.1.11) would imply that

$$\dot{f}_{\alpha}(\boldsymbol{\sigma}, \boldsymbol{q}) = \partial_{\sigma} f_{\alpha} : \mathbf{C} : \dot{\varepsilon} > 0\,, \tag{5.1.15}$$

which contradicts (5.1.12). Thus **ii** holds.

□

It should be noted that if plastic loading takes place at $(\boldsymbol{\sigma}, \boldsymbol{q}) \in \partial\mathbb{E}_{\sigma}$ and *several* yield surfaces are active, then the condition $\partial_{\sigma} f_{\alpha} : \mathbf{C} : \dot{\varepsilon} > 0$ *does not guarantee that f_{α} is ultimately active.* This observation is central to our subsequent developments and is examined in detail in Section **5.1.4** below.

BOX 5.1. Infinitesimal Multisurface Plasticity.

(i) Elastic stress-strain relationships

$$\boldsymbol{\sigma} = \nabla W(\boldsymbol{\varepsilon} - \boldsymbol{\varepsilon}^{p}),$$

where $\boldsymbol{\varepsilon} := \nabla^{s}\boldsymbol{u}$.

(ii) Associative flow rule

$$\dot{\boldsymbol{\varepsilon}}^{p} = \sum_{\alpha=1}^{m} \gamma^{\alpha} \partial_{\sigma} f_{\alpha}(\boldsymbol{\sigma}, \boldsymbol{q}).$$

(iii) Hardening law

$$\dot{\boldsymbol{\alpha}} = -\mathbf{D}^{-1}\dot{\boldsymbol{q}} = \sum_{\beta=1}^{m} \gamma^{\beta} \partial_{q} f_{\beta}(\boldsymbol{\sigma}, \boldsymbol{q}).$$

(iv) Yield and loading/unloading conditions

$$\gamma^{\alpha} \geq 0$$

$$f_{\alpha}(\boldsymbol{\sigma}, \boldsymbol{q}) \leq 0$$

$$\gamma^{\alpha} f_{\alpha}(\boldsymbol{\sigma}, \boldsymbol{q}) \equiv 0$$

$$\gamma^{\alpha} \dot{f}_{\alpha}(\boldsymbol{\sigma}, \boldsymbol{q}) \equiv 0.$$

(v) Tangent elastoplastic moduli

$$\mathbf{C}^{\text{ep}} = \mathbf{C} - \sum_{\alpha,\, \beta \,\in\, \mathbb{J}_{\text{act}}} g^{\alpha\beta}\left[\mathbf{C} : \partial_{\sigma} f_{\alpha}\right] \otimes \left[\mathbf{C} : \partial_{\sigma} f_{\beta}\right],$$

$$g_{\alpha\beta}(\boldsymbol{\sigma}, \boldsymbol{q}) := \left[\partial_{\sigma} f_{\alpha} : \mathbf{C} : \partial_{\sigma} f_{\beta} + \partial_{q} f_{\alpha} \cdot \mathbf{D}\partial_{q} f_{\beta}\right],$$

$$\sum_{\beta \,\in\, \mathbb{J}_{\text{act}}} g^{\alpha\beta}(\boldsymbol{\sigma}, \boldsymbol{q})\, g_{\beta\gamma}(\boldsymbol{\sigma}, \boldsymbol{q}) = \delta^{\alpha}_{\gamma}.$$

5.1.3 Consistency Condition. Elastoplastic Tangent Moduli

As in the preceding chapters, we assume that the strain history is given so that the strain rate $\dot{\varepsilon}$ is known at time t. Then, conditions (5.1.5) and (5.1.6) determine whether a constraint is active. In expanded form, these conditions read as follows:
i. if $f_\alpha(\sigma, q) < 0$ or $f_\alpha(\sigma, q) = 0$ and $\dot{f}_\alpha(\sigma, q) < 0$, then $\gamma^\alpha = 0$;
ii. if $f_\alpha(\sigma, q) = 0$ and $\dot{f}_\alpha(\sigma, q) = 0$, then $\gamma^\alpha \geq 0$.

Conditions **i** and **ii** are a restatement for multisurface plasticity of classical conditions; see Koiter[1960, equation (2.19)]. Now let m_{act} be the number of constraints at a given point for which **ii** holds. Set

$$\boxed{\mathbb{J}_{\text{act}} := \left\{\beta \in \mathbb{J}_{\text{adm}} \mid \dot{f}_\beta(\sigma, q) = 0\right\}.} \tag{5.1.16}$$

Then, since γ^α is nonzero only for $\alpha \in \mathbb{J}_{\text{act}}$, it follows from (5.1.11) that

$$\dot{f}_\alpha(\sigma, q) = 0 \;\Rightarrow\; \sum_{\beta \,\in\, \mathbb{J}_{\text{adm}}} g_{\alpha\beta}(\sigma, q)\gamma^\beta = \partial_\sigma f_\alpha : \mathbf{C} : \dot{\varepsilon}\,, \tag{5.1.17}$$

for all $\alpha \in \mathbb{J}_{\text{act}}$. This leads to a system of m_{act} equations with $m_{\text{adm}} \geq m_{\text{act}}$ unknowns. Then conditions $\gamma^\beta = 0$ if $\dot{f}_\beta(\sigma, q) < 0$ provide the remaining $m_{\text{adm}} - m_{\text{act}}$ equations that render (5.1.17) a determinate system.

In summary,

$$\left.\begin{aligned} \gamma^\beta &= 0\,, \text{ if } \beta \notin \mathbb{J}_{\text{act}}\,, \\ \gamma^\alpha &= \sum_{\beta \,\in\, \mathbb{J}_{\text{act}}} g^{\alpha\beta}(\sigma, q)\big[\partial_\sigma f_\beta(\sigma, q) : \mathbf{C} : \dot{\varepsilon}\big]\,, \text{ if } \alpha \in \mathbb{J}_{\text{act}}\,, \end{aligned}\right\} \tag{5.1.18}$$

where $g^{\alpha\beta}(\sigma, q) = [g_{\alpha\beta}(\sigma, q)]^{-1}$ and $g_{\alpha\beta}(\sigma, q)$ is defined by (5.1.10). By substituting (5.1.18) in the rate form of the stress-strain relationships $(5.1.1)_1$, we obtain $\dot{\sigma} = \mathbf{C}^{ep} : \dot{\varepsilon}$, where $\mathbf{C}^{ep}$ is the tensor of elastoplastic tangent moduli given by the expression

$$\boxed{\mathbf{C}^{ep} = \begin{cases} \mathbf{C} - \displaystyle\sum_{\alpha,\, \beta \,\in\, \mathbb{J}_{\text{act}}} g^{\alpha\beta}\big[\mathbf{C} : \partial_\sigma f_\alpha\big] \otimes \big[\mathbf{C} : \partial_\sigma f_\beta\big] & \text{iff } \mathbb{J}_{\text{act}} \neq \emptyset\,, \\ \mathbf{C} & \text{iff } \mathbb{J}_{\text{act}} = \emptyset\,. \end{cases}} \tag{5.1.19}$$

For convenience, the basic equations governing classical multisurface, rate-independent plasticity are summarized in BOX **5.1**.

5.1.4 Geometric Interpretation

We give a geometric interpretation of the loading conditions (5.1.6) and illustrate the fact, alluded to above, that *a constraint f_α may be active; i.e., $\gamma^\alpha > 0$ and, nevertheless, one may have* $\partial_\sigma f_\alpha : \mathbf{C} : \dot{\varepsilon} < 0$. For simplicity, we consider perfect plasticity. At each $\boldsymbol{\sigma} \in \partial\mathbb{E}_\sigma$, the vector space

$$\mathbb{M} := \text{span}\left[\mathbf{g}_\alpha := \partial_\sigma f_\alpha\,, \ \text{for } \alpha \in \mathbb{J}_{\text{adm}}\right]. \tag{5.1.20}$$

We equip $\mathbb{M}$ with the inner product induced by $\mathbf{C}^*$ according to (5.1.10) and define the *dual vectors* (*co-vectors*) $\left\{\mathbf{g}^\alpha\right\}_{\alpha \in \mathbb{J}_{\text{adm}}}$ in the standard fashion, i.e.,

$$g_{\alpha\beta} := \mathbf{g}_\alpha : \mathbf{C} : \mathbf{g}_\beta\,, \ \text{and } \mathbf{g}^\alpha = \sum_{\beta \in \mathbb{J}_{\text{adm}}} g^{\alpha\beta}\mathbf{g}_\beta\,. \tag{5.1.21}$$

Given $\dot{\boldsymbol{\varepsilon}}$, conditions (5.1.13) define the *accessible elastic region* as the *cone*

$$\boxed{\mathbb{M}^- = \left\{\boldsymbol{\xi} \in \mathbb{S} \mid \boldsymbol{\xi} : \mathbf{C} : \mathbf{g}_\alpha \leq 0\right\},} \tag{5.1.22}$$

whereas the *plastic region* is $\mathbb{M} - \mathbb{M}^-$. The *normal cone* $\mathbb{M}^+$ is given by (see Figure **5.2**)

$$\boxed{\mathbb{M}^+ = \left\{\boldsymbol{\xi} \in \mathbb{S} \mid \boldsymbol{\xi} = \textstyle\sum_{\alpha \in \mathbb{J}_{\text{adm}}} \lambda^\alpha \mathbf{g}_\alpha \text{ for } \lambda^\alpha \geq 0\right\}.} \tag{5.1.23}$$

A straightforward computation yields the values of the coefficients λ^α in (5.1.23) as

$$\lambda^\alpha = \sum_{\beta \in \mathbb{J}_{\text{act}}} g^{\alpha\beta}\partial_\sigma f_\beta : \mathbf{C} : \boldsymbol{\xi} > 0\,. \tag{5.1.24}$$

Therefore, for $\dot{\boldsymbol{\varepsilon}} \in \mathbb{M}^+$, γ^α and $\partial_\sigma f_\alpha : \mathbf{C} : \dot{\boldsymbol{\varepsilon}}$ may be interpreted as the *contravariant* and *covariant* components of $\dot{\boldsymbol{\varepsilon}}$ relative to $\{\mathbf{g}_\alpha\}$, respectively. The fact that

$$\boxed{\gamma^\alpha > 0 \not\Leftrightarrow \partial_\sigma f_\alpha : \mathbf{C} : \dot{\boldsymbol{\varepsilon}} > 0} \tag{5.1.25}$$

is illustrated in Figure **5.3**.

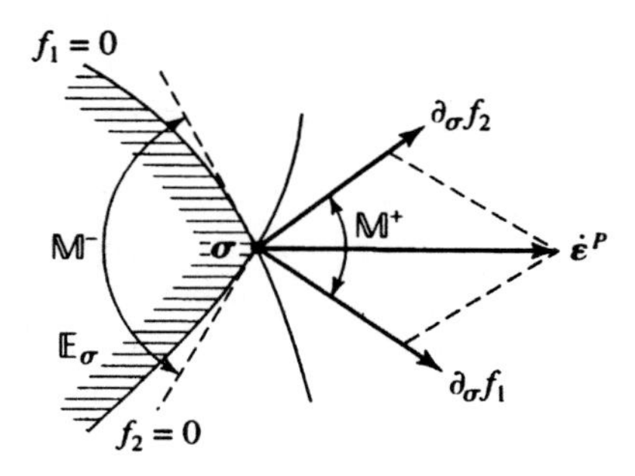

FIGURE 5.2. Illustration of the geometry at a singular point $\boldsymbol{\sigma} \in \partial\mathbb{E}_\sigma$, the intersection of two yield surfaces ($\mathbb{J}_{\text{act}} = \{1, 2\}$).

*In the presence of internal plastic variables, the inner product is induced by DIAG $[\mathbf{C}, \mathbf{D}]$.

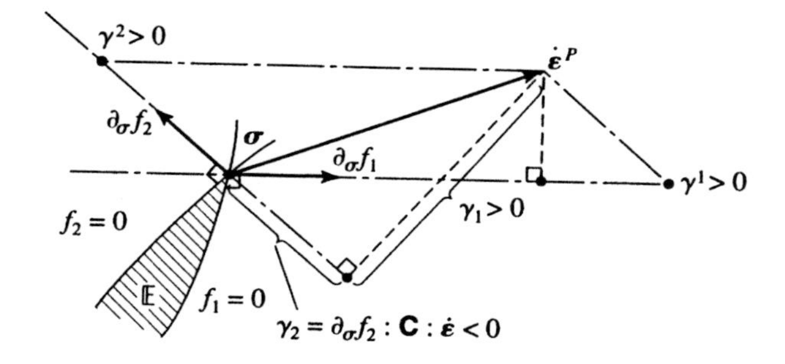

FIGURE 5.3. Positiveness of the contravariant components $\gamma^\alpha > 0$ does not guarantee positiveness of the covariant components $\partial_\sigma f_2 : \mathbf{C} : \dot{\boldsymbol{\varepsilon}} = g_{\alpha\beta}\gamma^\beta$.

5.2 Discrete Formulation. Rate-Independent Elastoplasticity

In this section we outline the extension of the return-mapping algorithms developed in Chapter **3** to multisurface plasticity. As shown below, the crucial difference with the developments in Chapter **3** concerns determining the active surfaces. For this purpose, the elastic trial state no longer suffices and an additional iterative procedure is required.

5.2.1 Closest Point Projection Algorithm for Multisurface Plasticity

As in Chapter **3**, we consider a time discretization of the interval $[0, T] \subset \mathbb{R}$ of interest, and let $\{\boldsymbol{\varepsilon}_n, \boldsymbol{\varepsilon}_n^p, \boldsymbol{\alpha}_n\}$ be the initial data at $t_n \in [0, T]$. Given an incremental displacement field $\Delta \boldsymbol{u} : \mathcal{B} \to \mathbb{R}^3$, application of an implicit backward Euler difference scheme to the evolutionary equations in BOX **5.1** results in the following nonlinear coupled system for the unknown state variables $\{\boldsymbol{\varepsilon}_{n+1}, \boldsymbol{\varepsilon}_{n+1}^p, \boldsymbol{\alpha}_{n+1}\}$ at time t_{n+1}

$$\boxed{\begin{aligned}
\boldsymbol{\varepsilon}_{n+1} &= \boldsymbol{\varepsilon}_n + \nabla^s(\Delta \boldsymbol{u}) \qquad \text{(trivial)}, \\
\boldsymbol{\sigma}_{n+1} &= \nabla W(\boldsymbol{\varepsilon}_{n+1} - \boldsymbol{\varepsilon}_{n+1}^p), \\
\boldsymbol{\varepsilon}_{n+1}^p &= \boldsymbol{\varepsilon}_n^p + \sum_{\alpha=1}^{m} \Delta\gamma^\alpha \partial_{\boldsymbol{\sigma}} f_\alpha(\boldsymbol{\sigma}_{n+1}, \boldsymbol{q}_{n+1}), \\
\boldsymbol{\alpha}_{n+1} &= \boldsymbol{\alpha}_n + \sum_{\alpha=1}^{m} \Delta\gamma^\alpha \partial_{\boldsymbol{q}} f_\alpha(\boldsymbol{\sigma}_{n+1}, \boldsymbol{q}_{n+1}), \\
\boldsymbol{q}_{n+1} &= -\mathbf{D}\boldsymbol{\alpha}_{n+1},
\end{aligned}} \tag{5.2.1}$$

where we have set $\Delta\gamma^\alpha := \Delta t \gamma^\alpha$. Now the discrete counterpart of the Kuhn–Tucker conditions which define the appropriate notion of loading/unloading takes the form

$$\boxed{\begin{aligned} f_\alpha(\boldsymbol{\sigma}_{n+1}, \boldsymbol{q}_{n+1}) &\leq 0\,, \\ \Delta\gamma^\alpha &\geq 0\,, \\ \Delta\gamma^\alpha f_\alpha(\boldsymbol{\sigma}_{n+1}, \boldsymbol{q}_{n+1}) &= 0\,, \end{aligned}} \tag{5.2.2}$$

for $\alpha = 1, 2, \ldots, m$. Finally, recall that the trial state is obtained formally by "freezing" plastic flow in the interval $[t_n, t_{n+1}]$. Accordingly, setting $\Delta\gamma^\alpha = 0$ in (5.2.1), we obtain

$$\left.\begin{aligned} \boldsymbol{\varepsilon}^{e^{\text{trial}}}_{n+1} &:= \boldsymbol{\varepsilon}_{n+1} - \boldsymbol{\varepsilon}^p_n\,, \\ \boldsymbol{\sigma}^{\text{trial}}_{n+1} &:= \nabla W\left(\boldsymbol{\varepsilon}^{e^{\text{trial}}}_{n+1}\right), \\ \boldsymbol{\varepsilon}^{p^{\text{trial}}}_{n+1} &:= \boldsymbol{\varepsilon}^p_n\,, \\ \boldsymbol{\alpha}^{\text{trial}}_{n+1} &:= \boldsymbol{\alpha}_n\,, \\ \boldsymbol{q}^{\text{trial}}_{n+1} &:= -\mathbf{D}\boldsymbol{\alpha}^{\text{trial}}_{n+1}\,, \\ f^{\text{trial}}_{\alpha,n+1} &:= f_\alpha\left(\boldsymbol{\sigma}^{\text{trial}}_{n+1}, \boldsymbol{q}^{\text{trial}}_{n+1}\right). \end{aligned}\right\} \tag{5.2.3}$$

The solution $\{\boldsymbol{\sigma}_{n+1}, \boldsymbol{q}_{n+1}\}$ to the discrete algorithmic problem (5.2.1)–(5.2.2) admits the same interpretation discussed in Section **3.2.2** as the solution of the following minimization problem:

Proposition 5.1. *The solution to problem* (5.2.1–2) *is unique and is characterized as the argument of the minimization problem*

$$\boxed{(\boldsymbol{\sigma}_{n+1}, \boldsymbol{q}_{n+1}) = \text{ARG}\left\{\underset{(\boldsymbol{\tau}, \boldsymbol{q}) \in \mathbb{E}_\sigma}{\text{MIN}} \left[\chi(\boldsymbol{\tau}, \boldsymbol{q})\right]\right\},} \tag{5.2.4}$$

where

$$\left.\begin{aligned} \chi(\boldsymbol{\tau}, \boldsymbol{q}) &:= \tfrac{1}{2}\left\|\boldsymbol{\sigma}^{\text{trial}}_{n+1} - \boldsymbol{\tau}\right\|^2_{\mathbf{C}^{-1}} + \tfrac{1}{2}\left\|\boldsymbol{q}_n - \boldsymbol{q}\right\|^2_{\mathbf{D}^{-1}}, \\ \left\|\boldsymbol{\sigma}^{\text{trial}}_{n+1} - \boldsymbol{\tau}\right\|^2_{\mathbf{C}^{-1}} &:= \left[\boldsymbol{\sigma}^{\text{trial}}_{n+1} - \boldsymbol{\tau}\right] : \mathbf{C}^{-1} : \left[\boldsymbol{\sigma}^{\text{trial}}_{n+1} - \boldsymbol{\tau}\right], \\ \left\|\boldsymbol{q}_n - \boldsymbol{q}\right\|^2_{\mathbf{D}^{-1}} &:= \left[\boldsymbol{q}_n - \boldsymbol{q}\right] : \mathbf{D}^{-1} : \left[\boldsymbol{q}_n - \boldsymbol{q}\right], \end{aligned}\right\} \tag{5.2.5}$$

where $\mathbf{C} := \nabla^2 W$ *and* $\mathbf{D}$ *are assumed constant and positive-definite.*

PROOF. Positive definiteness of $\mathbf{C}$ and $\mathbf{D}$ implies that $\chi : \mathbb{E}_\sigma \to \mathbb{R}$ is a strictly convex function. Since $\mathbb{E}_\sigma$ is a closed convex set as a result of the convexity assumption on $f_\alpha(\bullet, \bullet)$, uniqueness of the minimizer $(\boldsymbol{\sigma}_{n+1}, \boldsymbol{q}_{n+1})$ follows by standard results in convex analysis (see, e.g., Pshenichny and Danilin [1978],

Section **3**). To prove the equivalence of (5.2.4) and (5.2.1-2), as in Section **3.2.2**, we consider the Lagrangian associated with the minimization problem (5.2.4):

$$\mathcal{L}(\boldsymbol{\tau}, \boldsymbol{q}, \lambda^\alpha) := \chi(\boldsymbol{\tau}, \boldsymbol{q}) + \sum_{\alpha=1}^{m} \lambda^\alpha f_\alpha(\boldsymbol{\tau}, \boldsymbol{q}) . \tag{5.2.6}$$

Now the corresponding optimality conditions now take the form (see Luenberger [1984, page 314] or Strang [1986, page 724])

$$\left.\begin{aligned}
\partial_\tau \mathcal{L} &= -\mathbf{C}^{-1} : \left[\boldsymbol{\sigma}_{n+1}^{\text{trial}} - \boldsymbol{\sigma}_{n+1}\right] + \sum_{\alpha=1}^{m} \Delta\gamma^\alpha \partial_\sigma f_\alpha(\boldsymbol{\sigma}_{n+1}, \boldsymbol{q}_{n+1}) = \mathbf{0} , \\
\partial_q \mathcal{L} &= \boldsymbol{\alpha}_{n+1} - \boldsymbol{\alpha}_n + \sum_{\alpha=1}^{m} \Delta\gamma^\alpha \partial_q f_\alpha(\boldsymbol{\sigma}_{n+1}, \boldsymbol{q}_{n+1}) = \mathbf{0} , \\
\partial_{\lambda^\alpha} \mathcal{L} &= f_\alpha(\boldsymbol{\sigma}_{n+1}, \boldsymbol{q}_{n+1}) \le 0 , \\
\Delta\gamma^\alpha &\ge 0, \quad \Delta\gamma^\alpha f_\alpha(\boldsymbol{\sigma}_{n+1}, \boldsymbol{q}_{n+1}) = 0 ,
\end{aligned}\right\} \tag{5.2.7}$$

which are equivalent to (5.2.1-2). □

Again the solution $\boldsymbol{\Sigma} := (\boldsymbol{\sigma}_{n+1}, \boldsymbol{q}_{n+1})$ to the minimization problem (5.2.4) can be interpreted geometrically as the closest point projection of the trial state $(\boldsymbol{\sigma}_{n+1}^{\text{trial}}, \boldsymbol{q}_{n+1}^{\text{trial}})$ onto the boundary of the elastic region $\partial\mathbb{E}_\sigma$ in the metric $\mathbf{G}$ defined by (3.2.13):

$$\boxed{\|\boldsymbol{\Sigma}\|_{\mathbf{G}}^2 := \boldsymbol{\Sigma} : \mathbf{G} : \boldsymbol{\Sigma} = \boldsymbol{\sigma} : \mathbf{C}^{-1} : \boldsymbol{\sigma} + \boldsymbol{q} \cdot \mathbf{D}^{-1}\boldsymbol{q}.} \tag{5.2.8}$$

This interpretation is consistent with the work of Matthies [1978] and Halphen and Nguyen [1975].

Next we turn our attention to the discrete characterization of plastic loading in the context of multisurface plasticity.

5.2.2 Loading/Unloading. Discrete Kuhn–Tucker Conditions

Whether plastic loading or elastic response occurs in a time increment $[t_n, t_{n+1}]$ can be concluded solely from the trial elastic state according to the following conditions.

Proposition 5.2. *Assuming that* $f_\alpha : \mathbb{S} \times \mathbb{R}^q \to \mathbb{R}$, $\alpha = 1, 2, \ldots, m$, *are convex, one has the following algorithmic characterization of plastic loading:*

$$\boxed{\begin{aligned}
f_{\alpha,n+1}^{\text{trial}} &\le 0 \text{ for } \textbf{all} \quad \alpha \in (1, 2, \ldots, m) \Rightarrow \text{elastic step} , \\
f_{\beta,n+1}^{\text{trial}} &> 0 \text{ for } \textbf{some} \quad \beta \in (1, 2, \ldots, m) \Rightarrow \text{plastic step} .
\end{aligned}} \tag{5.2.9}$$

PROOF. (i) If $f^{\text{trial}}_{\alpha,n+1} < 0$ for all $\alpha \in \{1, 2, \ldots, m\}$, then $(\boldsymbol{\sigma}^{\text{trial}}_{n+1}, \boldsymbol{q}_n)$ *is admissible*. Thus, $\boldsymbol{\sigma}_{n+1} = \boldsymbol{\sigma}^{\text{trial}}_{n+1}$, $\boldsymbol{q}_{n+1} = \boldsymbol{q}_n$, $\Delta\gamma^\alpha = 0$, for all $\alpha \in \{1, 2, \ldots, m\}$ solves (5.2.1–2). Since the solution to (5.2.4) is unique, this constitutes the actual solution, and the step is elastic.

(ii) Suppose that there exists at least one $\beta \in \{1, 2, \ldots, m\}$ such that $f^{\text{trial}}_{\beta,n+1} > 0$. Then, $\boldsymbol{\sigma}^{\text{trial}}_{n+1}$ is not admissible; hence, the step is plastic. □

Remarks 5.2.1.

1. Observe that, if *only one* yield surface is active (i.e., $\Delta\gamma^\beta > 0$ for only one $\beta \in \{1, 2, \ldots, m\}$), then the condition $f^{\text{trial}}_{\beta,n+1} > 0$ implies that $\Delta\gamma^\beta > 0$; i.e., the β-constraint is active, agreeing with the conclusions drawn in single-surface plasticity.
2. However, if *several* yield conditions are active, then the condition $f^{\text{trial}}_{\alpha,n+1} > 0$ does not necessarily imply that $\Delta\gamma^\alpha > 0$. Equivalently, it is possible to have $f^{\text{trial}}_{\alpha,n+1} > 0$ and, at the same time, $f_{\alpha,n+1} < 0$. This situation is illustrated in Figure **5.4** and can be explained as follows. In view of relationship $(5.2.7)_1$, we can regard $\Delta\gamma^\alpha$ as the contravariant components of $[\boldsymbol{\sigma}^{\text{trial}}_{n+1} - \boldsymbol{\sigma}_{n+1}]$ in $\mathbb{S}$ along the vectors $\{\mathbf{C} : \partial_\sigma f_{\alpha,n+1}\}$. Thus, conditions $\Delta\gamma^\alpha > 0$, $\alpha = 1, 2$, define a corner region in stress space, denoted by Γ_{12}, and spanned by $\{\mathbf{C} : \partial_\sigma f_{\alpha,n+1}\}$. Within this cone, $f^{\text{trial}}_{1,n+1} > 0$ and $f^{\text{trial}}_{2,n+1} > 0$. If $\boldsymbol{\sigma}^{\text{trial}}_{n+1} \in \Gamma_{12}$, then $\boldsymbol{\sigma}_{n+1}$ is at the intersection of the two surfaces which defines the corner point. On the other hand, within regions Γ_1, and Γ_2, conditions $f^{\text{trial}}_{1,n+1} > 0$ and $f^{\text{trial}}_{2,n+1} > 0$ also hold, but $\Delta\gamma^2 < 0$ in region Γ_1, and $\Delta\gamma^1 < 0$ in region Γ_2.

5.2.3 Solution Algorithm and Implementation

Extending the closest point projection algorithm in BOX **3.5** to multisurface plasticity relies on its interpretation as an iterative solution technique for the constrained minimization problem (5.2.4), as discussed next.

5.2.3.1 Motivation. Convex programming.

To motivate the general solution strategy, we reformulate the algorithm outlined in Section **3.3.1** as a minimization procedure for problem (5.2.4) restricted to single-surface perfect plasticity. Then the Lagrangian (5.2.6) reduces to

$$\mathcal{L}(\boldsymbol{\tau}, \lambda) := \chi(\boldsymbol{\tau}) + \lambda f(\boldsymbol{\tau}). \tag{5.2.10}$$

Observe that the derivatives of $\mathcal{L}(\boldsymbol{\tau}, \lambda)$ are given by

$$\left.\begin{aligned} \partial_{\tau}\mathcal{L}(\boldsymbol{\tau}, \lambda) &:= \nabla\chi(\boldsymbol{\tau}) + \lambda\nabla f(\boldsymbol{\tau})\,, \\ \partial_{\tau\tau}\mathcal{L}(\boldsymbol{\tau}, \lambda) &= \mathbf{C}^{-1} + \lambda\nabla^2 f(\boldsymbol{\tau})\,, \\ \partial_{\tau\lambda}\mathcal{L}(\boldsymbol{\tau}, \lambda) &= \nabla f(\boldsymbol{\tau})\,. \end{aligned}\right\} \tag{5.2.11}$$

Then we consider the following Newton algorithm:

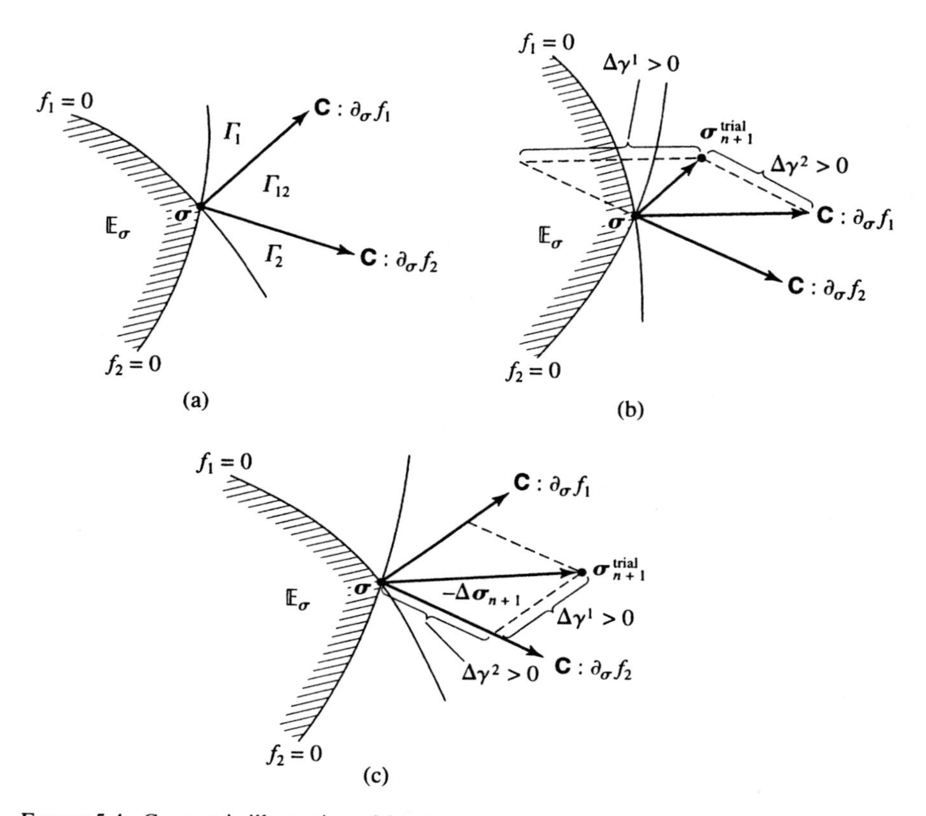

FIGURE 5.4. Geometric illustration of the geometry at a corner point $\boldsymbol{\sigma} \in \partial\mathbb{E}_\sigma$ the intersection of two yield surfaces ($\mathbb{J}_{\text{act}} = \{1, 2\}$). (a) Definition of regions Γ_1, Γ_2, and Γ_{12}, (b) region Γ_1 is characterized by $\gamma^1 > 0$, $\gamma^2 < 0$, (c) region Γ_{12} is characterized by $\gamma^1 > 0$, $\gamma^2 > 0$.

1. Define the residual at iteration (k) by

$$\nabla\mathcal{L}^{(k)} := \left\{ \begin{array}{c} \partial_\tau\mathcal{L}^{(k)} \\ \partial_\lambda\mathcal{L}^{(k)} \end{array} \right\} \equiv \left\{ \begin{array}{c} \mathbf{C}^{-1} : \left[\boldsymbol{\sigma}^{(k)}_{n+1} - \boldsymbol{\sigma}^{\text{trial}}_{n+1}\right] + \Delta\gamma^{(k)}\nabla f^{(k)}_{n+1} \\ f\left(\boldsymbol{\sigma}^{(k)}_{n+1}\right) \end{array} \right\}$$

2. Check whether convergence is attained.
   IF$\|\nabla\mathcal{L}^{(k)}\| <$ TOL, THEN: *Exit.*
   ELSE: *Continue.*
3. Compute Hessian matrix.

$$\nabla^2\mathcal{L}^{(k)} = \begin{bmatrix} \left[\mathbf{C}^{-1} + \Delta\gamma^{(k)}\nabla^2 f^{(k)}_{n+1}\right] & \nabla f^{(k)}_{n+1} \\ \left(\nabla f^{(k)}_{n+1}\right)^T & \mathbf{0} \end{bmatrix}$$

4. Solve the linearized system:

$$\nabla^2 \mathcal{L}^{(k)} \begin{Bmatrix} \Delta \boldsymbol{\sigma}_{n+1}^{(k)} \\ \Delta^2 \gamma^{(k)} \end{Bmatrix} = -\nabla \mathcal{L}^{(k)},$$

that is,

$$\left. \begin{aligned} \Delta^2 \gamma^{(k)} &= \frac{f_{n+1}^{(k)} - [\nabla f_{n+1}^{(k)}]^T [\partial_{\tau\tau} \mathcal{L}^{(k)}]^{-1} \partial_{\tau} \mathcal{L}^{(k)}}{[\nabla f_{n+1}^{(k)}]^T [\partial_{\tau\tau} \mathcal{L}^{(k)}]^{-1} \nabla f_{n+1}^{(k)}} \\ \Delta \boldsymbol{\sigma}_{n+1}^{(k)} &= -[\partial^2_{\tau\tau} \mathcal{L}^{(k)}]^{-1} [\Delta^2 \gamma^{(k)} \nabla f_{n+1}^{(k)} + \partial_{\tau} \mathcal{L}^{(k)}] \end{aligned} \right\}$$

5. Update the solution and GO TO 2.

$$\left. \begin{aligned} \Delta \gamma^{(k+1)} &= \Delta \gamma^{(k)} + \Delta^2 \gamma_{n+1}^{(k)} \\ \boldsymbol{\sigma}_{n+1}^{(k+1)} &= \boldsymbol{\sigma}_{n+1}^{(k)} + \Delta \boldsymbol{\sigma}_{n+1}^{(k)} \end{aligned} \right\}$$

Remarks 5.2.2.

1. Since $\mathbb{E}_{\sigma}$ is *convex*, the preceding algorithm is *unconditionally convergent, regardless of the initial starting point* $\boldsymbol{\sigma}_{n+1}^{\text{trial}}$. (Technically, step-size adjustment is needed, see, e.g., Luenberger [1984, page 212], Dennis and Schnabel [1983, page 116], or Pshenichny and Danilin [1978, page 60]).
2. Recall that the matrix

$$\boxed{\boldsymbol{\Xi} := [\mathbf{C}^{-1} + \Delta\gamma \nabla^2 f]^{-1},} \tag{5.2.12}$$

   the inverse of which appears in the Hessian matrix, is called the *elastic algorithmic tangent moduli.*
3. Different optimization algorithms can be used to solve the constrained minimization problem (5.2.4); (see e.g. Luenberger [1984, Part **II**]), in particular nonlinear versions of the conjugate gradient algorithm, such as the Polak–Ribiere method. See also Karmanov [1977].

The extension of the preceding algorithm to multisurface plasticity is straightforward and is based on the following Lagrangian:

$$\boxed{\mathcal{L}(\boldsymbol{\tau}, \boldsymbol{q}, \lambda^{\alpha}) := \chi(\boldsymbol{\tau}, \boldsymbol{q}) + \sum_{\alpha \,\in\, \mathbb{J}_{\text{act}}} \lambda^{\alpha} f_{\alpha}(\boldsymbol{\tau}, \boldsymbol{q}),} \tag{5.2.13}$$

where $\chi(\boldsymbol{\tau}, \boldsymbol{q})$ is defined in (5.2.5) and $\mathbb{J}_{\text{act}} \subseteq \{1, 2, \ldots, m\}$ is the set of indices associated with the *active constraints* at the *unknown solution point* $(\boldsymbol{\sigma}_{n+1}, \boldsymbol{q}_{n+1})$:

$$\boxed{\mathbb{J}_{\text{act}} := \left\{\alpha \in \{1, 2, \ldots, m\} \mid f_{\alpha}(\boldsymbol{\sigma}_{n+1}, \boldsymbol{q}_{n+1}) = 0\right\}.} \tag{5.2.14}$$

An added difficulty in multisurface plasticity lies in the fact that the set of active constraints defined by $\mathbb{J}_{\text{act}}$ is *not known in advance* since, as noted in Remarks **5.2.1**, the condition $f_{\alpha,n+1}^{\text{trial}} > 0$ *does not guarantee* that the surface $f_{\alpha,n+1}$ is ultimately active. We address this question next.

5.2.3.2 Determination of the active constraints.

A yield surface $f_{\alpha,n+1}$ is termed *active* if $\Delta\gamma^{\alpha} > 0$. The set of active constraints defined by $\mathbb{J}_{\text{act}}$ is determined by an iterative procedure which starts from an initial set of *trial* constraints defined as

$$\boxed{\mathbb{J}_{\text{act}}^{\text{trial}} := \left\{\alpha \in \{1, 2, \ldots, m\} \mid f_{\alpha,n+1}^{\text{trial}} > 0\right\}.} \tag{5.2.15}$$

Clearly $\mathbb{J}_{\text{act}} \subseteq \mathbb{J}_{\text{act}}^{\text{trial}}$. The basic idea here is to modify this initial set by enforcing, iteratively, the Kuhn–Tucker conditions. At each step of the iterative solution procedure one constructs a *working set of constraints*, denoted by $\mathbb{J}_{\text{act}}^{(k)}$, which is updated according to the following two alternative strategies:

i. **Procedure 1 (conceptual).** In this method, the working set of constraints $\mathbb{J}_{\text{act}}^{(k)}$ is held fixed during the iteration that restores consistency. Once the constraints $f_{n+1}^{(k)} = 0$ are satisfied for all $\alpha \in \mathbb{J}_{\text{act}}^{(k)}$, admissibility of the solution is checked by testing whether conditions $\Delta\gamma^{(k)} \geq 0$ hold for *all* $\alpha \in \mathbb{J}_{\text{act}}^{(k)}$. Accordingly, the iteration proceeds as follows:

   1. For $k = 0$ solve the closest point projection iteration with $\mathbb{J}_{\text{act}}^{(0)} := \mathbb{J}_{\text{act}}^{\text{trial}}$ to obtain $(\bar{\boldsymbol{\sigma}}_{n+1}, \bar{\boldsymbol{\varepsilon}}_{n+1}, \bar{\boldsymbol{q}}_{n+1})$, along with $\Delta\bar{\gamma}^{\alpha}$, $\alpha \in \mathbb{J}_{\text{act}}^{\text{trial}}$.
   2. Check the sign of $\Delta\bar{\gamma}^{\alpha}$. If $\Delta\bar{\gamma}^{\beta} < 0$, for some $\beta \in \mathbb{J}_{\text{act}}^{\text{trial}}$, obtain a new working set of constraints $\mathbb{J}_{\text{act}}^{(1)}$ by dropping the β-constraint from $\mathbb{J}_{\text{act}}^{\text{trial}}$. Set $k \leftarrow k + 1$ and repeat step 1. Otherwise stop.

ii. **Procedure 2.** In this method, the working set $\mathbb{J}_{\text{act}}^{\text{trial}}$ *is updated during the iterative process*. Consistency is restored by enforcing the admissibility constraint $\Delta\gamma^{\alpha^{(k)}}$ be nonnegative for all $\alpha \in \mathbb{J}_{\text{act}}^{(k)}$. Accordingly, the iteration proceeds as follows:

   1. Let $\mathbb{J}_{\text{act}}^{(k)}$ be the working set at the kth iteration of the return mapping. Compute increments $\Delta^2\gamma^{\alpha^{(k)}}$, $\alpha \in \mathbb{J}_{\text{act}}^{(k)}$, by solving the linearized return-mapping algorithm.
   2. Update $\Delta\gamma^{\alpha^{(k)}}$ by setting $\Delta\bar{\gamma}^{\alpha^{(k+1)}} = \Delta\gamma^{\alpha^{(k)}} + \Delta^2\gamma^{\alpha^{(k)}}$, and check the sign of $\Delta\bar{\gamma}^{\alpha^{(k+1)}}$. If $\Delta\bar{\gamma}^{\beta^{(k+1)}} < 0$ for some $\beta \in \mathbb{J}_{\text{act}}^{(k)}$, drop the β-constraint from the set $\mathbb{J}_{\text{act}}^{(k)}$, and restart the iteration. Otherwise, set $\Delta\gamma^{\beta^{(k+1)}} = \Delta\bar{\gamma}^{\beta^{(k+1)}}$, and proceed to the next iterate.

A summary of the algorithm associated with Procedure **2** is given in BOX **5.2a–c**.

5.2.4 Linearization: Algorithmic Tangent Moduli

The exact linearization of the algorithm developed above is accomplished by the same procedure discussed in detail in Section **3.6.2.** For completeness we sketch below the main steps involved in the derivation for the case of perfect plasticity.

BOX 5.2a. Elastic Predictor.

1. Compute elastic predictor

$$\boldsymbol{\sigma}_{n+1}^{\text{trial}} = \nabla W(\boldsymbol{\varepsilon}_{n+1} - \boldsymbol{\varepsilon}_{n+1}^{p})$$

$$f_{\alpha,n+1}^{\text{trial}} := f_{\alpha}\big(\boldsymbol{\sigma}_{n+1}^{\text{trial}}, \boldsymbol{q}_{n+1}^{\text{trial}}\big), \quad \text{for } \alpha \in \{1, 2, \dots, m\}$$

2. Check for plastic process

IF $f_{\alpha,n+1}^{\text{trial}} \leq 0$ for **all** $\alpha \in \{1, 2, \dots, m\}$ THEN:

Set $(\bullet)_{n+1} = (\bullet)_{n+1}^{\text{trial}}$ and *EXIT*

ELSE:

$$\mathbb{J}_{\text{act}}^{(0)} := \big\{\alpha \in \{1, 2, \dots, m\} \mid f_{\alpha,n+1}^{\text{trial}} > 0\big\}$$

$$\boldsymbol{\varepsilon}_{n+1}^{p^{(0)}} = \boldsymbol{\varepsilon}_{n}^{p}$$

$$\boldsymbol{\alpha}_{n+1}^{(0)} = \boldsymbol{\alpha}_{n}$$

$$\Delta\gamma^{\alpha^{(0)}} = 0$$

Goto BOX **5.2b**

ENDIF.

First, differentiating the elastic stress-strain relationships $(5.2.1)_2$ and the discrete flow rule $(5.2.1)_3$ yields

$$\left.\begin{aligned} d\boldsymbol{\sigma}_{n+1} &= \mathbf{C}_{n+1} : \big(d\boldsymbol{\varepsilon}_{n+1} - d\boldsymbol{\varepsilon}_{n+1}^{p}\big) \\ d\boldsymbol{\varepsilon}_{n+1}^{p} &= \sum_{\alpha=1}^{m} \big[\Delta\gamma^{\alpha}\partial_{\sigma\sigma}^{2} f_{\alpha}(\boldsymbol{\sigma}_{n+1}) : d\boldsymbol{\sigma}_{n+1} + d\Delta\gamma^{\alpha}\partial_{\sigma} f_{\alpha}(\boldsymbol{\sigma}_{n+1})\big]. \end{aligned}\right\} \tag{5.2.16}$$

By combining these two equations, one obtains the relationship

$$d\boldsymbol{\sigma}_{n+1} = \boldsymbol{\Xi}_{n+1} : \left[d\boldsymbol{\varepsilon}_{n+1} - \sum_{\alpha=1}^{m} d\Delta\gamma^{\alpha}\partial_{\sigma} f_{\alpha}(\boldsymbol{\sigma}_{n+1})\right], \tag{5.2.17}$$

where $\boldsymbol{\Xi}_{n+1}$ are *algorithmic* moduli now given by the expression

$$\boxed{\boldsymbol{\Xi}_{n+1} := \big[\mathbf{C}_{n+1}^{-1} + \sum_{\alpha=1}^{m} \Delta\gamma^{\alpha}\partial_{\sigma\sigma}^{2} f_{\alpha}(\boldsymbol{\sigma}_{n+1})\big]^{-1}.} \tag{5.2.18}$$

Next, the coefficients $d\Delta\gamma^{\alpha}$ are determined from the algorithmic version of the consistency condition obtained by differentiating $f_{\alpha}(\boldsymbol{\sigma}_{n+1}) = 0$. Explicitly, consider

$$\partial_{\sigma} f_{\alpha}(\boldsymbol{\sigma}_{n+1}) : d\boldsymbol{\sigma}_{n+1} = 0, \quad \alpha \in \mathbb{J}_{\text{act}}. \tag{5.2.19}$$

BOX 5.2b. General Multisurface Closest Point Projection Iteration.

3. Evaluate flow rule/hardening law residuals

$$\boldsymbol{\sigma}_{n+1}^{(k)} := \nabla W\big(\boldsymbol{\varepsilon}_{n+1} - \boldsymbol{\varepsilon}_{n+1}^{p^{(k)}}\big)$$

$$\boldsymbol{q}_{n+1}^{(k)} := -\mathbf{D}\boldsymbol{\alpha}_{n+1}^{(k)}$$

$$\boldsymbol{R}_{n+1}^{(k)} := \begin{Bmatrix} -\boldsymbol{\varepsilon}_{n+1}^{p^{(k)}} + \boldsymbol{\varepsilon}_n^p \\ -\boldsymbol{\alpha}_{n+1}^{(k)} + \boldsymbol{\alpha}_n \end{Bmatrix} + \sum_{\beta\,\in\,\mathbb{J}_{\text{act}}^{(k)}}^{m} \Delta\gamma^{\beta^{(k)}} \begin{Bmatrix} \partial_{\boldsymbol{\sigma}} f_\beta \\ \partial_{\boldsymbol{q}} f_\beta \end{Bmatrix}_{n+1}^{(k)}$$

4. Check convergence

$$f_{\alpha,n+1}^{(k)} := f_\alpha\big(\boldsymbol{\sigma}_{n+1}^{(k)}, \boldsymbol{q}_{n+1}^{(k)}\big), \quad \text{for } \alpha \in \mathbb{J}_{\text{act}}^{(k)}$$

IF: $f_{\alpha,n+1}^{(k)} < \text{TOL}_1$, for all $\alpha \in \mathbb{J}_{\text{act}}^{(k)}$ and $\|\boldsymbol{R}_{n+1}^{(k)}\| < \text{TOL}_2$ THEN:

EXIT

ENDIF

5. Compute elastic moduli and consistent tangent moduli

$$\big[G_{\alpha\beta}\big]_{n+1}^{(k)} := \big[\,\partial_{\boldsymbol{\sigma}} f_\alpha \quad \partial_{\boldsymbol{q}} f_\alpha\,\big]_{n+1}^{(k)} : \mathbf{A}_{n+1}^{(k)} : \begin{Bmatrix} \partial_{\boldsymbol{\sigma}} f_\beta \\ \partial_{\boldsymbol{q}} f_\beta \end{Bmatrix}_{n+1}^{(k)}$$

$$\big[G^{\alpha\beta}\big]_{n+1}^{(k)} := \big[G_{\alpha\beta}\big]_{n+1}^{(k)\;-1}$$

$$\big[\mathbf{C}_{n+1}^{(k)}\big]^{-1} := \Big[\nabla^2 W\big(\boldsymbol{\varepsilon}_{n+1} - \boldsymbol{\varepsilon}_{n+1}^{p^{(k)}}\big)\Big]^{-1}$$

$$\big[\mathbf{A}_{n+1}^{(k)}\big]^{-1} := \begin{bmatrix} \mathbf{C}^{-1} & \mathbf{0} \\ \mathbf{0} & \mathbf{D}^{-1} \end{bmatrix}_{n+1}^{(k)} + \sum_{\beta\,\in\,\mathbb{J}_{\text{act}}^{(k)}} \Delta\gamma^\beta \begin{bmatrix} \partial^2_{\boldsymbol{\sigma\sigma}} f_\beta & \partial^2_{\boldsymbol{\sigma q}} f_\beta \\ \partial^2_{\boldsymbol{q\sigma}} f_\beta & \partial^2_{\boldsymbol{qq}} f_\beta \end{bmatrix}_{n+1}^{(k)}$$

6. Obtain increment to consistency parameter

$$\Delta^2\gamma^{\alpha^{(k)}} := \sum_{\beta\,\in\,\mathbb{J}_{\text{act}}^{(k)}} \big[G^{\alpha\beta}\big]_{n+1}^{(k)} \big\{f_\beta - \big[\,\partial_{\boldsymbol{\sigma}} f_\beta \quad \partial_{\boldsymbol{q}} f_\beta\,\big] : \mathbf{A} : \boldsymbol{R}\big\}_{n+1}^{(k)}$$

$$\Delta\bar{\gamma}^{\alpha^{(k+1)}} = \Delta\gamma^{\alpha^{(k)}} + \Delta^2\gamma_{n+1}^{\alpha^{(k)}}$$

IF: $\Delta\bar{\gamma}_{n+1}^{\alpha^{(k+1)}} < 0, \quad \alpha \in \mathbb{J}_{\text{act}}^{(k)}$, THEN:

Reset $\mathbb{J}_{\text{act}}^{(k+1)} = \big\{\alpha \in \mathbb{J}_{\text{act}}^{(k)} \mid \Delta\bar{\gamma}^{\alpha^{(k+1)}} > 0\big\}$

Goto **3.**

ELSE:

Goto BOX **5.2c**

ENDIF

BOX 5.2c. (Cont'd) Closest Point Projection Iteration.

7. Obtain incremental plastic strains and internal variables

$$\left\{ \begin{array}{c} \Delta \varepsilon^p \\ \Delta \alpha \end{array} \right\}_{n+1}^{(k)} = \begin{bmatrix} \mathbf{C}^{-1} & \mathbf{0} \\ \mathbf{0} & \mathbf{D}^{-1} \end{bmatrix}_{n+1}^{(k)} : \mathbf{A}_{n+1}^{(k)} : \left[\mathbf{R} + \sum_{\beta \in \mathbb{J}_{\text{act}}^{(k)}}^{m} \Delta^2 \gamma^\beta \left\{ \begin{array}{c} \partial_\sigma f_\beta \\ \partial_q f_\beta \end{array} \right\} \right]_{n+1}^{(k)}$$

8. Update state variables and consistency parameter

$$\varepsilon_{n+1}^{p^{(k+1)}} = \varepsilon_{n+1}^{p^{(k)}} + \Delta \varepsilon_{n+1}^{p^{(k)}}$$

$$\boldsymbol{\alpha}_{n+1}^{(k+1)} = \boldsymbol{\alpha}_{n+1}^{(k)} + \Delta \boldsymbol{\alpha}_{n+1}^{(k)}$$

$$\Delta \gamma^{\alpha^{(k+1)}} = \Delta \gamma_{n+1}^{\alpha^{(k)}} + \Delta^2 \gamma_{n+1}^{\alpha^{(k)}} \quad \alpha \in \mathbb{J}_{\text{act}}^{(k+1)}$$

Set $k \leftarrow k + 1$ and Goto **3**

Then substituting (5.2.17) in (5.2.19) yields

$$d\Delta \gamma_{n+1}^{\beta} = \sum_{\alpha \in \mathbb{J}_{\text{act}}} \left[g_{n+1}^{\beta\alpha} \right] \left[\partial_\sigma f_{\alpha,n+1} : \boldsymbol{\Xi}_{n+1} : d\boldsymbol{\varepsilon}_{n+1} \right], \tag{5.2.20}$$

where $g_{n+1}^{\beta\alpha} := [g_{\beta\alpha,n+1}]^{-1}$, and $g_{\beta\alpha,n+1}$ is defined by (5.1.10) with $\mathbf{C}$ replaced by the algorithmic moduli $\boldsymbol{\Xi}_{n+1}$:

$$g_{n+1}^{\beta\alpha} := \left[g_{\beta\alpha,n+1} \right]^{-1} := \left[\partial_\sigma f_{\beta,n+1} : \boldsymbol{\Xi}_{n+1} : \partial_\sigma f_{\alpha,n+1} \right]^{-1}. \tag{5.2.21}$$

Finally, substituting (5.2.20) in (5.2.17) gives the desired expression for the algorithmic elastoplastic tangent moduli:

$$\left. \frac{d\boldsymbol{\sigma}}{d\boldsymbol{\varepsilon}} \right|_{n+1} = \boldsymbol{\Xi}_{n+1} - \sum_{\beta \in \mathbb{J}_{\text{act}}} \sum_{\alpha \in \mathbb{J}_{\text{act}}} g_{n+1}^{\beta\alpha} \boldsymbol{N}_{\beta,n+1} \otimes \boldsymbol{N}_{\alpha,n+1}, \qquad \boldsymbol{N}_{\alpha,n+1} := \boldsymbol{\Xi}_{n+1} : \partial_\sigma f_\alpha(\boldsymbol{\sigma}_{n+1}). \tag{5.2.22}$$

Note that the structure of (5.2.22) is analogous to expression (3.6.10). All that is needed to obtain the *algorithmic* tangent moduli is to replace the elastic moduli $\mathbf{C}_{n+1}$ in the expression for the *continuum* elastoplastic moduli by the *algorithmic* moduli $\boldsymbol{\Xi}_{n+1}$ defined by (5.2.18). An analogous but more elaborate calculation applies to the case of hardening plasticity.

5.3 Extension to Viscoplasticity

In this section we consider extending the preceding developments to accommodate rate-dependent response governed by a suitable generalization of the Duvaut–Lions viscoplastic model in Section **2.7.4.** First, we show that models of the Perzyna type considered in Section **2.7.3**, in general, do not admit meaningful generalizations to elastic domains bounded by multiple surfaces.

5.3.1 Motivation. Perzyna-Type Models

It appears that a straightforward extension of inviscid plasticity to the rate-dependent case is obtained by postulating a flow rule of the form

$$\dot{\boldsymbol{\varepsilon}}^{\mathrm{vp}} = \sum_{\alpha=1}^{m} \frac{\langle f_\alpha(\boldsymbol{\sigma})\rangle}{\eta} \partial_{\boldsymbol{\sigma}} f_\alpha(\boldsymbol{\sigma}) , \tag{5.3.1}$$

where $\eta \in (0, \infty)$ is a fluidity parameter, and $\langle \bullet \rangle$ is the ramp function defined as $\langle x \rangle := (x + |x|)/2$. Unfortunately, as $\eta \to 0$, this model *does not reduce to the rate-independent formulation* in BOX **5.1**, as the following example illustrates.

EXAMPLE: 5.3.1. Consider the case in which two convex functions $f_1(\boldsymbol{\sigma})$ and $f_2(\boldsymbol{\sigma})$ intersect nonsmoothly as shown in Figure **5.5**. In the limit, as $\eta \to 0$, since both $f_1 > 0$ and $f_2 > 0$, equation (5.3.1) gives the relaxation path and viscoplastic strain rate shown in Figure **5.5a** which corresponds to $\boldsymbol{\sigma} \to \bar{\boldsymbol{\sigma}}$. However, in the actual rate-independent solution shown in Figure **5.5b**, *only* $f_1 = 0$ *is active.* Therefore we conclude that viscoplastic flow rules of the type (5.3.1) may not produce the correct rate-independent solution in the limit, as $\eta \to 0$.

The example above illustrates the fact, alluded to in Remark **5.2.1**, that condition $f_\alpha(\boldsymbol{\sigma}) > 0$ does not imply that the α–constraint is necessarily active. By contrast, the evolution equation (5.3.1) activates the α–constraint whenever $f_\alpha > 0$. The model discussed below precludes this difficulty and properly reduces to the inviscid limit.

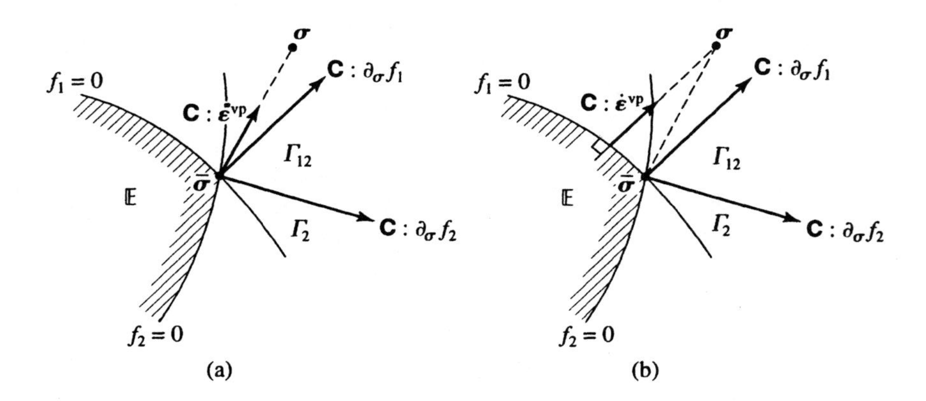

FIGURE 5.5. (a) Inviscid limit return path for Perzyna-type multisurface models, (b) actual inviscid return path.

5.3.2 Extension of the Duvaut–Lions Model

A straightforward extension of the model discussed in Section **2.7.4** to the case of an elastic domain $\mathbb{E}$ defined by multiple yield surfaces $f_\alpha : \mathbb{S} \times \mathbb{R}^q \to \mathbb{R}$ and including internal hardening variables $\boldsymbol{q}$ may be constructed as follows.

1. Let $(\bar{\sigma}, \bar{q})$ be the *inviscid solution* of the elastoplastic problem in BOX **5.1**.
2. Postulate constitutive equations of the form

$$\boxed{\begin{aligned} \dot{\varepsilon}^{\mathrm{vp}} &= \frac{1}{\eta}\mathbf{C}^{-1}[\sigma - \bar{\sigma}], \\ \dot{\alpha} &= \frac{1}{\eta}\mathbf{D}^{-1}[q - \bar{q}]. \end{aligned}} \tag{5.3.2}$$

Equations (5.3.2) are amenable to a straightforward numerical implementation as suggested by the algorithm discussed below.

Remark 5.3.1. More elaborate models may be constructed similarly. For example, let $g : \mathbb{R}_+ \to \mathbb{R}_+$ be a monotonic C^1 function. Then, one may consider

$$\left.\begin{aligned} \dot{\varepsilon}^{\mathrm{vp}} &= \frac{1}{\eta} g(\|\sigma - \bar{\sigma}\|_{\mathbf{C}^{-1}})\mathbf{C}^{-1} : [\sigma - \bar{\sigma}], \\ \dot{\alpha} &= \frac{1}{\eta} g(\|q - \bar{q}\|_{\mathbf{D}^{-1}})\mathbf{D}^{-1}[q - \bar{q}], \end{aligned}\right\} \tag{5.3.3}$$

where $\|\sigma\|^2_{\mathbf{C}^{-1}} := \sigma : \mathbf{C}^{-1} : \sigma$ is the energy norm and $\|q\|^2_{\mathbf{D}^{-1}} := q \cdot \mathbf{D}^{-1}q$. For metals, typical forms for g are exponentials and power laws.

5.3.3 Discrete Formulation

Let $[t_n, t_{n+1}] \subset \mathbb{R}_+$ be the time-step interval of interest. Then the stress rate

$$\dot{\sigma} = \mathbf{C} : [\dot{\varepsilon} - \dot{\varepsilon}^{\mathrm{vp}}] \equiv \mathbf{C} : \dot{\varepsilon} - \frac{1}{\eta}(\sigma - \bar{\sigma}), \tag{5.3.4}$$

may be integrated in closed form to obtain

$$\sigma_{n+1} = \exp(-\Delta t/\eta)\sigma_n + \int_{t_n}^{t_{n+1}} \exp[-(t_{n+1} - s)/\eta]\left(\frac{\bar{\sigma}}{\eta} + \mathbf{C} : \dot{\varepsilon}\right) ds\,. \tag{5.3.5}$$

Using the approximations

$$\begin{aligned} &\int_{t_n}^{t_{n+1}} \exp[-(t_{n+1} - s)/\eta]\mathbf{C} : \dot{\varepsilon}\, ds \\ &\qquad \approx \left\{\int_{t_n}^{t_{n+1}} \exp[-(t_{n+1} - s)/\eta]\, ds\right\}\mathbf{C} : \frac{\Delta\varepsilon_{n+1}}{\Delta t} \\ &\qquad = \frac{1 - \exp(-\Delta t/\eta)}{\Delta t/\eta}\mathbf{C} : \Delta\varepsilon_{n+1}, \end{aligned} \tag{5.3.6}$$

$$\frac{1}{\eta}\int_{t_n}^{t_{n+1}} \exp[-(t_{n+1}-s)/\eta]\bar{\sigma}(s)\,ds$$
$$\approx \left[1-\exp(-\Delta t/\eta)\right]\bar{\sigma}_{n+1}, \tag{5.3.7}$$

and proceeding in the same manner with equation $(5.3.2)_2$, one obtains the algorithm for viscoplasticity summarized in BOX **5.3**. Note that, with $\boldsymbol{q} = -\mathbf{D}\boldsymbol{\alpha}$, $(5.3.2)_2$ takes the form

$$\dot{\boldsymbol{q}} = \frac{-1}{\eta}\left[\boldsymbol{q}-\bar{\boldsymbol{q}}\right]. \tag{5.3.8}$$

BOX 5.3. Closed-Form Algorithm for Viscoplasticity.

1. Compute the closest-point projection $(\bar{\sigma}_{n+1}, \bar{q}_{n+1})$ by BOX **5.2a–5.2c**

2. Obtain the viscoplastic solution by the formulas

$$\begin{aligned}
\sigma_{n+1} = {} & \exp(-\Delta t/\eta)\sigma_n + \left[1-\exp(-\Delta t/\eta)\right]\bar{\sigma}_{n+1} \\
+ {} & \frac{1-\exp(-\Delta t/\eta)}{\Delta t/\eta}\,\mathbf{C} : \Delta\varepsilon_{n+1} \\
q_{n+1} = {} & \exp(-\Delta t/\eta)q_n + \left[1-\exp(-\Delta t/\eta)\right]\bar{q}_{n+1} \\
\alpha_{n+1} = {} & -\mathbf{D}^{-1}q_{n+1}
\end{aligned}$$

Remarks 5.3.1.

1. The *elastic and inviscid* cases are recovered from the preceding algorithm in the following limiting situations:
   a. Let $\Delta t/\eta \to 0$. It follows that $\exp(-\Delta t/\eta) \to 1$ and
   $$[1-\exp(-\Delta t/\eta)]/(\Delta t/\eta) \to 1.$$
   Hence $\sigma_{n+1} \to \sigma_n + \mathbf{C} : \Delta\varepsilon_{n+1}$, and $q_{n+1} \to q_n$. Therefore, one obtains the elastic case.
   b. Let $\Delta t/\eta \to \infty$. It follows that $\exp(-\Delta t/\eta) \to 0$ and
   $$[1-\exp(-\Delta t/\eta)]/(\Delta t/\eta) \to 0.$$
   Hence, $\sigma_{n+1} \to \bar{\sigma}_{n+1}$, $q_{n+1} \to \bar{q}_{n+1}$, and one recovers the inviscid plastic case.
2. Alternatively, from (5.3.2), by applying an implicit backward Euler algorithm, we obtain the first-order accurate formulas

$$\boxed{\begin{aligned}
\sigma_{n+1} &= \frac{\sigma_{n+1}^{\text{trial}} + (\Delta t/\eta)\bar{\sigma}_{n+1}}{1+\Delta t/\eta} \\
q_{n+1} &= \frac{q_n + (\Delta t/\eta)\bar{q}_{n+1}}{1+\Delta t/\eta}.
\end{aligned}} \tag{5.3.9}$$

Note that these expressions are identical to those in BOX **1.7** of Chapter **1**.

6

Numerical Analysis of General Return Mapping Algorithms

In this chapter, we present a rigorous, nonlinear, stability analysis of a class of time discretizations of the weak form of the initial boundary-value problem for both rate-independent and rate-dependent infinitesimal elastoplasticity. To motivate the notion of nonlinear stability appropriate for elastoplasticity, first we consider the simpler model problem of nonlinear heat conduction and give a rigorous proof of nonlinear stability for the generalized midpoint rule. The analysis that follows differs in several aspects from previous treatments of algorithmic stability; in particular:

i. The stability analysis is performed directly on the system of variational equations discretized in time, not on the algebraic system arising from both temporal and spatial discretizations. The results carry over immediately to the finite-dimensional problem obtained via a Galerkin (spatial) discretization.
ii. Previous stability analyses employ either the notion of A-stability, introduced by Dahlquist [1963] in the context of linear, multistep methods for systems of ODEs or the concept of *linearized stability*; see, e.g., Hughes [1983] and references therein. The results given below prove *nonlinear stability* in the sense that arbitrary perturbations in the initial data are attenuated by the algorithm relative to a certain *algorithmic-independent* norm associated with the continuum problem called the *natural norm*. See also Dahlquist [1975].

For nonlinear systems of ODEs, this notion of nonlinear stability reduces to the concept of A-contractivity or B-stability introduced by Butcher [1975] in the context of implicit Runge–Kutta methods. A-contractivity is widely accepted now as the proper definition of nonlinear stability; see, e.g., Burrage and Butcher [1979,1980]; and Dahlquist and Jeltsch [1979]. For *linear* semigroups, the definition of stability employed in this chapter coincides with the notion of Lax stability; see Richtmyer and Morton [1967].

A key step in the stability analysis given below is identifying the natural norm for the continuum problem relative to which the crucial contractivity property holds. For nonlinear heat conduction, this norm is a weighted $\mathbb{L}_2$–norm whose weighting factor is the specific heat capacity times the density. For the semidiscrete version of this problem obtained via a Galerkin spatial discretization, the natural norm reduces to the matrix norm induced by the mass matrix. For infinitesimal elastoplasticity, the natural norm is the norm induced by the complementary Helmholtz free energy function. A given algorithm is then said to be A-contractive if it inherits the contractivity property present in the continuum problem relative to the natural norm. For nonlinear heat conduction, the generalized midpoint rule is shown to be contractive (for $\alpha \geq \frac{1}{2}$) by crucially exploiting the *convexity* property of the heat-flux potential. It is well known that A-contractivity cannot hold for the generalized trapezoidal rule (see Wanner [1976] for a counterexample).

For infinitesimal elastoplasticity and viscoplasticity, it is shown that the system of variational inequalities associated with a class of return mapping algorithms based on the generalized midpoint rule, proposed in Simo and Govindjee [1991], is nonlinearly stable (A-contractive) for $\alpha \geq \frac{1}{2}$. This class of algorithms encompasses the widely used return maps based on the implicit backward Euler method ("catching-up" algorithms in the terminology of Moreau [1977]), in particular, the classical radial return method of Wilkins [1964] and its generalizations to linear kinematic/isotropic hardening (Krieg and Key [1976]; Balmer et al. [1974]) and plane stress (Simo and Taylor [1986]). The nonlinear stability proof given below applies to the class of generalized midpoint rule algorithms in Hughes and Taylor [1978] (perfect viscoplasticity) and Simo and Taylor [1986] (plane stress elastoplasticity), but does not cover the class of methods in Ortiz and Popov [1985], whose stability properties remain an open question.

6.1 Motivation: Nonlinear Heat Conduction

The steps in the stability analysis of the time discretization of the IBVP for nonlinear heat conduction by a generalized midpoint rule are analogous to those employed in analyzing the elastoplastic problem and can be summarized as follows:

i. The nonlinear heat conduction equation is formulated in weak form as a variational problem of evolution. Assuming convexity of the heat-flux potential, the solutions of this problem are contractive relative to the *natural norm* defined as a weighted $\mathbb{L}_2$–norm whose weighting factor is the density times the specific heat capacity.

ii. The time-dependent variational equality is discretized by a one–parameter family of generalized midpoint rule algorithms depending on the parameter $\alpha \in [0, 1]$. Then it is shown that this time-discrete problem of evolution inherits the contractivity property of the continuum problem provided that $\alpha \geq \frac{1}{2}$, hence proving unconditional stability relative to the natural norm.

As pointed out in the introduction, the preceding analysis is performed without introducing any spatial discretization. An identical result holds for the semidiscrete problem of evolution obtained by a Galerkin finite-element projection. For this problem, the natural norm is the matrix norm defined by the *positive-definite* mass matrix (see Hughes [1983]).

6.1.1 The Continuum Problem

Let $\Omega \subset \mathbb{R}^{n_{\rm dim}}$, with $n_{\rm dim} \le 3$, be the reference configuration of a nonlinear heat conductor with smooth boundary $\partial\Omega$ and particles labeled by $\boldsymbol{x} \in \overline{\Omega}$. Further, let $\mathbb{I} \subset \mathbb{R}_+$ be the time interval of interest and denote by

$$\vartheta : \overline{\Omega} \times \mathbb{I} \to \mathbb{R}_+$$

and (6.1.1)

$$\boldsymbol{q} : \overline{\Omega} \times \mathbb{I} \to \mathbb{R}^{n_{\rm dim}}$$

the absolute temperature and the heat-flux vector, respectively. Assume that $\vartheta(\boldsymbol{x}, t)$ and $\boldsymbol{q}(\boldsymbol{x}, t)$ are specified on parts of the boundary Γ_ϑ and Γ_q as

$$\vartheta = \bar{\vartheta} \text{ on } \Gamma_\vartheta \times \mathbb{I}$$

and (6.1.2)

$$\boldsymbol{q} \cdot \boldsymbol{n} = -\bar{q} \text{ on } \Gamma_q \times \mathbb{I},$$

respectively. Here $\bar{\vartheta} : \Gamma_\vartheta \times \mathbb{I} \to \mathbb{R}_+$ is the prescribed temperature, $\bar{q} : \Gamma_q \times \mathbb{I} \to \mathbb{R}$ is the prescribed heat-flux, and $\boldsymbol{n}$ is the unit outward normal to the boundary. As usual, it is assumed that the conditions

$$\Gamma_\vartheta \cap \Gamma_q = \emptyset,$$

and (6.1.3)

$$\overline{\Gamma_\vartheta \cup \Gamma_q} = \overline{\partial\Omega},$$

hold. Further, denote the reference density by $\rho_0 : \Omega \to \mathbb{R}_+$, assume that the *specific heat capacity* c is *constant*, and let $f : \Omega \times \mathbb{I} \to \mathbb{R}$ be the heat source per unit volume. Finally, the constitutive equation for the heat-flux vector is specified in terms of a *heat-flux potential*

$$\mathcal{H} : \Omega \times \mathbb{R}^{n_{\rm dim}} \to \mathbb{R}, \tag{6.1.4}$$

depending on position and temperature gradient, by the potential relationships

$$\boldsymbol{q}(\boldsymbol{x}, t) = -\nabla\mathcal{H}\left(\boldsymbol{x}, \nabla\vartheta(\boldsymbol{x}, t)\right), \quad \text{in } \Omega \times \mathbb{I}, \tag{6.1.5}$$

where $\nabla\vartheta = \frac{\partial\vartheta}{\partial x_i}\boldsymbol{e}_i$ denotes the temperature gradient relative to a Cartesian reference system with orthonormal basis $\{\boldsymbol{e}_i\}$. In components, constitutive equation (6.1.5) reads

$$q_i(\boldsymbol{x}, t) = -\frac{\partial}{\partial\vartheta_{,i}}\mathcal{H}\left(\boldsymbol{x}, \nabla\vartheta(\boldsymbol{x}, t)\right), \quad \text{in } \Omega \times \mathbb{I}, \tag{6.1.6}$$

with $\vartheta_{,i} := \frac{\partial\vartheta}{\partial x_i}$. One makes the crucial assumption that the heat-flux potential $\mathcal{H}(\boldsymbol{x}, \cdot) : \mathbb{R}^{n_{\rm dim}} \to \mathbb{R}$ is a smooth **convex** function for all $\boldsymbol{x} \in \Omega$. Accordingly, the following relationship holds

$$\mathcal{H}(\cdot, \boldsymbol{u}) - \mathcal{H}(\cdot, \boldsymbol{v}) \ge [\boldsymbol{u} - \boldsymbol{v}] \cdot \nabla\mathcal{H}(\cdot, \boldsymbol{v}), \quad \forall\, \boldsymbol{u}, \boldsymbol{v} \in \mathbb{R}^{n_{\rm dim}}. \tag{6.1.7}$$

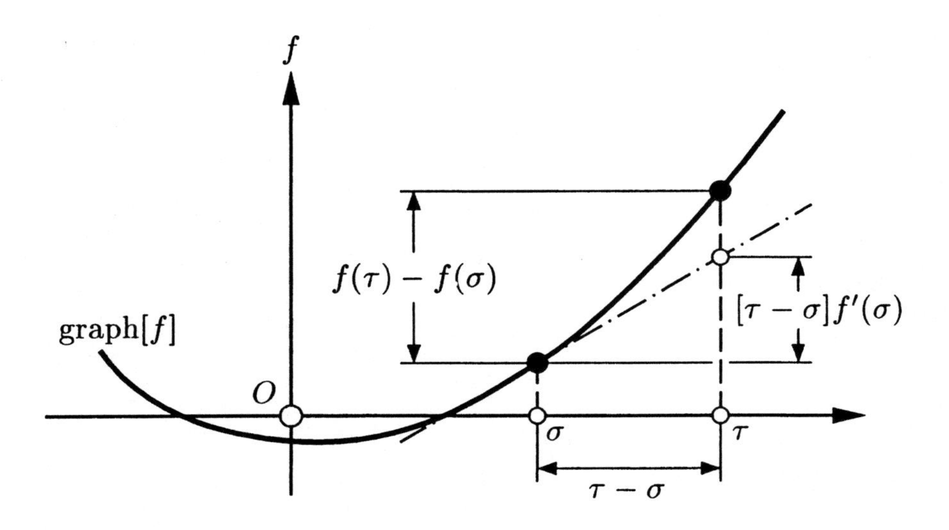

FIGURE 6-1. The convexity property (6.2.7) for a smooth, one-dimensional function $f : \mathbb{R} \to \mathbb{R}$ ($n_{\text{dim}} = 1$).

This property is illustrated in Figure **6.1.**

6.1.1.1 Strong form and weak form of the IBVP

With the preceding notation in hand, the strong form of the initial boundary-value problem (IBVP) is formulated as follows:

Problem S_t: Find $\vartheta : \overline{\Omega} \times \mathbb{I} \to \mathbb{R}_+$ such that

$$\left.\begin{aligned} \rho_0 c\, \dot{\vartheta} &= -\text{div}[\boldsymbol{q}] + f \\ \boldsymbol{q} &= -\nabla \mathcal{H} \end{aligned}\right\} \quad \text{in } \Omega \times \mathbb{I}, \tag{6.1.8a}$$

subject to the boundary conditions

$$\left.\begin{aligned} \vartheta &= \bar{\vartheta} \quad \text{on} \quad \Gamma_\vartheta \times \mathbb{I} \\ \boldsymbol{q} \cdot \boldsymbol{n} &= -\bar{q} \quad \text{on} \quad \Gamma_q \times \mathbb{I} \end{aligned}\right\} \tag{6.1.8b}$$

and the initial condition

$$\vartheta(\cdot, t)|_{t=0} = \vartheta_0(\cdot) \quad \text{in} \quad \Omega. \tag{6.1.9}$$

To formulate the weak form of this classical IBVP, one introduces a space $\mathcal{V}$ of admissible test functions defined as follows:

$$\mathcal{V} := \{\, \eta \in \mathbb{W}^{1,p}(\Omega) \; : \quad \eta = 0 \quad \text{on} \quad \Gamma_\vartheta \,\}, \tag{6.1.10a}$$

where $\mathbb{W}^{1,p}(\Omega)$ is the standard notation for Sobolev spaces. The choice of the appropriate exponent $p > 1$ is dictated by growth and coercivity conditions on the heat-flux potential of the type

$$|\mathcal{H}[\boldsymbol{x}, \nabla\vartheta(\boldsymbol{x})]| \leq C_1[1 + |\nabla\vartheta|^p]$$

and (6.1.10b)

$$\mathcal{H}[\boldsymbol{x}, \nabla\vartheta(\boldsymbol{x})] \geq C_2|\nabla\vartheta|^p,$$

where C_1 and C_2 are constants; see e.g., Zeidler [1985, p.260]. In addition, for fixed time $t \in \mathbb{I}$, the admissible solution space $\mathcal{S}_t$ is given by

$$\mathcal{S}_t := \{\, \vartheta(\cdot, t) \in \mathbb{W}^{1,p}(\Omega) \; : \quad \vartheta(\cdot, t) = \overline{\vartheta}(\cdot, t) \quad \text{on} \quad \Gamma_\vartheta \,\}. \tag{6.1.11}$$

Then standard arguments in the calculus of variations lead to the following variational problem:

Problem W_t: Find $\vartheta(\cdot, t) \in \mathcal{S}_t$ such that

$$\langle \rho_0 c\, \dot{\vartheta}, \eta\rangle = -\langle \nabla\mathcal{H}, \nabla\eta\rangle + \langle f, \eta\rangle + \langle \bar{q}, \eta\rangle_\Gamma, \quad \forall\, \eta \in \mathcal{V}, \tag{6.1.12}$$

subject to the initial condition

$$\langle \rho_0 c\vartheta(\cdot, 0), \eta\rangle = \langle \rho_0 c\vartheta_0, \eta\rangle, \quad \forall\, \eta \in \mathcal{V}. \tag{6.1.13}$$

Here, $\nabla\mathcal{H} := \frac{\partial\mathcal{H}}{\partial\vartheta_{,i}}\boldsymbol{e}_i$, $\langle \cdot, \cdot\rangle$ denotes the standard $\mathbb{L}_2$-pairing in Ω, and $\langle \cdot, \cdot\rangle_\Gamma$ is the $\mathbb{L}_2$-pairing on the boundary $\partial\Omega$.

The following property of the solution of the IBVP (6.1.12)–(6.1.13) is crucial to the subsequent algorithmic analysis and motivates the notion of nonlinear stability for the problem at hand.

6.1.1.2 Contractivity of the flow

Given the variational IBVP (6.1.12)–(6.1.13), fix the source term $f(\boldsymbol{x}, t)$ and the boundary conditions in (6.1.12) while considering *two* different initial conditions

$$\vartheta_0 : \Omega \to \mathbb{R}_+,$$

and (6.1.14)

$$\widetilde{\vartheta}_0 : \Omega \to \mathbb{R}_+.$$

The solutions generated by the IBVP (6.1.12)–(6.1.13) for these two initial conditions are denoted by

$$t \in \mathbb{I} \mapsto \vartheta(\cdot, t) \in \mathcal{S}_t$$

and (6.1.15)

$$t \in \mathbb{I} \mapsto \widetilde{\vartheta}(\cdot, t) \in \mathcal{S}_t,$$

respectively. Under these conditions, the IBVP is said to be *contractive* if there is an inner product denoted by $\langle\!\langle \cdot, \cdot \rangle\!\rangle$, with associated norm (called the *natural norm*) given by

$$|||\cdot||| := \sqrt{\langle\!\langle \cdot, \cdot \rangle\!\rangle}, \tag{6.1.16}$$

such that the solutions (6.1.15) satisfy the following inequality:

$$\boxed{|||\vartheta(\cdot, t) - \tilde{\vartheta}(\cdot, t)||| \leq |||\vartheta_0(\cdot) - \tilde{\vartheta}_0(\cdot)||| \quad \forall t \in \mathbb{I}.} \tag{6.1.17}$$

For nonlinear heat conduction, the following result is a direct consequence of the convexity assumption on the heat-flux potential (for completeness a proof is included)

Proposition 1.1. *The flow* $t \in \mathbb{I} \mapsto \vartheta(\cdot, t) \in \mathcal{S}_t$ *generated by (6.1.12)–(6.1.13) is contractive relative to the weighted* $\mathbb{L}_2$ *inner product* $\langle\!\langle \cdot, \cdot \rangle\!\rangle := \langle \rho_0 c \cdot, \cdot \rangle$.

PROOF. For convenience, introduce the following notation

$$\nabla\mathcal{H} := \frac{\partial}{\partial \vartheta_{,i}} \mathcal{H}(\boldsymbol{x}, \nabla\vartheta)\, \boldsymbol{e}_i,$$

and (6.1.18)

$$\nabla\tilde{\mathcal{H}} := \frac{\partial}{\partial \vartheta_{,i}} \mathcal{H}(\boldsymbol{x}, \nabla\tilde{\vartheta})\, \boldsymbol{e}_i.$$

By hypothesis, $\vartheta(\cdot, t)$ and $\tilde{\vartheta}(\cdot, t)$ satisfy variational equation (6.1.12). Therefore, the difference $\xi(\cdot, t) := \vartheta(\cdot, t) - \tilde{\vartheta}(\cdot, t)$ satisfies the variational equation

$$\langle \rho_0 c\, \dot{\xi}(\cdot, t), \eta \rangle = -\langle \nabla\mathcal{H} - \nabla\tilde{\mathcal{H}}, \eta \rangle, \quad \forall \eta \in \mathcal{V}. \tag{6.1.19}$$

In particular, for *fixed* but arbitrary $t \in \mathbb{I}$, one can choose $\eta = \xi(\cdot, t)$ in (6.1.19) since $\xi(\cdot, t) \in \mathcal{V}$. Using the identity

$$\langle\!\langle \dot{\xi}(\cdot, t), \xi(\cdot, t) \rangle\!\rangle = \tfrac{1}{2} \frac{d}{dt} |||\xi(\cdot, t)|||^2, \tag{6.1.20}$$

equation (6.1.19) with $\eta = \xi(\cdot, t)$ implies the following result:

$$\tfrac{1}{2} \frac{d}{dt} |||\xi(\cdot, t)|||^2 = -\langle \nabla\mathcal{H} - \nabla\tilde{\mathcal{H}}, \xi(\cdot, t) \rangle. \tag{6.1.21}$$

Now, use the convexity condition (6.1.7) on $\mathcal{H}(\boldsymbol{x}, \cdot)$ to obtain the estimate

$$\begin{aligned} \tfrac{1}{2} \frac{d}{dt} &|||\xi(\cdot, t)|||^2 \\ &= \int_\Omega \Big\{ [\nabla\vartheta - \nabla\tilde{\vartheta}] \cdot \nabla\mathcal{H}(\boldsymbol{x}, \nabla\tilde{\vartheta}) + [\nabla\tilde{\vartheta} - \nabla\vartheta] \cdot \nabla\mathcal{H}(\boldsymbol{x}, \nabla\vartheta) \Big\}\, dx \\ &\leq \int_\Omega \Big\{ [\mathcal{H}(\boldsymbol{x}, \nabla\vartheta) - \mathcal{H}(\boldsymbol{x}, \nabla\tilde{\vartheta})] + [\mathcal{H}(\boldsymbol{x}, \nabla\tilde{\vartheta}) - \mathcal{H}(\boldsymbol{x}, \nabla\vartheta)] \Big\}\, dx \\ &= 0, \quad \forall t \in \mathbb{I}, \end{aligned} \tag{6.1.22}$$

which implies that $\frac{d}{dt}|||\vartheta(\cdot,t) - \tilde{\vartheta}(\cdot,t)||| \leq 0$. Therefore,

$$|||\vartheta(\cdot,t)-\tilde{\vartheta}(\cdot,t)||| - |||\vartheta_0(\cdot)-\tilde{\vartheta}_0(\cdot)||| = \int_0^t \frac{d}{d\tau}|||\xi(\cdot,\tau)|||\,dt \leq 0, \quad (6.1.23)$$

which proves the result. □

A proof identical to that given above applies to the semidiscrete problem of evolution, obtained via a spatial Galerkin projection of the continuum problem onto a finite dimensional subspace $\mathcal{V}^h \subset \mathcal{V}$, with the natural norm induced by the mass matrix (see Hughes [1983]).

6.1.2 The Algorithmic Problem

Consider the time integration of the IBVP (6.1.12)–(6.1.13) for nonlinear heat conduction by the generalized midpoint rule algorithm, depending on the parameter $\alpha \in [0, 1]$. The goal is to prove nonlinear stability of the scheme for $\alpha \geq \frac{1}{2}$.

6.1.2.1 The generalized midpoint rule

Let $[t_n, t_{n+1}] \subset \mathbb{I}$, with $\Delta t := t_{n+1} - t_n > 0$, be a typical time subinterval. Assume that the following initial data is known at time $t = t_n$:

$$\vartheta_n : \overline{\Omega} \to \mathbb{R}_+,$$

and (6.1.24)

$$v_n : \overline{\Omega} \to \mathbb{R},$$

where $\vartheta_n(\boldsymbol{x})$ and $v_n(\boldsymbol{x})$ are algorithmic approximations of the temperature and temperature "velocity" $\vartheta(\boldsymbol{x}, t_n)$ and $v(\boldsymbol{x}, t_n) = \dot{\vartheta}(\boldsymbol{x}, t_n)$, respectively. The objective is to obtain algorithmic approximations $\vartheta_{n+1}(\boldsymbol{x})$ and $v_{n+1}(\boldsymbol{x})$ to the actual fields $\vartheta(\boldsymbol{x}, t_{n+1})$ and $v(\boldsymbol{x}, t_{n+1})$ at time t_{n+1}, respectively, for prescribed source term $f(\boldsymbol{x}, t)$ and boundary data $\bar{\vartheta}(\boldsymbol{x}, t)$ and $\bar{q}(\boldsymbol{x}, t)$ in the interval $[t_n, t_{n+1}]$. To this end, consider the following algorithmic problem:

Problem $W_{\Delta t}$: Find $\vartheta_{n+1} \in \mathcal{S}_{n+1}$ such that

$$\langle \rho_0 c\, v_{n+\alpha}, \eta\rangle = -\langle \nabla\mathcal{H}(\cdot, \nabla\vartheta_{n+\alpha}), \nabla\eta\rangle + \langle f(\cdot, t_{n+\alpha}), \eta\rangle + \langle \bar{q}(\cdot, t_{n+\alpha}), \eta\rangle_\Gamma, \quad (6.1.25)$$

for all $\eta \in \mathcal{V}$, where

$$\left.\begin{aligned} v_{n+\alpha} &= (\vartheta_{n+1} - \vartheta_n)/\Delta t \\ \vartheta_{n+\alpha} &= \alpha\vartheta_{n+1} + (1-\alpha)\vartheta_n \end{aligned}\right\} \quad \text{for } \alpha \in (0, 1] \text{ fixed.} \quad (6.1.26)$$

It is assumed that the forcing functions in the algorithmic problem above are given without approximation in the interval $[t_n, t_{n+1}]$. Using standard arguments in the calculus of variations, problem $W_{\Delta t}$ is formally equivalent to the following local problem:

$$\left.\begin{aligned} \rho_0 c\, v_{n+\alpha} &= \operatorname{div}[\nabla\mathcal{H}(\cdot, \nabla\vartheta_{n+\alpha})] + f(\cdot, t_{n+\alpha}) \quad \text{in } \Omega \\ \vartheta_{n+\alpha} &= \bar{\vartheta}(\cdot, t_{n+\alpha}) \quad \text{on } \Gamma_\vartheta \\ \nabla\mathcal{H}(\cdot, \nabla\vartheta_{n+\alpha}) \cdot \boldsymbol{n} &= \bar{q}(\cdot, t_{n+\alpha}) \quad \text{on } \Gamma_q, \end{aligned}\right\} \tag{6.1.27}$$

with $\vartheta_{n+\alpha}$ and $v_{n+\alpha}$ defined by (6.1.26).

Remark 1.1. In view of (6.1.27), it is clear that the temperature evolution equation in the algorithmic problem is *enforced at* $t_{n+\alpha} := \alpha t_{n+1} + (1-\alpha)t_n$. This point is crucial to the stability proof given below. Observe that, in general, the algorithmic counterpart of the temperature evolution equation does not hold at t_n or at t_{n+1}, unless the heat-flux potential function $\mathcal{H}(\boldsymbol{x}, \cdot)$ is quadratic.

6.1.3 Nonlinear Stability Analysis

Roughly speaking, an algorithm for the nonlinear heat conduction equation is said to be *nonlinearly stable* if the algorithm inherits the contractivity property of the continuum problem. More precisely, let

$$\{\vartheta_n\},$$

and (6.1.28)

$$\{\widetilde{\vartheta}_n\},$$

be two sequences generated by a given algorithm for two initial conditions

$$\vartheta_0 : \Omega \to \mathbb{R}_+,$$

and (6.1.29)

$$\widetilde{\vartheta}_0 : \Omega \to \mathbb{R}_+,$$

respectively. The algorithm is nonlinearly stable (or *A-contractive*) if the following inequality holds relative to the natural norm $|||\cdot|||$ for the continuum problem:

$$\boxed{|||\vartheta_n - \widetilde{\vartheta}_n||| \leq |||\vartheta_0 - \widetilde{\vartheta}_0||| \quad \forall\, n \geq 1.} \tag{6.1.30}$$

The following result holds for the generalized midpoint rule algorithm:

Proposition 1.2. *The algorithmic problem $W_{\Delta t}$ defined by (6.1.25)–(6.1.26) is nonlinearly stable for $\alpha \geq \frac{1}{2}$.*

PROOF. By hypothesis, the following two variational equalities hold:

$$\left.\begin{aligned} \frac{1}{\Delta t}\langle\!\langle \vartheta_{n+1} - \vartheta_n, \eta\rangle\!\rangle &= -\langle \nabla\mathcal{H}_{n+\alpha}, \nabla\eta\rangle + \langle f_{n+\alpha}, \eta\rangle + \langle \bar{q}_{n+\alpha}, \eta\rangle_\Gamma \\ \frac{1}{\Delta t}\langle\!\langle \widetilde{\vartheta}_{n+1} - \widetilde{\vartheta}_n, \eta\rangle\!\rangle &= -\langle \nabla\widetilde{\mathcal{H}}_{n+\alpha}, \nabla\eta\rangle + \langle f_{n+\alpha}, \eta\rangle + \langle \bar{q}_{n+\alpha}, \eta\rangle_\Gamma \end{aligned}\right\} \forall \eta \in \mathcal{V} \tag{6.1.31}$$

where, as in the continuum case, the following notation has been used:

$$\begin{aligned} \nabla\mathcal{H}_{n+\alpha} &:= \nabla\mathcal{H}(\cdot, \nabla\vartheta_{n+\alpha}), \\ \nabla\widetilde{\mathcal{H}}_{n+\alpha} &:= \nabla\mathcal{H}(\cdot, \nabla\widetilde{\vartheta}_{n+\alpha}). \end{aligned} \tag{6.1.32}$$

Set $\xi_{n+\alpha} := \vartheta_{n+\alpha} - \widetilde{\vartheta}_{n+\alpha}$, and observe that $\xi_{n+\alpha} \in \mathcal{V}$ for any $\alpha \in [0, 1]$. Subtracting $(6.1.31)_2$ from $(6.1.31)_1$ gives

$$\frac{1}{\Delta t} \langle\!\langle \xi_{n+1} - \xi_n, \eta \rangle\!\rangle = -\langle \nabla\mathcal{H}_{n+\alpha} - \nabla\widetilde{\mathcal{H}}_{n+\alpha}, \nabla\eta \rangle, \quad \forall \eta \in \mathcal{V}. \tag{6.1.33}$$

In particular, since $\xi_{n+\frac{1}{2}} = \frac{1}{2}[\xi_{n+1} + \xi_n]$ is in $\mathcal{V}$, choosing $\eta = \xi_{n+\frac{1}{2}}$ in (6.1.33), expanding the left-hand side and using definition (6.1.16) for the natural norm $|||\cdot|||$ gives

$$\frac{1}{2\Delta t}\left[|||\xi_{n+1}|||^2 - |||\xi_n|||^2\right] = -\langle \nabla\mathcal{H}_{n+\alpha} - \nabla\widetilde{\mathcal{H}}_{n+\alpha}, \nabla\xi_{n+\frac{1}{2}} \rangle. \tag{6.1.34}$$

Now insert the following identity on the right-hand side of (6.1.34):

$$\xi_{n+\frac{1}{2}} = \xi_{n+\alpha} - (\alpha - \tfrac{1}{2})[\xi_{n+1} - \xi_n], \quad \forall \alpha \in [0, 1], \tag{6.1.35}$$

and make use of the fact that, in particular, (6.1.33) also holds with $\eta = \xi_{n+1} - \xi_n \in \mathcal{V}$ to obtain the result

$$\begin{aligned} |||\xi_{n+1}|||^2 - |||\xi_n|||^2 = &-2\Delta t \langle \nabla\mathcal{H}_{n+\alpha} - \nabla\widetilde{\mathcal{H}}_{n+\alpha}, \nabla\xi_{n+\alpha} \rangle \\ &- (2\alpha - 1)|||\xi_{n+1} - \xi_n|||^2. \end{aligned} \tag{6.1.36}$$

Finally, substitute $\xi_{n+\alpha} = \vartheta_{n+\alpha} - \widetilde{\vartheta}_{n+\alpha}$ in (6.1.36), and make crucial use of the convexity condition (6.1.7) on the function $\mathcal{H}(\boldsymbol{x}, \cdot)$. Proceeding exactly as in the continuum problem, one obtains the following estimate:

$$\begin{aligned} |||\xi_{n+1}|||^2 - |||\xi_n|||^2 = &-(2\alpha - 1)|||\xi_{n+1} - \xi_n|||^2 \\ &+ 2\Delta t \int_\Omega \nabla\mathcal{H}(\boldsymbol{x}, \nabla\widetilde{\vartheta}_{n+\alpha}) \cdot (\nabla\vartheta_{n+\alpha} - \nabla\widetilde{\vartheta}_{n+\alpha})\, d\boldsymbol{x} \\ &+ 2\Delta t \int_\Omega \nabla\mathcal{H}(\boldsymbol{x}, \nabla\vartheta_{n+\alpha}) \cdot (\nabla\widetilde{\vartheta}_{n+\alpha} - \nabla\vartheta_{n+\alpha})\, d\boldsymbol{x} \\ \le &-(2\alpha - 1)|||\xi_{n+1} - \xi_n|||^2 \le 0 \quad \text{for } \alpha \ge \tfrac{1}{2}, \end{aligned} \tag{6.1.37}$$

so that $|||\xi_{n+1}||| \le |||\xi_n|||$ for any $n \ge 0$. Then a straightforward induction argument completes the stability proof. □

Once more, observe that the preceding proof depends critically on enforcing the algorithmic version of the temperature evolution equation at $t_{n+\alpha}$. In fact, this proof breaks down if the equilibrium equation is enforced, as is customary, at the end of the time step, i.e., at $t = t_{n+1}$.

6.2 Infinitesimal Elastoplasticity

The stability analysis carried out in the preceding section is readily extended to the case of rate-independent plasticity and viscoplasticity. The approach adopted depends crucially on a formulation of plasticity as a variational inequality and proceeds as follows.

i. One exploits the fact that the continuum problem is contractive relative to the norm induced by the complementary Helmholtz free energy function to identify the natural norm for the elastoplastic problem.

ii. Then it is shown that a one-parameter family of return-mapping algorithms inspired by the generalized midpoint rule inherits the contractivity property of the continuum problem, when $\alpha \geq \frac{1}{2}$. Hence, this class of algorithms is nonlinearly B-stable.

As in the heat conduction equation, the stability proof relies critically on enforcing the equilibrium condition at $t_{n+\alpha}$. The analysis presented below extends and generalizes the results in Hughes and Taylor [1978] and Simo and Govindjee [1991].

6.2.1 The Continuum Problem for Plasticity and Viscoplasticity

Once more, let $\Omega \subset \mathbb{R}^{n_{\text{dim}}}$, with $1 \leq n_{\text{dim}} \leq 3$, be the reference placement of an elastoplastic body with smooth boundary $\partial\Omega$; let $\mathbb{I} \subset \mathbb{R}_+$ the time interval of interest, and denote the displacement field and the stress tensor by

$$\boldsymbol{u} : \overline{\Omega} \times \mathbb{I} \rightarrow \mathbb{R}^{n_{\text{dim}}},$$

and (6.2.1)

$$\boldsymbol{\sigma} : \overline{\Omega} \times \mathbb{I} \rightarrow \mathbb{S}$$

respectively. Here $\mathbb{S} = \mathbb{R}^{(n_{\text{dim}}+1)\cdot n_{\text{dim}}/2}$ is the vector space of symmetric rank-two tensors. Assume that $\boldsymbol{u}(\boldsymbol{x}, t)$ and $\boldsymbol{\sigma}(\boldsymbol{x}, t)$ are specified on parts of the boundary Γ_u and Γ_σ as

$$\boldsymbol{u} = \bar{\boldsymbol{u}} \quad \text{on} \quad \Gamma_u \times \mathbb{I},$$

and (6.2.2)

$$\boldsymbol{\sigma n} = \bar{\boldsymbol{t}} \quad \text{on} \quad \Gamma_\sigma \times \mathbb{I},$$

respectively, where $\boldsymbol{n}$ is the unit outward normal vector to the boundary, $\bar{\boldsymbol{u}} : \Gamma_u \times \mathbb{I} \rightarrow \mathbb{R}^{n_{\text{dim}}}$ is a prescribed boundary displacement, and $\bar{\boldsymbol{t}} : \Gamma_\sigma \times \mathbb{I} \rightarrow \mathbb{R}^{n_{\text{dim}}}$ is the prescribed boundary traction vector. As usual, one assumes that

$$\overline{\Gamma_u \cup \Gamma_\sigma} = \overline{\partial\Omega},$$

and (6.2.3)

$$\Gamma_u \cap \Gamma_\sigma = \emptyset$$

(with the conventional interpretation). In addition, denote the body force (per unit of volume) by $\boldsymbol{f} : \Omega \times \mathbb{I} \to \mathbb{R}^{n_{\text{dim}}}$, so that the local form of the equilibrium equation becomes

$$-\text{div}[\boldsymbol{\sigma}] = \boldsymbol{f}, \qquad \text{in} \qquad \Omega \times \mathbb{I}. \tag{6.2.4}$$

This equation is linear. The source of nonlinearity in this problem arises from the constitutive equation that relates the stress field and the displacement field, as discussed below.

6.2.1.1 Classical rate-independent plasticity

In addition to the stress tensor $\boldsymbol{\sigma}(\boldsymbol{x}, t)$, one introduces an n_{int}-dimensional vector field $\boldsymbol{q}: \Omega \times \mathbb{I} \to \mathbb{R}^{n_{\text{int}}}$ ($n_{\text{int}} \geq 1$) of phenomenological internal variables which, from a physical standpoint, characterize strain hardening in the material. For convenience, the following notation is adopted:

$$\boldsymbol{\Sigma}(\boldsymbol{x}, t) := [\boldsymbol{\sigma}(\boldsymbol{x}, t), \ \boldsymbol{q}(\boldsymbol{x}, t)], \qquad \text{for } (\boldsymbol{x}, t) \in \Omega \times \mathbb{I}. \tag{6.2.5}$$

One refers to $\boldsymbol{\Sigma}$ as the *generalized stress* which is constrained to lie within a *convex domain*, called the elastic domain and denoted by $\mathbb{E}$. Typically, $\mathbb{E}$ is defined in terms of smooth functions $\phi_\mu : \mathbb{S} \times \mathbb{R}^{n_{\text{int}}} \to \mathbb{R}$, with $\mu \in \{1, \cdots, m\}$, as the constrained *convex* set

$$\mathbb{E} := \{\boldsymbol{\Sigma} := (\boldsymbol{\sigma}, \boldsymbol{q}) \in \mathbb{S} \times \mathbb{R}^{n_{\text{int}}} : \ \phi_\mu(\boldsymbol{\sigma}, \boldsymbol{q}) \leq 0, \ \text{for } \mu = 1, \cdots, m\}. \tag{6.2.6}$$

The boundary $\partial\mathbb{E}$ of $\mathbb{E} \subset \mathbb{S} \times \mathbb{R}^{n_{\text{int}}}$ need not be smooth; in fact, in applications, $\partial\mathbb{E}$ typically exhibits "corners". A classical example is provided by the Tresca yield condition.

Let $\mathbf{C}$ be the *elasticity tensor*, which is assumed constant in what follows. Further, let $\mathbf{D}$ denote a $n_{\text{int}} \times n_{\text{int}}$ given matrix, which is assumed to be positive-definite and constant and is called the *generalized hardening moduli*. Under these assumptions, the complementary Helmholtz free energy function defined by

$$\chi(\boldsymbol{\Sigma}) := \tfrac{1}{2}\boldsymbol{\sigma} \cdot \mathbf{C}^{-1}\boldsymbol{\sigma} + \tfrac{1}{2}\boldsymbol{q} \cdot \mathbf{D}^{-1}\boldsymbol{q}. \tag{6.2.7}$$

is strictly convex on $\mathbb{S} \times \mathbb{R}^{n_{\text{int}}}$. By definition, the *local dissipation function* is the total stress power minus the change in complementary Helmholtz free energy:

$$\mathcal{D} := \boldsymbol{\sigma} : \boldsymbol{\varepsilon}[\dot{\boldsymbol{u}}] - \frac{d}{dt}\chi(\boldsymbol{\Sigma}); \quad \text{in} \quad \Omega \times \mathbb{I}, \tag{6.2.8}$$

where $\boldsymbol{\varepsilon}[\cdot] = \text{sym}[\nabla(\cdot)]$ is the strain operator and $\dot{\boldsymbol{u}} := \frac{\partial}{\partial t}\boldsymbol{u}$ is the velocity field.

The local form of the second law requires that $\mathcal{D} \geq 0$. The classical rate-independent plasticity model is obtained by postulating the following *local principle of maximum dissipation*: For fixed rates $(\boldsymbol{\varepsilon}[\dot{\boldsymbol{u}}], \dot{\boldsymbol{\sigma}})$, the actual state $(\boldsymbol{\sigma}, \boldsymbol{q}) \in \mathbb{E}$ maximizes the dissipation:

$$[\boldsymbol{\sigma} - \boldsymbol{\tau}] : \boldsymbol{\varepsilon}[\dot{\boldsymbol{u}}] - \frac{d}{dt}\chi(\boldsymbol{\Sigma}) + \langle\langle \boldsymbol{T}, \dot{\boldsymbol{\Sigma}} \rangle\rangle \geq 0, \quad \forall\, \boldsymbol{T} = (\boldsymbol{\tau}, \boldsymbol{p}) \in \mathbb{E}. \tag{6.2.9}$$

Using the method of Lagrange multipliers, it is easily concluded that, for the quadratic complementary Helmholtz free energy function (6.2.7) and the elastic domain (6.2.6), the standard Kuhn–Tucker optimality conditions associated with the maximum principle (6.2.9) yield the local flow rule and hardening law:

$$\left.\begin{aligned} \boldsymbol{\varepsilon}[\dot{\boldsymbol{u}}] - \mathbf{C}^{-1}\dot{\boldsymbol{\sigma}} &= \sum_{\mu=1}^{m} \gamma^{\mu} \frac{\partial}{\partial \boldsymbol{\sigma}} \phi_{\mu}(\boldsymbol{\sigma}, \boldsymbol{q}), \\ -\mathbf{D}^{-1}\dot{\boldsymbol{q}} &= \sum_{\mu=1}^{m} \gamma^{\mu} \frac{\partial}{\partial \boldsymbol{q}} \phi_{\mu}(\boldsymbol{\sigma}, \boldsymbol{q}), \end{aligned}\right\} \tag{6.2.10a}$$

where

$$\gamma^{\mu} \geq 0, \quad \phi_{\mu}(\sigma, q) \leq 0 \quad \text{and} \quad \gamma^{\mu}\phi_{\mu}(\sigma, q) = 0, \quad \text{for } \mu = 1, \cdots, m, \tag{6.2.10b}$$

are the Kuhn–Tucker complementarity conditions. $\gamma^{\mu} \geq 0$ are the *plastic consistency parameters.*

6.2.1.2 The weak form of the constitutive equations

The weak form of the constitutive equations is simply the global formulation of the principle of maximum (plastic) dissipation (6.2.9) over the entire body. Accordingly, let (σ_{ij}, q_i) be components of $\boldsymbol{\Sigma} = (\boldsymbol{\sigma}, \boldsymbol{q})$ relative to a Cartesian orthonormal frame, and let

$$\mathcal{T} := \{\boldsymbol{\Sigma} = (\boldsymbol{\sigma}, \boldsymbol{q}) : \Omega \to \mathbb{S} \times \mathbb{R}^{n_{\text{int}}} : \sigma_{ij} \in L_2(\Omega) \text{ and } q_i \in L_2(\Omega)\}. \tag{6.2.11}$$

Define bilinear forms $a(\cdot, \cdot)$ and $b(\cdot, \cdot)$ by the expressions

$$a(\boldsymbol{\sigma}, \boldsymbol{\tau}) := \int_{\Omega} \boldsymbol{\sigma} : \mathbf{C}^{-1}\boldsymbol{\tau}\, d\Omega,$$

and (6.2.12a)

$$b(\boldsymbol{q}, \boldsymbol{p}) := \int_{\Omega} \boldsymbol{q} \cdot \mathbf{D}^{-1}\boldsymbol{p}\, d\Omega.$$

Since the elastic moduli $\mathbf{C}$ and plastic hardening moduli $\mathbf{D}$ are positive-definite on $\mathbb{S}$ and $\mathbb{R}^{n_{\text{int}}}$, respectively, it follows that $a(\cdot\,,\,\cdot)$, and $b(\cdot\,,\,\cdot)$ are coercive. Consequently, the bilinear form $\langle\!\langle \cdot\,,\,\cdot \rangle\!\rangle : \mathcal{T} \times \mathcal{T} \to \mathbb{R}$ defined by

$$\langle\!\langle \boldsymbol{\Sigma}, \boldsymbol{T} \rangle\!\rangle := a(\boldsymbol{\sigma}, \boldsymbol{\tau}) + b(\boldsymbol{q}, \boldsymbol{p}), \tag{6.2.12b}$$

induces an inner product on $\mathcal{T}$. It is shown below that the associated norm, denoted by $|||\cdot|||:= \sqrt{\langle\!\langle \cdot\,,\,\cdot \rangle\!\rangle}$, is the *natural norm* for the elastoplastic problem. Observe that relationships (6.2.7) and (6.2.11) imply that the norm squared $|||\cdot|||^2$ is precisely twice the integral over the body of the complementary Helmholtz free energy function.

With the preceding notation in hand, the dissipation over the entire body, denoted by $\mathcal{D}_{\Omega}$ and obtained by integration of (6.2.8) over the reference placement Ω, can be written as

$$\mathcal{D}_\Omega = \langle\!\langle \boldsymbol{\Sigma}, \dot{\boldsymbol{\Sigma}}^{\text{trial}} \rangle\!\rangle - \tfrac{1}{2}\frac{d}{dt}|||\boldsymbol{\Sigma}|||^2 = \langle\!\langle \dot{\boldsymbol{\Sigma}}^{\text{trial}} - \dot{\boldsymbol{\Sigma}}, \boldsymbol{\Sigma} \rangle\!\rangle \geq 0, \tag{6.2.13}$$

where

$$\dot{\boldsymbol{\Sigma}}^{\text{trial}} := \big(\mathbf{C}\,\boldsymbol{\varepsilon}[\dot{\boldsymbol{u}}]\,,\ \mathbf{0}\big)\,. \tag{6.2.14}$$

Note that $\dot{\boldsymbol{\Sigma}}^{\text{trial}}$ is the stress rate about $\boldsymbol{\Sigma} \in \mathbb{E}$ obtained by *freezing plastic flow* (i.e., by setting $\dot{\boldsymbol{\varepsilon}}^p = \mathbf{0}$ and $\dot{\boldsymbol{q}} = \mathbf{0}$) and, therefore, it is called the rate of trial elastic stress. In view of (6.2.13), the global version of the principle of maximum dissipation leads to the variational inequality

$$\langle\!\langle \dot{\boldsymbol{\Sigma}}^{\text{trial}} - \dot{\boldsymbol{\Sigma}}, \boldsymbol{T} - \boldsymbol{\Sigma} \rangle\!\rangle \leq 0, \quad \forall \boldsymbol{T} \in \mathbb{E} \cap \mathcal{T}, \tag{6.2.15}$$

which gives the weak formulation of the constitutive equation for rate-independent (hardening) elastoplasticity.

6.2.1.3 The viscoplastic regularization

Following Duvaut and Lions [1972], classical viscoplasticity is regarded as a (Yosida) regularization of rate-independent plasticity, constructed as follows. Define the functional $J\colon \mathbb{S} \times \mathbb{R}^{n_{\text{int}}} \to \mathbb{R}$ by the constrained minimization problem

$$J(\boldsymbol{\Sigma}) := \min\{2\,\chi(\boldsymbol{\Sigma} - \boldsymbol{T}),\ \text{for all } \boldsymbol{T} \in \mathbb{E}\}. \tag{6.2.16}$$

Thus, $J(\boldsymbol{\Sigma})$ gives the (unique) distance measured in the complementary Helmholtz free energy $\chi(\cdot)$ between any $\boldsymbol{\Sigma} \in \mathbb{S} \times \mathbb{R}^{n_{\text{int}}}$ and the convex set $\mathbb{E}$. Clearly, $J(\boldsymbol{\Sigma}) \geq 0$, and $J(\boldsymbol{\Sigma}) = 0$ iff $\boldsymbol{\Sigma} \in \mathbb{E}$. Now consider the following regularization of the dissipation function (6.2.13):

$$\mathcal{D}^\eta_\Omega := \mathcal{D}_\Omega + \frac{1}{\eta}\int_\Omega g[J(\boldsymbol{\Sigma})]\,d\Omega, \tag{6.2.17a}$$

where $\eta \in (0, \infty)$ is the regularization parameter and $g(\cdot)$ is a *nonnegative* convex function with the property

$$g(x) \geq 0,$$

and (6.2.17b)

$$g(x) = 0 \iff x = 0.$$

By standard results in convex optimization (see, e.g., Luenberger [1984]), it follows that the problem that maximizes the regularized dissipation $\mathcal{D}^\eta_\Omega$ over all *unconstrained* stresses $\boldsymbol{\Sigma} \in \mathcal{T}$ is simply the penalty regularization of the classical *constrained* principle of maximum dissipation, with dissipation function $\mathcal{D}_\Omega$. Then the optimality condition associated with the regularized principle of maximum dissipation yields the following inequality which characterizes the constitutive response of classical viscoplasticity:

$$\langle\!\langle \dot{\boldsymbol{\Sigma}}^{\text{trial}} - \dot{\boldsymbol{\Sigma}}, \boldsymbol{T} - \boldsymbol{\Sigma} \rangle\!\rangle \leq \frac{1}{\eta}\int_\Omega \{g[J(\boldsymbol{T})] - g[J(\boldsymbol{\Sigma})]\}\,d\Omega, \quad \forall \boldsymbol{T} \in \mathcal{T}. \tag{6.2.18}$$

Observe that in contrast with rate-independent plasticity, neither the actual stress $\boldsymbol{\Sigma} \in \mathcal{T}$ nor the admissible stress variations $\boldsymbol{T} \in \mathcal{T}$ are constrained to lie in the elastic domain $\mathbb{E}$. However, it is well known that, as $\eta \to 0$, the stress $\boldsymbol{\Sigma}$ is constrained to lie in the elastic domain, and (6.2.18) reduces to inequality (6.2.15); see Duvaut and Lions [1972] and Johnson [1976,1978].

6.2.1.4 The weak formulation of the IBVP

Let $\mathcal{S}_t$ denote the displacement solution space for all $t \in \mathbb{I}$:

$$\mathcal{S}_t = \{\boldsymbol{u}(\cdot, t) \in [\mathbb{H}^1(\Omega)]^{n_{\text{dim}}} : \boldsymbol{u}(\cdot, t) = \bar{\boldsymbol{u}}(\cdot, t) \quad \text{on} \quad \Gamma_u\}. \tag{6.2.19}$$

In addition, let $\mathcal{V}$ be the space of displacement test functions:

$$\mathcal{V} = \{\boldsymbol{\eta} \in [\mathbb{H}^1(\Omega)]^{n_{\text{dim}}} : \boldsymbol{\eta} = \boldsymbol{0} \quad \text{on} \quad \Gamma_u\}. \tag{6.2.20}$$

The weak form of the equilibrium equations (6.2.4) along with the dissipation inequality (6.2.15) [or (6.2.18) for viscoplasticity] leads to the following variational problem:

Problem W_t: For all $t \in \mathbb{I}$, find $\boldsymbol{u} \in \mathcal{S}_t$ and $\boldsymbol{\Sigma} = (\boldsymbol{\sigma}, \boldsymbol{q}) \in \mathbb{E} \cap \mathcal{T}$ such that

$$\left.\begin{aligned} \langle \boldsymbol{\sigma}, \boldsymbol{\varepsilon}[\boldsymbol{\eta}]\rangle - \langle \boldsymbol{f}, \boldsymbol{\eta}\rangle - \langle \bar{\boldsymbol{t}}, \boldsymbol{\eta}\rangle_\Gamma = 0, &\quad \forall \boldsymbol{\eta} \in \mathcal{V}, \\ \langle\!\langle \dot{\boldsymbol{\Sigma}}^{\text{trial}} - \dot{\boldsymbol{\Sigma}}, \boldsymbol{T} - \boldsymbol{\Sigma} \rangle\!\rangle \le 0, &\quad \forall \boldsymbol{T} \in \mathbb{E} \cap \mathcal{T}, \end{aligned}\right\} \tag{6.2.21a}$$

with $\boldsymbol{T} = (\boldsymbol{\tau}, \boldsymbol{p})$ subject to the initial condition

$$\langle\!\langle \boldsymbol{\Sigma}(\cdot, 0), \boldsymbol{T}\rangle\!\rangle = \langle\!\langle \boldsymbol{\Sigma}_0, \boldsymbol{T}\rangle\!\rangle, \quad \forall \boldsymbol{T} \in \mathbb{E} \cap \mathcal{T}. \tag{6.2.21b}$$

In what follows, with a slight abuse in notation, the same symbol $\langle \cdot, \cdot \rangle$ is used to denote the standard $L_2(\Omega)$ inner product of functions, vectors, or tensors depending on the specific context.

Remarks 6.2.1.

1. The geometric interpretation of inequality $(6.2.21a)_2$ is illustrated in Figure **6.2**. The actual stress rate $\dot{\boldsymbol{\Sigma}}$ is the projection of the trial stress rate $\dot{\boldsymbol{\Sigma}}^{\text{trial}}$ onto the tangent plane at $\boldsymbol{\Sigma} \in \partial\mathbb{E}$. Then convexity of $\mathbb{E}$ implies that the angle measured in the natural norm between $[\dot{\boldsymbol{\Sigma}}^{\text{trial}} - \dot{\boldsymbol{\Sigma}}]$ and $[\boldsymbol{\Sigma} - \boldsymbol{T}]$ is greater or equal to $\pi/2$, a condition equivalent to $(6.2.21a)_2$.
2. The variational formulation of plasticity given by equations (6.2.21) is standard and has been considered by a number of authors, in particular, Johnson [1976,1978], extending early work of Duvaut & Lions [1972]. The assumption of hardening plasticity, i.e., the presence of the bilinear form $b(\cdot, \cdot)$ defined by $(6.2.12a)_2$, is crucial if the functional analysis framework outlined above is to remain physically meaningful.
3. The situation afforded by perfect plasticity is significantly more complicated than hardening plasticity since the regularity implicit in the choice of $\mathcal{S}_t$ in (6.2.11) no longer holds. The underlying physical reason for this lack of regularity is the presence of strong discontinuities in the displacement field, the

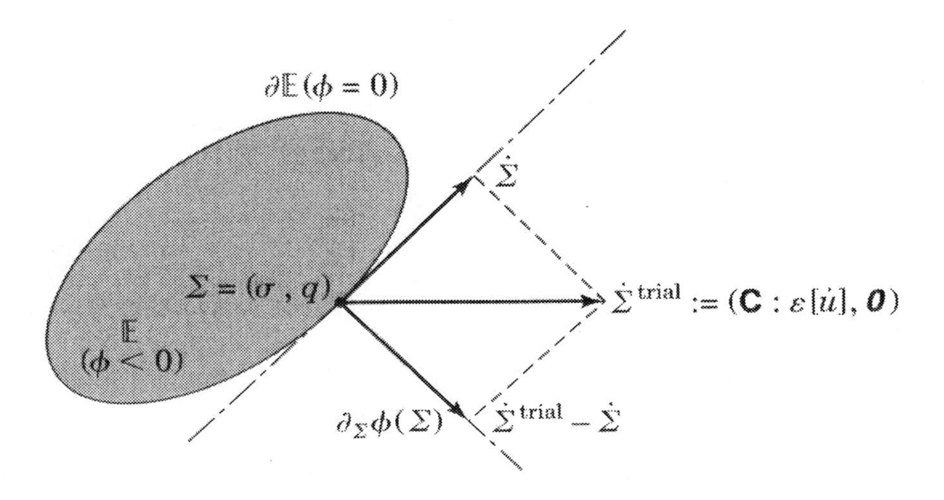

FIGURE 6-2. Geometric interpretation of the variational inequality $\left\langle \dot{\boldsymbol{\Sigma}}^{\text{trial}} - \dot{\boldsymbol{\Sigma}}, \boldsymbol{\Sigma} - \boldsymbol{T} \right\rangle \leq 0$ for $\mathbb{E} := \{\boldsymbol{\Sigma} : \phi(\boldsymbol{\Sigma}) \leq 0\}$.

so-called *slip lines*, which rule out the use of standard Sobolev spaces. For perfect plasticity the appropriate choice appears to be $\mathcal{S}_t = BD(\Omega)$, i.e. the space of bounded deformations [displacements in $\mathbb{L}_2(\Omega)$ with strain field a bounded measure] introduced by Matthies, Strang, and Christiansen [1979], and further analyzed in Temam and Strang [1980]. See Matthies [1978,1979]; Suquet [1979]; Strang, Matthies, and Temam [1980]; and the recent summary in Demengel [1989] for a detailed elaboration on this and related issues.

6.2.1.5 Contractivity of elastoplastic flow.

The following contractivity property inherent to problem W_t identifies the norm $|||\cdot|||$ defined by (6.2.12*b*) as the natural norm and, therefore, plays a crucial role in the subsequent algorithmic stability analysis. Let

$$\boldsymbol{\Sigma}_0 = (\boldsymbol{\sigma}_0, \boldsymbol{q}_0) \in \mathbb{E} \cap \mathcal{T},$$

and (6.2.22)

$$\tilde{\boldsymbol{\Sigma}}_0 = (\tilde{\boldsymbol{\sigma}}_0, \tilde{\boldsymbol{q}}_0) \in \mathbb{E} \cap \mathcal{T},$$

be two *arbitrary* initial conditions for the problem of evolution defined by (6.2.21) and denote the corresponding flows generated by (6.2.21) by

$$t \in \mathbb{I} \mapsto \boldsymbol{\Sigma} = (\boldsymbol{\sigma}, \boldsymbol{q}) \in \mathbb{E} \cap \mathcal{T},$$

and (6.2.23)

$$t \in \mathbb{I} \mapsto \widetilde{\boldsymbol{\Sigma}} = (\widetilde{\boldsymbol{\sigma}}, \widetilde{\boldsymbol{q}}) \in \mathbb{E} \cap \mathcal{T},$$

respectively. The following result is implicitly contained in Moreau [1977]; see also Nguyen [1977].

Proposition 2.1.. Relative to the norm $|||\cdot|||$ induced by (6.2.12), the following contractivity property holds:

$$\boxed{|||\boldsymbol{\Sigma}(\cdot, t) - \widetilde{\boldsymbol{\Sigma}}(\cdot, t)||| \leq |||\boldsymbol{\Sigma}_0 - \widetilde{\boldsymbol{\Sigma}}_0|||, \quad \forall t \in \mathbb{I}.} \tag{6.2.24}$$

PROOF. By hypothesis, the flows in (6.2.23) satisfy the variational inequality $(6.2.21)_2$. In particular,

$$\left.\begin{aligned} \langle\!\langle \dot{\boldsymbol{\Sigma}}, \boldsymbol{\Sigma} - \widetilde{\boldsymbol{\Sigma}} \rangle\!\rangle &\leq \ \ \langle\!\langle \dot{\boldsymbol{\Sigma}}^{\text{trial}}, \boldsymbol{\Sigma} - \widetilde{\boldsymbol{\Sigma}} \rangle\!\rangle, \\ -\langle\!\langle \dot{\widetilde{\boldsymbol{\Sigma}}}, \boldsymbol{\Sigma} - \widetilde{\boldsymbol{\Sigma}} \rangle\!\rangle &\leq -\langle\!\langle \dot{\widetilde{\boldsymbol{\Sigma}}}^{\text{trial}}, \boldsymbol{\Sigma} - \widetilde{\boldsymbol{\Sigma}} \rangle\!\rangle. \end{aligned}\right\} \tag{6.2.25}$$

Adding these inequalities and using bilinearity along with definitions $(6.2.12a)_1$ and (6.2.14) yields

$$\langle\!\langle \dot{\boldsymbol{\Sigma}} - \dot{\widetilde{\boldsymbol{\Sigma}}}, \boldsymbol{\Sigma} - \widetilde{\boldsymbol{\Sigma}} \rangle\!\rangle \leq \langle\!\langle \dot{\boldsymbol{\Sigma}}^{\text{trial}} - \dot{\widetilde{\boldsymbol{\Sigma}}}^{\text{trial}}, \boldsymbol{\Sigma} - \widetilde{\boldsymbol{\Sigma}} \rangle\!\rangle = \langle \boldsymbol{\varepsilon}[\dot{\boldsymbol{u}} - \dot{\widetilde{\boldsymbol{u}}}], \boldsymbol{\sigma} - \widetilde{\boldsymbol{\sigma}} \rangle. \tag{6.2.26}$$

Now observe that, for fixed (but arbitrary) $t \in \mathbb{I}$, $\dot{\boldsymbol{u}}(\cdot, t) - \dot{\widetilde{\boldsymbol{u}}}(\cdot, t) \in \mathcal{V}$. Using the fact that the flows (6.2.23) satisfy $(6.2.21a)_1$ gives

$$\langle \boldsymbol{\sigma}, \boldsymbol{\varepsilon}[\dot{\boldsymbol{u}} - \dot{\widetilde{\boldsymbol{u}}}] \rangle - \langle \widetilde{\boldsymbol{\sigma}}, \boldsymbol{\varepsilon}[\dot{\boldsymbol{u}} - \dot{\widetilde{\boldsymbol{u}}}] \rangle = 0, \tag{6.2.27}$$

so that the right-hand side of (6.2.26) vanishes. Combining (6.2.26) and (6.2.27) yields

$$\tfrac{1}{2} \frac{d}{dt} |||\boldsymbol{\Sigma} - \widetilde{\boldsymbol{\Sigma}}|||^2 \equiv \langle\!\langle \dot{\boldsymbol{\Sigma}} - \dot{\widetilde{\boldsymbol{\Sigma}}}, \boldsymbol{\Sigma} - \widetilde{\boldsymbol{\Sigma}} \rangle\!\rangle \leq 0. \tag{6.2.28}$$

Therefore, for any $t \in \mathbb{I}$,

$$|||\boldsymbol{\Sigma} - \widetilde{\boldsymbol{\Sigma}}||| - |||\boldsymbol{\Sigma}_0 - \widetilde{\boldsymbol{\Sigma}}_0||| = \int_0^t \frac{d}{d\tau} |||\boldsymbol{\Sigma}(\cdot, \tau) - \widetilde{\boldsymbol{\Sigma}}(\cdot, \tau)||| \, d\tau \leq 0, \tag{6.2.29}$$

which proves the result. □

Remarks 6.2.2.. An identical contractivity result holds for classical viscoplasticity. The corresponding IBVP is obtained by replacing inequality $(6.2.21a)_2$ with (6.2.18), and the flows $t \in \mathbb{I} \mapsto \boldsymbol{\Sigma} \in \mathcal{T}$ and $t \in \mathbb{I} \mapsto \widetilde{\boldsymbol{\Sigma}} \in \mathcal{T}$ associated with the two initial conditions (6.2.22) satisfy

$$\left.\begin{aligned} \langle\!\langle \dot{\boldsymbol{\Sigma}}, \boldsymbol{\Sigma} - \widetilde{\boldsymbol{\Sigma}} \rangle\!\rangle &\leq \ \ \langle\!\langle \dot{\boldsymbol{\Sigma}}^{\text{trial}}, \boldsymbol{\Sigma} - \widetilde{\boldsymbol{\Sigma}} \rangle\!\rangle + \frac{1}{\eta} \int_\Omega [g(J(\widetilde{\boldsymbol{\Sigma}})) - g(J(\boldsymbol{\Sigma}))] \, d\Omega, \\ -\langle\!\langle \dot{\widetilde{\boldsymbol{\Sigma}}}, \boldsymbol{\Sigma} - \widetilde{\boldsymbol{\Sigma}} \rangle\!\rangle &\leq -\langle\!\langle \dot{\widetilde{\boldsymbol{\Sigma}}}^{\text{trial}}, \boldsymbol{\Sigma} - \widetilde{\boldsymbol{\Sigma}} \rangle\!\rangle - \frac{1}{\eta} \int_\Omega [g(J(\widetilde{\boldsymbol{\Sigma}})) - g(J(\boldsymbol{\Sigma}))] \, d\Omega. \end{aligned}\right\} \tag{6.2.30}$$

Adding these inequalities again yields (6.2.26) and the contractivity result for viscoplasticity follows the proof of Proposition 2.1 exactly.

As in the nonlinear heat conduction equation, the contractivity property relative to the natural norm $|||\cdot|||$ motivates the algorithmic definition of stability employed subsequently.

6.2.2 The Algorithmic Problem

Let $\mathbb{I} = \cup_{n=0}^{N}[t_n, t_{n+1}]$ be a partition of the time interval $\mathbb{I} \subset \mathbb{R}_+$ of interest. To develop an algorithmic approximation to the IBVP (6.2.21), it suffices to consider a typical subinterval $[t_n, t_{n+1}] \subset \mathbb{I}$ and assume that the initial conditions $\boldsymbol{\Sigma}_n \in \mathcal{T}$ and $\boldsymbol{u}_n \in \mathcal{S}_n$ are given. Then the incremental algorithmic problem reduces to finding $\boldsymbol{\Sigma}_{n+1} \in \mathcal{T}$ and $\boldsymbol{u}_{n+1} \in \mathcal{S}_{n+1}$ for prescribed forcing function $\boldsymbol{f}(\cdot, t)$ and prescribed boundary conditions $\bar{\boldsymbol{u}}(\cdot, t), \bar{\boldsymbol{t}}(\cdot, t)$ for $t \in [t_n, t_{n+1}]$.

The construction of the algorithmic counterpart to the IBVP (6.2.21), which gives the approximations $\boldsymbol{\Sigma}_{n+1}$ and $\boldsymbol{u}_{n+1}$ to the exact values $\boldsymbol{\Sigma}(\cdot, t_{n+1})$ and $\boldsymbol{u}(\cdot, t_{n+1})$, is motivated as follows.

6.2.2.1 Algorithmic approximation

For a typical time interval $[t_n, t_{n+1}]$, define the generalized midpoint stress by the expression

$$\boldsymbol{\Sigma}_{n+\alpha} := \alpha\boldsymbol{\Sigma}_{n+1} + (1-\alpha)\boldsymbol{\Sigma}_n, \quad \alpha \in (0, 1]. \tag{6.2.31}$$

Let $\Delta\boldsymbol{u} := \boldsymbol{u}_{n+1} - \boldsymbol{u}_n$ be the incremental displacement field. Consistent with the generalized midpoint rule algorithm and in view of (6.2.14), set

$$\Delta t\, \dot{\boldsymbol{\Sigma}}^{\text{trial}}_{n+\alpha} = \big(\mathbf{C}\boldsymbol{\varepsilon}[\Delta\boldsymbol{u}], \mathbf{0}\big),$$

and

$$\Delta t\, \dot{\boldsymbol{\Sigma}}_{n+\alpha} = \boldsymbol{\Sigma}_{n+1} - \boldsymbol{\Sigma}_n. \tag{6.2.32}$$

Then, the algorithmic counterpart of variational inequality $(6.2.21a)_2$ becomes

$$\begin{aligned}
&\langle\!\langle \dot{\boldsymbol{\Sigma}}^{\text{trial}}_{n+\alpha} - \dot{\boldsymbol{\Sigma}}_{n+\alpha}, \boldsymbol{T} - \boldsymbol{\Sigma}_{n+\alpha}\rangle\!\rangle \\
&\quad = \frac{1}{\alpha\Delta t}\langle\!\langle \big(\alpha\mathbf{C}\boldsymbol{\varepsilon}[\Delta\boldsymbol{u}], \mathbf{0}\big) - \alpha[\boldsymbol{\Sigma}_{n+1} - \boldsymbol{\Sigma}_n], \boldsymbol{T} - \boldsymbol{\Sigma}_{n+\alpha}\rangle\!\rangle \\
&\quad = \frac{1}{\alpha\Delta t}\langle\!\langle \big(\boldsymbol{\sigma}_n + \alpha\mathbf{C}\boldsymbol{\varepsilon}[\Delta\boldsymbol{u}], \boldsymbol{q}_n\big) - \boldsymbol{\Sigma}_{n+\alpha}, \boldsymbol{T} - \boldsymbol{\Sigma}_{n+\alpha}\rangle\!\rangle.
\end{aligned} \tag{6.2.33}$$

Thus, if one defines $\boldsymbol{\Sigma}^{\text{trial}}_{n+\alpha}$ by the expression

$$\boldsymbol{\Sigma}^{\text{trial}}_{n+\alpha} := (\boldsymbol{\sigma}_n + \alpha\mathbf{C}\boldsymbol{\varepsilon}[\Delta\boldsymbol{u}], \boldsymbol{q}_n), \tag{6.2.34}$$

then inequality $(6.2.21a)_2$, along with approximation (6.2.33), leads to the variational inequality

$$\langle\!\langle \boldsymbol{\Sigma}^{\text{trial}}_{n+\alpha} - \boldsymbol{\Sigma}_{n+\alpha}, \boldsymbol{T} - \boldsymbol{\Sigma}_{n+\alpha}\rangle\!\rangle \le 0, \quad \forall \boldsymbol{T} \in \mathbb{E} \cap \mathcal{T}. \tag{6.2.35}$$

An identical argument for the viscoplastic problem yields

$$\langle\!\langle \boldsymbol{\Sigma}_{n+\alpha}^{\text{trial}} - \boldsymbol{\Sigma}_{n+\alpha}, \boldsymbol{T} - \boldsymbol{\Sigma}_{n+\alpha} \rangle\!\rangle \leq \frac{\alpha \Delta t}{\eta} \int_{\Omega} \{ g[J(\boldsymbol{T})] - g[J(\boldsymbol{\Sigma}_{n+\alpha})] \} \, d\Omega, \quad \forall \boldsymbol{T} \in \mathcal{T}, \tag{6.2.36}$$

which is the algorithmic counterpart of variational inequality (6.2.18).

6.2.2.2 The incremental algorithmic problem

As pointed out above, the generalized stresses $\boldsymbol{\Sigma} \in \mathcal{T}$ for continuum viscoplastic problems need not lie within the elastic domain $\mathbb{E}$, a fact also reflected in the algorithmic inequality (6.2.36). In addition, inequality (6.2.35) shows that, in the algorithmic version of rate-independent plasticity, except for the initial condition $\boldsymbol{\Sigma}_0$, in general, $\boldsymbol{\Sigma}_n$ and $\boldsymbol{\Sigma}_{n+1}$ need not be in the elastic domain. Only the algorithmic approximation $\boldsymbol{\Sigma}_{n+\alpha}$ is in $\mathbb{E} \cap \mathcal{T}$. This suggests the enforcement of the algorithmic counterpart of $(6.2.21a)_1$ also at $t_{n+\alpha}$, and leads to the following algorithmic problem:

Problem $W_{\Delta t}$: Find $\Delta \boldsymbol{u}$ and $\boldsymbol{\Sigma}_{n+\alpha} = (\boldsymbol{\sigma}_{n+\alpha}, \boldsymbol{q}_{n+\alpha})$ such that

$$\left.\begin{aligned} \langle \boldsymbol{\sigma}_{n+\alpha}, \boldsymbol{\varepsilon}[\boldsymbol{\eta}] \rangle - \langle \boldsymbol{f}_{n+\alpha}, \boldsymbol{\eta} \rangle - \langle \bar{\boldsymbol{t}}_{n+\alpha}, \boldsymbol{\eta} \rangle_{\Gamma} &= 0, \quad \forall \boldsymbol{\eta} \in \mathcal{V}, \\ \langle\!\langle \boldsymbol{\Sigma}_{n+\alpha}^{\text{trial}} - \boldsymbol{\Sigma}_{n+\alpha}, \boldsymbol{T} - \boldsymbol{\Sigma}_{n+\alpha} \rangle\!\rangle &\leq 0, \quad \forall \boldsymbol{T} \in \mathbb{E} \cap \mathcal{T}. \end{aligned}\right\} \tag{6.2.37}$$

The variational equations (6.2.37) determine the generalized stress $\boldsymbol{\Sigma}_{n+\alpha} \in \mathbb{E} \cap \mathcal{T}$ and the displacement increment $\Delta \boldsymbol{u}$. Then the initial condition $\boldsymbol{\Sigma}_n$ and the displacement field are updated by the formulas

$$\left.\begin{aligned} \boldsymbol{\Sigma}_{n+1} &= \frac{1}{\alpha} \boldsymbol{\Sigma}_{n+\alpha} + [1 - \frac{1}{\alpha}] \boldsymbol{\Sigma}_n, \\ &\text{and} \\ \boldsymbol{u}_{n+1} &= \boldsymbol{u}_n + \Delta \boldsymbol{u}. \end{aligned}\right\} \tag{6.2.38}$$

For viscoplasticity, equation $(6.2.37)_2$ is replaced by (6.2.36). The preceding algorithm was introduced in Simo and Govindjee [1991].

Remarks 6.2.3.

1. In general, the stress $\boldsymbol{\sigma}_{n+1}$ does not satisfy the equilibrium equation at t_{n+1} unless the forcing terms are linear in time. Hence, as in nonlinear heat conduction, the equilibrium equation $(6.2.21a)_1$ is enforced at $t_{n+\alpha}$.
2. The geometric interpretation of variational inequality $(6.2.37)_2$ is illustrated in Figure **6.3.** Using standard results in convex analysis, one concludes that the point $\boldsymbol{\Sigma}_{n+\alpha} \in \mathbb{E} \cap \mathcal{T}$ is the closest point projection in the natural norm $|||\cdot|||$ of the trial elastic state $\boldsymbol{\Sigma}_{n+\alpha}^{\text{trial}}$ onto $\mathbb{E} \cap \mathcal{T}$. In the context of an isoparametric finite-element approximation with numerical quadrature, the stresses need to be evaluated only at the quadrature points. Consequently, the infinite-dimensional, constrained, optimization problem defined by inequality $(6.2.37)_2$

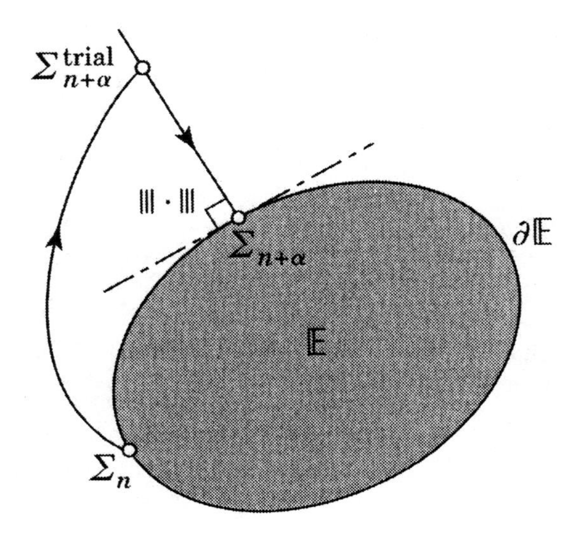

FIGURE 6-3. Geometric interpretation of the variational inequality $\langle\!\langle \boldsymbol{\Sigma}^{\text{trial}}_{n+\alpha} - \boldsymbol{\Sigma}_{n+\alpha}, \boldsymbol{T} - \boldsymbol{\Sigma}_{n+\alpha} \rangle\!\rangle \leq 0$.

reduces to a collection of *finite-dimensional*, constrained, optimization problems at quadrature points. This is the crucial observation exploited in actual numerical implementations; see Simo, Kennedy, and Govindjee [1988], Simo and Hughes [1987] and references therein.

3. The value $\alpha = \frac{1}{2}$ results in second-order accuracy and leads to an algorithm that preserves the second law exactly; see Simo and Govindjee [1991] for further details and numerical simulations.

The main objective of the following discussion is to study the nonlinear stability properties of problem $W_{\Delta t}$. It is shown below that the algorithmic solutions generated by $W_{\Delta t}$ are nonlinearly stable for $\alpha \geq \frac{1}{2}$.

6.2.3 Nonlinear Stability Analysis

Let $\boldsymbol{\Sigma}_n \in T$ and $\tilde{\boldsymbol{\Sigma}}_n \in T$ be two initial conditions at time t_n. Further, let

$$\begin{aligned} &\{\boldsymbol{\Sigma}_n\}, \\ \text{and} \qquad & \qquad\qquad\qquad\qquad (6.2.39) \\ &\{\tilde{\boldsymbol{\Sigma}}_n\}, \end{aligned}$$

be sequences generated by the preceding algorithm. As in nonlinear heat conduction, nonlinear stability holds if the algorithm inherits the contractivity property of the continuum problem relative to the natural norm $|||\cdot|||$; a property also known as A-contractivity or B-stability. Our main result is given in the following proposition

Proposition 2.2. The algorithm defined by variational problem $W_{\Delta t}$ and the update formulas (6.2.38) is A-contractive for $\alpha \geq \frac{1}{2}$, i.e., the following inequality holds:

$$\boxed{|||\boldsymbol{\Sigma}_n - \widetilde{\boldsymbol{\Sigma}}_n||| \leq |||\boldsymbol{\Sigma}_0 - \widetilde{\boldsymbol{\Sigma}}_0|||, \quad \forall n \geq 1,} \tag{6.2.40}$$

provided that $\alpha \geq \frac{1}{2}$.

PROOF. By hypothesis, (6.2.39) satisfy (6.2.37). Choosing $\boldsymbol{T} = \boldsymbol{\Sigma}_{n+\alpha}$ and $\boldsymbol{T} = \widetilde{\boldsymbol{\Sigma}}_{n+\alpha}$ and using the definition of $\boldsymbol{\Sigma}^{\text{trial}}_{n+\alpha}$,

$$\left.\begin{aligned} a(\alpha\mathbf{C}\varepsilon[\Delta\widetilde{\boldsymbol{u}}], \boldsymbol{\sigma}_{n+\alpha} - \widetilde{\boldsymbol{\sigma}}_{n+\alpha}) + \langle\!\langle \widetilde{\boldsymbol{\Sigma}}_n - \widetilde{\boldsymbol{\Sigma}}_{n+\alpha}, \boldsymbol{\Sigma}_{n+\alpha} - \widetilde{\boldsymbol{\Sigma}}_{n+\alpha}\rangle\!\rangle \leq 0, \\ -a(\alpha\mathbf{C}\varepsilon[\Delta\boldsymbol{u}], \boldsymbol{\sigma}_{n+\alpha} - \widetilde{\boldsymbol{\sigma}}_{n+\alpha}) - \langle\!\langle \boldsymbol{\Sigma}_n - \boldsymbol{\Sigma}_{n+\alpha}, \boldsymbol{\Sigma}_{n+\alpha} - \widetilde{\boldsymbol{\Sigma}}_{n+\alpha}\rangle\!\rangle \leq 0. \end{aligned}\right\} \tag{6.2.41}$$

Now set

$$\Delta\boldsymbol{\Sigma}_{n+\alpha} := \boldsymbol{\Sigma}_{n+\alpha} - \widetilde{\boldsymbol{\Sigma}}_{n+\alpha} \quad \text{for any } \alpha \in [0, 1]. \tag{6.2.42}$$

Add equations $(6.2.41)_{1,2}$, and use the notation in (6.2.42) to get

$$\langle\!\langle \Delta\boldsymbol{\Sigma}_{n+\alpha} - \Delta\boldsymbol{\Sigma}_n, \Delta\boldsymbol{\Sigma}_{n+\alpha}\rangle\!\rangle \leq \alpha a(\mathbf{C}\varepsilon[\Delta\boldsymbol{u} - \Delta\widetilde{\boldsymbol{u}}], \boldsymbol{\sigma}_{n+\alpha} - \widetilde{\boldsymbol{\sigma}}_{n+\alpha}). \tag{6.2.43}$$

Since the difference $\Delta\boldsymbol{u} - \Delta\widetilde{\boldsymbol{u}}$ is in $\mathcal{V}$, definition $(6.2.12a)_1$ along with the algorithmic weak form of the equilibrium equations yields

$$\begin{aligned} a(\mathbf{C}\varepsilon[\Delta\boldsymbol{u} - \Delta\widetilde{\boldsymbol{u}}], \boldsymbol{\sigma}_{n+\alpha} - \widetilde{\boldsymbol{\sigma}}_{n+\alpha}) = &\ \langle \boldsymbol{\sigma}_{n+\alpha}, \varepsilon[\Delta\boldsymbol{u} - \Delta\widetilde{\boldsymbol{u}}]\rangle \\ &- \langle \widetilde{\boldsymbol{\sigma}}_{n+\alpha}, \varepsilon[\Delta\boldsymbol{u} - \Delta\widetilde{\boldsymbol{u}}]\rangle = 0. \end{aligned} \tag{6.2.44}$$

Thus, the right-hand side of inequality (6.2.43) vanishes. By using the identity $\Delta\boldsymbol{\Sigma}_{n+\alpha} - \Delta\boldsymbol{\Sigma}_n = \alpha[\Delta\boldsymbol{\Sigma}_{n+1} - \Delta\boldsymbol{\Sigma}_n]$ along with (6.2.43) and (6.2.44), one concludes that

$$\langle\!\langle \Delta\boldsymbol{\Sigma}_{n+1} - \Delta\boldsymbol{\Sigma}_n, \Delta\boldsymbol{\Sigma}_{n+\alpha}\rangle\!\rangle \leq 0, \tag{6.2.45}$$

provided that $\alpha \in (0, 1]$. The proof is concluded by exploiting the identity

$$\Delta\boldsymbol{\Sigma}_{n+\alpha} = \Delta\boldsymbol{\Sigma}_{n+\frac{1}{2}} + (\alpha - \tfrac{1}{2})[\Delta\boldsymbol{\Sigma}_{n+1} - \Delta\boldsymbol{\Sigma}_n]. \tag{6.2.46}$$

Since $\Delta\boldsymbol{\Sigma}_{n+\frac{1}{2}} = \frac{1}{2}(\Delta\boldsymbol{\Sigma}_{n+1} + \Delta\boldsymbol{\Sigma}_n)$, combining (6.2.45) and (6.2.46) yields

$$\langle\!\langle \Delta\boldsymbol{\Sigma}_{n+1} + \Delta\boldsymbol{\Sigma}_n, \Delta\boldsymbol{\Sigma}_{n+1} - \Delta\boldsymbol{\Sigma}_n\rangle\!\rangle \leq -(2\alpha - 1)|||\Delta\boldsymbol{\Sigma}_{n+1} - \Delta\boldsymbol{\Sigma}_n, |||^2. \tag{6.2.47}$$

Using bilinearity along with definition (6.2.12) of the natural norm $|||\cdot|||$ gives the estimate

$$\begin{aligned} |||\Delta\boldsymbol{\Sigma}_{n+1}|||^2 - |||\Delta\boldsymbol{\Sigma}_n|||^2 &\leq -(2\alpha - 1)|||\Delta\boldsymbol{\Sigma}_{n+1} - \Delta\boldsymbol{\Sigma}_n|||^2 \\ &\leq 0, \quad \text{if} \quad \alpha \geq \tfrac{1}{2}, \end{aligned} \tag{6.2.48}$$

which implies (6.2.40) in conjunction with a straightforward induction argument. □

An argument analogous to that given above also shows that the incremental algorithmic problem for viscoplasticity is A-contractive relative to the natural norm $|||\cdot|||$.

6.3 Concluding Remarks

The analysis presented in this chapter settles the question of nonlinear stability of widely used algorithms for nonlinear heat conduction, infinitesimal plasticity, and classical viscoplasticity in the form considered by Duvaut and Lions [1972]. The accuracy of this class of return-mapping algorithms is examined in Simo and Govindjee [1991]. As expected, the midpoint rule ($\alpha = \frac{1}{2}$) is second-order accurate. The analysis for plasticity and viscoplasticity proves nonlinear stability for the generalized stresses $\boldsymbol{\Sigma} = (\boldsymbol{\sigma}, \boldsymbol{q})$. It appears, however, that an analogous result for the displacement field cannot be proved by the methods employed in this chapter. It should be noted that the contractivity property is certainly false for the case of perfect plasticity.

7

Nonlinear Continuum Mechanics and Phenomenological Plasticity Models

First in this chapter we review some basic results of nonlinear continuum mechanics needed for our subsequent development of nonlinear plasticity. Our account provides a brief overview of the relevant aspects of the theory and should, by *no* means, be considered exhaustive. Detailed expositions of the subject are found in the classical treatises of Truesdell and Toupin [1960]; Truesdell and Noll [1965]; and the more recent accounts of Gurtin [1981]; Marsden and Hughes [1994]; Ogden [1984]; and Ciarlet [1988]. However, we emphasize some aspects relevant to the formulation and implementation of widely used plasticity models.

In our presentation we have chosen to ignore any reference to the geometric structure which underlies continuum mechanics. For comprehensive expositions we refer to Marsden and Hughes [1994]; and Simo, Marsden, and Krishnaprasad [1988].

With the preceding background at hand, in Section **3** we examine a class of models of phenomenological plasticity which are widely used. Despite their shortcomings, we believe a discussion of these models provides a useful perspective on the current state of the art in computational plasticity. In addition, this class of models motivates our subsequent discussion, deferred to Chapter **9**, of more sound formulations of plasticity. The algorithmic treatment of the two representative models presented in Section **3** is considered in detail in the next chapter.

Readers familiar with the standard notation summarized in Sections **1** and **2** may wish to proceed directly to Section **3**.

7.1 Review of Some Basic Results in Continuum Mechanics

Below we summarize some basic results of nonlinear continuum mechanics relevant to our subsequent developments. For further details we refer to Ciarlet [1988] and Marsden and Hughes [1994].

7.1.1 Configurations. Basic Kinematics.

We let $\mathcal{B} \subset \mathbb{R}^3$ be the *reference configuration* of a continuum body with particles labeled by $\boldsymbol{X} \in \mathcal{B}$. For our purposes, it suffices to regard $\mathcal{B}$ as an open bounded set in $\mathbb{R}^3$. A smooth deformation is a one-to-one mapping:

$$\varphi : \mathcal{B} \longrightarrow \mathcal{S} \subset \mathbb{R}^3. \tag{7.1.1}$$

We refer to $\boldsymbol{x} \in \mathcal{S}$ as a point in the *current configuration* $\mathcal{S} = \varphi(\mathcal{B})$. The *deformation gradient* is the derivative of the deformation. We use the notation

$$\boldsymbol{F}(\boldsymbol{X}) = D\varphi(\boldsymbol{X}) = \frac{\partial \varphi(\boldsymbol{X})}{\partial \boldsymbol{X}}. \tag{7.1.2a}$$

The *local* condition of impenetrability of matter requires that

$$J(\boldsymbol{X}) := \det\left[\boldsymbol{F}(\boldsymbol{X})\right] > 0. \tag{7.1.2b}$$

In addition, the *right and left Cauchy–Green tensors* are defined as

$$\boldsymbol{C} = \boldsymbol{F}^T\boldsymbol{F}$$

and (7.1.3)

$$\boldsymbol{b} = \boldsymbol{F}\boldsymbol{F}^T,$$

respectively. According to the polar decomposition theorem, we recall that the deformation gradient at any $\boldsymbol{X} \in \mathcal{B}$ can be decomposed as

$$\boldsymbol{F}(\boldsymbol{X}) = \boldsymbol{R}(\boldsymbol{X})\boldsymbol{U}(\boldsymbol{X}) = \boldsymbol{V}\left[\varphi(\boldsymbol{X})\right]\boldsymbol{R}(\boldsymbol{X}), \tag{7.1.4}$$

where $\boldsymbol{R}(\boldsymbol{X})$ is a proper orthogonal tensor, called the *rotational tensor*, and $\boldsymbol{U}(\boldsymbol{X})$, $\boldsymbol{V}[\varphi(\boldsymbol{X})]$ are symmetric positive-definite tensors called the *right and left stretch tensors*, respectively. Omitting explicit indication of the argument,

$$\begin{aligned} \boldsymbol{R}\boldsymbol{R}^T &= \mathbf{1}, \\ \boldsymbol{U} &= \boldsymbol{C}^{\frac{1}{2}}, \\ \boldsymbol{V} &= \boldsymbol{b}^{\frac{1}{2}}. \end{aligned} \tag{7.1.5}$$

7.1.1.1 Spectral decomposition of the strain and rotation tensors.

Let I_A, $(A = 1, 2, 3)$ be the principal invariants of $\boldsymbol{C}$ (or $\boldsymbol{b}$), defined as

$$\begin{aligned} I_1 &:= \operatorname{tr} \boldsymbol{C}, \\ I_2 &:= \frac{1}{2}\left(I_1^2 - \operatorname{tr}\left[\boldsymbol{C}^2\right]\right), \\ I_3 &:= \det \boldsymbol{C}. \end{aligned} \tag{7.1.6}$$

Since $\boldsymbol{C}$ is symmetric and positive-definite, by the spectral theorem,

$$\boldsymbol{C} = \sum_{A=1}^{3} \lambda_A^2 \boldsymbol{N}^{(A)} \otimes \boldsymbol{N}^{(A)}, \quad \| \boldsymbol{N}^{(A)} \| = 1, \tag{7.1.7}$$

where $\lambda_A^2 > 0$ are the eigenvalues of $\boldsymbol{C}$, and $\boldsymbol{N}^{(A)}$ are the associated principal directions:

$$\boldsymbol{C}\boldsymbol{N}^{(A)} = \lambda_A^2 \boldsymbol{N}^{(A)} \quad ; \quad (A = 1, 2, 3). \tag{7.1.8}$$

Sometimes one calls the triad $\{\boldsymbol{N}^{(1)}, \boldsymbol{N}^{(2)}, \boldsymbol{N}^{(3)}\}$ the Lagrangian axes at $\boldsymbol{X} \in \mathcal{B}$ or the *principal directions* at $\boldsymbol{X}$. We also recall that

$$\boldsymbol{F}\boldsymbol{N}^{(A)} = \lambda_A \boldsymbol{n}^{(A)} \quad , \quad \| \boldsymbol{n}^{(A)} \| = 1. \tag{7.1.9}$$

The triad $\{\boldsymbol{n}^{(1)}, \boldsymbol{n}^{(2)}, \boldsymbol{n}^{(3)}\}$ is called the Eulerian axes at $\boldsymbol{x} = \varphi(\boldsymbol{X}) \in \mathcal{S}$, so that the spectral decomposition of $\boldsymbol{F}$ takes the form

$$\boldsymbol{F} = \sum_{A=1}^{3} \lambda_A \boldsymbol{n}^{(A)} \otimes \boldsymbol{N}^{(A)}. \tag{7.1.10}$$

λ_A, $(A = 1, 2, 3)$, are called as the *principal stretches* along the *principal directions* $\boldsymbol{N}^{(A)}$. Their squares (i.e., the eigenvalues of $\boldsymbol{C}$) are the solutions of the characteristic polynomial

$$p\left(\lambda^2\right) := \lambda^6 - I_1\lambda^4 + I_2\lambda^2 - I_3 = 0. \tag{7.1.11}$$

From (7.1.5) and (7.1.8), the spectral decompositions of the right and left stretch tensor are given by

$$\boldsymbol{U} = \sum_{A=1}^{3} \lambda_A \boldsymbol{N}^{(A)} \otimes \boldsymbol{N}^{(A)},$$

and (7.1.12)

$$\boldsymbol{V} = \sum_{A=1}^{3} \lambda_A \boldsymbol{n}^{(A)} \otimes \boldsymbol{n}^{(A)},$$

and the spectral decomposition of the rotation tensor takes the form

$$\boldsymbol{R} = \sum_{A=1}^{3} \boldsymbol{n}^{(A)} \otimes \boldsymbol{N}^{(A)}. \tag{7.1.13}$$

These decompositions are crucial in the closed-form numerical implementation of the polar decomposition discussed below.

7.1.1.2 Closed-form algorithm for the polar decomposition.

Recall that the roots of the characteristic polynomial (7.1.11) are found *explicitly* in *closed form* according to well-known formulas (see, e.g., Malvern [1969, pp. 91,92]). Only the principal directions remain to computed. The fact that these directions are obtained in *closed form* is a computationally important result which follows easily from Serrin's representation theorem, see Morman [1987].

Let $\{A, B, C\}$ be an even permutation of the indices $\{1, 2, 3\}$. We have the following three possibilities

i. *Three different principal stretches*: $\lambda_A \neq \lambda_B \neq \lambda_C$

$$\boxed{V = \sum_{A=1}^{3} \lambda_A \left[\frac{(\boldsymbol{b} - \lambda_B^2 \mathbf{1})(\boldsymbol{b} - \lambda_C^2 \mathbf{1})}{(\lambda_A^2 - \lambda_B^2)(\lambda_A^2 - \lambda_C^2)} \right].} \tag{7.1.14a}$$

ii. *Two equal principal stretches*: $\lambda := \lambda_1 = \lambda_2 \neq \lambda_3$

$$\boxed{V = \lambda \mathbf{1} + (\lambda_3 - \lambda) \left[\frac{(\boldsymbol{b} - \lambda^2 \mathbf{1})^2}{(\lambda_3^2 - \lambda^2)^2} \right].} \tag{7.1.14b}$$

iii. *Three equal principal stretches*: $\lambda := \lambda_1 = \lambda_2 = \lambda_3$

$$\boxed{V = \lambda \mathbf{1}.} \tag{7.1.14c}$$

Similar expressions are derived for $\boldsymbol{C}$ and $\boldsymbol{U}$ by noting the relationships

$$\boldsymbol{C} = \boldsymbol{R}^T \boldsymbol{b} \boldsymbol{R} \;\Rightarrow\; \boldsymbol{U} = \boldsymbol{R}^T \boldsymbol{V} \boldsymbol{R}, \tag{7.1.15}$$

so that, because $\boldsymbol{R}^T \boldsymbol{R} = \mathbf{1}$, (7.1.14) becomes

$$\boxed{\boldsymbol{U} = \begin{cases} \sum_{A=1}^{3} \lambda_A \dfrac{\left(\boldsymbol{C} - \lambda_B^2 \mathbf{1}\right)\left(\boldsymbol{C} - \lambda_C^2 \mathbf{1}\right)}{\left(\lambda_A^2 - \lambda_B^2\right)\left(\lambda_A^2 - \lambda_C^2\right)}, & \lambda_1 \neq \lambda_2 \neq \lambda_3, \\ \lambda \mathbf{1} + (\lambda_3 - \lambda) \dfrac{\left(\boldsymbol{C} - \lambda^2 \mathbf{1}\right)^2}{\left(\lambda_3^2 - \lambda^2\right)^2}, & \lambda := \lambda_1 \equiv \lambda_2 \neq \lambda_3, \\ \lambda \mathbf{1}, & \lambda := \lambda_1 \equiv \lambda_2 \equiv \lambda_3. \end{cases}} \tag{7.1.16}$$

The preceding formulas are used to obtain explicit expressions leading to a *closed-form algorithm* for *any isotropic function* of $\boldsymbol{C}$ and $\boldsymbol{b}$; see Simo and Taylor [1991]. These expressions, although explicit, depend on the number of repeated eigenvalues. However, for the case of the square root, the preceding formulas are combined into a *unified, singularity-free* expression which encompasses all three different cases. $\boldsymbol{U}$ can be written as (see Hoger and Carlson [1984*a*,*b*]; and Ting [1985])

$$\boxed{\boldsymbol{U} = \frac{1}{i_1 i_2 - i_3} \left[-\boldsymbol{C}^2 + (i_1^2 - i_2)\boldsymbol{C} + i_1 i_3 \mathbf{1} \right],} \tag{7.1.17a}$$

where i_A, $(A = 1, 2, 3)$ are the principal invariants of $\boldsymbol{U}$. A similar expression can be derived for the inverse tensor $\boldsymbol{U}^{-1}$, as given in BOX **7.1** where the closed-form algorithm for the polar decomposition is summarized. The derivation of this type of formula involves systematic use of the Cayley–Hamilton theorem. See also Franca [1989] for a robust algorithm for computing the square root of a 3×3 positive-definite matrix.

BOX 7.1. Algorithm for the Polar Decomposition.

i. Compute the squares of the principal stretches λ_A^2, $(A = 1, 2, 3)$ (the eigenvalues of $\boldsymbol{C}$) by solving (in closed form) the characteristic polynomial:

Set

$$b = I_2 - I_1^2/3$$

$$c = -\frac{2}{27}I_1^3 + \frac{I_1 I_2}{3} - I_3$$

IF ($|b| \leq \text{TOL}_3$) THEN:

$$x_A = -c^{1/3}$$

ELSE:

$$m = 2\sqrt{-b/3}$$

$$n = \frac{3c}{mb}$$

$$t = \arctan\left[\sqrt{1 - n^2}/n\right]^* /3$$

$$x_A = m \cos\left[t + 2(A-1)\pi/3\right]$$

ENDIF

$$\lambda_A^2 = x_A + I_1/3$$

ii. Compute the stretch tensor $\boldsymbol{U}$

Compute the invariants of $\boldsymbol{U}$

$$i_1 := \lambda_1 + \lambda_2 + \lambda_3$$

$$i_2 := \lambda_1\lambda_2 + \lambda_1\lambda_3 + \lambda_2\lambda_3$$

$$i_3 := \lambda_1\lambda_2\lambda_3$$

Set

$$D = i_1 i_2 - i_3$$

$$= (\lambda_1 + \lambda_2)(\lambda_1 + \lambda_3)(\lambda_2 + \lambda_3) > 0$$

$$\boxed{\begin{aligned} \boldsymbol{U} &= \frac{1}{D}\left[-\boldsymbol{C}^2 + (i_1^2 - i_2)\boldsymbol{C} + i_1 i_3 \mathbf{1}\right] \\ \boldsymbol{U}^{-1} &= \frac{1}{i_3}\left[\boldsymbol{C} - i_1\boldsymbol{U} + i_2\mathbf{1}\right] \end{aligned}}$$

iii. Compute the rotation tensor $\boldsymbol{R}$

$$\boldsymbol{R} = \boldsymbol{F}\boldsymbol{U}^{-1}$$

* The FORTRAN function datan2 $(\sqrt{1-n^2}, n) = \arctan(\sqrt{1-n^2}/n)$ is used instead of the arccosine function dacos (n) to avoid the ill conditioning of the latter function near the origin.

7.1.2 Motions. Lagrangian and Eulerian Descriptions.

A *motion* of a continuum body is a one-parameter family of configurations indexed by time. Explicitly, let $[0, T] \subset \mathbb{R}_+$ be the time interval of interest. Then, *for each* $t \in [0, T]$, the mapping

$$\varphi_t : \mathcal{B} \longrightarrow \mathcal{S}_t \subset \mathbb{R}^3 \tag{7.1.18}$$

is a deformation which maps the *reference configuration* $\mathcal{B}$ onto the configuration $\mathcal{S}_t \subset \mathbb{R}^3$ at time t. We write

$$\boldsymbol{x} = \varphi_t(\boldsymbol{X}) = \varphi(\boldsymbol{X}, t), \tag{7.1.19}$$

for the position of $\boldsymbol{X} \in \mathcal{B}$ at time t.

7.1.2.1 Lagrangian description.

The *material velocity*, denoted by $\boldsymbol{V}(\boldsymbol{X}, t)$, is the time derivative of the motion:*

$$\boldsymbol{V}(\boldsymbol{X}, t) = \frac{\partial \varphi(\boldsymbol{X}, t)}{\partial t}. \tag{7.1.20}$$

For fixed $t \in [0, T]$, we call $\boldsymbol{V}_t = \boldsymbol{V}(\bullet, t)$ the material velocity field at time t. Similarly, the *material acceleration* is defined as the time derivative of the material velocity:

$$\boldsymbol{A}(\boldsymbol{X}, t) = \frac{\partial \boldsymbol{V}(\boldsymbol{X}, t)}{\partial t} = \frac{\partial^2 \varphi(\boldsymbol{X}, t)}{\partial t^2}. \tag{7.1.21}$$

Again, for fixed $t \in [0, T]$, $\boldsymbol{A}_t = \boldsymbol{A}(\bullet, t)$ denotes the material acceleration field at time t. Note that the motion, the material velocity, and the material acceleration fields are associated with *material points* $\boldsymbol{X} \in \mathcal{B}$, and hence parameterized by *material coordinates*. In components we write

$$V_a(X_A, t) = \frac{\partial \varphi_a(X_A, t)}{\partial t}, \quad A_a(X_A, t) = \frac{\partial^2 \varphi_a(X_A, t)}{\partial t^2}, \tag{7.1.22}$$

where $\{X_A\} = (X_1, X_2, X_3)$ are the Cartesian coordinates of a *material point* $\boldsymbol{X} \in \mathcal{B}$, relative to an inertial frame, and $\{x_a\} = (x_1, x_2, x_3)$ are the coordinates of a point $\boldsymbol{x} \in \mathcal{S}_t$. Because the material coordinates $\{X_A\}$ in this description are the *independent* variables, one speaks of the *Lagrangian* or *material* description of the motion. See Figure 7.1.

For fixed $\boldsymbol{X} \in \mathcal{B}$, one refers to the mapping

$$t \in [0, T] \longmapsto \varphi_t(\boldsymbol{X})\big|_{X = \text{ fixed}} \tag{7.1.23}$$

as the trajectory of the material point $\boldsymbol{X}$ in the time interval $[0, T]$. Clearly, the material velocity field at a point is tangent to the trajectory of this point through time.

---

*Do not confuse the material velocity with the left stretch tensor (see (7.1.4)); the context will make clear which object is being considered.

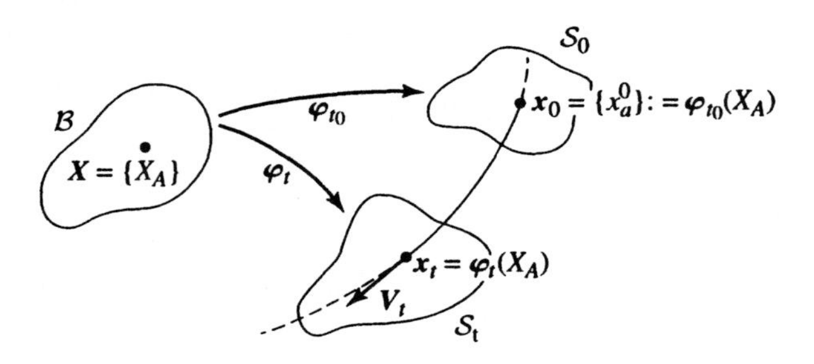

FIGURE 7.1. Lagrangian description of motion.

7.1.2.2 The spatial description of motion.

The *spatial or Eulerian description* can be viewed as obtained from the material description by changing the independent variables from material coordinates (of a particle) to positions in Euclidean space. Accordingly, at any time $t \in [0, T]$ one defines the *spatial velocity* and *acceleration fields*, denoted by $\boldsymbol{v}(\boldsymbol{x}, t)$ and $\boldsymbol{a}(\boldsymbol{x}, t)$, respectively, by the change of variables formulas

$$\left.\begin{aligned} \boldsymbol{x} &= \varphi(\boldsymbol{X}, t) \ \text{ i.e., } \ x_a = \varphi_a(X_A, t) \\ V_a(X_A, t) &= v_a\left(\varphi_b\left(X_A, t\right), t\right), \\ A_a(X_A, t) &= a_a\left(\varphi_b\left(X_A, t\right), t\right). \end{aligned}\right\} \tag{7.1.24}$$

Using direct notation, since $\varphi_t : \mathcal{B} \to \mathbb{R}^3$ maps particles $\boldsymbol{X} \in \mathcal{B}$ onto positions $\boldsymbol{x} = \varphi_t(\boldsymbol{X})$ at time $t \in [0, T]$, relationships (7.1.24) are written in compact form simply as

$$\boxed{\boldsymbol{V}_t = \boldsymbol{v}_t \circ \varphi_t}$$

and (7.1.25)

$$\boxed{\boldsymbol{A}_t = \boldsymbol{a}_t \circ \varphi_t,}$$

where $\circ$ denotes composition. Equivalently, since φ_t is the deformation for all t, condition (7.1.26), written below as

$$J_t := \det \boldsymbol{F}_t = \det [D\varphi_t] > 0 \ \text{ in } \mathcal{B}, \tag{7.1.26}$$

ensures that φ_t is *invertible*, i.e., the map $\varphi_t^{-1} : \mathcal{S}_t \subset \mathbb{R}^3 \to \mathcal{B}$ is well defined. Consequently, relationships (7.1.25) are written as

$$\boxed{\boldsymbol{v}_t = \boldsymbol{V}_t \circ \varphi_t^{-1},}$$

and (7.1.27)

$$\boxed{\boldsymbol{a}_t = \boldsymbol{A}_t \circ \varphi_t^{-1}.}$$

EXAMPLE: 7.1.1. As an illustration of the preceding relationships, let us compute the spatial acceleration in terms of the spatial velocity. From $(7.1.24)_{1,2}$ and (7.1.22),

$$\begin{aligned}
A_a(X_A, t) &= \frac{\partial V_a}{\partial t}(X_A, t) \\
&= \frac{\partial}{\partial t}\left\{v_a\left[\varphi_b(X_A, t), t\right]\right\} \\
&= \frac{\partial v_a\left[\varphi_b(X_A, t], t\right)}{\partial t} + \frac{\partial v_a\left[\varphi_b(X_A, t), t\right]}{\partial x_c}\frac{\partial \varphi_c(X_A, t)}{\partial t} \\
&= \left[\frac{\partial v_a(x_b, t)}{\partial t} + \frac{\partial v_a(x_b, t)}{\partial x_c} v_c(x_b, t)\right]_{x_b=\varphi_b(X_A, t)},
\end{aligned}$$

or, in direct notation,

$$A(X, t) = \left[\frac{\partial v(x, t)}{\partial t} + v(x, t) \cdot \nabla v(x, t)\right]_{x=\varphi(X, t)}. \tag{7.1.28b}$$

However, by (7.1.25),

$$A(X, t) = a(x, t)\big|_{x=\varphi(X, t)}. \tag{7.1.29}$$

Consequently, from $(7.1.27)_2$ and (7.1.29),

$$\boxed{a_t = \frac{\partial v_t}{\partial t} + v_t \cdot \nabla v_t,} \tag{7.1.30}$$

which is the desired expression. The tensor $\nabla v(x, t)$ with components $\partial v_a(x_b, t)/\partial x_b$ is called the *spatial velocity gradient*.

The *material time* derivative of a spatial object, such as the spatial velocity, the spatial acceleration or any other tensor function of the variables $(x, t) \in \mathbb{R}^3 \times [0, T]$, is the *time derivative holding the particle* (**not** its current position) *fixed*. For example, for the spatial velocity, we denote its material time derivative by $\dot{v}(x, t)$. Then, by definition,

$$\begin{aligned}
\dot{v}(x, t)\big|_{x=\varphi(X, t)} &:= \frac{\partial}{\partial t}\left\{v[\varphi(X, t), t]\right\}\big|_{X=\text{ Fixed}} \\
&= \frac{\partial}{\partial t} V(X, t)\big|_{X=\text{Fixed}} \\
&= A(X, t).
\end{aligned} \tag{7.1.31}$$

Therefore, by definition of spatial acceleration,

$$\dot{v}(x, t) = A(X, t)\big|_{X=\varphi_t^{-1}(x)} = a(x, t), \tag{7.1.32}$$

i.e., the material time derivative of the spatial velocity field is the spatial acceleration. In general if $\sigma(x, t)$ is a spatial tensor field, by definition, its material time derivative, denoted by $\dot{\boldsymbol{\sigma}}(\boldsymbol{x}, t)$, is obtained by the formula

$$\boxed{\dot{\boldsymbol{\sigma}} := \left[\frac{\partial}{\partial t}\left(\boldsymbol{\sigma} \circ \varphi_t\right)\right] \circ \varphi_t^{-1}.} \tag{7.1.33}$$

7.1.3 Rate of Deformation Tensors.

Below we summarize the expression for the rates of deformation tensors in several alternative descriptions of the motion.

7.1.3.1 Material and spatial descriptions.

We start our developments by computing the rate of change of the deformation gradient. From (7.1.2),

$$\frac{\partial \boldsymbol{F}}{\partial t} = \frac{\partial}{\partial t}\left[D\varphi_t\right] = D\frac{\partial \varphi_t}{\partial t} = D\boldsymbol{V}_t, \tag{7.1.34}$$

where $\boldsymbol{V}_t$ is the material velocity field given by (7.1.22). $D\boldsymbol{V}_t$ with components $\dfrac{\partial V_a}{\partial X_B}$ is called the *material velocity gradient*. By the chain rule,

$$D\boldsymbol{V}_t = D\left[\boldsymbol{v}_t \circ \varphi_t\right] = \nabla \boldsymbol{v}_t D\varphi_t = \nabla \boldsymbol{v}_t \boldsymbol{F}_t. \tag{7.1.35}$$

Combining (7.1.34) and (7.1.35), we arrive at the following expression for the *spatial velocity gradient* $\nabla \boldsymbol{v}_t$:

$$\boxed{\nabla \boldsymbol{v}_t = \frac{\partial \boldsymbol{F}_t}{\partial t}\boldsymbol{F}_t^{-1}.} \tag{7.1.36}$$

The symmetric part of $\nabla \boldsymbol{v}_t$, denoted by $\boldsymbol{d}_t$, is called the *spatial rate of deformation tensor*, and its skew-symmetric part is called the *spin, or vorticity, tensor*, denoted by $\boldsymbol{w}_t$. Thus

$$\boxed{\boldsymbol{d}_t = \frac{1}{2}\left[\nabla \boldsymbol{v}_t + \nabla \boldsymbol{v}_t^T\right],}$$

and (7.1.37)

$$\boxed{\boldsymbol{w}_t = \frac{1}{2}\left[\nabla \boldsymbol{v}_t - \nabla \boldsymbol{v}_t^T\right].}$$

The following relationship is useful in our algorithmic treatment of plasticity. By (7.1.35) and (7.1.3),

$$\begin{aligned} \frac{\partial}{\partial t}\boldsymbol{C}_t &= \frac{\partial \boldsymbol{F}_t^T}{\partial t}\boldsymbol{F}_t + \boldsymbol{F}_t^T\frac{\partial \boldsymbol{F}_t}{\partial t} \\ &= \boldsymbol{F}_t^T\left[(\nabla \boldsymbol{v}_t)^T + \nabla \boldsymbol{v}_t\right]\boldsymbol{F}_t. \end{aligned} \tag{7.1.38}$$

Consequently,

$$\boxed{\frac{\partial}{\partial t} \boldsymbol{C}_t = 2\boldsymbol{F}_t^T \boldsymbol{d}_t \boldsymbol{F}_t.} \tag{7.1.39}$$

Expression (7.1.39) justifies the name *material rate of deformation tensor* often given to $\frac{1}{2} \frac{\partial}{\partial t} \boldsymbol{C}_t$.

7.1.3.2 Rotated rate of deformation tensor.

We call the *local* configuration, obtained by applying the rotational tensor to a neighborhood $\mathcal{O}_x$ of a point $\boldsymbol{x}$ in the current configuration $\mathcal{S}_t$, the *rotated configuration*. Such a configuration is local in the sense that the "rotated neighborhoods" do not "fit together" unless the deformation of the body is homogeneous, see Figure 7.2.

Then we define the *rotated rate of deformation tensor* by the expression

$$\mathbf{D}(\boldsymbol{X}, t) = \boldsymbol{R}^T(\boldsymbol{X}, t)\boldsymbol{d}[\varphi(\boldsymbol{X}, t), t]\boldsymbol{R}(\boldsymbol{X}, t). \tag{7.1.40a}$$

Symbolically,

$$\boxed{\mathbf{D}_t = \boldsymbol{R}_t^T(\boldsymbol{d}_t \circ \varphi_t)\boldsymbol{R}_t.} \tag{7.1.40b}$$

An alternative expression is derived as follows. From (7.1.39) and the polar decomposition,

$$\frac{\partial}{\partial t} \boldsymbol{C}_t = 2\boldsymbol{U}_t \boldsymbol{R}_t^T \left(\boldsymbol{d}_t \circ \varphi_t\right) \boldsymbol{R}_t \boldsymbol{U}_t. \tag{7.1.41}$$

Therefore,

$$\boxed{\mathbf{D}_t = \frac{1}{2} \boldsymbol{C}_t^{-\frac{1}{2}} \frac{\partial \boldsymbol{C}_t}{\partial t} \boldsymbol{C}_t^{-\frac{1}{2}}.} \tag{7.1.42}$$

Finally, we introduce a skew-symmetric tensor that gives the rate of change of the rotation tensor. We set

$$\boxed{\boldsymbol{\Omega}_t := \left(\frac{\partial}{\partial t} \boldsymbol{R}_t\right) \boldsymbol{R}_t^T \;\Rightarrow\; \boldsymbol{\Omega}_t + \boldsymbol{\Omega}_t^T = \boldsymbol{0}.} \tag{7.1.43}$$

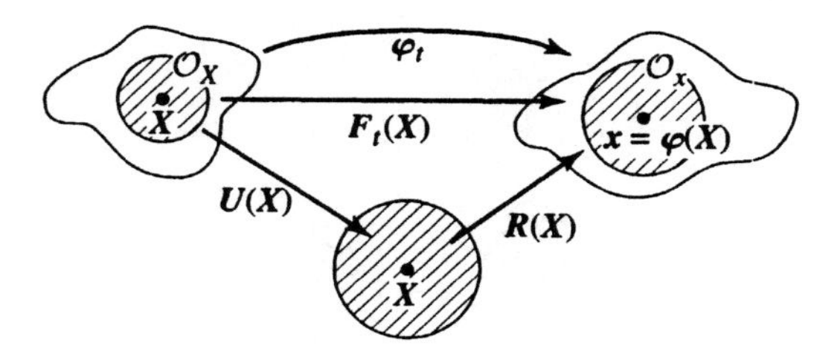

FIGURE 7.2. The rotated configuration of a *local* neighborhood $\mathcal{O}_x \subset \mathcal{B}$.

We observe that $\boldsymbol{\Omega}_t$ does not coincide with the spin tensor $\boldsymbol{w}_t$. In fact, a direct calculation shows that

$$\begin{aligned}\frac{\partial \boldsymbol{F}_t}{\partial t} &= \frac{\partial \boldsymbol{R}_t}{\partial t}\boldsymbol{R}_t^T\boldsymbol{R}_t\boldsymbol{U}_t + \boldsymbol{R}_t\frac{\partial \boldsymbol{U}_t}{\partial t}\boldsymbol{U}_t^{-1}\boldsymbol{R}_t^T\boldsymbol{F}_t \\ &= \left[\boldsymbol{\Omega}_t + \boldsymbol{R}_t\left(\frac{\partial \boldsymbol{U}_t}{\partial t}\boldsymbol{U}_t^{-1}\right)\boldsymbol{R}_t^T\right]\boldsymbol{F}_t.\end{aligned} \tag{7.1.44}$$

Combining (7.1.36), $(7.1.37)_2$, and (7.1.44), we find that

$$\boxed{\boldsymbol{w}_t = \boldsymbol{\Omega}_t + \boldsymbol{R}_t\text{skew}\left[\left(\frac{\partial}{\partial t}\boldsymbol{U}_t\right)\boldsymbol{U}_t^{-1}\right]\boldsymbol{R}_t^T,} \tag{7.1.45}$$

where skew[·] indicates the skew-symmetric part of the second-order tensor [·].

7.1.4 Stress Tensors. Equations of Motion.

Here, we summarize some of the basic stress tensors and their relationships relevant to the alternative descriptions of continuum mechanics.

We let $\boldsymbol{\sigma}$ be the *Cauchy* stress tensor and $\boldsymbol{\tau} := J\boldsymbol{\sigma}$ be the *Kirchhoff* stress tensor. These objects are *symmetric tensors* defined on the *current configuration* of the body.

In addition we denote by $\boldsymbol{P}$ the *nonsymmetric nominal* stress tensor, also known as the first Piola-Kirchhoff tensor. This tensor is relevant to a Lagrangian description of continuum mechanics. We have the relationships

$$\boxed{\boldsymbol{\tau}\circ\boldsymbol{\varphi} = \boldsymbol{P}\boldsymbol{F}^T,}$$

and (7.1.46)

$$\boxed{\boldsymbol{\tau} = \left(J\circ\boldsymbol{\varphi}^{-1}\right)\boldsymbol{\sigma}.}$$

Finally, we let $\boldsymbol{S}$ be the *symmetric (or second) Piola–Kirchhoff* tensor defined as

$$\boxed{\boldsymbol{S} = \boldsymbol{F}^{-1}\boldsymbol{P} = \boldsymbol{F}^{-1}\left(\boldsymbol{\tau}\circ\boldsymbol{\varphi}\right)\boldsymbol{F}^{-T}.} \tag{7.1.47}$$

In index notation, the component expressions are

$$\tau_{ab} = F_{aA}S_{AB}F_{bB} = P_{aA}F_{bA}. \tag{7.1.48}$$

Relative to the rotated configuration introduced above, we define the *rotated stress tensor* by the expression

$$\boxed{\boldsymbol{\Sigma} = \boldsymbol{R}^T\boldsymbol{\tau}\boldsymbol{R},}$$

i.e., (7.1.49)

$$\Sigma_{AB} = R_{aA}\tau_{ab}R_{bB}.$$

All of the stress tensors introduced above are conjugate to associated rate of deformation tensors through the following important stress-power relationships:

$$\tau_{ab} d_{ab} = P_{aA} \dot{F}_{aA} = \Sigma_{AB} D_{AB} = \tfrac{1}{2} S_{AB} \dot{C}_{AB}, \tag{7.1.50a}$$

or, in direct notation,

$$\boxed{\boldsymbol{\tau} : \boldsymbol{d} = \boldsymbol{P} : \dot{\boldsymbol{F}} = \boldsymbol{\Sigma} : \mathbf{D} = \tfrac{1}{2} \boldsymbol{S} : \dot{\boldsymbol{C}}.} \tag{7.1.50b}$$

Remark 7.1.1. In elasticity, one often introduces a nonsymmetric stress tensor defined by the relationship

$$\boxed{\boldsymbol{T} := \boldsymbol{R}^T \boldsymbol{P} = \boldsymbol{U}\boldsymbol{S},} \tag{7.1.51}$$

called the *Biot stress*. By the polar decomposition theorem,

$$\begin{aligned} \boldsymbol{P} : \dot{\boldsymbol{F}} &= \boldsymbol{P} : \left(\dot{\boldsymbol{R}} \boldsymbol{R}^T \boldsymbol{R} \boldsymbol{U} + \boldsymbol{R} \dot{\boldsymbol{U}} \right) \\ &= \boldsymbol{P} \boldsymbol{F}^T : \boldsymbol{\Omega} + \boldsymbol{R}^T \boldsymbol{P} : \dot{\boldsymbol{U}} \\ &= \boldsymbol{\tau} : \boldsymbol{\Omega} + \operatorname{sym}[\boldsymbol{T}] : \dot{\boldsymbol{U}}, \end{aligned} \tag{7.1.52}$$

where sym[·] indicates the symmetric part. Since $\boldsymbol{\tau}$ is symmetric and $\boldsymbol{\Omega} := \dot{\boldsymbol{R}} \boldsymbol{R}^T$ is skew-symmetric, it follows that $\boldsymbol{\tau} : \boldsymbol{\Omega} = \boldsymbol{0}$. Consequently,

$$\boxed{\boldsymbol{P} : \dot{\boldsymbol{F}} = \boldsymbol{T}^s : \dot{\boldsymbol{U}},} \tag{7.1.53}$$

that is, the (symmetric part) of $\boldsymbol{T}$ is conjugate to the right stretch tensor. The skew-symmetric part of the Biot stress is a *reaction force* because it does not enter in the expression for the stress power. Other stress tensors and conjugate strain measures are introduced by a formalism essentially due to Hill; see Ogden [1984].

7.1.4.1 Equations of motion.

Next, we summarize the equations of motion in local form relevant to numerical implementation by the finite-element method.

i. *Lagrangian description.* We let $\partial\mathcal{B}$ be the boundary of $\mathcal{B}$ and assume that the deformation is prescribed on $\partial_\varphi \mathcal{B} \subset \partial\mathcal{B}$ as

$$\varphi = \bar{\varphi} \text{ (prescribed) on } \partial_\varphi \mathcal{B}, \tag{7.1.54}$$

whereas the *nominal traction vector* $\boldsymbol{t}^N$ is prescribed on the part of the boundary $\partial_t \mathcal{B} \subset \partial\mathcal{B}$, with unit normal $\hat{\boldsymbol{N}}$ as

$$\boldsymbol{t}^N := \boldsymbol{P}\hat{\boldsymbol{N}} = \bar{\boldsymbol{t}}^N \text{ (prescribed) on } \partial_t \mathcal{B}. \tag{7.1.55}$$

Of course, one assumes that $\partial_\varphi \mathcal{B} \cap \partial_t \mathcal{B} = \emptyset$ and $\overline{\partial_\varphi \mathcal{B} \cup \partial_t \mathcal{B}} = \overline{\partial\mathcal{B}}$. Then the local equations of motion take the form

$$\boxed{\operatorname{DIV} \boldsymbol{P} + \rho_0 \boldsymbol{B} = \rho_0 \boldsymbol{A}_t, \quad \text{in} \quad \mathcal{B},} \tag{7.1.56}$$

where $\rho_0 : \mathcal{B} \to \mathbb{R}$ is the reference density and $\boldsymbol{B}$ the body force. Here,

$$(\mathrm{DIV}\, \boldsymbol{P})_a = \frac{\partial P_{aA}}{\partial X_A}. \qquad (7.1.57).$$

ii. *Eulerian description.* The counterpart of equation (7.1.56) in the Eulerian description takes the form

$$\boxed{\mathrm{div}\, \boldsymbol{\sigma} + \rho_t \boldsymbol{b}_t = \rho_t \boldsymbol{a}_t \quad \text{in} \quad \mathcal{S}_t = \varphi_t(\mathcal{B}),} \qquad (7.1.58)$$

where $\rho_t = (\frac{\rho_0}{J}) \circ \varphi_t^{-1}$ is the density in the current placement $\mathcal{S}_t$ and $\boldsymbol{b}_t = \boldsymbol{B} \circ \varphi_t^{-1}$ is the body force per unit of volume in the current placement $\mathcal{S}_t$. In a Cartesian coordinate system,

$$(\mathrm{div}\, \boldsymbol{\sigma})_a = \frac{\partial \sigma_{ab}}{\partial x_b}. \qquad (7.1.59)$$

In what follows, attention is focused on quasi-static loading so that the inertial terms on the right-hand side of (7.1.56) and (7.1.58) are neglected.

7.1.5 Objectivity. Elastic Constitutive Equations.

We continue our overview of continuum mechanics with a brief discussion of the notion of objectivity, one of the most fundamental principles of mechanics, along with the statement of the classical constitutive equations for elasticity.

7.1.5.1 Superposed rigid body motions. Objective transformation.

Let $\varphi : \mathcal{B} \times [0, T] \to \mathbb{R}$ be a given motion, and let $\boldsymbol{x} = \varphi(\boldsymbol{X}, t)$ be the position of a particle $\boldsymbol{X} \in \mathcal{B}$ at time $t \in [0, T]$ in the current placement $\mathcal{S}_t = \varphi_t(\mathcal{B})$. Consider a *superposed rigid body* motion, i.e., a map

$$\boldsymbol{x} \in \mathcal{S}_t \longmapsto \boldsymbol{x}^+ = \boldsymbol{c}(t) + \boldsymbol{Q}(t)\boldsymbol{x} \in \mathbb{R}^3, \qquad (7.1.60)$$

where $\boldsymbol{c}(t)$ is function of time and $\boldsymbol{Q}(t)$ is a proper orthogonal transformation depending only on time. We write $\boldsymbol{Q}(t) \in \mathrm{SO}(3)$, the *proper orthogonal group*. The superposed motion is called rigid because, given any two points $\boldsymbol{x}_1, \boldsymbol{x}_2 \in \mathcal{S}_t$, since $\boldsymbol{Q}(t)$ is orthogonal,

$$\boldsymbol{x}_1^+ - \boldsymbol{x}_2^+ = \boldsymbol{Q}(t)\,[\boldsymbol{x}_1 - \boldsymbol{x}_2] \Rightarrow \| \boldsymbol{x}_1^+ - \boldsymbol{x}_2^+ \|^2 = \| \boldsymbol{x}_1 - \boldsymbol{x}_2 \|^2, \qquad (7.1.61)$$

where $\| \boldsymbol{x}_1 - \boldsymbol{x}_2 \|^2 = (\boldsymbol{x}_1 - \boldsymbol{x}_2) \cdot (\boldsymbol{x}_1 - \boldsymbol{x}_2)$ is the square of the Euclidean distance. Thus, (7.1.60) *preserves distances* and, therefore, is rigid (i.e., an isometry). A **spatial** tensor field is said to *transform objectively* under superposed rigid body motions if it transforms according to the standard rules of tensor analysis.

EXAMPLE: 7.1.2. The total motion obtained by composition of (7.1.60) with the given motion is

$$\boldsymbol{x}^+ = \varphi^+(\boldsymbol{X}, t) := \boldsymbol{c}(t) + \boldsymbol{Q}(t)\varphi(\boldsymbol{X}, t). \qquad (7.1.62)$$

Thus the deformation gradient becomes

$$\boldsymbol{F}_t^+ = D\varphi_t^+ = \boldsymbol{Q}(t)D\varphi(\boldsymbol{X}, t) \equiv \boldsymbol{Q}(t)\boldsymbol{F}_t. \tag{7.1.63}$$

Therefore, the spatial velocity gradient is given by

$$\nabla^+\boldsymbol{v}_t^+ = \left(\frac{\partial \boldsymbol{F}_t^+}{\partial t}\right)(\boldsymbol{F}_t^+)^{-1} = \boldsymbol{Q}(t)\nabla\boldsymbol{v}_t\boldsymbol{Q}^T(t) + \dot{\boldsymbol{Q}}(t)\boldsymbol{Q}^T(t), \tag{7.1.64}$$

which *does not* transform objectively because of the additional *skew-symmetric* term $\dot{\boldsymbol{Q}}(t)\boldsymbol{Q}^T(t)$. However, its symmetric part, the rate of deformation tensor, transforms objectively since, from (7.1.64) and $(7.1.37)_1$,

$$\boxed{\boldsymbol{d}_t^+ = \boldsymbol{Q}(t)\boldsymbol{d}_t\boldsymbol{Q}^T(t).} \tag{7.1.65}$$

Note that the spin tensor, defined by $(7.1.37)_2$, transforms according to the nonobjective rule

$$\boxed{\boldsymbol{w}_t^+ = \boldsymbol{Q}(t)\boldsymbol{w}_t\boldsymbol{Q}^T(t) + \dot{\boldsymbol{Q}}(t)\boldsymbol{Q}^T(t)\,.} \tag{7.1.66}$$

Remark 7.1.2. Material objects, that is tensor fields on the reference configuration, remain *unaltered under spatially superposed rigid body motions*. For example, from (7.1.63),

$$\boldsymbol{C}_t^+ = (\boldsymbol{F}_t^+)^T\boldsymbol{F}_t^+ = \boldsymbol{F}_t^T\boldsymbol{Q}^T(t)\boldsymbol{Q}(t)\boldsymbol{F}_t \equiv \boldsymbol{C}_t. \tag{7.1.67}$$

Similarly, $\dot{\boldsymbol{C}}_t^+ = \dot{\boldsymbol{C}}_t$ is unaltered by spatially superposed rigid body motions. Figure 7.3 may prove useful for understanding this result.

7.1.5.2 Objective stress rates.

Assuming that the Cauchy stress tensor is objective, its material time derivative is *not objective*. To see this, first we compute $\dot{\boldsymbol{\sigma}}_t$ as follows. By definition (7.1.33)

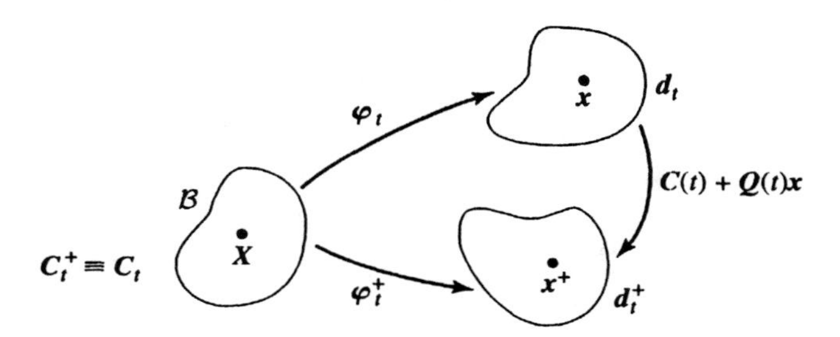

FIGURE 7.3. Illustration of superposed rigid body motions.

and the chain rule,

$$\begin{aligned}\dot{\boldsymbol{\sigma}}_t &= \left[\frac{\partial}{\partial t}\left(\boldsymbol{\sigma}_t \circ \varphi_t\right)\right] \circ \varphi_t^{-1} \\ &= \left[\frac{\partial \boldsymbol{\sigma}_t}{\partial t} \circ \varphi_t + \left(\frac{\partial \boldsymbol{\sigma}_t}{\partial x} \circ \varphi_t\right) \frac{\partial \varphi_t}{\partial t}\right] \circ \varphi_t^{-1} \\ &= \frac{\partial \boldsymbol{\sigma}_t}{\partial t} + \nabla \boldsymbol{\sigma}_t \left(\boldsymbol{V}_t \circ \varphi_t^{-1}\right),\end{aligned} \tag{7.1.68}$$

so that, by (7.1.27),

$$\boxed{\dot{\boldsymbol{\sigma}}_t = \frac{\partial \boldsymbol{\sigma}_t}{\partial t} + \boldsymbol{v}_t \cdot \nabla \boldsymbol{\sigma}_t .} \tag{7.1.69a}$$

In component form, expression (7.1.69a) reads

$$\boxed{\dot{\sigma}_{ab}\left(\boldsymbol{x}, t\right) = \frac{\partial \sigma_{ab}(\boldsymbol{x}, t)}{\partial t} + \frac{\partial \sigma_{ab}(\boldsymbol{x}, t)}{\partial x_c} v_c(\boldsymbol{x}, t).} \tag{7.1.69b}$$

Next, assume that $\boldsymbol{\sigma}_t$ transforms objectively, that is

$$\boldsymbol{\sigma}_t^+ = \boldsymbol{Q}(t)\boldsymbol{\sigma}_t\boldsymbol{Q}^T(t). \tag{7.1.70}$$

Then, from (7.1.69) and (7.1.70), one easily finds that

$$\dot{\boldsymbol{\sigma}}_t^+ = \boldsymbol{Q}(t)\dot{\boldsymbol{\sigma}}_t\boldsymbol{Q}^T(t) + \left[\dot{\boldsymbol{Q}}(t)\boldsymbol{Q}^T(t)\right]\boldsymbol{\sigma}_t^+ - \boldsymbol{\sigma}_t^+\left[\dot{\boldsymbol{Q}}(t)\boldsymbol{Q}^T(t)\right] \tag{7.1.71}$$

which is clearly nonobjective.

Objective rates are essentially modified time derivatives of the Cauchy stress tensor constructed to preserve objectivity. A large body of literature is concerned with the development of objective rates which, remarkably, extends to recent dates. Concerning the practically infinite number of proposals made, we remark, following Truesdell and Noll [1965, p. 404], that "· · · Despite claims and whole papers to the contrary, any advantage claimed for one such rate over another is pure illusion."

In fact, one can show that any possible objective stress rate is a particular case of a fundamental geometric object known as the *Lie derivative*; see Marsden and Hughes [1994, Chapter **1**]; Arnold [1978]; and Simo and Marsden [1984]. Below we illustrate some of the proposals found in the literature.

i. *The Lie derivative*, of the Kirchhoff stress tensor, also known (up to a factor of J_t) as the Truesdell stress rate, is defined as

$$\begin{aligned}L_v\boldsymbol{\tau}_t :&= \left\{\boldsymbol{F}_t \frac{\partial}{\partial t}\left[\boldsymbol{F}_t^{-1}\left(\boldsymbol{\tau}_t \circ \varphi_t\right)\boldsymbol{F}_t^{-T}\right]\boldsymbol{F}_t^T\right\} \circ \varphi_t^{-1} \\ &= \left\{\boldsymbol{F}_t\left[\frac{\partial}{\partial t}\boldsymbol{S}_t\right]\boldsymbol{F}_t^T\right\} \circ \varphi_t^{-1}.\end{aligned} \tag{7.1.72}$$

Using the expression for the derivative of the inverse,

$$\frac{\partial}{\partial t}\left(\boldsymbol{F}_t^{-1}\right) = -\boldsymbol{F}_t^{-1}\frac{\partial \boldsymbol{F}_t}{\partial t}\boldsymbol{F}_t^{-1}, \tag{7.1.73}$$

along with the definition of material time derivative and spatial velocity gradient,

$$\boxed{L_v \tau_t = \dot{\tau}_t - (\nabla v_t)\, \tau_t - \tau_t\, (\nabla v_t)^T .} \tag{7.1.74}$$

One can easily show that (7.1.74) is objective (in fact, $L_v \tau_t$ meets the much stronger condition of *covariance*, see Marsden and Hughes [1994]).

ii. *The Jaumann–Zaremba* stress rate of the Kirchhoff stress is essentially a corotated derivative relative to spatial axes with instantaneous velocity given by the spin tensor (i.e., the *vorticity*):

$$\boxed{\overset{\nabla}{\tau} = \dot{\tau}_t - w_t \tau_t + \tau_t w_t .} \tag{7.1.75}$$

iii. *The Green–McInnis–Naghdi* stress rate of the Kirchhoff stress is defined by an expression similar to (7.1.72), but with F_t replaced by R_t:

$$\begin{aligned} \overset{\square}{\tau}_t &= \left\{ R_t \frac{\partial}{\partial t} \left[R_t^T \left(\tau_t \circ \varphi_t \right) R_t \right] R_t^T \right\} \circ \varphi_t^{-1} \\ &= \left\{ R_t \frac{\partial}{\partial t} \left[\Sigma_t \circ \varphi_t \right] R_t^T \right\} \circ \varphi_t^{-1}. \end{aligned} \tag{7.1.76}$$

Recalling that $\dot{R}_t = (\Omega_t \circ \varphi_t) R_t$,

$$\boxed{\overset{\square}{\tau}_t = \dot{\tau}_t - \Omega_t \tau_t + \tau_t \Omega_t .} \tag{7.1.77}$$

Although (7.1.77) is similar in structure to (7.1.75), we remark that $\Omega_t \neq w_t$ unless $\dot{U}_t = 0$ (i.e., an instantaneously rigid motion); see (7.1.45).

For a catalog of the many proposed objective stress rates, up to the early 60s, see Truesdell and Toupin [1960, Section 48, page 151].

7.1.5.3 Objectivity of the constitutive response; frame indifference.

We illustrate this fundamental notion within the context of elasticity restricted to the purely mechanical theory; see Truesdell and Noll [1965, Section 17–19] for a detailed discussion in a general context.

Recall that the constitutive equation for a hyperelastic material is defined in terms of a stored-energy function which depends on the deformation *locally only through the deformation gradient*, that is, a function $W(X, F(X, t))$ is given such that

$$\boxed{P(X, t) = \frac{\partial W[X, F(X, t)]}{\partial F} .} \tag{7.1.78}$$

The stored-energy function $W(X, F)$ is said to be *objective or frame indifferent* if the following condition holds. Let $\varphi(X, t)$ be an arbitrary motion and consider a superposed rigid body motion of the form (7.1.60), so that the resulting deformation gradient is given by (7.1.63), i.e., we let

$$F^+(X, t) := Q(t) F(X, t), \tag{7.1.79}$$

for any proper, orthogonal, time-dependent transformation $\boldsymbol{Q}(t) \in \mathrm{SO}(3)$. If no restrictions are placed on the stored-energy function $W(\boldsymbol{X}, \boldsymbol{F})$, in general, $W(\boldsymbol{X}, \boldsymbol{F}^+)$ need not coincide with $W(\boldsymbol{X}, \boldsymbol{F})$. The requirement of objectivity demands that $W(\boldsymbol{X}, \boldsymbol{F})$ be (left) rotationally invariant in the sense that

$$\boxed{W(\boldsymbol{X}, \boldsymbol{QF}) = W(\boldsymbol{X}, \boldsymbol{F}),} \tag{7.1.80}$$

for all possible proper orthogonal transformations $\boldsymbol{Q} \in \mathrm{SO}(3)$. From restriction (7.1.80), it easily follows that W depends on $\boldsymbol{F}$ only through $\boldsymbol{C} = \boldsymbol{F}^T\boldsymbol{F}$, i.e.,

$$\boxed{W(\boldsymbol{X}, \boldsymbol{F}) = \bar{W}(\boldsymbol{X}, \boldsymbol{C}).} \tag{7.1.81}$$

Note that (7.1.81) automatically satisfies the restriction placed by objectivity since, as noted in (7.1.67), $\boldsymbol{C}_t^+ \equiv \boldsymbol{C}_t$ under superposed rigid body motions.

From the reduced constitutive function (7.1.81) and (7.1.78), one obtains the following classical constitutive equations for elasticity

$$\boxed{\begin{aligned} \boldsymbol{S} &= 2\frac{\partial \bar{W}}{\partial \boldsymbol{C}}, \\ \boldsymbol{P} &= 2\boldsymbol{F}\frac{\partial \bar{W}}{\partial \boldsymbol{C}}, \\ &\text{and} \\ \boldsymbol{\tau} &= 2\boldsymbol{F}\frac{\partial \bar{W}}{\partial \boldsymbol{C}}\boldsymbol{F}^T. \end{aligned}} \tag{7.1.82}$$

where for clarity, explicit indication of the argument is omitted. We often follow this practice in subsequent developments.

Objectivity of the stored-energy function is closely related to balance of angular momentum, i.e., to the symmetry of the Cauchy stress tensor. In fact, one can easily show that $W(\boldsymbol{X}, \boldsymbol{F})$ is *objective* (i.e., (7.1.80) holds) *if and only if* the balance of angular momentum condition $\boldsymbol{PF}^T = \boldsymbol{FP}^T$ holds (with $\boldsymbol{P}$ given by (7.1.78)).

7.1.5.4 Hyperelastic rate constitutive equations.

Time differentiation of relationship $(7.1.82)_1$ yields

$$\boxed{\dot{\boldsymbol{S}} = \mathbf{C} : \tfrac{1}{2}\dot{\boldsymbol{C}},}$$

i.e. (7.1.83)

$$\dot{S}_{AB} = \mathsf{C}_{ABCD}\, \tfrac{1}{2}\dot{C}_{CD},$$

where $\mathbf{C}(\boldsymbol{X}, \boldsymbol{C}_t)$ is the material elasticity tensor given by

$$\boxed{\mathbf{C} = 4\frac{\partial^2 \bar{W}}{\partial \boldsymbol{C} \partial \boldsymbol{C}},}$$

i.e. (7.1.84)

$$\boxed{\mathsf{C}_{ABCD} = 4\frac{\partial^2 \bar{W}}{\partial C_{AB}\partial C_{CD}}}.$$

By recalling that $L_v \boldsymbol{\tau}_t = [\boldsymbol{F}_t \dot{\boldsymbol{S}}_t \boldsymbol{F}_t^T] \circ \varphi_t^{-1}$, along with the relationship $\dot{\boldsymbol{C}}_t = 2\boldsymbol{F}_t^T \boldsymbol{d}_t \boldsymbol{F}_t$, expression (7.1.83) leads to

$$\begin{aligned}
(L_v \boldsymbol{\tau}_t)_{ab} &= F_{aA} \dot{S}_{AB} F_{bB} \\
&= F_{aA} \left[\mathsf{C}_{ABCD} F_{cC} d_{cd} F_{dD}\right] F_{bB} \\
&= \left[F_{aA} F_{bB} F_{cC} F_{dD} \mathsf{C}_{ABCD}\right] d_{cd} \\
&= \mathsf{c}_{abcd} d_{cd}, \qquad (7.1.85)
\end{aligned}$$

where $\mathbf{c}$ with components c_{abcd} is the spatial elasticity tensor related to $\mathbf{C}$ by the (push-forward) transformation†

$$\boxed{\mathsf{c}_{abcd} \circ \varphi_t = F_{aA} F_{bB} F_{cC} F_{dD} \mathsf{C}_{ABCD}.} \qquad (7.1.86)$$

Note that (7.1.85) results in the spatial rate-constitutive equation

$$\boxed{L_v \boldsymbol{\tau}_t = \mathbf{c} : \boldsymbol{d}_t,}$$

i.e. (7.1.87)

$$(L_v \boldsymbol{\tau})_{ab} = \mathsf{c}_{abcd} d_{cd}.$$

Also note that analogous expressions can be derived for any objective stress rate other than the Lie derivative. The derivation of hyperelastic rate equations for other stress and strain measures constitutes an exercise (often nontrivial) involving the application of the chain rule (see Ogden [1984, Chapter **5**] and Remark **1.3** below). Here, we simply quote one further result useful in numerical implementations. From (7.1.78),

$$\dot{\boldsymbol{P}} = \mathbf{A} : \dot{\boldsymbol{F}},$$

i.e., (7.1.88)

$$\dot{P}_{aA} = \mathsf{A}_{aAbB} \dot{F}_{bB},$$

where $\mathbf{A}$ is called the first elasticity tensor, given by

$$\mathbf{A} = \frac{\partial^2 W}{\partial \boldsymbol{F} \partial \boldsymbol{F}},$$

i.e., (7.1.89)

$$\mathsf{A}_{aAbB} = \frac{\partial^2 W}{\partial F_{aA} \partial F_{bB}}.$$

---

†The definition of the spatial elasticity tensor often includes a factor of $\frac{1}{J}$, as in Marsden and Hughes [1994].

Alternatively, starting from $(7.1.82)_2$, we arrive at (7.1.88) and the following important relationship connecting $\mathbf{A}$ and $\mathbf{c}$

$$\boxed{\mathsf{A}_{aBcD} = F_{Bb}^{-1}\left[\mathsf{c}_{abcd} + \tau_{ac}\delta_{bd}\right] F_{Dd}^{-1}.} \tag{7.1.90}$$

Remark 7.1.3. Starting from properly invariant hyperelastic relationships of the form (7.1.82), one can derive spatial rate-constitutive equations of the form (7.1.87) which are also properly invariant. This equation can be expressed in terms of any objective rate; for instance, in terms of the Jaumann stress rate, on account of the relationship

$$\begin{aligned} L_v\boldsymbol{\tau} &= \dot{\boldsymbol{\tau}} - (\boldsymbol{d} + \boldsymbol{w})\boldsymbol{\tau} - \boldsymbol{\tau}(\boldsymbol{d} + \boldsymbol{w})^T \\ &= \dot{\boldsymbol{\tau}} - \boldsymbol{w}\boldsymbol{\tau} + \boldsymbol{\tau}\boldsymbol{w} - \boldsymbol{d}\boldsymbol{\tau} - \boldsymbol{\tau}\boldsymbol{d} \\ &= \overset{\nabla}{\boldsymbol{\tau}} - \boldsymbol{d}\boldsymbol{\tau} - \boldsymbol{\tau}\boldsymbol{d}, \end{aligned} \tag{7.1.91}$$

equation (7.1.87) can be rephrased as

$$\boxed{\begin{aligned} \overset{\nabla}{\boldsymbol{\tau}} &= \mathbf{a} : \boldsymbol{d}, \\ \mathsf{a}_{abcd} &:= \mathsf{c}_{abcd} + \delta_{ac}\tau_{bd} + \delta_{bd}\tau_{ac}. \end{aligned}} \tag{7.1.92}$$

Conversely, the following question arises. Given any rate constitutive equation of the form $(7.1.92)_1$, is there a stored-energy function such that $\boldsymbol{\tau}$ is given by $(7.1.82)_3$? In general, the answer to this question is negative. In addition to the full symmetry of the moduli $\mathbf{a}$, a set of *compatibility relationships* essentially due to Bernstein must hold for a rate equation of the form $\overset{\nabla}{\boldsymbol{\tau}} = \mathbf{a} : \boldsymbol{d}$ to be derivable from a stored-energy function. We refer to Truesdell and Noll [1965, Chapter **IV**], for a detailed account of the relevant results. Rate equations of the form (7.1.92) which are not derivable from a stored-energy function lead to the notion of *hypoelasticity*. We recall two basic results.

i. In general, hypoelastic materials produce *nonzero dissipation* in a closed cycle. See Truesdell and Noll [1965, page 401] for the precise statements.

ii. The assumption of $\overset{\nabla}{\boldsymbol{\tau}} = \mathbf{a} : \boldsymbol{d}$, $\mathbf{a}$ being the constant isotropic tensor of the infinitesimal theory of elasticity, in general, is *incompatible* with hyperelasticity (Simo and Pister [1984]). We note that such an assumption is typically made in phenomenological theories of plasticity.

7.1.5.5 A simple model of hyperelastic response.

Consider the following stored-energy function (Ciarlet [1988]):

$$\bar{W} = \lambda\frac{(J^2 - 1)}{4} - \left(\frac{\lambda}{2} + \mu\right)\ln J + \tfrac{1}{2}\mu\left(\operatorname{tr}\boldsymbol{C} - 3\right), \tag{7.1.93}$$

where $\lambda, \mu > 0$ are interpreted as Lamé constants. Since

$$\frac{\partial J}{\partial \boldsymbol{C}} = \tfrac{1}{2}J\boldsymbol{C}^{-1}, \tag{7.1.94}$$

from (7.1.82),

$$\left.\begin{aligned} S &= \lambda \frac{(J^2-1)}{2} C^{-1} + \mu\left(1 - C^{-1}\right), \\ \boxed{\tau = \lambda \frac{J^2-1}{2} \mathbf{1} + \mu\left(FF^T - \mathbf{1}\right).} \end{aligned}\right\} \tag{7.1.95}$$

Using (7.1.84) and (7.1.86) we find the following expression for the spatial elasticity tensor:

$$\boxed{\mathbf{c} = \lambda J^2 \mathbf{1} \otimes \mathbf{1} + 2\mu\left[1 + \frac{\lambda}{2\mu}\left(1 - J^2\right)\right]\mathbf{I},} \tag{7.1.96}$$

where $\mathbf{I}$ with components $I_{abcd} = [\delta_{ac}\delta_{bd} + \delta_{ad}\delta_{bc}]/2$ is the fourth-order symmetric unit tensor. We note the following important properties:

i. As $J \to 0$ or $J \to \infty$, $\bar{W} \to \infty$.

ii. For $F = \mathbf{1} \Rightarrow \bar{W} = 0$, $\tau = \mathbf{0}$ and $\mathbf{c}$ reduces to the elasticity tensor of the linear theory, i.e., $\mathbf{c}|_{J=1} = \lambda \mathbf{1} \otimes \mathbf{1} + 2\mu \mathbf{I}$.

iii. $\bar{W}$ can be written as $\bar{W} = U(J) + \frac{1}{2}\mu(\operatorname{tr} C - 3)$, where

$$U''(J) = \frac{\lambda}{2}\left(1 + \frac{1}{J^2}\right) + \frac{\mu}{J^2} > 0, \quad \text{for } J \in (0, \infty). \tag{7.1.97}$$

Hence, $U(J)$ is a *convex function* of J.

iv. $\bar{W}$ is a *polyconvex* function of F; see Ball [1977] and Ciarlet [1988, Section **4.9**]. Thus the only known *global* existence results for elasticity, which are based on the existence of minimizers of the total potential energy for polyconvex stored-energy functions, apply to this model (see, e.g., Ciarlet [1988, Chapter **7**] or Marsden and Hughes [1994, Chapter **6**]).

7.1.6 The Notion of Isotropy. Isotropic Elastic Response.

We conclude our brief overview of some basic aspects of continuum mechanics with a discussion of the notion of *isotropy*. This notion is concerned with the possible invariance of the constitutive response of a material under *certain superposed rigid body motions of the reference configuration* and should not be confused with the notion of frame indifference. As we shall see, the former notion refers to a particular property which the material response may enjoy, whereas the latter notion is a fundamental principle of mechanics, which holds for *all* possible response functions. Again, in the following elementary discussion we illustrate the basic ideas within the context of elasticity.

7.1.6.1 Isotropic group at a point in the reference state.

Let $X \in \mathcal{B}$ be a point in the reference placement $\mathcal{B}$ of an elastic body. Consider a superposed rigid deformation of $\mathcal{B}$:

$$X \in \mathcal{B} \longmapsto X^+ = \psi(X) := c + QX \in \mathcal{B}, \qquad Q \in \mathrm{SO}(3), \tag{7.1.98}$$

which transforms $\mathcal{B}$ onto $\mathcal{B}^+$ as shown in Figure 7.4. Therefore, from Figure 7.4, $\varphi_t^+ \circ \psi = \varphi_t$, i.e.,

$$\varphi(X, t) = \varphi^+(QX + c, t). \tag{7.1.99}$$

Consequently, by the chain rule,

$$F_t = (F_t^+ Q) \Rightarrow \boxed{F_t^+ = FQ^T,} \tag{7.1.100}$$

which says that-under superposed rigid body motions of the reference state $\mathcal{B}$, the deformation gradient transforms according to (7.1.100). This implies the transformation

$$\boxed{C_t^+ = QCQ^T,} \tag{7.1.101}$$

for the right Cauchy–Green tensor. In general, the values of the stored-energy function at $X \in \mathcal{B}$, associated with C_t and C_t^+, are different. The *isotropic group at* $X \in \mathcal{B}$ is precisely the set of proper orthogonal transformations that leave the stored-energy function unchanged:

$$\boxed{G_X := \left\{Q \in \mathrm{SO}(3) \mid \bar{W}\left(X, QC_tQ^T\right) = \bar{W}(X, C_t)\right\}.} \tag{7.1.102}$$

It can be easily shown that $G_X \subset \mathrm{SO}(3)$ is indeed a group. Note the following:

i. G_X is a *local set* associated with a point $X \in \mathcal{B}$, unless the material is homogeneous, i.e., $\bar{W}$ is independent of X.

ii. The isotropy group G_X is defined *relative to a particular reference configuration.*

If $G_X \equiv \mathrm{SO}(3)$, the material is said to be *isotropic* (relative to $\mathcal{B}$ and $X \in \mathcal{B}$), otherwise, the material is said to be *anisotropic*.

7.1.6.2 Isotropic functions.

The condition of isotropic response places strong restrictions on the admissible forms of the response function. Here we recall only one important result which is used below.

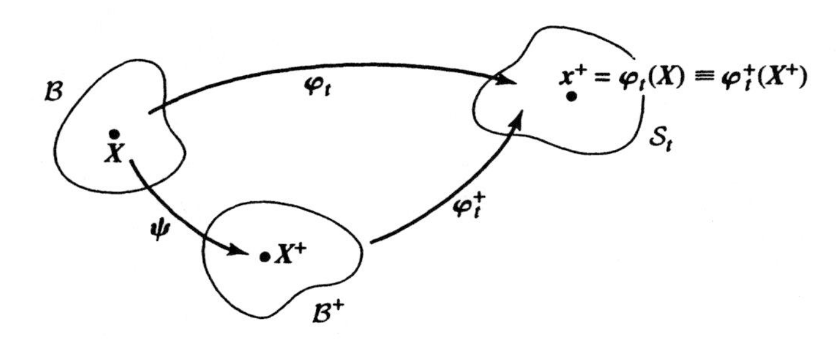

FIGURE 7.4. Superposed rigid body deformation of the reference state.

From our preceding discussion, *a function $f : \mathbb{S} \to \mathbb{R}$ of symmetric tensors $\boldsymbol{H} \in \mathbb{S}$ is isotropic if and only if*

$$\boxed{f(\boldsymbol{Q}\boldsymbol{H}\boldsymbol{Q}^T) = f(\boldsymbol{H}) \ \text{ for all } \boldsymbol{Q} \in \mathrm{SO}(3).} \tag{7.1.103}$$

We may think of f as the free energy or any other response function (e.g., the yield condition as discussed below). Depending on the context, $\boldsymbol{H} \in \mathbb{S}$ denotes the right Cauchy–Green tensor or any other *symmetric* tensor (e.g., the Cauchy stress tensor). One has the following well-known basic result.

Representation theorem. *(for isotropic functions). A function $f : \mathbb{S} \to \mathbb{R}$ is isotropic if and only if $f(\boldsymbol{H})$ depends on $\boldsymbol{H} \in \mathbb{S}$ through its principal invariants, i.e., if and only if there exists a function $f : \mathbb{R}^3 \to \mathbb{R}$ such that*

$$\boxed{f(\boldsymbol{H}) = \tilde{f}(I_{\boldsymbol{H}}, \mathbb{I}2_{\boldsymbol{H}}, III_{\boldsymbol{H}}) \textit{ for all } \boldsymbol{H} \in \mathbb{S},} \tag{7.1.104}$$

where $I_{\boldsymbol{H}} := tr\boldsymbol{H}$, $\mathbb{I}2_{\boldsymbol{H}} := \frac{1}{2}[I_{\boldsymbol{H}}^2 - tr\, \boldsymbol{H}^2]$, $III_{\boldsymbol{H}} = det\, \boldsymbol{H}$ are the principal invariants of $\boldsymbol{H}$. (We also use the alternative notation I_1, I_2 and I_3 for the principal invariants.)

This result is an immediate consequence of the spectral theorem for symmetric tensors; see, e.g., Gurtin [1981, page 230].

EXAMPLE: 7.1.3. For homogeneous isotropic elasticity, the stored energy is a function only of the principal invariants of $\boldsymbol{C} = \boldsymbol{F}^T\boldsymbol{F}$:

$$\bar{W}(\boldsymbol{C}) = \tilde{W}(I_1, I_2, I_3), \tag{7.1.105}$$

where I_A, $(A = 1, 2, 3)$ are given by (7.1.6). An example of an isotropic stored-energy function was given by (7.1.93). Using the relationships

$$\begin{aligned} \frac{\partial I_1}{\partial \boldsymbol{C}} &= \mathbf{1}, \\ \frac{\partial I_2}{\partial \boldsymbol{C}} &= I_1\mathbf{1} - \boldsymbol{C}, \\ \frac{\partial I_3}{\partial \boldsymbol{C}} &= I_3\boldsymbol{C}^{-1}; \end{aligned} \tag{7.1.106}$$

and the constitutive equations (7.1.82), the symmetric Piola–Kirchhoff tensor becomes

$$\boxed{\boldsymbol{S} = 2\left[\frac{\partial \tilde{W}}{\partial I_1} + \frac{\partial \tilde{W}}{\partial I_2} I_1\right]\mathbf{1} - 2\frac{\partial \tilde{W}}{\partial I_2}\boldsymbol{C} + 2\frac{\partial \tilde{W}}{\partial I_3} I_3\boldsymbol{C}^{-1}.} \tag{7.1.107}$$

Using the relation $\boldsymbol{\tau} = \boldsymbol{F}\boldsymbol{S}\boldsymbol{F}^T$, one obtains the following constitutive equation for the Kirchhoff stress:

$$\boxed{\boldsymbol{\tau} = 2\left[\frac{\partial \tilde{W}}{\partial I_3} I_3\right]\mathbf{1} + 2\left[\frac{\partial \tilde{W}}{\partial I_1} + \frac{\partial \tilde{W}}{\partial I_2} I_1\right]\boldsymbol{b} - 2\frac{\partial \tilde{W}}{\partial I_2}\boldsymbol{b}^2,} \tag{7.1.108}$$

where $\boldsymbol{b} = \boldsymbol{F}\boldsymbol{F}^T$.

Remarks 7.1.4.

1. Note that, for isotropic response and only for this case, the stored-energy function depends on the motion through the left Cauchy-Green tensor, $\boldsymbol{b}$, i.e., $\bar{W}(\boldsymbol{X}, \boldsymbol{C}) = \hat{W}(\boldsymbol{X}, \boldsymbol{b})$, see Chadwick [1976] for a direct proof of this well-known result.
2. The obvious way to automatically ensure satisfaction of frame indifference without precluding anisotropic response is to formulate the constitutive response function in terms of objects *associated with the reference state*. For elasticity this amounts to formulating the constitutive response in terms of the right Cauchy-Green tensor $\boldsymbol{C}$ and the second Piola–Kirchhoff tensor $\boldsymbol{S}$.
3. The formulation of elasticity in terms of $(\boldsymbol{S}, \boldsymbol{C})$ is called the *convective representation of elasticity*. The reason for this name is that the coordinates of $(\boldsymbol{S}, \boldsymbol{C})$ relative to the inertial reference frame coincide with the coordinates of the spatial object $(\boldsymbol{\tau}, \boldsymbol{g})$ (here, $\boldsymbol{g}$ is the spatial metric tensor) in the *convected coordinate system*. Zaremba [1903] understood that the use of convected coordinates (the convected representation) automatically ensures frame indifference; see Truesdell and Noll [1965, p. 45]. For a discussion of the convected representation (including the corresponding reduced Lie–Poisson–Hamiltonian structure), see Simo, Marsden, and Krishnaprasad [1988].
4. In a computational context, the convective representation in terms of $(\boldsymbol{S}, \boldsymbol{C})$ (or, equivalently, use of convected coordinates) has been used by several authors; see, e.g., Needleman [1982], in the context of *plasticity* theories.

7.2 Variational Formulation. Weak Form of Momentum Balance

In this section we develop the weak formulation of momentum balance equations as a first step toward numerically implementating the models discussed subsequently within the framework of the finite-element method. For a discussion of functional analysis issues entirely omitted in the present introductory account, we refer to Ciarlet [1988] and Marsden and Hughes [1994, Chapter **6**]. A few comments are made in Remark **2.2**.

7.2.1 Configuration Space and Admissible Variations.

Motivated by our discussion of the admissible deformations in a continuum body, with reference configuration $\mathcal{B} \subset \mathbb{R}^3$ and prescribed deformations $\bar{\varphi} : \partial_\varphi \mathcal{B} \to \mathbb{R}^3$ on a portion $\partial_\varphi \mathcal{B}$ of the boundary $\partial \mathcal{B}$, we define its configuration space as the set

$$\mathcal{C} := \left\{ \varphi : \mathcal{B} \to \mathbb{R}^3 \mid \det\left[D\varphi\right] > 0 \text{ in } \mathcal{B} \text{ and } \varphi\big|_{\partial_\varphi \mathcal{B}} = \bar{\varphi} \right\}. \tag{7.2.1}$$

As before, we denote by $\mathcal{S} = \varphi(\mathcal{B})$ the current configuration of the body under a deformation $\varphi \in \mathcal{C}$. Displacements superposed on $\varphi(\mathcal{B})$, which do not violate the prescribed boundary conditions of place, are called *admissible variations* of φ. They span a linear space denoted by $\mathcal{V}_\varphi$ and defined as (see Figure 7.5 for an illustration)

$$\mathcal{V}_\varphi := \{\eta : \varphi(\mathcal{B}) \to \mathbb{R}^3 \mid \eta(\varphi(\boldsymbol{X})) = \mathbf{0} \text{ for } \boldsymbol{X} \in \partial_\varphi \mathcal{B}\}. \tag{7.2.2}$$

Note that $\mathcal{V}_\varphi$ changes with $\varphi \in \mathcal{C}$.‡ We call $\eta \in \mathcal{V}_\varphi$ a *spatial variation*. To remove the dependence on φ, one defines *material variations* by the change of variables

$$\eta_0(\boldsymbol{X}) = \eta[\varphi(\boldsymbol{X})] \text{ for } \eta \in \mathcal{V}_\varphi. \tag{7.2.3}$$

Thus, material variations span a *fixed* linear space, denoted by $\mathcal{V}_0$ and defined as

$$\mathcal{V}_0 = \left\{\eta_0 : \mathcal{B} \to \mathbb{R}^3 \mid \eta_0(\boldsymbol{X}) = \mathbf{0} \text{ for } \boldsymbol{X} \in \partial_\varphi \mathcal{B}\right\}. \tag{7.2.4}$$

Note that it is often more convenient to think of $\mathcal{V}_0$ as *given* and then construct $\eta \in \mathcal{V}_\varphi$ by setting $\eta = \eta_0 \circ \varphi^{-1}$ for each $\eta_0 \in \mathcal{V}_0$. By the chain rule, from (7.2.3), in components

$$\frac{\partial \eta_{0a}}{\partial X_A} = \frac{\partial \eta_a}{\partial x_b}\frac{\partial \varphi_b}{\partial X_A} \Rightarrow \boxed{\eta_{0_a,A} = \eta_{a,b} F_{bA}.} \tag{7.2.5}$$

Alternatively, using direct notation, we set

$$\boxed{\text{GRAD}\, \eta_0 = (\nabla \eta \circ \varphi)\, \boldsymbol{F}.} \tag{7.2.6}$$

With this notation in hand, we can formulate the weak form of the momentum balance equation as follows.

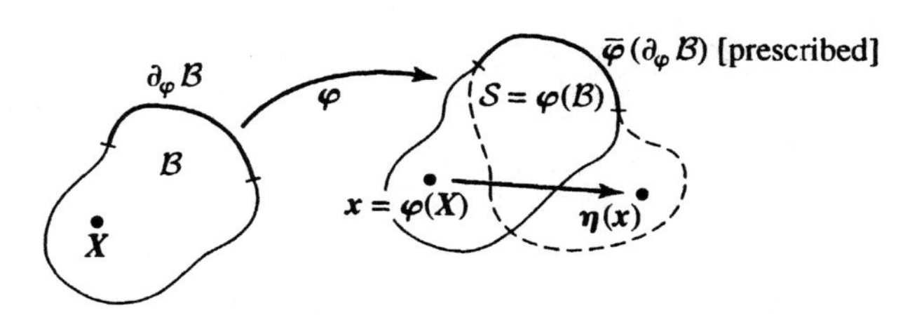

FIGURE 7.5. A superposed spatial variation $\eta : \varphi(\mathcal{B}) \to \mathbb{R}^3$ defined on the configuration $\varphi \in \mathcal{C}$. Note that $\eta(\boldsymbol{x}) = \mathbf{0}$ for $\boldsymbol{x} \in \bar{\varphi}(\partial_\varphi \mathcal{B})$. Also note that *material* variations $\eta_0 : \mathcal{B} \to \mathbb{R}^3$ are obtained by changing to material coordinates through $\boldsymbol{x} = \varphi(\boldsymbol{X})$, as $\eta_0(\boldsymbol{X}) = \eta[\varphi(\boldsymbol{X})]$, for $\eta \in \mathcal{V}_\varphi$.

‡In a more geometric setting, $\mathcal{V}_\varphi$ is called the *tangent space to $\mathcal{C}$ at the configuration $\varphi \in \mathcal{C}$*.

7.2.2 The Weak Form of Momentum Balance.

Recall that on $\partial_t\mathcal{B} \subset \partial\mathcal{B}$ we assume that the nominal traction vector is prescribed as

$$\bar{\boldsymbol{t}}^{\hat{N}} = \boldsymbol{P}\hat{\boldsymbol{N}}\big|_{\partial_t\mathcal{B}}, \tag{7.2.7}$$

where $\hat{\boldsymbol{N}}$ is the outer normal to $\partial_t\mathcal{B}$ and $\partial_t\mathcal{B} \cap \partial_\varphi\mathcal{B} = \emptyset$, with $\overline{\partial_t\mathcal{B} \cup \partial_\varphi\mathcal{B}} = \overline{\partial\mathcal{B}}$. A standard construction of the weak form of equation (7.1.56) proceeds as follows.

7.2.2.1 Material (Lagrangian) description.

By taking the dot product of (7.1.56) with any $\boldsymbol{\eta}_0 \in \mathcal{V}_0$, integrating over the reference volume, and using the divergence theorem,

$$\begin{aligned} G_{\text{dyn}}(\boldsymbol{P}, \boldsymbol{\varphi}_t; \boldsymbol{\eta}_0) := & -\int_{\mathcal{B}} \text{DIV}\,\boldsymbol{P}\cdot\boldsymbol{\eta}_0\,dV - \int_{\mathcal{B}} \rho_0\boldsymbol{B}\cdot\boldsymbol{\eta}_0\,dV + \int_{\mathcal{B}} \rho_0\boldsymbol{A}_t\cdot\boldsymbol{\eta}_0\,dV \\ = & \int_{\mathcal{B}} \boldsymbol{P} : \text{GRAD}\,\boldsymbol{\eta}_0\,dV - \int_{\partial\mathcal{B}} \boldsymbol{P}\hat{\boldsymbol{N}}\cdot\boldsymbol{\eta}_0\,d\Gamma \\ & - \int_{\mathcal{B}} \rho_0\boldsymbol{B}\cdot\boldsymbol{\eta}_0\,dV + \int_{\mathcal{B}} \rho_0\boldsymbol{A}_t\cdot\boldsymbol{\eta}_0\,dV. \end{aligned} \tag{7.2.8}$$

Since by construction $\boldsymbol{\eta}_0\big|_{\partial_\varphi\mathcal{B}} = \boldsymbol{0}$, from (7.2.8) and (7.2.7), we arrive at the following expression:

$$\boxed{G_{\text{dyn}}\left(\boldsymbol{P}, \boldsymbol{\varphi}_t; \boldsymbol{\eta}_0\right) := \int_{\mathcal{B}} \rho_0 \frac{\partial^2\boldsymbol{\varphi}_t}{\partial t^2}\cdot\boldsymbol{\eta}_0 dV + G\left(\boldsymbol{P}, \boldsymbol{\eta}_0\right) = 0,} \tag{7.2.9}$$

which holds for *any* (material) variation $\boldsymbol{\eta}_0 \in \mathcal{V}_0$. Here, $G(\boldsymbol{P}, \boldsymbol{\eta}_0)$ denotes the (static) weak form of the equilibrium equations given by

$$\boxed{G\left(\boldsymbol{P}, \boldsymbol{\eta}_0\right) := \int_{\mathcal{B}} \boldsymbol{P} : \text{GRAD}\,\boldsymbol{\eta}_0 dV - \int_{\mathcal{B}} \rho_0\boldsymbol{B}\cdot\boldsymbol{\eta}_0 dV - \int_{\partial\mathcal{B}} \bar{\boldsymbol{t}}^{\hat{N}}\cdot\boldsymbol{\eta}_0 d\Gamma.}\ {}^{\S} \tag{7.2.10}$$

Conversely, by assuming enough smoothness of $\boldsymbol{\varphi} \in \mathcal{C}$ and $\boldsymbol{\eta}_0 \in \mathcal{V}_0$, (7.2.9) yields the local form of the momentum equations along with the stress boundary conditions (7.2.7). We note, however, that statement (7.2.9) is considerably more general than the local form (7.1.56), since the configurations $\boldsymbol{\varphi}_t$ are subject to less restrictive continuity requirements.

---

§Note that the boundary integral in (7.2.10) can be also written over $\partial_t\mathcal{B}$ because $\boldsymbol{\eta}_0$ vanishes on $\partial_\varphi\mathcal{B}$.

In what follows we restrict our discussion to the static problem. Consequently, the variational formulation of the boundary value problem may be stated as follows.

$$\boxed{\begin{array}{c}\text{Find the stress field } \boldsymbol{P} \text{ and the configuration } \varphi \in C \text{ such that} \\ G\left(\boldsymbol{P}, \eta_0\right) = 0, \text{ for all } \eta_0 \in \mathcal{V}_0.\end{array}} \tag{7.2.11}$$

Remark 7.2.1. For elasticity, $\boldsymbol{P}$ is a local functional of the deformation through the right Cauchy–Green tensor, i.e.,

$$\boldsymbol{P} = \hat{\boldsymbol{P}}(\varphi) := \boldsymbol{F} \left. \frac{\partial \overline{W}(\boldsymbol{X}, \boldsymbol{F}^T\boldsymbol{F})}{\partial \boldsymbol{C}} \right|_{\boldsymbol{F}=D\varphi} . \tag{7.2.12}$$

If this constitutive equation is enforced locally, then $G(\boldsymbol{P}, \eta_0) = G(\hat{\boldsymbol{P}}(\varphi), \eta_0)$ becomes a functional of $\varphi \in C$. Thus, the only unknown variable in the variational problem is the deformation φ.

7.2.2.2 Spatial (Eulerian) description.

The weak form (7.2.11) is phrased entirely in the spatial description by the following change of variables. First, note that

$$\begin{aligned} \boldsymbol{P} : \text{GRAD}\, \eta_0 &= (\boldsymbol{\tau} \circ \varphi)\boldsymbol{F}^{-T} : \text{GRAD}\, \eta_0 \\ &= (\boldsymbol{\tau} \circ \varphi) : \text{GRAD}\, \eta_0 \boldsymbol{F}^{-1} \\ &= (\boldsymbol{\tau} \circ \varphi) : \nabla\eta \circ \varphi \\ &= (\boldsymbol{\tau} : \nabla\eta) \circ \varphi, \end{aligned} \tag{7.2.13a}$$

where $\eta = \eta_0 \circ \varphi^{-1}$, and we have used relationships (7.2.6) and (7.1.48). In components,

$$P_{aA}(\boldsymbol{X})\eta_{0a,A}(\boldsymbol{X}) = \left[\tau_{ab}(\boldsymbol{x})\eta_{a,b}(\boldsymbol{x})\right]_{\boldsymbol{x}=\varphi(\boldsymbol{X})} . \tag{7.2.13b}$$

Second, we recall the change of variable formulas

$$\left.\begin{aligned} \boldsymbol{B}(\boldsymbol{X}) &= \boldsymbol{b}(\boldsymbol{x})\big|_{\boldsymbol{x}=\varphi(\boldsymbol{X})}, \\ \rho_0(\boldsymbol{X})J^{-1}(\boldsymbol{X}) &= \rho(\boldsymbol{x})\big|_{\boldsymbol{x}=\varphi(\boldsymbol{X})}, \\ \text{and} \qquad J(\boldsymbol{X})dV(\boldsymbol{X}) &= dv(\boldsymbol{x})\big|_{\boldsymbol{x}=\varphi(\boldsymbol{X})}, \end{aligned}\right\} \tag{7.2.14}$$

Therefore, substituting (7.2.13) and (7.2.14) in (7.2.10) produces the result

$$\boxed{G(\boldsymbol{\tau}, \eta) := \int_{\varphi(\mathcal{B})} \left(J \circ \varphi^{-1}\right)^{-1} \boldsymbol{\tau} : \nabla\eta dv - \int_{\varphi(\mathcal{B})} \rho\boldsymbol{b} \cdot \eta dv - \int_{\partial\varphi_t(\mathcal{B})} \boldsymbol{t}^{\hat{\boldsymbol{n}}} \cdot \eta d\gamma,} \tag{7.2.15}$$

where the traction vector is transformed with the aid of the change in area formula $d\gamma\hat{\boldsymbol{n}} = [J\boldsymbol{F}^{-T}\hat{\boldsymbol{N}}]d\Gamma \circ \varphi^{-1}$ as

$$\boldsymbol{t}^{\hat{\boldsymbol{N}}} d\Gamma =: \left(\boldsymbol{t}^{\hat{\boldsymbol{n}}} d\gamma\right) \circ \varphi. \tag{7.2.16}$$

Note that the Jacobian factor in (7.2.15) disappears by replacing the Kirchhoff stress with the Cauchy stress according to the formula $\boldsymbol{\tau} = (J \circ \boldsymbol{\varphi}^{-1})\boldsymbol{\sigma}$.

Remarks 7.2.2.

1. The variational formulation of the equations of continuum mechanics discussed above is formal in the sense that little is known about the appropriate mathematical structure for specific models. For instance, the only known satisfactory theory in elasticity is based on *minimizers* of the potential energy with a polyconvex stored-energy function. Existence can be proved in a suitable Sobolev space (typically $W^{1,p}$ with p large enough as dictated by certain growth conditions), but it is not clear in what sense the minimizers satisfy the weak form (7.2.11). For a recent review of known results, we refer to Ciarlet [1988]. Practically nothing is known for plasticity models, except in the context of the infinitesimal theory, where a complete existence theory exists; see Temam [1985] for a comprehensive review.
2. To simplify the notation, we often omit explicit indication of the composition operation involved in the change of variables from material to spatial coordinates, with the understanding that all the quantities involved are evaluated at appropriate points. This is usually clear from the context. For instance, we write

$$dv = J dV, \boldsymbol{\tau} = J\boldsymbol{\sigma},$$

and (7.2.17)

$$\rho = \frac{\rho_0}{J}$$

in place of the more precise notation $dv \circ \boldsymbol{\varphi} = J dV$, $\boldsymbol{\tau} = (J \circ \boldsymbol{\varphi}^{-1})\boldsymbol{\sigma}$, and $\rho \circ \boldsymbol{\varphi} = \rho_0 / J$.

7.2.3 The Rate Form of the Weak Form of Momentum Balance.

To illustrate some of the difficulties involved in the numerical implementation of certain hypoelastic plasticity models discussed below in the simplest possible context, we examine the structure underlying the rate formulation of the weak form of momentum balance equations. Traditionally, this rate form has played an important role in the incremental solution of quasi-static elastoplasticity and remains widely used in many numerical solution schemes (see, e.g., McMeeking and Rice [1975]; Needleman [1982]). In finite-element formulations of elastoplasticity, it leads to the "continuum" tangent stiffness matrix on the basis of which an iterative solution procedure is constructed. This tangent matrix, however, is not consistent with the discrete form of the equations obtained through a return-mapping integration algorithm, as observed by Simo and Taylor [1985] who employed the notion of a "consistent tangent" obtained by direct linearization of the discrete equations. The origin of this notion is found in the work of Hughes and Taylor [1978] on implicit algorithms for viscoplasticity and Nagtegaal [1982].

In any event, one requires that the (tangent) operator emanating from the variational formulation of the rate form of the momentum balance equations be symmetric, in the sense described below, for rate-constitutive equations. This symmetry condition leads to the notion of rate potentials, in the sense of Hill [1958], and results in finite-element symmetric (continuum) tangent stiffness matrices. Such a result is the direct consequence of a well-known theorem often credited to Vainberg [1964], and bears no relationship whatsoever to the question of existence of global potentials (In this connection, see, e.g., Marsden and Hughes [1994, Chapter **5**, problem 1.3].

7.2.3.1 Admissible velocity fields.

The crucial observation to be made is that the spatial velocity field for fixed time $t \in [0, T]$ is an admissible spatial variation, i.e., $\boldsymbol{v}(\bullet, t) \in \mathcal{V}_{\varphi_t}$. This can be seen by noting that

i. $\boldsymbol{v}_t = \boldsymbol{v}(\bullet, t)$ is a mapping that assigns the velocity vector $\boldsymbol{v}(\boldsymbol{x}, t) \in \mathbb{R}^3$ to each $\boldsymbol{x} \in \varphi_t(\mathcal{B})$, that is, $\boldsymbol{v}(\bullet, t) : \varphi_t(\mathcal{B}) \to \mathbb{R}^3$; and
ii. $\boldsymbol{v}(\bullet, t)$ vanishes on the prescribed boundary $\bar{\varphi}(\partial_\varphi \mathcal{B})$, i.e., $\boldsymbol{v}(\boldsymbol{x}, t) = \boldsymbol{0}$ for $\boldsymbol{x} \in \bar{\varphi}(\partial_\varphi \mathcal{B})$.

Consequently, $\boldsymbol{v}_t = \boldsymbol{v}(\bullet, t)$ satisfies the conditions in (7.2.2). In addition, the material velocity field satisfies relationship (7.2.3) since, by definition, $\boldsymbol{V}_t(\boldsymbol{X}) = \boldsymbol{v}_t[\varphi_t(\boldsymbol{X})]$. Hence, $\boldsymbol{V}_t = \boldsymbol{V}(\bullet, t)$ is in $\mathcal{V}_0$.

In this context, $\mathcal{V}_{\varphi_t}$ is called as the space of admissible spatial velocity fields tangent to the motion φ_t. Then the *actual* velocity field $\boldsymbol{v}_t = \boldsymbol{v}(\bullet, t)$ at time t is an element of $\mathcal{V}_{\varphi_t}$.

7.2.3.2 Variational form of the rate equilibrium equations.

We develop the appropriate variational form under the usual assumption of *dead loading*, i.e., we assume that $\boldsymbol{B}$ in $\mathcal{B}$ and $\hat{\boldsymbol{t}^N}$ on $\partial_t \mathcal{B}$ are *configurationally independent*. We also assume that the process is quasi-static so that the inertial contributions are neglected. With this assumption at hand, let φ_t define a configuration in equilibrium at time t. For a rate of change in the loading given by $\dot{\boldsymbol{B}}$ and $\dot{\boldsymbol{t}}^N$, the equilibrium equations become

$$\text{DIV}\, \dot{\boldsymbol{P}}_t + \rho_0 \dot{\boldsymbol{B}} = 0, \ \text{in } \mathcal{B},$$

i.e., (7.2.18)

$$\dot{P}_{aA.A} + \rho_0 \dot{B}_a = 0, \quad \text{in } \mathcal{B}.$$

Let $\boldsymbol{\eta}_t \in \mathcal{V}_{\varphi_t}$ be an *admissible* spatial velocity field tangent to the deformation φ_t, and let $\boldsymbol{\eta}_0 = \boldsymbol{\eta}_t \circ \varphi_t^{-1}$ be the corresponding material field. Using the divergence theorem, we obtain

$$\int_{\mathcal{B}} \dot{\boldsymbol{P}}_t : \text{GRAD}\, \boldsymbol{\eta}_0 dV - \int_{\mathcal{B}} \rho_0 \dot{\boldsymbol{B}} \cdot \boldsymbol{\eta}_0 dV - \int_{\partial_t \mathcal{B}} \dot{\boldsymbol{t}}^{\hat{N}} \cdot \boldsymbol{\eta}_0 d\Gamma = \boldsymbol{0}, \qquad (7.2.19)$$

for all $\eta_0 \in \mathcal{V}_0$. Note that this equation is the *rate-incremental* version of (7.2.10). To transform (7.2.19) to the spatial description, we proceed as follows. Since $\boldsymbol{P} = \boldsymbol{F}_t \boldsymbol{S}_t = \boldsymbol{\tau}_t \boldsymbol{F}_t^{-T}$,

$$\begin{aligned}
\dot{\boldsymbol{P}}_t : \text{GRAD}\, \eta_0 &= \left[\dot{\boldsymbol{F}}_t \boldsymbol{S}_t + \boldsymbol{F}_t \dot{\boldsymbol{S}}_t\right] : \text{GRAD}\, \eta_0 \\
&= \left[\nabla \boldsymbol{v}_t \boldsymbol{F}_t \boldsymbol{S}_t + \boldsymbol{F}_t \dot{\boldsymbol{S}}_t\right] : \text{GRAD}\, \eta_0 \\
&= \left[\nabla \boldsymbol{v}_t \boldsymbol{F}_t \boldsymbol{S}_t \boldsymbol{F}_t^T + \boldsymbol{F}_t \dot{\boldsymbol{S}}_t \boldsymbol{F}_t^T\right] \boldsymbol{F}^{-T} : \text{GRAD}\, \eta_0 \\
&= \left[\nabla \boldsymbol{v}_t \boldsymbol{\tau}_t + L_v \boldsymbol{\tau}_t\right] : \text{GRAD}\, \eta_0 \boldsymbol{F}^{-1} \\
&= \left[\nabla \boldsymbol{v}_t \boldsymbol{\tau}_t + L_v \boldsymbol{\tau}_t\right] : \nabla \eta_t,
\end{aligned} \tag{7.2.20}$$

where we have used the Lie derivative formula (7.1.72). Therefore, (7.2.19) becomes

$$\boxed{\int_{\mathcal{B}} \left[\nabla \boldsymbol{v}_t \boldsymbol{\tau}_t + L_v \boldsymbol{\tau}_t\right] : \nabla \eta_t dV = \int_{\mathcal{B}} \dot{\boldsymbol{B}} \cdot \eta_0 \rho_o dV + \int_{\partial_t \mathcal{B}} \dot{\boldsymbol{t}}^{\hat{N}} \cdot \eta_0 d\Gamma,} \tag{7.2.21}$$

for all $\eta_t \in \mathcal{V}_{\varphi_t}$. Next consider a rate-constitutive equation of the form

$$\boxed{L_v \boldsymbol{\tau} = \mathbf{a} : \boldsymbol{d} = \mathbf{a} : \nabla \boldsymbol{v}_t,} \tag{7.2.22}$$

where replacement of $\boldsymbol{d}$ by $\nabla \boldsymbol{v}_t$ is justified since $\mathbf{a}$ is assumed to enjoy the symmetry conditions

$$\mathsf{a}_{abcd} = \mathsf{a}_{cdab} = \mathsf{a}_{bacd} = \mathsf{a}_{abdc}. \tag{7.2.23}$$

Then, for a given right-hand side in (7.2.21), a given configuration at time t defined by φ_t, and a given stress field $\boldsymbol{\tau}_t$ in equilibrium at time t, the *actual spatial velocity* $\boldsymbol{v}_t$ is given from (7.2.21) and (7.2.22) by the linear variational equation

$$\boxed{\int_{\varphi_t(\mathcal{B})} \nabla \eta_t : \left[\nabla \boldsymbol{v}_t \boldsymbol{\tau}_t + \mathbf{a} : \nabla \boldsymbol{v}_t\right] \frac{dv}{J_t} = \int_{\varphi_t(\mathcal{B})} \dot{\boldsymbol{b}} \cdot \eta_t \rho dv + \int_{\partial \varphi_t(\mathcal{B})} \dot{\boldsymbol{t}}^{\hat{n}} \cdot \eta_t d\gamma.} \tag{7.2.24}$$

Since $\boldsymbol{\tau}_t$ is given and $\mathbf{a}$ is known, the left-hand side of (7.2.24) defines a *bilinear form*, denoted by $B_{\varphi_t}(\eta_t, \boldsymbol{v}_t)$ which, by virtue of (7.2.23) and the symmetry of $\boldsymbol{\tau}$, is easily seen to be symmetric, i.e.,

$$\begin{aligned}
B_{\varphi_t}(\eta_t, \boldsymbol{v}_t) :&= \int_{\mathcal{B}} \nabla \eta_t : \left[\nabla \boldsymbol{v}_t \boldsymbol{\tau}_t + \mathbf{a} : \nabla \boldsymbol{v}_t\right] \frac{dv}{J_t} \\
&= \int_{\mathcal{B}} \nabla \boldsymbol{v}_t : \left[\nabla \eta_t \boldsymbol{\tau}_t + \mathbf{a} : \nabla \eta_t\right] \frac{dv}{J_t} \\
&= B_{\varphi_t}(\boldsymbol{v}_t, \eta_t).
\end{aligned} \tag{7.2.25}$$

The analysis of linear variational problems of the form (7.2.24) is completely standard; see, e.g., Johnson [1987] for an introductory account.

The point we wish to emphasize with the preceding analysis is that result (7.2.24) depends crucially on the structure of the rate-constitutive equation (7.2.22). We show below that a widely used formulation of rate-independent plasticity does not conform to the format of (7.2.22).

7.3 Ad Hoc Extensions of Phenomenological Plasticity Based on Hypoelastic Relationships

In this section, we examine a class of models of phenomenological, rate-independent plasticity obtained by an ad hoc extension of the infinitesimal theory which relies on a *hypoelastic* characterization of the "elastic response". Despite the number of serious objections from a physical and a fundamental perspective, which the structure of these models raises, they remain widely used in many circles of the computational and applied mechanics community. As we shall see, the advantage of these models relies on the conceptual simplicity of their formulation. From our perspective, however, this class of models suffers from two major drawbacks.

i. Hypoelastic characterizations of the elastic response postulated at the outset are objectionable on fundamental grounds.

ii. As we shall see, extending the numerical integration algorithms for the linearized theory to the present class of models requires an important additional step: an integration algorithm which preserves objectivity and defines the trial elastic step. The formulation of such an algorithm is by no means a trivial task.

In spite of the aforementioned objections, this class of models remains widely used, and, thus, it is worthwhile discussing in some detail. Furthermore, this discussion helps to put in perspective the advantages of some formulations recently proposed and examined in the last chapter of this monograph.

7.3.1 Formulation in the Spatial Description.

The basic elements of the model are patterned after those of the infinitesimal theory. First we consider the formulation of the theory in the *spatial* (Eulerian) description.

7.3.1.1 Constitutive model in the spatial description.

The first step concerns the decomposition of the strain measures into elastic and plastic parts as follows.

i. *Additive decomposition.* Motivated by the infinitesimal theory, at the outset one introduces an additive decomposition of the spatial rate of deformation tensor into elastic and plastic parts:

$$\boxed{\boldsymbol{d} = \boldsymbol{d}^e + \boldsymbol{d}^p.} \tag{7.3.1}$$

ii. *Stress response.* Next, the "elastic" response is characterized by a *hypoelastic* rate constitutive equation of the form

$$\boxed{\overset{\circ}{\boldsymbol{\tau}} = \mathbf{a} : [\boldsymbol{d} - \boldsymbol{d}^p],} \tag{7.3.2}$$

where $\overset{\circ}{(\bullet)}$ denotes *any* objective stress rate.

Remark 7.3.1. Since $\boldsymbol{d}^e$ need not be the rate of deformation of any "elastic" measure of strain, it appears hopeless to attempt to relate (7.3.2) to any truly hyperelastic response function. Consequently, according to the standard set-up of hypoelasticity, one assumes that $\mathbf{a}(\boldsymbol{\tau})$ is a function of the stress tensor.

However, frame indifference then requires that the tensor $\mathbf{a}(\boldsymbol{\tau})$ be an *isotropic function* of $\boldsymbol{\tau}$, see Truesdell and Noll [1965, p. 405]. In addition to the very nature of hypoelasticity, this conclusion raises questions on the generality and significance of a model like (7.3.2).

iii. *Elastic domain in stress space.* As in the linear theory, one introduces an elastic domain which restricts the admissible stress fields. We set

$$\boxed{\mathbb{E}_\sigma := \left\{(\boldsymbol{\tau}, \boldsymbol{q}) \in \mathbb{S} \times \mathbb{R}^m \mid f(\boldsymbol{\tau}, \boldsymbol{q}) \leq 0\right\},} \tag{7.3.3}$$

where, again, as in the infinitesimal theory, $\boldsymbol{q}$ denotes a set of m internal variables that characterize the hardening of the material, with evolution equations given below.

Remark 7.3.2. Frame indifference places a strong restriction on the admissible functional forms of the yield condition. To see this, consider the case of perfect plasticity, so that the *yield condition* is given simply as $f(\boldsymbol{\tau}) \leq 0$.

Since $\boldsymbol{\tau}$ transforms objectively under superposed rigid body motions, the requirement that the function $f : \mathbb{S} \to \mathbb{R}$ be frame indifferent demands that

$$f(\boldsymbol{Q}\boldsymbol{\tau}\boldsymbol{Q}^T) = f(\boldsymbol{\tau}), \tag{7.3.4}$$

that is, the *yield condition must be an isotropic function of* $\boldsymbol{\tau}$. An identical result holds for the case in which f depends on internal variables $\boldsymbol{q}$, provided that these variables transform objectively. In our opinion, the restriction to isotropy is too strong a requirement.

iv. *Flow rule and hardening law.* Restricting our attention to the case of an associative flow rule, the evolution equations are the following:

$$\boxed{\begin{aligned} \boldsymbol{d}^p &= \gamma \frac{\partial f(\boldsymbol{\tau}, \boldsymbol{q})}{\partial \boldsymbol{\tau}}, \\ \overset{\circ}{\boldsymbol{q}} &= -\gamma \boldsymbol{h}(\boldsymbol{\tau}, \boldsymbol{q}). \end{aligned}} \tag{7.3.5}$$

Again frame indifference restricts these evolution equations to *isotropic* response.

v. *Loading/unloading conditions*. The formulation of the model is completed by introducing loading/unloading conditions in the classical Kuhn–Tucker form

$$\boxed{\gamma \geq 0\,, \;\; f(\boldsymbol{\tau}, \boldsymbol{q}) \leq 0\,, \;\; \gamma f(\boldsymbol{\tau}, \boldsymbol{q}) = 0\,,} \tag{7.3.6a}$$

along with the *consistency* requirement

$$\boxed{\gamma \dot{f}(\boldsymbol{\tau}, \boldsymbol{q}) = 0.} \tag{7.3.6b}$$

7.3.1.2 Criticism of the model.

Models of the type outlined above have been extensively used in the computational literature, particularly with reference to metal plasticity, in conjunction with the classical von Mises yield criterion. Representative formulations include Key and Krieg [1982]; Pinsky, Ortiz, and Pister [1983]; Nagtegaal and DeJong [1981]; Nagtegaal [1982]; Nagtegaal and Veldpaus [1984]; Needleman [1982]; and Rolph and Bathe [1984]. Concerning the particular form of the hypoelastic constitutive equation (7.3.2), the moduli $\mathbf{a}$ are typically assumed constant and isotropic with the same form as the elastic moduli of linear isotropic elasticity. As pointed out in Remark **1.3** such a choice is incompatible with hyperelasticity; see Simo and Pister [1984]. However, it can be argued on a physical basis that, for small elastic strains, this objection may not become crucial. Such a situation is characteristic of metal plasticity where elastic strains are orders of magnitude less than plastic strains in well-developed plastic flow. Finally, the structure of the hardening law is typically borrowed from that of classical infinitesimal models, e.g., some form of kinematic and isotropic hardening rule. The choice of the kinematic hardening law has become a somewhat controversial issue because of results reported by Nagtegaal and DeJong [1981].

Although the class of constitutive models outlined above may be adequate for predicting the response of some metal at finite plastic strains, we believe that the number of aforementioned drawbacks and the restriction to isotropy, make this class of models too restrictive to qualify as a general theory of plasticity at finite strains. As we show in the next chapter, the numerical implementation is somewhat cumbersome. Further comments on the model are deferred to Section **7.3.2.4**.

7.3.2 Formulation in the Rotated Description.

Motivated by the strong restriction to isotropy inherent in the formulation outlined in the preceding section, it is suggested to rephrase this model in the rotated description discussed in Section **7.1.3.2** keeping its basic structure essentially unchanged. An additional motivation (perhaps the essential one) is provided by the form that the classical kinematic hardening law takes in this description. The structure of the Prager–Ziegler law remains unchanged with the time derivative of the back stress now replaced by the Green–McInnis–Naghdi rate. It appears that the anomalies reported in Nagtegaal and DeJong [1981] for the simple shear test with kinematic hardening disappear for this evolution law. The resulting formulation is used in several large scale simulation codes, such as NIKE and DYNA (Hallquist

[1979]; [1988]) and HONDO (Key [1974]), and has been advocated by a number of authors, notably Dienes [1979]; and Johnson and Bamman [1984].

In spite of the aforementioned improvements, in particular, the removal of the restriction to isotropy, we believe that the formulation discussed below remains unsatisfactory. First, as originally stated, it is not clear how to bypass the hypoelastic characterization of the elastic response. Second, the linearized variational equations and equivalently the weak form of the rate of momentum balance are not symmetric.

7.3.2.1 Local constitutive equation.

The structure of the theory is identical to that discussed in Section **3.1**, but with spatial variables now replaced by rotated variables, as summarized in BOX **7.2**. The J_2 flow theory version of the model is given in Hughes [1984].

It is of interest to examine briefly the structure of the hypoelastic equation for the stress response given by

$$\dot{\Sigma} = \mathbf{A} : [\mathbf{D} - \mathbf{D}^p], \tag{7.3.7}$$

where $\mathbf{A}$ are the rotated moduli and $\mathbf{D}^p := \boldsymbol{R}^T \boldsymbol{d}^p \boldsymbol{R}$. Upon defining spatial moduli a by the relationship

$$\mathsf{a}_{abcd} = R_{aA} R_{bB} R_{cC} R_{dD} \mathsf{A}_{ABCD}, \tag{7.3.8}$$

and noting relationships (7.1.40*b*) along with (7.1.76), equation (7.2.7) is recast in the spatial description as

$$\overset{\square}{\boldsymbol{\tau}} = \mathsf{a} : [\boldsymbol{d} - \boldsymbol{d}^p], \tag{7.3.9}$$

where $\overset{\square}{\boldsymbol{\tau}}$ is the Green–McInnis–Naghdi rate of the Kirchhoff stress tensor. Thus, (7.3.9) is of the form (7.3.2).

7.3.2.2 Symmetry of the spatial elastic moduli.

At this point, a basic difficulty involved in the present formulation should be noted. As shown in Section **2**, the Lie derivative of the Kirchhoff stress tensor $L_v\boldsymbol{\tau}$ is the object that appears in the linearized weak form (or rate form) of the momentum balance equation. Then symmetry of the tangent operator is guaranteed if $L_v\boldsymbol{\tau}$ is related to the rate of deformation tensor through symmetric moduli. This condition does not hold for an equation of the form (7.3.9), as the following argument shows.

For simplicity, we consider the case for which $\boldsymbol{d}^p \equiv \boldsymbol{0}$ vanishes identically. From (7.1.91) and (7.1.92), recall the relationships

$$L_v\boldsymbol{\tau} = \mathsf{c} : \boldsymbol{d} \Rightarrow \overset{\nabla}{\boldsymbol{\tau}} = \mathsf{a} : \boldsymbol{d}, \tag{7.3.10}$$

where a is given by (7.1.92) $_2$ and thus is completely symmetric; i.e., $\mathsf{a}_{abcd} = \mathsf{a}_{cdab} = \mathsf{a}_{bacd} = \mathsf{a}_{abdc}$. Hence, rate equations of the form (7.3.10) lead to symmetric tangent operators. In the alternative terminology of Hill [1958], one speaks of the existence of rate potentials. Now, from (7.1.75), (7.1.77) and (7.3.9), we obtain

$$\begin{aligned}\overset{\triangledown}{\tau} &= \overset{\square}{\tau} - (w - \Omega)\tau + \tau(w - \Omega) \\ &= \mathsf{a} : d - (w - \Omega)\tau + \tau(w - \Omega).\end{aligned} \tag{7.3.11a}$$

However, from (7.1.45) and (7.1.39),

$$\left.\begin{aligned} w - \Omega &= R\left[\dot{U}U^{-1} - U^{-1}\dot{U}\right]R^T/2 \\ d &= R\left[\dot{U}U^{-1} + U^{-1}\dot{U}\right]R^T/2. \end{aligned}\right\} \tag{7.3.11b}$$

Therefore, we obtain

$$\overset{\square}{V} := R\dot{U}R^T = dV + (w - \Omega)V. \tag{7.3.11c}$$

Now, it is possible to solve this equation explicitly for $(w - \Omega)$ by repeatedly using the Cayley–Hamilton theorem; see Mehrabadi and Nemat–Nasser [1987]. The end result is expressed as

$$\boxed{w - \Omega = \Lambda(V) : d,} \tag{7.3.12a}$$

where $\Lambda(V)$ is a fourth-order tensor defined by

$$\boxed{\Lambda : d = \frac{1}{I_V \mathrm{II}_V - III_V}\left\{I_V^2(Vd - dV) - I_V(bd - db) + bdV - Vdb\right\}.} \tag{7.3.12b}$$

Note that Λ_{abcd} *does not* have the major symmetries but possesses minor symmetry and skew-symmetry, i.e.,

$$\left.\begin{aligned} &\Lambda_{abcd} \neq \Lambda_{cdab} \\ &\Lambda_{abcd} = \Lambda_{abdc} = -\Lambda_{badc}. \end{aligned}\right\} \tag{7.3.12c}$$

This conclusion is undesirable. In effect by substituting (7.3.12a) in (7.3.11a), we obtain

$$\begin{aligned}\overset{\square}{\tau}_{ab} &= [\mathsf{a}_{abcd} - \Lambda_{aecd}\tau_{eb} + \tau_{ae}\Lambda_{ebcd}]\, d_{cd} \\ &=: \tilde{\mathsf{a}}_{abcd}(V) d_{cd}.\end{aligned} \tag{7.3.13a}$$

However, in view of (7.3.12c) and (7.3.12b),

$$\tilde{\mathsf{a}}_{abcd} - \tilde{\mathsf{a}}_{cdab} \neq 0. \tag{7.3.13b}$$

In other words, from (7.3.13a) and (7.3.13b) it follows that the constitutive equation (7.3.10) possesses *nonsymmetric moduli*. As a result, fundamental issues aside, this formulation leads necessarily to nonsymmetric tangent stiffness matrices in numerical implementations. We view this result as an undesirable feature.

7.3.2.3 Consistency condition. Elastoplastic moduli.

The development of elastoplastic moduli for the model summarized in BOX **7.2** follows the familiar lines discussed in detail in the context of the linear theory.

BOX 7.2. Ad Hoc Phenomenological Model in the Rotated Description.

i. Hypoelastic stress-strain relationships

$$\dot{\boldsymbol{\Sigma}} = \mathbf{A} : [\mathbf{D} - \mathbf{D}^p];$$

$$\boldsymbol{\Sigma} := \boldsymbol{R}^T \boldsymbol{\tau} \boldsymbol{R} \quad ; \quad \mathbf{D} := \boldsymbol{R}^T \boldsymbol{d} \boldsymbol{R}.$$

ii. Elastic domain and yield condition

$$\mathbb{E}_\Sigma := \left\{ (\boldsymbol{\Sigma}, \boldsymbol{Q}) \in \mathbb{S} \times \mathbb{R}^m \mid \mathcal{F}(\boldsymbol{\Sigma}, \boldsymbol{Q}) \leq 0 \right\}.$$

iii. Flow rule and hardening law

$$\mathbf{D}^p = \lambda \frac{\partial \mathcal{F}(\boldsymbol{\Sigma}, \boldsymbol{Q})}{\partial \boldsymbol{\Sigma}},$$

$$\dot{\boldsymbol{Q}} = -\lambda \mathsf{H}(\boldsymbol{\Sigma}, \boldsymbol{Q}).$$

iv. Kuhn–Tucker loading/unloading conditions

$$\lambda \geq 0, \quad \mathcal{F}(\boldsymbol{\Sigma}, \boldsymbol{Q}) \leq 0, \quad \lambda \mathcal{F}(\boldsymbol{\Sigma}, \boldsymbol{Q}) = 0.$$

v. Consistency condition

$$\lambda \dot{\mathcal{F}}(\boldsymbol{\Sigma}, \boldsymbol{Q}) = 0.$$

Time differentiation of the yield condition gives

$$\begin{aligned} \dot{\mathcal{F}} &= \frac{\partial \mathcal{F}}{\partial \boldsymbol{\Sigma}} : \dot{\boldsymbol{\Sigma}} + \frac{\partial \mathcal{F}}{\partial \boldsymbol{Q}} \cdot \dot{\boldsymbol{Q}} \\ &= \frac{\partial \mathcal{F}}{\partial \boldsymbol{\Sigma}} : \mathbf{A} : \left[\mathbf{D} - \lambda \frac{\partial \mathcal{F}}{\partial \boldsymbol{\Sigma}} \right] - \lambda \frac{\partial \mathcal{F}}{\partial \boldsymbol{Q}} \cdot \mathbf{H}, \end{aligned} \tag{7.3.14}$$

where we have used the hypoelastic relationships with the flow rule and hardening law. Then the Kuhn–Tucker conditions and the consistency condition require that

$$\mathcal{F} = \dot{\mathcal{F}} = 0, \tag{7.3.15}$$

for λ to be nonzero. Therefore,

$$\boxed{\lambda = \frac{\frac{\partial \mathcal{F}}{\partial \boldsymbol{\Sigma}} : \mathbf{A} : \mathbf{D}}{\frac{\partial \mathcal{F}}{\partial \boldsymbol{\Sigma}} : \mathbf{A} : \frac{\partial \mathcal{F}}{\partial \boldsymbol{\Sigma}} + \frac{\partial \mathcal{F}}{\partial \boldsymbol{Q}} \cdot \mathbf{H}} \geq 0,} \tag{7.3.16}$$

and we arrive at the expression

$$\dot{\boldsymbol{\Sigma}} = \mathbf{A}_{\text{ep}} : \mathbf{D}, \tag{7.3.17a}$$

where

$$\boxed{\mathbf{A}_{\text{ep}} = \begin{cases} \mathbf{A} & \text{if } \lambda = 0, \\ \mathbf{A} - \dfrac{(\mathbf{A} : \frac{\partial \mathcal{F}}{\partial \boldsymbol{\Sigma}}) \otimes (\mathbf{A} : \frac{\partial \mathcal{F}}{\partial \boldsymbol{\Sigma}})}{\frac{\partial \mathcal{F}}{\partial \boldsymbol{\Sigma}} : \mathbf{A} : \frac{\partial \mathcal{F}}{\partial \boldsymbol{\Sigma}} + \frac{\partial \mathcal{F}}{\partial \boldsymbol{Q}} \cdot \mathbf{H}} & \text{if } \lambda > 0. \end{cases}} \tag{7.3.17b}$$

A rate equation in the spatial description identical to (7.2.9) but with $\mathbf{a}$ now replaced by $\mathbf{a}_{ep}$ is obtained by defining $\mathbf{a}_{ep}$ in terms of $\mathbf{A}_{ep}$ by the transformation rule (7.3.8). Note that the resulting rate equation is subjected to the same shortcomings alluded to in Remarks **7.3.1** and **7.3.2**.

7.3.2.4 Criticism of the model.

The constitutive model outlined in this section is subject to a number of objections which make the model too specific to be considered a suitable extension of the linearized theory with general applicability. In particular,

i. the hypoelastic nature of the stress response renders the model questionable on fundamental grounds, and limits its applicability to "small" elastic strains (which, by the way, are never defined; only the *rate* of deformation $\mathbf{D}^e$ appears in the theory);
ii. the entire construction seems somewhat unmotivated. One wonders what is the special feature that points to the rotated description as the canonical set-up for the formulation of the theory;
iii. the difficulties alluded to in Remarks **7.3.1** and **7.3.2** appear more substantial than a mere computational inconvenience;
iv. computationally, the implementation of the model is involved and requires repeated use of the polar decomposition.

8

Objective Integration Algorithms for Rate Formulations of Elastoplasticity

In this chapter, we address the formulation of integration algorithms for the class of plasticity models at finite strains outlined in Chapter 7. The distinctive feature of these widely used models lies in the fact that the elastic response is formulated in the Eulerian or spatial description by a hypoelastic constitutive equation. Despite the number of shortcomings alluded to in Section 7.3, this class of models is currently widely used in large-scale inelastic calculations, as in Key [1974]; Key, Stone, and Krieg [1981]; Nagtegaal and Veldpaus [1984]; Goudreau and Hallquist [1982]; Rolph and Bathe [1984]; and Hallquist [1979,1988]. Therefore, in view of their practical relevance, a careful analysis of the issues involved in their numerical implementation is worthwhile.

From a computational standpoint, the central issue concerns the numerical integration of the constitutive model so that the resulting discrete equations identically satisfy the principle of material frame indifference. Satisfaction of this fundamental restriction leads to the so-called *incrementally objective algorithms*, a notion formalized in Hughes and Winget [1980] and subsequently considered by a number of authors; e.g., Rubinstein and Atluri [1983]; and Pinsky, Ortiz, and Pister [1983]. The condition of incremental objectivity precludes the generation of spurious stresses in rigid body motions. Although it appears that such a requirement furnishes a rather obvious criterion, a number of algorithmic treatments of plasticity at finite strain exist which violate this condition (e.g., McMeeking and Rice[1975]).

Conceptually, the procedure underlying the formulation of objective constitutive algorithms is rather straightforward. The given spatial, rate-constitutive equations are mapped to a local configuration which is now *unaffected* by superposed spatial rigid body motions. Then a time-stepping algorithm is performed in this local configuration, and the discrete equations are mapped back to the Eulerian description. In the actual implementation of this idea, two basic methodologies can be adopted.

i. *Convective representation.* In this approach one exploits the fact that objects in the convective representation, such as the right Cauchy–Green tensor $\boldsymbol{C}$ or the symmetric second Piola–Kirchhoff tensor $\boldsymbol{S}$, remain *unaltered* under superposed spatial rigid body motions. We recall that the convective representation is obtained from the Eulerian or spatial representation by appropriate tensorial transformations with the deformation gradient. Equivalently, by using convected coordinates, the governing equations are automatically transformed to the convective representation (recall that in convected coordinates, the deformation gradient is the identity tensor).

In rigid body mechanics, the convective representation is called the *body representation*, and the convected coordinates become the familiar body coordinates; see e.g., Goldstein [1981]. In solid mechanics, convected coordinates have been used by a number of authors, notably Green and Zerna [1960] in the context of nonlinear elasticity and Hill [1978] in the wider context of inelasticity. In computational applications, convected coordinates have been extensively used by Needleman [1984, and references therein].

Computationally, the convective representation plays a crucial role in the derivation of an algorithmic approximation to the rate of deformation tensor based on the generalized midpoint rule. In contrast with approximations constructed by several authors, e.g., Pinsky, Ortiz, and Pister [1983], in Section **8.1** we show that objectivity of the algorithmic approximation over an interval $[t_n, t_{n+1}]$ can be retained for any choice of the configuration at $t_{n+\alpha} \in [t_n, t_{n+1}]$, with $\alpha \in [0, 1]$ not necessarily restricted to the value $\alpha = \frac{1}{2}$. On the negative side, however, we show in Section **8.3** that, for J_2 flow theory, the simple structure of the radial return algorithm can be retained only for a very specific choice of objective rate: the Lie derivative of the Kirchhoff stress.

ii. *Local rotated representation.* In this approach, examined in detail Section **8.3**, the evolution equations are transformed to a locally Cartesian rotating coordinate system which is constructed precisely to ensure that the "rotated objects" remain *unaltered* under superposed spatial rigid body motions. Then the constitutive integrative algorithm is performed in the rotated description, and subsequently the discrete equations are transformed back to the spatial configuration. This methodology is ideally suited for "rotational-like" objective stress rates obtained by modifying the material time derivative of the Kirchhoff stress with additional terms involving spin-like tensors. Typical examples include the Jaumann derivative and the Green–McInnis–Naghdi rate. For J_2 flow theory, we show that this approach results in a return mapping essentially identical to the classical radial return method of the infinitesimal theory.

The crucial computational aspect in this second methodology addressed in Section **8.3.2**, concerns determinating the rotated local configuration. As pointed out in Hughes [1984], the problem can be reduced to numerically integrating an initial-value problem that generates a one-parameter subgroup of proper orthogonal transformations. In Hughes and Winget [1980] and Hughes [1984], integration over the interval $[t_n, t_{n+1}]$ is accomplished by a midpoint rule algorithm, which produces an incremental orthogonal transformation when the midpoint configuration coincides with the *half-step* configuration, i.e., for the value $\alpha = \frac{1}{2}$. Here, we improve on these results in several aspects, in particular,

a. the generalized midpoint rule algorithm in Section 8.3.2 generates an incremental orthogonal transformation regardless of the value assigned to the parameter $\alpha \in [0, 1]$ which defines the $t_{n+\alpha}$ configuration; and

b. the algorithm is singularity-free, placing no restriction on the value of the admissible incremental rotation, and remains objective for all $\alpha \in [0, 1]$.

A detailed account of the implementation of this algorithm, which is inspired by Simo and Quoc [1986], is given in Section **8.3.2.**

8.1 Objective Time-Stepping Algorithms

In this section we consider formulating integration algorithms for objects defined in the spatial description, which identically satisfy the fundamental requirement of frame indifference. These types of algorithms are commonly called *incrementally objective*, a nomenclature first introduced in Hughes and Winget [1980].

We are concerned with the generalized midpoint rule type of algorithms formulated to exactly preserve objectivity. As outlined above, one proceeds as follows:

i. Given an *objective rate quantity* in the spatial description, such as the rate of deformation tensor or an objective stress rate, *tensorially transform this object to the reference configuration*. Geometrically, this transformation is called a *pull-back* whereas, from a mechanical point of view, it can be interpreted *as going from the spatial to the convected description.*

ii. Time step with the generalized midpoint rule (or any other algorithm) *in the convected description.*

iii. Transform the result to the spatial description. Geometrically, this operation is called a push-forward.

By construction, this procedure, illustrated in Figure **8.1**, preserves objectivity exactly.

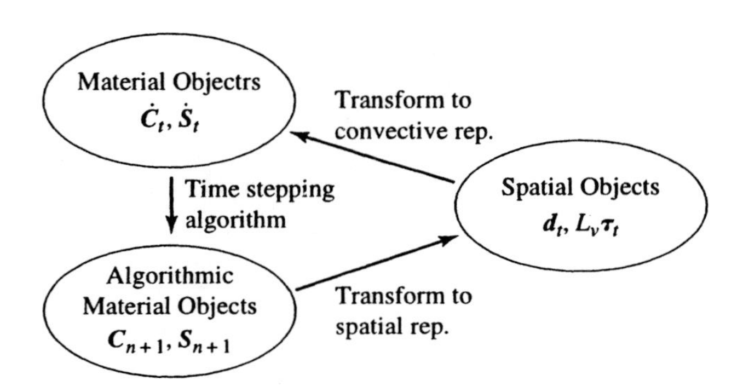

FIGURE 8.1. General algorithmic scheme for developing incrementally objective algorithms.

8.1.1 The Geometric Setup

As usual, we let $\mathcal{B} \subset \mathbb{R}^{n_{\text{dim}}}$ be the reference configuration of the continuum body of interest and let $\varphi : \mathcal{B} \times [0, T] \to \mathbb{R}^{n_{\text{dim}}}$ be the motion. We assume that the configuration is known at time $t_n \in [0, T]$. Equivalently, we regard the set

$$\varphi_n(\mathcal{B}) = \{x_n = \varphi_n(X) \mid X \in \mathcal{B}\} \tag{8.1.1}$$

as *given* data. Further, let $\varphi_{n+1}: \mathcal{B} \to \mathbb{R}^{n_{\text{dim}}}$ be the configuration at time $t_{n+1} = t_n + \Delta t$, where $\Delta t > 0$, which is also assumed to be given by

$$\boxed{\begin{aligned} \varphi_{n+1}(X) &= \varphi_n(X) + U(X) \\ &= \varphi_n(X) + u\left[\varphi_n(X)\right]. \end{aligned}} \tag{8.1.2}$$

Here $u : \varphi_n(\mathcal{B}) \to \mathbb{R}^{n_{\text{dim}}}$ is the *given* incremental displacement of $\varphi_n(\mathcal{B})$. Now let $\varphi_{n+\alpha}(\mathcal{B})$ denote a one-parameter family of configurations defined by the expression (see Figure **8.2**):

$$\boxed{\varphi_{n+\alpha} = \alpha\varphi_{n+1} + (1-\alpha)\varphi_n\,, \quad \alpha \in [0, 1].} \tag{8.1.3}$$

Our objective is to *find algorithmic approximations for spatial rate-like objects*, in terms of the incremental displacement $u(x_n)$ and the time increment Δt, *which exactly preserve proper invariance under superposed rigid body motions*, i.e. objectivity. In particular, we consider the following objects:

i. the *rate of deformation tensor* d, defined in Section **7.1.3**.
ii. the following two representative objective stress rates:
   a. the Lie derivative of the Kirchhoff stress tensor, $L_v\tau$, defined by (7.1.72); and
   b. the Jaumann derivative of the Kirchhoff stress tensor $\overset{\nabla}{\tau}$, defined by (7.1.75).

We use these results to develop integration algorithms for the hypoelastic equations considered in Section **7.3**.

8.1.1.1 Kinematic relationsships.

From definition (8.1.3), it follows that the deformation gradient

$$F_{n+\alpha} = \frac{\partial\varphi_{n+\alpha}}{\partial X} = D\varphi_{n+\alpha}(X) \tag{8.1.4a}$$

is given by the relationship

$$\boxed{F_{n+\alpha} = \alpha F_{n+1} + (1-\alpha)F_n\,, \quad \alpha \in [0, 1].} \tag{8.1.4b}$$

In addition, for subsequent developments, it proves convenient to introduce relative deformation gradients with respect to the configurations $\varphi_n(\mathcal{B})$. Thus, by definition, we set (see Figure **8.3**)

$$\boxed{\begin{aligned} f_{n+\alpha} &:= F_{n+\alpha}F_n^{-1}, \\ f_{n+1} &:= F_{n+1}F_n^{-1}, \\ \tilde{f}_{n+\alpha} &:= f_{n+1}f_{n+\alpha}^{-1}. \end{aligned}} \tag{8.1.5}$$

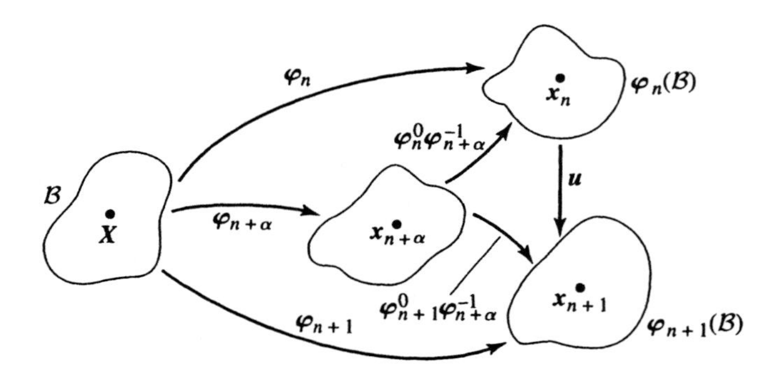

FIGURE 8.2. Geometric setup for integrating spatial rate-like quantities.

The expressions below will be useful in the developments that follow.

Lemma 8.1. *Define the relative incremental displacement gradient by the expression*

$$\boxed{h_{n+\alpha}(x_{n+\alpha}) := \nabla_{n+\alpha}\tilde{u}(x_{n+\alpha}) \equiv \frac{\partial \tilde{u}(x_{n+\alpha})}{\partial x_{n+\alpha}},} \tag{8.1.6}$$

where $\tilde{u}(x_{n+\alpha}) = U(X)\big|_{X=\varphi_{n+\alpha}^{-1}(x_{n+\alpha})}$ *is the incremental displacement field referred to the configuration* $\varphi_{n+\alpha}(\mathcal{B})$. *Then*

$$\boxed{\begin{aligned} h_{n+\alpha} &= [GRAD\, U]F_{n+\alpha}^{-1} = [\nabla_n u]f_{n+\alpha}^{-1}, \\ f_{n+\alpha} &= 1 + \alpha\nabla_n u, \end{aligned}} \tag{8.1.7}$$

where $\nabla_n u(x_n) = \partial u(x_n)/\partial x_n$ *and* $\alpha \in [0, 1]$.

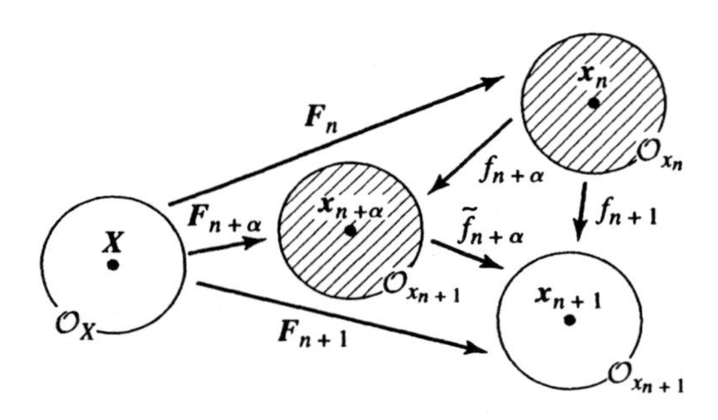

FIGURE 8.3. Deformation gradients and relative deformation gradients mapping neighborhoods $\mathcal{O}_x$, $\mathcal{O}_{x_n}$, and $\mathcal{O}_{x_{n+1}}$.

PROOF. It follows by a straightforward application of the chain rule. For instance, for the relationship $\boldsymbol{U}(\boldsymbol{X}) = \boldsymbol{u}[\varphi_n(\boldsymbol{X})]$,

$$\text{GRAD}\, \boldsymbol{U} = \left[\nabla_n \boldsymbol{u}\right] \boldsymbol{F}_n. \tag{8.1.8}$$

On the other hand, rearranging (8.1.4*b*),

$$\begin{aligned} \boldsymbol{F}_{n+\alpha} &= \boldsymbol{F}_n + \alpha(\boldsymbol{F}_{n+1} - \boldsymbol{F}_n) \\ &= \boldsymbol{F}_n + \alpha\, \text{GRAD}\, \boldsymbol{U}. \end{aligned} \tag{8.1.9}$$

Combining (8.1.8) and (8.1.9),

$$\boldsymbol{f}_{n+\alpha} = \boldsymbol{F}_{n+\alpha} \boldsymbol{F}_n^{-1} = \boldsymbol{1} + \alpha \nabla_n \boldsymbol{u}, \tag{8.1.10}$$

which proves $(8.1.7)_2$. Similarly, from $\boldsymbol{U}(\boldsymbol{X}) = \tilde{\boldsymbol{u}}\left[\varphi_{n+\alpha}(\boldsymbol{X})\right]$,

$$\begin{aligned} \nabla_{n+\alpha} \tilde{\boldsymbol{u}} &= (\text{GRAD}\, \boldsymbol{U}) \boldsymbol{F}_{n+\alpha}^{-1} \\ &= (\nabla_n \boldsymbol{u})\, \boldsymbol{F}_n \boldsymbol{F}_{n+\alpha}^{-1} \\ &= (\nabla_n \boldsymbol{u})(\boldsymbol{F}_{n+\alpha} \boldsymbol{F}_n^{-1})^{-1}, \end{aligned} \tag{8.1.11}$$

which, in view of $(8.1.5)_1$, proves $(8.1.7)_1$. □

8.1.2 *Approximation for the Rate of Deformation Tensor*

We construct an incrementally objective approximation for the rate of deformation tensor following the steps outlined above. The crucial identity to be exploited is given by (7.1.39) as

$$\boxed{\dot{\boldsymbol{C}} = 2\boldsymbol{F}^T \boldsymbol{d} \boldsymbol{F};}$$

i.e. (8.1.12)

$$\boxed{\dot{C}_{AB} = 2F_{aA} d_{ab} F_{bB},}$$

which transforms $\boldsymbol{d}$, a spatial tensor field, into $\dot{\boldsymbol{C}}$, a tensor field defined on $\mathcal{B}$ (convective description).

Proposition 8.1. *An objective approximation for the rate of deformation tensor in $[t_n, t_{n+1}]$, where $t_{n+1} = t_n + \Delta t$, is given by the formulas*

$$\boxed{\begin{aligned} \boldsymbol{d}_{n+\alpha} &= \frac{1}{\Delta t} \boldsymbol{f}_{n+\alpha}^{-T} \boldsymbol{E}_{n+1} \boldsymbol{f}_{n+\alpha}^{-1}, \\ \boldsymbol{E}_{n+1} &:= \tfrac{1}{2}\left[\boldsymbol{f}_{n+1}^T \boldsymbol{f}_{n+1} - \boldsymbol{1}\right], \\ \boldsymbol{f}_{n+\alpha} &:= \boldsymbol{1} + \alpha \nabla_n \boldsymbol{u}, \quad \alpha \in [0, 1], \end{aligned}} \tag{8.1.13}$$

where $\boldsymbol{E}_{n+1}$ is the relative Lagrangian strain tensor of configuration $\varphi_{n+1}(\mathcal{B})$ with respect to configuration $\varphi_n(\mathcal{B})$.

PROOF. Step **i** (see the preamble of Section 8.1) requires transforming $\boldsymbol{d}_{n+\alpha}$ to the convective description. From (8.1.12),

$$\dot{\boldsymbol{C}}_{n+\alpha} = 2\boldsymbol{F}_{n+\alpha}^T \boldsymbol{d}_{n+\alpha} \boldsymbol{F}_{n+\alpha}; \quad \alpha \in [0, 1]. \tag{8.1.14}$$

According to step **ii**, the time-stepping algorithm takes place in the convective description. By the generalized midpoint rule,

$$\boldsymbol{C}_{n+1} - \boldsymbol{C}_n = \dot{\boldsymbol{C}}_{n+\alpha} \Delta t, \quad \alpha \in [0, 1]. \tag{8.1.15}$$

Next, we recall that

$$\boldsymbol{F}_{n+1} = \boldsymbol{f}_{n+1} \boldsymbol{F}_n \Rightarrow \boldsymbol{C}_{n+1} = \boldsymbol{F}_n^T \big[\boldsymbol{f}_{n+1}^T \boldsymbol{f}_{n+1}\big] \boldsymbol{F}_n. \tag{8.1.16}$$

Combining (8.1.14), (8.1.15), and (8.1.16),

$$\boldsymbol{F}_{n+\alpha}^T \boldsymbol{d}_{n+\alpha} \boldsymbol{F}_{n+\alpha} = \frac{1}{2\Delta t} \boldsymbol{F}_n^T \big[\boldsymbol{f}_{n+1}^T \boldsymbol{f}_{n+1} - \mathbf{1}\big] \boldsymbol{F}_n. \tag{8.1.17}$$

Finally, in view of the relationship $\boldsymbol{f}_{n+\alpha} = \boldsymbol{F}_{n+\alpha} \boldsymbol{F}_n^{-1}$, (8.1.17) yields

$$\boldsymbol{d}_{n+\alpha} = \frac{1}{2\Delta t} \boldsymbol{f}_{n+\alpha}^{-T} \big[\boldsymbol{f}_{n+1}^T \boldsymbol{f}_{n+1} - \mathbf{1}\big] \boldsymbol{f}_{n+\alpha}^{-1}, \tag{8.1.18}$$

which proves the result. □

Remarks 8.1.1.

1. Suppose that the *deformation in* $[t_n, t_{n+1}]$ *is rigid*. Then

$$\boldsymbol{x}_{n+1} = \boldsymbol{Q} \boldsymbol{x}_n + \boldsymbol{c}, \quad \text{for } \boldsymbol{Q} \in \mathrm{SO}(3), \tag{8.1.19}$$

   and $\boldsymbol{c} \in \mathbb{R}^3$. It follows that

$$\boldsymbol{f}_{n+1} = \boldsymbol{Q} \in \mathrm{SO}(3) \Rightarrow \boldsymbol{E}_{n+1} = \tfrac{1}{2}\big[\boldsymbol{Q}^T \boldsymbol{Q} - \mathbf{1}\big] \equiv \mathbf{0}, \tag{8.1.20}$$

   which, in view of proposition 8.1, implies

$$\boxed{\boldsymbol{f}_{n+1} \in \mathrm{SO}(3) \iff \boldsymbol{d}_{n+\alpha} = 0\,.} \tag{8.1.21}$$

   Therefore $\boldsymbol{d}_{n+\alpha}$, *as given by (8.1.13), vanishes identically for incremental rigid motions*. This result is a manifestation of the property of incremental objectivity, which holds for all $\alpha \in [0, 1]$.
2. Note that (8.1.13) makes geometric sense. $\boldsymbol{E}_{n+1}$ is a tensor defined on $\varphi_n(\mathcal{B})$ which is transformed tensorially (by push-forward) to $\varphi_{n+\alpha}(\mathcal{B})$ by the right-hand side of (8.1.13). On the other hand, $\boldsymbol{d}_{n+\alpha}$ is a tensor on $\varphi_{n+\alpha}(\mathcal{B})$, and therefore formula (8.1.13) makes sense.

A convenient alternative expression for $\boldsymbol{d}_{n+\alpha}$ is contained in the following.

Corollary 8.1. *The following expression is equivalent to (8.1.13):*

$$\boxed{\begin{aligned} \boldsymbol{d}_{n+\alpha} &= \frac{1}{2\Delta t}\big[\boldsymbol{h}_{n+\alpha} + \boldsymbol{h}_{n+\alpha}^T + (1 - 2\alpha)\boldsymbol{h}_{n+\alpha}^T \boldsymbol{h}_{n+\alpha}\big], \\ \boldsymbol{h}_{n+\alpha} &:= \nabla_{n+\alpha} \widetilde{\boldsymbol{u}} \quad \textit{(see Lemma } \mathbf{8.1}\textit{)}. \end{aligned}} \tag{8.1.22}$$

PROOF. From (8.1.4) and $(8.1.7)_1$,

$$\begin{aligned} F_{n+1} &= F_{n+\alpha} + (1-\alpha)[F_{n+1} - F_n] \\ &= F_{n+\alpha} + (1-\alpha)\ \mathrm{GRAD}\, U \\ &= F_{n+\alpha} + (1-\alpha) h_{n+\alpha} F_{n+\alpha}. \end{aligned} \tag{8.1.23a}$$

Then an analogous calculation for F_n yields

$$\left.\begin{aligned} F_{n+1} &= [1 + (1-\alpha) h_{n+\alpha}] F_{n+\alpha}, \\ F_n &= [1 - \alpha h_{n+\alpha}] F_{n+\alpha}. \end{aligned}\right\} \tag{8.1.23b}$$

From these expressions we obtain

$$C_{n+1} - C_n = F^T_{n+\alpha}\left[h_{n+\alpha} + h^T_{n+\alpha} + (1-2\alpha) h^T_{n+\alpha} h_{n+\alpha}\right] F_{n+\alpha}\,. \tag{8.1.24}$$

Now the result follows by inserting (8.1.24) into (8.1.15) and using (8.1.14). □

Remark 8.1.2. For $\alpha = \frac{1}{2}$, expression (8.1.22) collapses to the simple formula

$$\boxed{d_{n+1/2} = \frac{1}{2\Delta t}\left(h_{n+1/2} + h^T_{n+1/2}\right),} \tag{8.1.25}$$

which is second-order accurate and *linear in* $\boldsymbol{u}$. Clearly, this is the most attractive expression from the viewpoint of implementation.

8.1.3 Approximation for the Lie Derivative

An objective algorithmic approximation for the Lie derivative of the Kirchhoff stress $L_v\tau$ is constructed by exploiting the identity

$$\boxed{L_v \tau = F \dot{S} F^T.} \tag{8.1.26}$$

The result is given in the following.

Proposition 8.2. *Within the interval* $[t_n, t_{n+1}]$, *the algorithmic approximation*

$$\boxed{L_v \tau_{n+\alpha} = \frac{1}{\Delta t} f_{n+\alpha}\left[f^{-1}_{n+1}\tau_{n+1} f^{-T}_{n+1} - \tau_n\right] f^T_{n+\alpha}} \tag{8.1.27}$$

is objective for all $\alpha \in [0, 1]$.

PROOF. Let $t_{n+\alpha} \in [t_n, t_{n+1}]$. From (8.1.26) and the generalized midpoint rule,

$$\begin{aligned} F^{-1}_{n+\alpha}(L_v \tau_{n+\alpha}) F^{-T}_{n+\alpha} &= \dot{S}_{n+\alpha} \\ &= \frac{1}{\Delta t}(S_{n+1} - S_n) \\ &= \frac{1}{\Delta t}\left(F^{-1}_{n+1}\tau_{n+1}F^{-T}_{n+1} - F^{-1}_n \tau_n F^{-T}_n\right) \\ &= \frac{1}{\Delta t} F^{-1}_n \left(f^{-1}_{n+1}\tau_{n+1} f^{-T}_{n+1} - \tau_n\right) F^{-T}_n, \end{aligned} \tag{8.1.28}$$

where we have used $(8.1.5)_2$. The result follows from $(8.1.5)_1$. □

Remarks 8.1.3.

1. Expression (8.1.27) can be formulated in the following alternative form. Let

$$\boxed{\tilde{\boldsymbol{f}}_{n+\alpha} := \boldsymbol{F}_{n+1}\boldsymbol{F}_{n+\alpha}^{-1} = \boldsymbol{f}_{n+1}\boldsymbol{f}_{n+\alpha}^{-1}} \tag{8.1.29}$$

   be the deformation gradient of configuration $\varphi_{n+1}(\mathcal{B})$ *relative* to configuration $\varphi_{n+\alpha}(\mathcal{B})$. Then, a straightforward computation yields

$$\boxed{L_v \boldsymbol{\tau}_{n+\alpha} = \frac{1}{\Delta t}\tilde{\boldsymbol{f}}_{n+\alpha}^{-1}\left(\boldsymbol{\tau}_{n+1} - \boldsymbol{f}_{n+1}\boldsymbol{\tau}_n \boldsymbol{f}_{n+1}^T\right)\tilde{\boldsymbol{f}}_{n+\alpha}^{-T}.} \tag{8.1.30}$$

2. Similarly, expression (8.1.13) can be recast in the form

$$\boxed{\begin{aligned} \boldsymbol{d}_{n+\alpha} &= \frac{1}{\Delta t}\tilde{\boldsymbol{f}}_{n+\alpha}^T \boldsymbol{e}_{n+1}\tilde{\boldsymbol{f}}_{n+\alpha}, \\ \boldsymbol{e}_{n+1} &:= \tfrac{1}{2}\left[\mathbf{1} - (\boldsymbol{f}_{n+1}\boldsymbol{f}_{n+1}^T)^{-1}\right]. \end{aligned}} \tag{8.1.31}$$

   Observe that $\boldsymbol{e}_{n+1}$ defined by $(8.1.31)_2$ is the *Eulerian strain tensor relative* to the configurations $\varphi_{n+1}(\mathcal{B})$ and $\varphi_n(\mathcal{B})$. This expression is consistent with the definition of $\boldsymbol{d}$ as the Lie derivative of the strain tensor; see Hughes [1984] and Simo, Marsden, and Krishnaprasad [1988].

8.1.3.1 The Jaumann derivative of the Kirchhoff stress.

As alluded to in Chapter 7, any objective stress rate is a specialization of the Lie derivative, as first observed in Marsden and Hughes [1994]. Therefore, with formula (8.1.27) at our disposal, we can construct objective approximations to any stress rate. In particular, for the Jaumann derivative, we recall that according to relationship (7.1.91),

$$\overset{\triangledown}{\boldsymbol{\tau}} = L_v\boldsymbol{\tau} + \boldsymbol{d\tau} + \boldsymbol{\tau d}. \tag{8.1.32}$$

Consequently, for any $t_{n+\alpha} \in [t_n, t_{n+1}]$,

$$\boxed{\overset{\triangledown}{\boldsymbol{\tau}}_{n+\alpha} = L_v\boldsymbol{\tau}_{n+\alpha} + \boldsymbol{d}_{n+\alpha}\boldsymbol{\tau}_{n+\alpha} + \boldsymbol{\tau}_{n+\alpha}\boldsymbol{d}_{n+\alpha}.} \tag{8.1.33}$$

An explicit expression for the algorithmic approximation to $\overset{\triangledown}{\boldsymbol{\tau}}_{n+\alpha}$ can be immediately constructed by substituting our expressions (8.1.13) and (8.1.27) for $\boldsymbol{d}_{n+\alpha}$ and $L_v\boldsymbol{\tau}_{n+\alpha}$, respectively, in (8.1.27) as follows. First, let

$$\boldsymbol{\tau}_{n+\alpha} = \tilde{\boldsymbol{f}}_{n+\alpha}^{-1}\boldsymbol{\tau}_{n+1}\tilde{\boldsymbol{f}}_{n+\alpha}^{-T}, \tag{8.1.34}$$

a relationship which which is justified by tensorially transforming τ_{n+1} (a tensor field on $\varphi_{n+1}(\mathcal{B})$) to $\tau_{n+\alpha}$ (a tensor field on $\varphi_{n+\alpha}(\mathcal{B})$). Using (8.1.31),

$$\begin{aligned} \tau_{n+\alpha}\boldsymbol{d}_{n+\alpha} &= \frac{1}{\Delta t}\widetilde{\boldsymbol{f}}^{-1}_{n+\alpha}\tau_{n+1}\widetilde{\boldsymbol{f}}^{-T}_{n+\alpha}\widetilde{\boldsymbol{f}}^{T}_{n+\alpha}\boldsymbol{e}_{n+1}\widetilde{\boldsymbol{f}}_{n+\alpha} \\ &= \frac{1}{\Delta t}\widetilde{\boldsymbol{f}}^{-1}_{n+\alpha}\left(\tau_{n+1}\boldsymbol{e}_{n+1}\widetilde{\boldsymbol{b}}_{n+1}\right)\widetilde{\boldsymbol{f}}^{-T}_{n+\alpha}, \end{aligned} \tag{8.1.35}$$

where $\widetilde{\boldsymbol{b}}_{n+1} := \widetilde{\boldsymbol{f}}_{n+\alpha}\widetilde{\boldsymbol{f}}^{T}_{n+\alpha}$ is the *relative left Cauchy–Green tensor*. Then a similar calculation for $\boldsymbol{d}_{n+\alpha}\tau_{n+\alpha}$ yields

$$\boxed{\tau_{n+\alpha}\boldsymbol{d}_{n+\alpha} + \boldsymbol{d}_{n+\alpha}\tau_{n+\alpha} = \frac{1}{\Delta t}\widetilde{\boldsymbol{f}}^{-1}_{n+\alpha}\left(\tau_{n+1}\boldsymbol{e}_{n+1}\widetilde{\boldsymbol{b}}_{n+1} + \widetilde{\boldsymbol{b}}_{n+1}\boldsymbol{e}_{n+1}\tau_{n+1}\right)\widetilde{\boldsymbol{f}}^{-T}_{n+\alpha}.} \tag{8.1.36}$$

From (8.1.30), (8.1.32), and (8.1.36), we arrive at the formula

$$\boxed{\overset{\triangledown}{\tau}_{n+\alpha} = \frac{1}{\Delta t}\widetilde{\boldsymbol{f}}^{-1}_{n+\alpha}\left(\tau_{n+1} + \tau_{n+1}\boldsymbol{e}_{n+1}\widetilde{\boldsymbol{b}}_{n+1} + \widetilde{\boldsymbol{b}}_{n+1}\boldsymbol{e}_{n+1}\tau_{n+1} - \boldsymbol{f}_{n+1}\tau_{n}\boldsymbol{f}^{T}_{n+1}\right)\widetilde{\boldsymbol{f}}^{-T}_{n+\alpha},} \tag{8.1.37}$$

which gives the algorithmic approximation to $\overset{\triangledown}{\tau}_{n+\alpha}$.

8.1.4 Application: Numerical Integration of Rate Constitutive Equations

As an illustration of the foregoing methodology, we consider the application of these objective integration algorithms to two representative rate constitutive equations:

8.1.4.1 Lie derivative of the Kirchhoff stress.

Assume a rate constitutive equation of the form (7.3.22):

$$\boxed{L_v\tau = \mathbf{a} : \boldsymbol{d}.} \tag{8.1.38}$$

Evaluation at $t_{n+\alpha} \in [t_n, t_{n+1}]$ and use of (8.1.30), (8.1.31) yield

$$\begin{aligned} \tau_{n+1} - \boldsymbol{f}_{n+1}\tau_n\boldsymbol{f}^{T}_{n+1} &= \widetilde{\boldsymbol{f}}_{n+\alpha}\left[\mathbf{a}_{n+\alpha} : \widetilde{\boldsymbol{f}}^{T}_{n+\alpha}\boldsymbol{e}_{n+1}\widetilde{\boldsymbol{f}}_{n+\alpha}\right]\widetilde{\boldsymbol{f}}^{T}_{n+\alpha} \\ &= \widetilde{\mathbf{a}}_{n+1} : \boldsymbol{e}_{n+1}, \end{aligned} \tag{8.1.39}$$

where

$$\boxed{\widetilde{\mathbf{a}}_{ijkl}|_{n+1} := \left(\widetilde{f}_{ia}\widetilde{f}_{jb}\widetilde{f}_{kc}\widetilde{f}_{ld}\mathbf{a}_{abcd}\right)_{n+\alpha}} \tag{8.1.40}$$

is the tensor of elastic moduli $\mathbf{a}$, *evaluated at configuration* $\varphi_{n+\alpha}(\mathcal{B})$ *and transformed to configuration* $\varphi_{n+1}(\mathcal{B})$ *with the relative deformation gradient* $\widetilde{\boldsymbol{f}}_{n+\alpha}$ (geometrically, the *push-forward* of $\mathbf{a}_{n+\alpha}$ to configuration $\varphi_{n+1}(\mathcal{B})$.)

From (8.1.29) and (8.1.30), we arrive at the update formula

$$\boxed{\begin{aligned} \boldsymbol{\tau}_{n+1} &= \boldsymbol{f}_{n+1}\boldsymbol{\tau}_n\boldsymbol{f}_{n+1}^T + \tilde{\mathbf{a}}_{n+1} : \boldsymbol{e}_{n+1}, \\ \boldsymbol{e}_{n+1} &= \tfrac{1}{2}\left[\mathbf{1} - \boldsymbol{f}_{n+1}^{-T}\boldsymbol{f}_{n+1}^{-1}\right], \end{aligned}} \tag{8.1.41}$$

for any $\alpha \in [0, 1]$.

Proposition 8.3. *Assume that the motion in* $[t_n, t_{n+1}]$ *is rigid. Then* $\boldsymbol{\tau}_n$ *and* $\boldsymbol{\tau}_{n+1}$ *are related through an objective transformation*

$$\boxed{\boldsymbol{f}_{n+1} = \boldsymbol{Q} \in \mathrm{SO}(3) \Longrightarrow \boldsymbol{\tau}_{n+1} = \boldsymbol{Q}\boldsymbol{\tau}_n\boldsymbol{Q}^T,} \tag{8.1.42}$$

for all $\alpha \in [0, 1]$.

PROOF. If the motion is rigid in $[t_n, t_{n+1}]$, by Remark 8.1.1, it follows that $\boldsymbol{e}_{n+1} = \mathbf{0}$ and $\boldsymbol{f}_{n+1} = \boldsymbol{Q} \in \mathrm{SO}(3)$. Hence, the second term in $(8.1.41)_1$ vanishes identically, and we obtain (8.1.42). □

Contrary to some statements found in the literature, the algorithm (8.1.41) is incrementally objective for *any* value of α in [0, 1], *not* necessarily restricted to $\alpha = \frac{1}{2}$.

8.1.4.2 Jaumann derivative of the Kirchhoff stress.

Next, we consider a constitutive equation of the form

$$\boxed{\overset{\nabla}{\boldsymbol{\tau}} = \mathbf{a} : \boldsymbol{d}.} \tag{8.1.43}$$

By evaluating this rate equation at $t_{n+\alpha} \in [t_n, t_{n+1}]$, transforming the result to $\varphi_{n+1}(\mathcal{B})$, and using (8.1.31) and (8.1.40),

$$\begin{aligned} \tilde{\boldsymbol{f}}_{n+\alpha}\overset{\nabla}{\boldsymbol{\tau}}_{n+\alpha}\tilde{\boldsymbol{f}}_{n+\alpha}^T &= \tilde{\boldsymbol{f}}_{n+\alpha}[\mathbf{a}_{n+\alpha} : \boldsymbol{d}_{n+\alpha}]\tilde{\boldsymbol{f}}_{n+\alpha}^T \\ &= \frac{1}{\Delta t}\tilde{\mathbf{a}}_{n+1} : \boldsymbol{e}_{n+1}, \end{aligned} \tag{8.1.44}$$

where $\tilde{\mathbf{a}}_{n+1}$ is given in terms of $\mathbf{a}_{n+\alpha}$ by (8.1.40). Substituting (8.1.37) in (8.1.44) results in the expression

$$\boxed{\boldsymbol{\tau}_{n+1} + \boldsymbol{\tau}_{n+1}\boldsymbol{e}_{n+1}\tilde{\boldsymbol{b}}_{n+1} + \tilde{\boldsymbol{b}}_{n+1}\boldsymbol{e}_{n+1}\boldsymbol{\tau}_{n+1} = \boldsymbol{f}_{n+1}\boldsymbol{\tau}_n\boldsymbol{f}_{n+1}^T + \tilde{\mathbf{a}}_{n+1} : \boldsymbol{e}_{n+1}.} \tag{8.1.45}$$

Assuming that $\mathbf{a}$ does not depend on $\boldsymbol{\tau}$, a hypothesis typically made in applications, equation (8.1.45) reduces to a relationship *linear in* $\boldsymbol{\tau}_{n+1}$, which is easily solved for a given deformation to obtain the updated stress $\boldsymbol{\tau}_{n+1}$. In fact, upon defining

$$k_{ijkl} = \delta_{ik}\delta_{jl} + \delta_{il}e_{km}\tilde{b}_{mj} + \tilde{b}_{im}e_{ml}\delta_{kj}, \tag{8.1.46}$$

which is known a priori for a given incremental deformation in $[t_n, t_{n+1}]$, we obtain the closed-form expression

$$\boxed{\boldsymbol{\tau}_{n+1} = \boldsymbol{k}_{n+1}^{-1}\left(\boldsymbol{f}_n \boldsymbol{\tau}_n \boldsymbol{f}_n^T + \tilde{\mathbf{a}} : \boldsymbol{e}_{n+1}\right).} \tag{8.1.47}$$

Remarks 8.1.4.

1. The algorithm outlined above is objective for *any* value of $\alpha \in [0, 1]$. This observation follows by recalling that $\boldsymbol{e}_{n+1} = \mathbf{0}$ for a rigid motion in $[t_n, t_{n+1}]$, so that (8.1.45) reduces to

$$\boldsymbol{\tau}_{n+1} = \boldsymbol{Q}\boldsymbol{\tau}_n \boldsymbol{Q}^T \quad \text{for } \boldsymbol{f}_{n+1} = \boldsymbol{Q} \in \mathrm{SO}(3) \iff \boldsymbol{e}_{n+1} = \mathbf{0}.$$

2. Observe that the algorithm proposed in Pinsky, Ortiz, and Pister [1983] requires a Newton iteration to solve for $\boldsymbol{\tau}_{n+1}$ for a given deformation. On the other hand, expression (8.1.47) furnishes a closed-form solution.

8.2 Application to J_2 Flow Theory at Finite Strains

From a computational standpoint, it is interesting to characterize the class of hypoelastic formulations of J_2 flow theory which result in a straightforward generalization of the classical radial return algorithm. It turns out that, within the algorithmic framework of the preceding section, such a generalization depends critically on two factors:

1. the structure of the rate equation governing hypoelastic response. In particular, for the class of objective integrators in Section **8.1**, a straightforward extension of the radial return method is possible provided that
   a. the objective stress rate is the Lie derivative of the Kirchhoff stress and
   b. the constitutive tensor $\mathbf{a}$ in (8.1.38) is constant and isotropic;
2. the choice of midpoint configuration in the objective integrator. This configuration and the configuration chosen to perform the radial return method must agree. In particular,
   a. the classical radial return, which is performed in the $\varphi_{n+1}(\mathcal{B})$–configuration requires use of the backward Euler method. Recall that this algorithm is only *first order accurate*;
   b. if the midpoint rule algorithm is adopted, the radial return must be performed in the midpoint configuration.

Thus, the preceding algorithmic framework excludes the use of J_2-plasticity models that employ the Jaumann derivative if the inherent simplicity of the radial return method is to be preserved. Note that the class of objective integrators in Section **8.1** is essentially equivalent to algorithms obtained by formulating the elastoplastic constitutive equations in convected coordinates, as in Needleman [1982].

As shown in the next section, the optimal algorithmic framework for hypoelastic formulations of plasticity that employ "rotational" objective rates, such as the Jaumann derivative, relies on the use of a *corotated coordinate* system. Within such a framework, the structure of the radial return method is preserved for J_2 flow theory.

8.2.1 A J_2 Flow Theory

Consider an extension to finite deformations of the classical J_2 flow theory governed by the following evolution equations:

$$\left.\begin{aligned} L_v\boldsymbol{\tau} &= \mathbf{a} : [\boldsymbol{d} - \boldsymbol{d}^p] \\ \boldsymbol{d}^p &= \gamma \boldsymbol{n} \\ \dot{\alpha} &= \sqrt{\tfrac{2}{3}}\gamma \\ f &:= \|\text{dev}[\boldsymbol{\tau}]\| - \sqrt{\tfrac{2}{3}}[\sigma_Y + K\alpha], \end{aligned}\right\} \tag{8.2.1a}$$

subject to the standard Kuhn–Tucker conditions. In particular, for plastic loading, $\dot{f} = 0$, and $\gamma > 0$. In addition, by analogy with the infinitesimal theory, we assume the conditions

$$\left.\begin{aligned} \mathbf{a} &= \lambda \mathbf{1} \otimes \mathbf{1} + 2\mu \mathbf{I}, \\ \boldsymbol{n} &= \text{dev}[\boldsymbol{\tau}] / \|\text{dev}[\boldsymbol{\tau}]\|. \end{aligned}\right\} \tag{8.2.1b}$$

Concerning the computation of $\dot{f}$, it should be carefully noted that

$$\frac{\partial}{\partial t}\|\text{dev}[\boldsymbol{\tau}]\| \neq \boldsymbol{n} : L_v\boldsymbol{\tau} \quad \text{and so} \quad 2\mu\gamma \neq \frac{\boldsymbol{n} : \mathbf{a} : \boldsymbol{d}}{1 + K/3\mu}. \tag{8.2.1c}$$

In other words, the standard expression for γ in the infinitesimal theory does not hold.

The goal of this section is to illustrate the algorithms discussed in Section **8.1.2** in the context of this model and show that, despite relationship (8.2.1c), the return mapping reduces to the radial return algorithm.

8.2.1.1 Discrete problem of evolution.

By evaluating (8.2.1a)$_1$ at the $\varphi_{n+\alpha}(\mathcal{B})$ configuration and using the flow rule (8.2.1a)$_2$, we find that

$$\Delta t L_v \boldsymbol{\tau}_{n+\alpha} = \mathbf{a} : [\Delta t \boldsymbol{d}_{n+\alpha} - \Delta\gamma \boldsymbol{n}_{n+\alpha}], \tag{8.2.2}$$

where we have set $\Delta\gamma = \gamma_{n+\alpha}\Delta t$. By recalling (8.1.34), using (8.1.27) and (8.1.30), and the fact that $\boldsymbol{f}_{n+\alpha} = \widetilde{\boldsymbol{f}}_{n+1}^{-1}\boldsymbol{f}_{n+1}$, equation (8.2.2) reduces to

$$\boldsymbol{\tau}_{n+\alpha} - \boldsymbol{f}_{n+\alpha}\boldsymbol{\tau}_n\boldsymbol{f}_{n+\alpha}^T = \mathbf{a} : (\Delta t \boldsymbol{d}_{n+\alpha} - \Delta\gamma \boldsymbol{n}_{n+\alpha}). \tag{8.2.3}$$

On the other hand, in view of (8.1.31), we set

$$\Delta t \boldsymbol{d}_{n+\alpha} = \widetilde{\boldsymbol{f}}_{n+\alpha}^T \boldsymbol{e}_{n+1} \widetilde{\boldsymbol{f}}_{n+\alpha} =: \widetilde{\boldsymbol{e}}_{n+\alpha}. \tag{8.2.4}$$

Since $\mathbf{a}$ is an isotropic tensor and tr $[\boldsymbol{n}_{n+\alpha}] = 0$, (8.2.1*b*), (8.2.3), and (8.2.4) lead to the following discrete system:

$$\boxed{\begin{aligned} \boldsymbol{\tau}_{n+\alpha}^{\text{trial}} &:= \boldsymbol{f}_{n+\alpha}\boldsymbol{\tau}_n\boldsymbol{f}_{n+\alpha}^T + \mathbf{a} : \widetilde{\boldsymbol{e}}_{n+\alpha}, \\ \boldsymbol{\tau}_{n+\alpha} &= \boldsymbol{\tau}_{n+\alpha}^{\text{trial}} - 2\mu\Delta\gamma\boldsymbol{n}_{n+\alpha}, \\ \alpha_{n+1} &= \alpha_n + \sqrt{\frac{2}{3}}\,\Delta\gamma. \end{aligned}} \tag{8.2.5}$$

Observe carefully that the *flow rule is also evaluated at the* $\varphi_{n+\alpha}(\mathcal{B})$*-configuration.*

8.2.1.2 Return-mapping algorithm.

Consistency remains to be imposed by *enforcing the Kuhn–Tucker conditions at the* $(n+\alpha)$*-configuration.* Accordingly, let

$$f_{n+\alpha} := \|\text{dev}[\boldsymbol{\tau}_{n+\alpha}^{\text{trial}}]\| - \sqrt{\tfrac{2}{3}}\,(\sigma_Y + K\alpha_n). \tag{8.2.6}$$

Then

$$\left.\begin{aligned} f_{n+\alpha}^{\text{trial}} < 0 \Rightarrow \text{ elastic step } &\iff \Delta\gamma = 0, \\ f_{n+\alpha}^{\text{trial}} > 0 \Rightarrow \text{ plastic step } &\iff \Delta\gamma > 0. \end{aligned}\right\} \tag{8.2.7}$$

Assuming that $f_{n+1}^{\text{trial}} > 0$, from $(8.2.5)_{2,3}$ and (8.2.6), we obtain

$$\boxed{\begin{aligned} \text{tr}[\boldsymbol{\tau}_{n+\alpha}] &= \text{tr}[\boldsymbol{\tau}_{n+\alpha}^{\text{trial}}], \\ \boldsymbol{n}_{n+\alpha} &= \frac{\text{dev}[\boldsymbol{\tau}_{n+\alpha}^{\text{trial}}]}{\|\text{dev}[\boldsymbol{\tau}_{n+\alpha}^{\text{trial}}]\|}, \\ \Delta\gamma &= \frac{\langle f_{n+1}^{\text{trial}}\rangle/2\mu}{1 + K/3\mu}. \end{aligned}} \tag{8.2.8}$$

Formulas (8.2.8) are identical to those appearing in the radial return algorithm discussed in Chapter **3**.

Remarks 8.2.1.

1. Kinematic hardening is easily incorporated into the theory by postulating an equation of evolution of the form

$$L_v\bar{\boldsymbol{q}} = \tfrac{2}{3}H\gamma\boldsymbol{n}; \qquad \boldsymbol{n} = \frac{\text{dev}[\boldsymbol{\tau}] - \bar{\boldsymbol{q}}}{\|\text{dev}[\boldsymbol{\tau}] - \bar{\boldsymbol{q}}\|}. \tag{8.2.9}$$

The integration algorithm becomes

$$\bar{\boldsymbol{q}}_{n+\alpha} = \boldsymbol{f}_{n+\alpha}\bar{\boldsymbol{q}}_n\boldsymbol{f}_{n+\alpha}^T + \tfrac{2}{3}\Delta\gamma H\boldsymbol{n}_{n+\alpha}, \tag{8.2.10}$$

and the return-mapping algorithm reduces to the radial return of the linearized theory, with

$$\boxed{\boldsymbol{\eta}_{n+\alpha}^{\text{trial}} = \text{dev}\left[\boldsymbol{\tau}_{n+\alpha}^{\text{trial}} - \boldsymbol{f}_{n+\alpha}\bar{\boldsymbol{q}}_n\boldsymbol{f}_{n+\alpha}^T\right].} \tag{8.2.11}$$

2. The algorithm is incrementally objective for $\alpha \in [0, 1]$.
3. The return map no longer reduces to the radial return if the Lie derivative is replaced by the Jaumann rate. To see this, observe that (8.1.33) yields

$$\begin{aligned} \Delta t L_v \boldsymbol{\tau}_{n+\alpha} &= \boldsymbol{\tau}_{n+\alpha} - \boldsymbol{f}_{n+\alpha}\boldsymbol{\tau}_n \boldsymbol{f}_{n+\alpha}^T \\ &= \overset{\nabla}{\boldsymbol{\tau}}_{n+\alpha}\Delta t - \widetilde{\boldsymbol{e}}_{n+\alpha}\boldsymbol{\tau}_{n+\alpha} - \boldsymbol{\tau}_{n+\alpha}\widetilde{\boldsymbol{e}}_{n+\alpha}. \end{aligned} \tag{8.2.12}$$

Since $\overset{\nabla}{\boldsymbol{\tau}}_{n+\alpha} = \mathbf{a} : [\boldsymbol{d}_{n+\alpha} - \boldsymbol{d}^p_{n+\alpha}]$, (8.2.12) and the flow rule yield the equation

$$\boxed{\boldsymbol{\tau}_{n+\alpha} + \widetilde{\boldsymbol{e}}_{n+\alpha}\boldsymbol{\tau}_{n+\alpha} + \boldsymbol{\tau}_{n+\alpha}\widetilde{\boldsymbol{e}}_{n+\alpha} = \boldsymbol{f}_{n+\alpha}\boldsymbol{\tau}_n \boldsymbol{f}_{n+\alpha}^T + \mathbf{a} : \widetilde{\boldsymbol{e}}_{n+\alpha} - 2\mu\Delta\gamma \boldsymbol{n}_{n+\alpha}.} \tag{8.2.13}$$

which does not lead to a solution of the form (8.2.8). In fact, now the consistency equation is nonlinear.

8.3 Objective Algorithms Based on the Notion of a Rotated Configuration

The construction of objective algorithms is also accomplished by modifying the procedure discussed in Section **8.1**. The idea is to use a *rotation neutralized description* of the spatial evolution equations in place of the convective description considered in Section **8.2**. To formalize this approach, let $\widehat{\boldsymbol{\omega}}$ be **any** spatial, skew-symmetric, second-order, tensor field. Consider the following evolution equation

$$\boxed{\begin{aligned} \dot{\boldsymbol{\Lambda}} &= (\widehat{\boldsymbol{\omega}} \circ \varphi)\boldsymbol{\Lambda} \\ \boldsymbol{\Lambda}\big|_{t=0} &= \mathbf{1}, \end{aligned}} \tag{8.3.1}$$

where $\boldsymbol{\Lambda}(\boldsymbol{X}, t)$ is a proper orthogonal tensor (in SO(3)) for fixed $\boldsymbol{X} \in \mathcal{B}$. For a given skew-symmetric $\widehat{\boldsymbol{\omega}}$, the solutions of (8.3.1) generate a *one-parameter subgroup of the special orthogonal group.*

EXAMPLE: 8.3.1. Typical choices for $\widehat{\boldsymbol{\omega}}$ include
i. the spin tensor $\boldsymbol{w} = \frac{1}{2}[\nabla \boldsymbol{v} - (\nabla \boldsymbol{v})^T]$; and
ii. the tensor $\boldsymbol{\Omega}$ defined as $\boldsymbol{\Omega} = \dot{\boldsymbol{R}}\boldsymbol{R}^T$ where $\boldsymbol{R}$ is the rotation tensor in the polar decomposition of $\boldsymbol{F}$.
Recall that $\boldsymbol{w}$ and $\boldsymbol{\Omega}$ are related through (7.3.12).

The one-parameter group of rotations generated by (8.3.1) leads to a one-parameter family of rotated local configurations. Let

$$\boxed{\begin{aligned} \boldsymbol{\Sigma} &:= \boldsymbol{\Lambda}^T \boldsymbol{\tau}\boldsymbol{\Lambda}, \\ \mathbf{D} &:= \boldsymbol{\Lambda}^T \boldsymbol{d}\boldsymbol{\Lambda}, \end{aligned}} \tag{8.3.2}$$

be the rotated Kirchhoff stress tensor and the rotated rate of deformation tensor, respectively. Observe that

$$\boxed{\dot{\boldsymbol{\Sigma}} = \boldsymbol{\Lambda}^T[\dot{\boldsymbol{\tau}} + \boldsymbol{\tau}\widehat{\boldsymbol{\omega}} - \widehat{\boldsymbol{\omega}}\boldsymbol{\tau}]\boldsymbol{\Lambda} =: \boldsymbol{\Lambda}^T\overset{\nabla}{\boldsymbol{\tau}}\boldsymbol{\Lambda},} \tag{8.3.3}$$

where $\overset{\nabla}{\boldsymbol{\tau}}$ is a corotated objective rate relative to axes with spin $\widehat{\boldsymbol{\omega}}$. For $\widehat{\boldsymbol{\omega}} = \boldsymbol{w}$, we recover the Jaumann derivative, whereas setting $\widehat{\boldsymbol{\omega}} = \boldsymbol{\Omega}$, we obtain the Green–McInnis–Naghdi rate.

Clearly, once a particular choice of $\widehat{\boldsymbol{\omega}}$ is made, a plasticity model can be formulated with a structure identical to the model summarized in BOX **7.2** of Chapter **7**, with rotation tensor $\boldsymbol{\Lambda}$ now generated by (8.3.1).

The objective of this section is the formulation of integration algorithms suitable for the class of plasticity models outlined above.

8.3.1 Objective Integration of Elastoplastic Models

As in Section **8.2**, the key to preserving objectivity of the discrete equations is to formulate the integration algorithm in a local configuration that remains unaltered by superposed spatial rigid body motions. We proceed as follows.

8.3.1.1 Integration algorithm for hypoelastic equations.

Consider rate-constitutive equations of the form

$$\dot{\boldsymbol{\Sigma}} = \mathbf{A} : \mathbf{D}, \tag{8.3.4}$$

where $\boldsymbol{\Sigma}$ and $\mathbf{D}$ are defined by (8.3.2). Applying the generalized midpoint rule yields

$$\begin{aligned} \boldsymbol{\Sigma}_{n+1} - \boldsymbol{\Sigma}_n &= \Delta t\, \dot{\boldsymbol{\Sigma}}_{n+\alpha} \\ &= \Delta t\, \mathbf{A}_{n+\alpha} : \mathbf{D}_{n+\alpha} \\ &= \mathbf{A}_{n+\alpha} : \boldsymbol{\Lambda}^T_{n+\alpha}[\Delta t \boldsymbol{d}_{n+\alpha}]\boldsymbol{\Lambda}_{n+\alpha}. \end{aligned} \tag{8.3.5}$$

As in Section **8.2**, in view of (8.2.4), we set

$$\boxed{\widetilde{\boldsymbol{e}}_{n+\alpha} := \widetilde{\boldsymbol{f}}^T_{n+\alpha}\boldsymbol{e}_{n+1}\widetilde{\boldsymbol{f}}_{n+1} = \Delta t \boldsymbol{d}_{n+\alpha},} \tag{8.3.6a}$$

Thus equation (8.3.5) yields the update formula

$$\boxed{\boldsymbol{\Sigma}_{n+1} = \boldsymbol{\Sigma}_n + \mathbf{A}_{n+\alpha} : \boldsymbol{\Lambda}^T_{n+\alpha}\widetilde{\boldsymbol{e}}_{n+\alpha}\boldsymbol{\Lambda}_{n+\alpha}.} \tag{8.3.6b}$$

Remarks 8.3.1.

1. Obviously, the algorithm (8.3.6) is incrementally objective for all $\alpha \in [0, 1]$, since

$$\boxed{\boldsymbol{f}_{n+1} = \boldsymbol{Q} \in \mathrm{SO}(3) \iff \boldsymbol{e}_{n=1} = \boldsymbol{0} \iff \boldsymbol{\Sigma}_{n+1} = \boldsymbol{\Sigma}_n.} \tag{8.3.7}$$

2. The algorithm (8.3.6) can be formulated entirely in the spatial description. Let $\mathbf{a}_{n+\alpha}$ be the moduli relative to the spatial $\varphi_{n+\alpha}$ configuration. By definition,

$$\mathsf{a}_{ijkl_{n+\alpha}} = \Lambda_{iA_{n+\alpha}} \Lambda_{jB_{n+\alpha}} \Lambda_{kC_{n+\alpha}} \Lambda_{lD_{n+\alpha}} \mathsf{A}_{ABCD_{n+\alpha}}. \tag{8.3.8}$$

Substituting (8.3.8) in (8.3.6b) yields

$$\boldsymbol{\Lambda}_{n+\alpha} \boldsymbol{\Sigma}_{n+1} \boldsymbol{\Lambda}_{n+\alpha}^T = \boldsymbol{\Lambda}_{n+\alpha} \boldsymbol{\Sigma}_n \boldsymbol{\Lambda}_{n+\alpha}^T + \mathsf{a}_{n+\alpha} : \tilde{\boldsymbol{e}}_{n+\alpha}, \tag{8.3.9}$$

Now let

$$\boxed{\tilde{\boldsymbol{\Lambda}}_{n+\alpha} := \boldsymbol{\Lambda}_{n+1} \boldsymbol{\Lambda}_{n+\alpha}^T} \tag{8.3.10}$$

be the relative rotation between $t_{n+\alpha}$ and t_{n+1} (recall that the solutions of (8.3.1) form a one-parameter subgroup; hence, $\boldsymbol{\Lambda}_{n+\alpha}$ is a well defined object). Set

$$\boxed{\tilde{\mathsf{a}}_{ijkl_{n+1}} = \tilde{\Lambda}_{ia_{n+\alpha}} \tilde{\Lambda}_{jb_{n+\alpha}} \tilde{\Lambda}_{kc_{n+\alpha}} \tilde{\Lambda}_{ld_{n+\alpha}} \mathsf{a}_{abcd_{n+\alpha}}.} \tag{8.3.11}$$

By premultiplying and postmultiplying (8.3.9) with $\tilde{\boldsymbol{\Lambda}}_{n+\alpha}$ and $\tilde{\boldsymbol{\Lambda}}_{n+\alpha}^T$, respectively, we obtain

$$\boxed{\begin{aligned} \boldsymbol{\tau}_{n+1} &= \left[\boldsymbol{\Lambda}_{n+1} \boldsymbol{\Lambda}_n^T\right] \boldsymbol{\tau}_n \left[\boldsymbol{\Lambda}_{n+1} \boldsymbol{\Lambda}_n^T\right]^T + \tilde{\mathsf{a}}_{n+1} : \tilde{\tilde{\boldsymbol{e}}}_{n+1}, \\ \tilde{\tilde{\boldsymbol{e}}}_{n+\alpha} &= \tilde{\boldsymbol{\Lambda}}_{n+\alpha} \tilde{\boldsymbol{e}}_{n+\alpha} \tilde{\boldsymbol{\Lambda}}_{n+\alpha}^T. \end{aligned}} \tag{8.3.12}$$

In actual implementations, form (8.3.6) is often more convenient.

3. An important property of the algorithms just developed should be noted. *If* a *is isotropic, the* $\tilde{\mathsf{a}}$ *defined by (8.3.11) is also isotropic* since the transformations (8.3.8) and (8.3.11) *involve rotations*. This property plays a crucial role in numerical implementations of J_2 flow theory.

8.3.1.2 Application to J_2 flow theory.

Applying the preceding class of algorithms to J_2 flow theory formulated in the rotated description leads to a straightforward generalization of the radial return method. The development of the algorithm follows along the same lines as discussed in Section **8.2** and in Chapter **3**, and thus details are omitted. A summary of the overall procedure is given in BOX **8.1**.

Remarks 8.3.2.

1. The radial return mapping is performed in the $\varphi_{n+1}(\mathcal{B})$-configuration, as in the classical radial return algorithm of the infinitesimal theory.
2. The rotation tensors $\boldsymbol{\Lambda}_n$, $\boldsymbol{\Lambda}_{n+1}$, $\boldsymbol{\Lambda}_{n+\alpha}$ are obtained according to the particular choice of rotational rate. We address this computation below.
3. A closed-form linearization of the algorithm is complex. Thus, the notion of consistent tangent moduli is not yet available for the class of algorithms considered in this chapter.

BOX 8.1. Rotationally Neutralized Objective Algorithm for J_2-Flow Theory.

1. Database: $\{\boldsymbol{\tau}_n, \alpha_n, \bar{\boldsymbol{q}}_n\}$, and $\varphi_n: \mathcal{B} \to \mathbb{R}^3$.

2. Given $\boldsymbol{u}: \varphi_n(\mathcal{B}) \to \mathbb{R}^3$ compute deformation gradients

$$\boldsymbol{f}_{n+1} = \mathbf{1} + \nabla_{\boldsymbol{x}_n}\boldsymbol{u}$$

$$\boldsymbol{f}_{n+\alpha} = \mathbf{1} + \alpha\nabla_{\boldsymbol{x}_n}\boldsymbol{u}$$

$$\widetilde{\boldsymbol{f}}_{n+\alpha} = \boldsymbol{f}_{n+1}\boldsymbol{f}_{n+\alpha}^{-1}$$

3. Compute incremental strain tensor

$$\widetilde{\boldsymbol{e}}_{n+\alpha} = \frac{1}{2}\widetilde{\boldsymbol{f}}_{n+\alpha}^{T}\left[\mathbf{1} - (\boldsymbol{f}_{n+1}\boldsymbol{f}_{n+1}^{T})^{-1}\right]\widetilde{\boldsymbol{f}}_{n+\alpha}$$

4. Compute relative rotation tensors
(according to the choice of "rotational" stress rate)

$$\boldsymbol{r}_{n+1} = \boldsymbol{\Lambda}_{n+1}\boldsymbol{\Lambda}_n^T, \quad \widetilde{\boldsymbol{r}}_{n+\alpha} = \boldsymbol{\Lambda}_{n+1}\boldsymbol{\Lambda}_{n+\alpha}^{-1}$$

5. Elastic predictor $[\mathbf{a} := \lambda\mathbf{1}\otimes\mathbf{1} + 2\mu\mathbf{I}]$

$$\boldsymbol{\tau}_{n+1}^{\text{trial}} = \boldsymbol{r}_{n+1}\boldsymbol{\tau}_n\boldsymbol{r}_{n+1}^T + \mathbf{a} : [\widetilde{\boldsymbol{r}}_{n+\alpha}\widetilde{\boldsymbol{e}}_{n+\alpha}\widetilde{\boldsymbol{r}}_{n+\alpha}^T]$$

$$\bar{\boldsymbol{q}}_{n+1}^{\text{trial}} = \boldsymbol{r}_{n+1}\bar{\boldsymbol{q}}_n\boldsymbol{r}_{n+1}^T$$

$$\boldsymbol{\eta}_{n+1}^{\text{trial}} = \operatorname{dev}[\boldsymbol{\tau}_{n+1}^{\text{trial}} - \bar{\boldsymbol{q}}_{n+1}^{\text{trial}}]$$

$$f_{n+1}^{\text{trial}} := \|\boldsymbol{\eta}_{n+1}^{\text{trial}}\| - \sqrt{\tfrac{2}{3}}(\sigma_Y + K\alpha_n)$$

7. Radial return method

IF $f_{n+1}^{\text{trial}} < 0$ THEN

Set $(\bullet)_{n+1} = (\bullet)_{n+1}^{\text{trial}}$ & *EXIT*

ELSE

$$\Delta\gamma = \frac{\langle f_{n+1}^{\text{trial}}\rangle/2\mu}{1 + \frac{H+K}{3\mu}}$$

$$\boldsymbol{n}_{n+1} = \boldsymbol{\eta}_{n+1}^{\text{trial}}/\|\boldsymbol{\eta}_{n+1}^{\text{trial}}\|$$

$$\boldsymbol{\tau}_{n+1} = \boldsymbol{\tau}_{n+1}^{\text{trial}} - 2\mu\Delta\gamma\boldsymbol{n}_{n+1}$$

$$\bar{\boldsymbol{q}}_{n+1} = \bar{\boldsymbol{q}}_{n+1}^{\text{trial}} + \tfrac{2}{3}\Delta\gamma H\boldsymbol{n}_{n+1}$$

$$\alpha_{n+1} = \alpha_n + \sqrt{\tfrac{2}{3}}\Delta\gamma$$

ENDIF

8.3.1.3 Determination of the rotation tensor.

As pointed out above, the rotation tensors $\boldsymbol{\Lambda}_n$, $\boldsymbol{\Lambda}_{n+\alpha}$, $\boldsymbol{\Lambda}_{n+1}$ are obtained as solutions of (8.3.1), which is defined once the skew-symmetric tensor is specified. We consider two illustrative possibilities:

i. $\boxed{\widehat{\omega} = \boldsymbol{\Omega}}$ *or, equivalently,* $\boldsymbol{\Lambda} := \boldsymbol{R}$ *is the rotation tensor in the polar decomposition of* $\boldsymbol{F}$. This choice leads to the Green–McInnis–Naghdi stress rate defined by (7.1.77). Then the simplest computational procedure is to use the algorithm in BOX **7.1** and directly perform polar decompositions, as summarized in BOX **8.2**.

BOX 8.2. Determination of $\boldsymbol{\Lambda}_n$, $\boldsymbol{\Lambda}_{n+\alpha}$, $\boldsymbol{\Lambda}_{n+1}$ for the Green–McInnis–Naghdi Rate.

1. Obtain total deformation gradients

$$\begin{aligned} \boldsymbol{F}_n &= D\varphi_n \\ \boldsymbol{F}_{n+1} &= \boldsymbol{f}_{n+1}\boldsymbol{F}_n \\ \boldsymbol{F}_{n+\alpha} &= \alpha\boldsymbol{F}_{n+1} + (1-\alpha)\boldsymbol{F}_n. \end{aligned}$$

2. Perform three polar decompositions using the algorithm in BOX **7.1** to get

$$\boldsymbol{\Lambda}_n = \boldsymbol{R}_n, \quad \boldsymbol{\Lambda}_{n+1} = \boldsymbol{R}_{n+1}, \quad \boldsymbol{\Lambda}_{n+\alpha} = \boldsymbol{R}_{n+\alpha}.$$

ii. $\boxed{\widehat{\omega} = \boldsymbol{w}}$, *where* $\boldsymbol{w}$ *is the spin tensor.* This choice leads to the Jaumann rate of the Kirchhoff stress. An algorithmic approximation to the spin tensor is given in the following.

Proposition 8.4. *Let* $\boldsymbol{u}\colon \varphi_n(\mathcal{B}) \to \mathbb{R}^3$ *be the incremental displacement field and* $\widetilde{\boldsymbol{u}} := \boldsymbol{u} \circ \varphi_n \circ \varphi_{n+\alpha}^{-1}$, *i.e.,* $\widetilde{\boldsymbol{u}}(\boldsymbol{x}_{n+\alpha}) = \boldsymbol{u}(\boldsymbol{x}_n)$. *Then*

$$\boxed{\begin{aligned} \boldsymbol{w}_{n+\alpha} &= \frac{1}{2\Delta t}\left[\boldsymbol{h}_{n+\alpha} - \boldsymbol{h}_{n+\alpha}^T\right]. \\ \boldsymbol{h}_{n+\alpha} &:= \nabla_{n+\alpha}\widetilde{\boldsymbol{u}}(\boldsymbol{x}_{n+\alpha}). \end{aligned}} \tag{8.3.13}$$

are consistent with a generalized midpoint rule approximation.

PROOF. Let $\boldsymbol{V}_t = \partial\varphi/\partial t$ be the material velocity field, and let $\boldsymbol{v}_t = \boldsymbol{V}_t \circ \varphi_t^{-1}$ be the spatial velocity field (see Section **7.1.2.2**). By the generalized midpoint rule,

$$\varphi_{n+1} - \varphi_n = \boldsymbol{V}_{n+\alpha}\Delta t = \boldsymbol{v}_{n+\alpha} \circ \varphi_{n+\alpha}\Delta t. \tag{8.3.14}$$

Using the chain rule and relationships (8.1.23*b*),

$$(\nabla_{n+\alpha}\boldsymbol{v}_{n+\alpha})\boldsymbol{F}_{n+\alpha}\Delta t = \boldsymbol{F}_{n+1} - \boldsymbol{F}_n = \boldsymbol{h}_{n+\alpha}\boldsymbol{F}_{n+\alpha}. \tag{8.3.15}$$

Thus,

$$\boldsymbol{w}_{n+\alpha}\Delta t := \frac{\Delta t}{2}\left[\nabla_{n+\alpha}\boldsymbol{v}_{n+\alpha} - (\nabla_{n+\alpha}\boldsymbol{v}_{n+\alpha})^T\right] = \tfrac{1}{2}\left(\boldsymbol{h}_{n+\alpha} - \boldsymbol{h}_{n+\alpha}^T\right), \tag{8.3.16}$$

and the result follows. □

With this result in hand, we turn our attention to integrating the evolution equation (8.3.1).

8.3.2 Time-Stepping Algorithms for the Orthogonal Group

Below we develop the proper extension to the special orthogonal group of the generalized midpoint rule time-stepping algorithm. The proposed algorithm remains valid for any value of $\alpha \in [0, 1]$. Before formulating the algorithm, we review some basic properties of the special orthogonal group. For a more detailed account, see, e.g., Abraham and Marsden [1978, Section 4.1].

8.3.2.1 Basic facts about the orthogonal group.

Recall that the special orthogonal group, denoted by SO(3), is the (compact Lie) subgroup of the general linear group, defined by

$$\mathrm{SO}(3) := \{\boldsymbol{\Lambda}: \mathbb{R}^3 \to \mathbb{R}^3 \mid \boldsymbol{\Lambda}^T\boldsymbol{\Lambda} = \mathbf{1} \ \ \det[\boldsymbol{\Lambda}] = 1\}. \tag{8.3.17}$$

On the other hand, the set of skew-symmetric matrices is the linear vector space defined by

$$\mathrm{so}(3) = \left\{\widehat{\omega} : \mathbb{R}^3 \to \mathbb{R}^3 \mid \widehat{\omega} + \widehat{\omega}^T = \mathbf{0}\right\}. \tag{8.3.18}$$

We recall that so(3) is a Lie algebra with the ordinary matrix commutator as its bracket. Associated with any skew-symmetric matrix $\widehat{\omega} \in \mathrm{so}(3)$, there is a unique vector $\boldsymbol{\omega} \in \mathbb{R}^3$ defined by the relationship

$$\widehat{\omega}\boldsymbol{h} = \boldsymbol{\omega} \times \boldsymbol{h}, \quad \text{for all} \quad \boldsymbol{h} \in \mathbb{R}^3. \tag{8.3.19a}$$

In components, the expressions are

$$\boldsymbol{\omega} = \begin{Bmatrix} \omega_1 \\ \omega_2 \\ \omega_3 \end{Bmatrix}; \qquad \widehat{\omega} = \begin{bmatrix} 0 & -\omega_3 & \omega_2 \\ \omega_3 & 0 & -\omega_1 \\ -\omega_2 & \omega_1 & 0 \end{bmatrix}. \tag{8.3.19b}$$

The *exponential map*, denoted by exp: so(3) $\to$ SO(3), transforms skew-symmetric matrices into orthogonal matrices according to the expression

$$\exp[\widehat{\omega}] = \sum_{n=0}^{\infty} \frac{1}{n!}[\widehat{\omega}]^n. \tag{8.3.21}$$

For the orthogonal group, one has the following *closed-form expression* which goes back at least to the middle of the nineteenth century and is attributed to Rodrigues (see, e.g., Whittaker[1944, Chapter I] and Goldstein[1981])

$$\boxed{\exp[\widehat{\omega}] = \mathbf{1} + \frac{\sin(\|\boldsymbol{\omega}\|)}{\|\boldsymbol{\omega}\|}\widehat{\omega} + \frac{1}{2}\left[\frac{\sin(\|\boldsymbol{\omega}\|/2)}{\|\boldsymbol{\omega}\|/2}\right]^2\widehat{\omega}^2.} \tag{8.3.22a}$$

Several parametrizations of this expression are possible. In particular, the following reformulation of (8.3.22*a*) in terms of the so-called *Euler parameters* will prove useful in our subsequent developments

$$\boxed{\begin{aligned}\exp[\hat{\omega}] &= \mathbf{1} + \frac{2}{1+\|\bar{\omega}\|^2}\left[\hat{\bar{\omega}} + \hat{\bar{\omega}}^2\right],\\ \bar{\omega} &:= \tan\left[\|\omega\|/2\right]\frac{\omega}{\|\omega\|}.\end{aligned}} \tag{8.3.22b}$$

As shown below, the exponential mapping is the crucial tool in formulating time-stepping algorithms.

8.3.2.2 The generalized midpoint rule in SO(3).

For the evolution problem (8.3.1), the following algorithm furnishes the appropriate extension of the generalized midpoint rule

$$\boxed{\boldsymbol{\Lambda}_{n+1} = \exp[\Delta t\hat{\omega}_{n+\alpha}]\boldsymbol{\Lambda}_n; \quad \alpha \in [0, 1],} \tag{8.3.23}$$

where $\exp[\Delta t\hat{\omega}_{n+\alpha}]$ is defined by (8.3.22*a*). Note that

$$\left.\begin{aligned}\alpha = 0 &\Rightarrow \text{explicit (forward) Euler,}\\ \alpha = \tfrac{1}{2} &\Rightarrow \text{midpoint rule,}\\ \alpha = 1 &\Rightarrow \text{implicit (backward) Euler.}\end{aligned}\right\} \tag{8.3.23b}$$

Remarks 8.3.3.

1. One can easily show that this algorithm is *consistent* with problem (8.3.1) for **all** $\alpha \in [0, 1]$, and *second-order accurate* for $\alpha = \frac{1}{2}$; see Simo and Quoc [1986].
2. Expression (8.3.22*b*) possesses a singularity at $\|\omega\| = \pi$. However, there are several alternative, singularity-free parametrizations. The *optimal* parametrization is obtained in terms of quaternion parameters, summarized in BOX **8.3**.
3. The algorithm (8.3.23) places *no restriction on the magnitude* of $\|\omega_{n+\alpha}\Delta t\|$.
4. As shown below, the algorithm proposed in Hughes and Winget [1980] is obtained from (8.3.23) by introducing the approximation

$$\tan\left[\|\omega_{n+\alpha}\Delta t/2\|\right] \cong \tfrac{1}{2}\Delta t\|\omega_{n+\alpha}\| \tag{8.3.24}$$

and setting $\alpha = 1/2$. However, this approximation restricts the algorithm to $\|\omega_{n+1/2}\Delta t\| < 180^\circ$ (the so-called "black-hole" condition; Hughes [1984]).

BOX 8.3. Optimal Evaluation of the Exponential Map.

1. Given $\widehat{\omega} \in so(3)$, compute quaternion parameters

$$q_0 = \cos(\|\omega\|/2)$$

$$q^* = \sin(\|\omega\|/2)$$

IF: $|q^*| > \text{TOL}$, THEN:

$$q^* = \frac{1}{2}\frac{\sin(\|\omega\|/2)}{(\|\omega\|/2)}$$

ELSE:

$$q^* = \tfrac{1}{2}\left[1 - \|\omega\|^2/24 + \|\omega\|^4/1920 + \ldots\right]$$

ENDIF.

$$\boldsymbol{q} = q^*\boldsymbol{\omega}$$

2. Compute the exponential by the formula

$$\begin{aligned}
\exp[\widehat{\omega}] &= 2(q_0^2 - \tfrac{1}{2})\mathbf{1} + 2q_0\widehat{\boldsymbol{q}} + 2\boldsymbol{q}\boldsymbol{q}^T \\
&= 2\begin{bmatrix} q_0^2 + q_1^2 - \frac{1}{2} & q_1q_2 - q_3q_0 & q_1q_3 + q_2q_0 \\ q_2q_1 + q_3q_0 & q_0^2 + q_2^2 - \frac{1}{2} & q_2q_3 - q_1q_0 \\ q_3q_1 - q_2q_0 & q_3q_2 + q_1q_0 & q_0^2 + q_3^2 - \frac{1}{2} \end{bmatrix}
\end{aligned}$$

8.3.2.3 Application to elastoplasticity formulated in terms of the Jaumann derivative.

The application of algorithm (8.3.23) to determining the rotation tensors $\boldsymbol{\Lambda}_{n+\alpha}$ and $\boldsymbol{\Lambda}_{n+1}$, when the objective stress rate is the Jaumann derivative, is summarized in BOX **8.4** for the usual value $\alpha = 1/2$.

The computation of $\boldsymbol{\Lambda}_{n+1/2}$ is remarkably simple within the framework of the algorithm (8.3.23) and is justified as follows. Let $\boldsymbol{Q} \in SO(3)$ be such that

$$\left.\begin{aligned} \boldsymbol{\Lambda}_{n+1/2} &= \boldsymbol{Q}\boldsymbol{\Lambda}_n \\ \boldsymbol{\Lambda}_{n+1} &= \boldsymbol{Q}\boldsymbol{\Lambda}_{n+1/2} \end{aligned}\right\} \Rightarrow \boldsymbol{\Lambda}_{n+1} = \boldsymbol{Q}^2\boldsymbol{\Lambda}_n\,. \tag{8.3.25}$$

On the other hand, setting $\widehat{\boldsymbol{\Theta}} = \Delta t\widehat{\omega}_{n+1/2}$ and noting that $\widehat{\omega}/2$ commutes with itself, from (8.3.23), we obtain

$$\begin{aligned}
\boldsymbol{\Lambda}_{n+1} &= \exp\big(\widehat{\boldsymbol{\Theta}}\big)\boldsymbol{\Lambda}_n \\
&= \exp\big(\widehat{\boldsymbol{\Theta}}/2 + \widehat{\boldsymbol{\Theta}}/2\big)\boldsymbol{\Lambda}_n \\
&= \exp\big(\widehat{\boldsymbol{\Theta}}/2\big)\exp\big(\widehat{\boldsymbol{\Theta}}/2\big)\boldsymbol{\Lambda}_n \\
&\equiv \big[\exp\big(\widehat{\boldsymbol{\Theta}}/2\big)\big]^2\boldsymbol{\Lambda}_n\,.
\end{aligned} \tag{8.3.26}$$

BOX 8.4. Determination of $\boldsymbol{\Lambda}_{n+1}$ and $\boldsymbol{\Lambda}_{n+1/2}$ for the Jaumann Derivative.

i. Compute the *spin* tensor

$$\boldsymbol{h}_{n+1/2}(\boldsymbol{x}_{n+1/2}) = \nabla_{n+1/2}\boldsymbol{u}(\boldsymbol{x}_{n+1/2})$$

$$\widehat{\boldsymbol{\Theta}} = \Delta t\widehat{\boldsymbol{\omega}}_{n+1/2} := \tfrac{1}{2}\left(\boldsymbol{h}_{n+1/2} - \boldsymbol{h}^T_{n+1/2}\right)$$

ii. Compute $\boldsymbol{\Lambda}_{n+1/2}$ and $\boldsymbol{\Lambda}_{n+1}$ using BOX **8.3**

$$\boldsymbol{\Lambda}_{n+1/2} = \exp\left[\widehat{\boldsymbol{\Theta}}/2\right]\boldsymbol{\Lambda}_n$$

$$\boldsymbol{\Lambda}_{n+1} = \exp\left[\widehat{\boldsymbol{\Theta}}\right]\boldsymbol{\Lambda}_n$$

By comparing (8.3.25) and (8.3.26), we conclude that

$$\boxed{\boldsymbol{Q} = \exp\left(\widehat{\boldsymbol{\Theta}}/2\right),} \tag{8.3.27}$$

which leads to the procedure in BOX **8.4**. Note that the present approach based on algorithm (8.3.23) bypasses the need for computing the square root of an orthogonal tensor, a necessary step in some approaches (see Hughes [1984]).

8.3.2.4 Relationship to the midpoint rule formula.

It is of interest to examine the relationship between the generalized midpoint algorithm in SO(3), as given by (8.3.23), and a widely used algorithm based on the midpoint rule formula proposed in Hughes and Winget [1980].

The derivation of the formula in Hughes and Winget proceeds as follows: Using the standard version of the generalized midpoint rule in (8.3.1) gives

$$\begin{aligned}\boldsymbol{\Lambda}_{n+1} - \boldsymbol{\Lambda}_n &= \Delta t\widehat{\boldsymbol{\omega}}_{n+\alpha}\boldsymbol{\Lambda}_{n+\alpha} \\ &= \Delta t\widehat{\boldsymbol{\omega}}_{n+\alpha}\left[\alpha\boldsymbol{\Lambda}_{n+1} + (1-\alpha)\boldsymbol{\Lambda}_n\right].\end{aligned} \tag{8.3.28}$$

Solving for $\boldsymbol{\Lambda}_{n+1}$ yields

$$\boxed{\boldsymbol{\Lambda}_{n+1} = \left[\mathbf{1} - \alpha\Delta t\widehat{\boldsymbol{\omega}}_{n+\alpha}\right]^{-1}\left[\mathbf{1} + \Delta t(1-\alpha)\widehat{\boldsymbol{\omega}}_{n+\alpha}\right]\boldsymbol{\Lambda}_n,} \tag{8.3.29}$$

which is the expression found in Hughes and Winget [1980]. Further simplification of (8.3.29) is obtained as follows. Let $\widehat{\boldsymbol{\Theta}} \in$ so(3) be a skew-symmetric matrix. Recall that, by the Neumann series expansion,

$$(\mathbf{1} - \widehat{\boldsymbol{\Theta}})^{-1} = \sum_{n=0}^{\infty}(\widehat{\boldsymbol{\Theta}})^n. \tag{8.3.30}$$

Now observe that since $\widehat{\boldsymbol{\Theta}}$ is skew-symmetric,

$$\left.\begin{aligned} \widehat{\Theta}^3 &= -\|\Theta\|^2\widehat{\Theta}\,, \\ \widehat{\Theta}^4 &= -\|\Theta\|^2\widehat{\Theta}^2\,, \\ \widehat{\Theta}^5 &= +\|\Theta\|^4\widehat{\Theta}\,, \end{aligned}\right\} \tag{8.3.31}$$

and so on. Consequently, (8.3.30) reduces to

$$\begin{aligned} \left(\mathbf{1} - \widehat{\Theta}\right)^{-1} &= \mathbf{1} + \left(1 - \|\Theta\|^2 + \|\Theta\|^4 - \|\Theta\|^6 + \cdots\right)\left(\widehat{\Theta} + \widehat{\Theta}^2\right) \\ &= \mathbf{1} + \frac{1}{1 + \|\Theta\|^2}\left(\widehat{\Theta} + \widehat{\Theta}^2\right). \end{aligned} \tag{8.3.32}$$

Setting $\widehat{\Theta} := \Delta t\widehat{\omega}_{n+\alpha}$, and substituting (8.3.32) in (8.3.29) yields the final result

$$\boxed{\begin{aligned} \boldsymbol{\Lambda}_{n+1} &= \mathbf{1} + \frac{1/\alpha}{1 + \|\alpha\Theta\|^2}\left[\left(\alpha\widehat{\Theta}\right) + \left(\alpha\widehat{\Theta}\right)^2\right]\boldsymbol{\Lambda}_n \\ \widehat{\Theta} &:= \Delta t\widehat{\omega}_{n+\alpha}\,. \end{aligned}} \tag{8.3.33}$$

Then a straightforward calculation leads to the relationship

$$\boldsymbol{\Lambda}_{n+1}^T\boldsymbol{\Lambda}_n\boldsymbol{\Lambda}_n^T\boldsymbol{\Lambda}_{n+1} = \mathbf{1} + \frac{1}{1 + \|\alpha\Theta\|^2}\widehat{\Theta}^2(2\alpha - 1). \tag{8.3.34}$$

Therefore, in view of (8.3.34), we conclude that

$$\boxed{\boldsymbol{\Lambda}_{n+1}^T\boldsymbol{\Lambda}_n\boldsymbol{\Lambda}_n^T\boldsymbol{\Lambda}_{n+1} = \mathbf{1} \iff \alpha = 1/2\,,} \tag{8.3.35}$$

a result obtained by Hughes and Winget [1980]. Observe further that, for $\alpha = 1/2$, expression (8.3.33) agrees with (8.3.23) under the approximation

$$\boxed{\tan\left[\|\Theta\|/2\right] \cong \|\Theta\|/2 \iff \bar{\Theta} \cong \Theta/2\,.} \tag{8.3.36}$$

As noted in Hughes and Winget [1980] and Hughes [1984], (8.3.29) (or equivalently, (8.3.33)) becomes singular for $\|\Theta\| = 180^{\circ}$. However, algorithm (8.3.23) yields a meaningful approximation for *any* value of $\|\bar{\Theta}\|$.

For generalization to dynamics see Simo and Wong [1991].

9

Phenomenological Plasticity Models Based on the Notion of an Intermediate Stress-Free Configuration

In this chapter we outline a formulation of nonlinear plasticity which, in contrast with the phenomenological models examined in Chapter 7, is motivated by a well-understood micromechanical picture of single-crystal metal plasticity. A comprehensive exposition of the current status of the micromechanical description of single-crystal metal plasticity is in the review article of Asaro [1983]. The basic ideas go back to the fundamental work of Taylor [1938] subsequently expanded upon in Hill [1966]; Hill and Rice [1972]; Asaro and Rice [1977]; and Asaro [1979].

An essential feature of this micromechanical description is the introduction of an *intermediate local configuration*, relative to which the elastic response of the material is characterized. From a phenomenological standpoint this notion leads to a local multiplicative decomposition of the deformation gradient of the form

$$\boldsymbol{F}(\boldsymbol{X}, t) = \boldsymbol{F}^e(\boldsymbol{X}, t)\boldsymbol{F}^p(\boldsymbol{X}, t) \tag{9.1.1}$$

for each material point $\boldsymbol{X} \in \mathcal{B}$, the reference configuration of the body. Multiplicative decompositions of this type have been considered by Lee and Liu [1967]; Lee [1969]; Kroner and Teodosiu [1972]; Mandel [1964,1974]; Kratochvil [1973]; Sidoroff [1974]; Nemat–Nasser[1982]; Agah–Tehrani et al. [1986]; Lubliner [1984,1986]; Simo and Ortiz [1985]; Simo [1988a]; and others.

The objective of this chapter is two-fold:

1. To outline the continuum basis of elastoplastic constitutive models at finite strains based on the notion of an intermediate (*local*) configuration, as embodied in the multiplicative decomposition (9.1.1). In particular, to motivate the general theory in a concrete setting, we consider in detail the formulation of a J_2 flow theory based on the multiplicative decomposition.

2. To demonstrate that the notion of an intermediate configuration leads to a remarkably simple class of integration algorithms which include, as a particular case, a straightforward extension of the radial return algorithm of infinitesimal J_2 flow theory. With the exceptions of Argyris and Doltsinis [1979,1980]; Argyris et al. [1979]; Simo and Ortiz [1985]; and Simo [1988a,b]; this approach had not been explored in the computational literature previously.

To this date plasticity at finite strains still remains a somewhat controversial subject. Because this monograph is largely concerned with numerical analysis, we have chosen to focus our attention on the computational implications of the multiplicative decomposition (9.1.1) in the simplest possible context afforded by J_2 flow theory. From this computational perspective, the following features are noteworthy.

i. The stress-strain relationships derive from a stored-energy function, formulated relative to the intermediate configuration, which decouples exactly into volumetric and deviatoric parts.

ii. The integration algorithm reduces to the classical, radial return algorithm discussed in detail in Chapter **3**, in which the elastic predictor is computed exactly by a *function evaluation* of the stress-strain relationship.

iii. In the absence of plastic flow, i.e., for elastic response, the algorithm is exact and reduces to the classical update procedure of finite elasticity.

iv. As in the infinitesimal theory, the entire algorithmic procedure can be linearized leading to a *closed-form expression* for the algorithmic (consistent) tangent elastoplastic moduli.

An outline of this chapter is as follows. First we summarize some kinematic relationships associated with multiplicative decomposition (9.1.1) and the notion of an intermediate configuration. Next, as an important illustration of the general theory, we consider the formulation of J_2 flow theory within the framework of the multiplicative decomposition. Subsequently, we provide a detailed description of the algorithmic aspects involved in numerical implementation. Finally, we assess the performance of J_2 flow theory in a selected number of examples that include comparisons with exact solutions and experimental results and also comparisons with numerical simulations reported in the literature.

9.1 Kinematic Preliminaries. The (Local) Intermediate Configuration

In this section we introduce the notion of intermediate configuration and summarize some basic kinematic relationships needed for our subsequent developments.

9.1.1 Micromechanical Motivation. Single-Crystal Plasticity

From a micromechanical standpoint, plastic flow in crystalline plasticity can be viewed as a *flow of material through the crystal lattice via dislocation motion*. This point of view goes back to the work of G.I. Taylor and his associates, e.g., Taylor and Elam [1923, 1925]; and Taylor [1938]; and provides the starting point for current micromechanical descriptions of plastic flow in metal plasticity, as in Asaro and Rice [1977] and Asaro [1979].

For a crystal with a single slip system denoted by $\{s, m\}$ with $m \cdot s = 0$ and $\|s\| = \|m\| = 1$, the point of view above leads to a micromechanical description illustrated in Figure **9.1**. The unit vectors $\{s, m\}$ are attached to the lattice, and plastic flow is characterized by the tensor F^p defined as

$$\boxed{F^p = 1 + \gamma s \otimes m,} \tag{9.1.2}$$

where γ is the plastic shearing on the crystallographic slip system defined by $\{s, m\}$. Then the total deformation of the crystal is decomposed as

$$\boxed{F = F^e F^p,} \tag{9.1.3}$$

where F^e is the deformation caused by stretching and rotation of the crystal lattice and F is the total deformation gradient. The plastic slip rate γ is defined according to the Schmidt-resolved shear law. The preceding kinematic description can be extended to the case of several slip systems. We refer to the review article of Asaro [1983] for further details including an up-to-date discussion of several micromechanical models. Here we simply note that proper account of the physical mechanisms underlying plastic flow in crystal plasticity leads rather naturally to a multiplicative decomposition of the deformation gradient of the form (9.1.3).

9.1.2 Kinematic Relationships Associated with the Intermediate Configuration

Motivated by the micromechanical picture of plastic deformation outlined above, for a continuum body with reference placement $\mathcal{B} \subset \mathbb{R}^3$, one postulates a local multiplicative decomposition of the form (9.1.1):

$$\boxed{F(X, t) = F^e(X, t) F^p(X, t).} \tag{9.1.4}$$

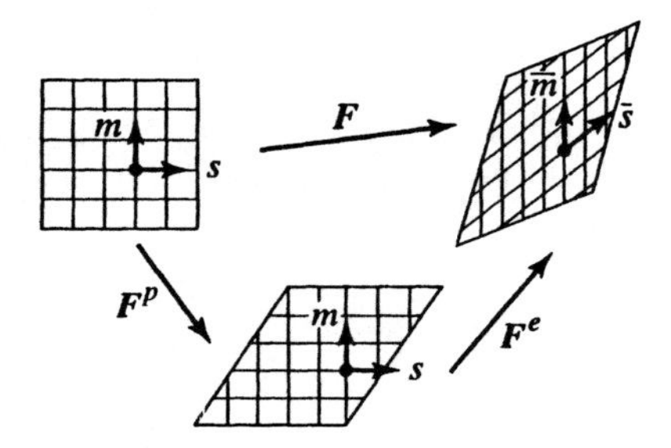

FIGURE 9.1. Micromechanical aspects of the deformation in a single slip crystal. F^p "moves" the material through the lattice via dislocation motion. F^e rotates and distorts the crystal lattice.

From a phenomenological standpoint, one interprets $\boldsymbol{F}^{e\,-1}$ as the local deformation that releases the stresses from each neighborhood $\mathcal{O}_x \subset \varphi(\mathcal{B})$ in the current placement of the body. Accordingly, it is implicitly assumed that the *local intermediate configuration* defined by $\boldsymbol{F}^{e\,-1}$ is *stress-free.* Note that, globally, the intermediate configuration is incompatible in the same sense as the rotated configuration introduced in Section **7.1.3.2**. See Figure **9.2** for an illustration.

9.1.2.1 Lagrangian and Eulerian strain tensors.

Following the standard conventions in continuum mechanics discussed in Chapter **7** relative to the reference placement of the body, the right Cauchy-Green tensors are defined as

$$\boxed{\boldsymbol{C} := \boldsymbol{F}^T\boldsymbol{F},}$$

and (9.1.5)

$$\boxed{\boldsymbol{C}^p = \boldsymbol{F}^{p\,T}\boldsymbol{F}^p.}$$

Then the corresponding Lagrangian strain tensors become

$$\boldsymbol{E} := \tfrac{1}{2}\left(\boldsymbol{C} - \boldsymbol{1}\right),$$

and (9.1.6)

$$\boldsymbol{E}^p = \tfrac{1}{2}\left(\boldsymbol{C}^p - \boldsymbol{1}\right),$$

where **1** denotes the symmetric unit tensor with components δ_{AB} in a Cartesian reference system. The fact that $\boldsymbol{C}, \boldsymbol{C}^p$ and $\boldsymbol{E}, \boldsymbol{E}^p$ are objects associated with the reference configuration is indicated schematically in Figure **9.3**.

Similarly, associated with the current configuration are the Eulerian tensors

$$\boxed{\boldsymbol{b} = \boldsymbol{F}\boldsymbol{F}^T,}$$

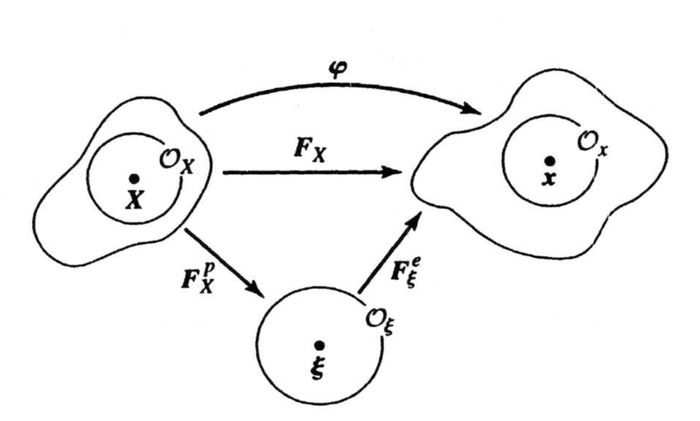

FIGURE 9.2. The multiplicative decomposition of the deformation gradient.

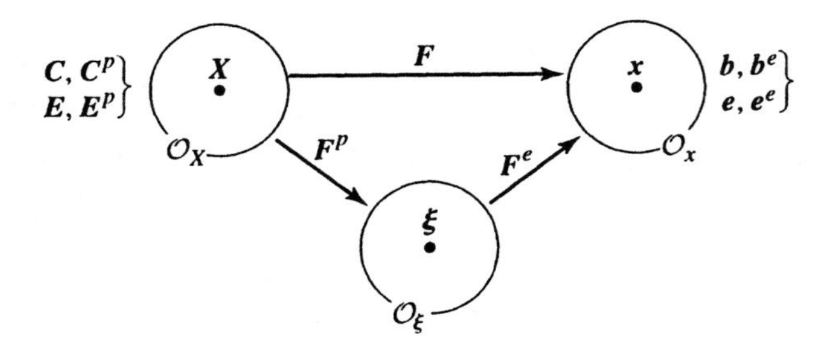

FIGURE 9.3. Examples of Lagrangian and Eulerian strain tensors.

and (9.1.7)

$$\boxed{\boldsymbol{b}^e = \boldsymbol{F}^e \boldsymbol{F}^{e\,T},}$$

called the total and elastic left Cauchy–Green tensors, respectively. The inverse form of the left Cauchy–Green tensor is called the Finger deformation tensor. In the present context, then*

$$\boldsymbol{c} := \boldsymbol{b}^{-1} \equiv \boldsymbol{F}^{-T}\boldsymbol{F}^{-1},$$

and (9.1.8)

$$\boldsymbol{c}^e := \boldsymbol{b}^{e\,-1} \equiv \boldsymbol{F}^{e\,-T}\boldsymbol{F}^{e\,-1}.$$

Finally, the Eulerian strain tensors take the form

$$\boldsymbol{e} = \tfrac{1}{2}\left(\mathbf{1} - \boldsymbol{c}\right),$$

and (9.1.9)

$$\boldsymbol{e}^e = \tfrac{1}{2}\left(\mathbf{1} - \boldsymbol{c}^e\right),$$

where $\mathbf{1}$ is the symmetric unit tensor in the current configuration with components δ_{ab}. Since Cartesian coordinates are used throughout our discussion, we use the same symbol for the unit tensor in both the reference and the current configurations.

The following relationship is crucial in our analysis. From $(9.1.7)_2$ and (9.1.4),

$$\boldsymbol{b}^e := \boldsymbol{F}\boldsymbol{F}^{p\,-1}\boldsymbol{F}^{p\,-T}\boldsymbol{F}^T = \boldsymbol{F}\left[\boldsymbol{F}^{p\,T}\boldsymbol{F}^p\right]^{-1}\boldsymbol{F}^T, \qquad (9.1.10)$$

which, in view of $(9.1.5)_2$, yields

$$\boxed{\boldsymbol{b}^e = \boldsymbol{F}\boldsymbol{C}^{p\,-1}\boldsymbol{F}^T.} \qquad (9.1.11)$$

Analogous expressions connecting other spatial and Lagrangian strain tensors defined above also hold; see, e.g., Simo and Ortiz [1985], Simo [1988*a*,*b*].

*Strictly speaking, one should write $\boldsymbol{b} = (\boldsymbol{F}\boldsymbol{F}^T)\circ\varphi^{-1}$ since $\boldsymbol{b}$ is a spatial object which depends on $\boldsymbol{x} = \varphi(\boldsymbol{X})$, and not $\boldsymbol{X} \in \mathcal{B}$. Following a standard abuse in notation, we often omit the composition with φ^{-1}.

9.1.2.2 Rates of deformation.

We recall only two relationships needed for our subsequent developments. First, note that the total rate of deformation tensor defined by (7.1.37) is alternatively expressed according to (7.1.39) as

$$d = F^{-T}\left[\tfrac{1}{2}\dot{C}\right]F^{-1} \equiv F^{-T}\dot{E}F^{-1}. \tag{9.1.12}$$

Now observe from (9.1.7) and (9.1.9) that

$$E = \tfrac{1}{2}F^{T}\left[1 - F^{-T}F^{-1}\right]F = F^{T}eF. \tag{9.1.13}$$

Combining (9.1.12) and (9.1.13), we write symbolically

$$\boxed{d = F^{-T}\left\{\frac{\partial}{\partial t}\left[F^{T}eF\right]\right\}F^{-1}.} \tag{9.1.14}$$

By analogy with (7.1.72), d may be viewed a Lie derivative; see Marsden and Hughes [1983, Chapter 1] or Simo, Marsden, and Krishnaprasad [1988] for a careful definition of this concept. Definition (9.1.14), however, suffices for our purposes.

By analogy with (9.1.14), in view of (9.1.11), we set

$$\boxed{L_{v}b^{e} := F\left\{\frac{\partial}{\partial t}\left[F^{-1}b^{e}F^{-T}\right]\right\}F^{T} \equiv F\dot{C}^{p\,-1}F^{T}.} \tag{9.1.15}$$

It can be shown that this formal definition rigorously agrees with the actual definition of the Lie derivative for the tensor b^{e}.

To formulate the J_2-plasticity model discussed in detail in the following section, we need to introduce one further kinematic notion.

9.1.3 Deviatoric-Volumetric Multiplicative Split

Within the context of the *infinitesimal theory*, the strain tensor ε is decomposed into *volumetric* and *deviatoric* parts according to the following standard *additive split*. Let

$$\mathrm{dev}[\varepsilon] := \varepsilon - \tfrac{1}{3}\,\mathrm{tr}[\varepsilon]\mathbf{1} \;\Rightarrow\; \mathrm{tr}\left\{\mathrm{dev}[\varepsilon]\right\} = 0 \tag{9.1.16}$$

be the strain deviator. Then

$$\varepsilon = \underbrace{\mathrm{dev}[\varepsilon]}_{\substack{\text{Deviatoric}\\ \text{part}}} + \underbrace{\tfrac{1}{3}\,\mathrm{tr}[\varepsilon]\mathbf{1}.}_{\substack{\text{Volumetric}\\ \text{part}}} \tag{9.1.17}$$

This additive decomposition, although formally valid, *loses its physical content in the nonlinear theory*. The correct split takes the following multiplicative form.

9.1.3.1 Multiplicative decomposition.

Let $\bar{F}$ denote the volume-preserving part of the deformation gradient. Accordingly, $\det[\bar{F}] = 1$. Further, recall that $J := \det[F]$ gives the local volume change. Then set

$$\boxed{\bar{F} := J^{-1/3}F \Rightarrow \det[\bar{F}] = 1}$$

and (9.1.18)

$$\boxed{F = J^{1/3}\bar{F}.}$$

This decomposition was originally introduced by Flory [1961] and has been used by several authors in different contexts: Sidoroff [1974]; Hughes, Taylor and Sackman [1975]; Ogden [1982, 1984]; Simo, Taylor, and Pister [1985]; and Atluri and Reissner [1988]. From (9.1.18),

$$\bar{C} := \bar{F}^T\bar{F} \equiv J^{-2/3}C \Rightarrow \det[\bar{C}] = 1. \tag{9.1.19}$$

Other volume-preserving tensors are defined similarly. To gain further insight into the nature of the decomposition (9.1.18), it is useful to examine the rate form of $(9.1.19)_1$. For this purpose, recall the standard formula

$$\dot{J} = J \operatorname{div} v \equiv J \operatorname{tr}[d]. \tag{9.1.20}$$

In view of (9.1.5), (9.1.12), and (9.1.20), time differentiation of (9.1.19) yields

$$\begin{aligned}\dot{\bar{C}} &= J^{-2/3}\dot{C} - \tfrac{2}{3} J^{-2/3}C \operatorname{div} v \\ &= 2J^{-2/3}\left[F^T d F - \tfrac{1}{3} F^T F \operatorname{tr}[d]\right] \\ &= 2J^{-2/3}F^T\left[d - \tfrac{1}{3}\operatorname{tr}[d]\mathbf{1}\right]F,\end{aligned} \tag{9.1.21}$$

which is rephrased as

$$\boxed{\dot{\bar{C}} = 2\bar{F}^T \operatorname{dev}[d]\bar{F}.} \tag{9.1.22}$$

This expression shows that $\dot{\bar{C}}$ is indeed a *material* deviatoric tensor in the correct sense, since $\dot{\bar{C}} : C^{-1} = 0$.

9.2 J_2 Flow Theory at Finite Strains. A Model Problem

In this section we consider the formulation of the simplest plasticity model: J_2 flow theory with isotropic hardening. Our objective is to motivate the main features of the general theory and examine its computational implications in the simplest possible context.

9.2.1 Formulation of the Governing Equations

We assume throughout that the *stress response is isotropic*. Accordingly, the free energy is independent of the orientation of the reference configuration. This assumption is introduced at the outset to avoid any controversy concerning the appropriate invariance restrictions on the intermediate configuration. This issue is addressed in the formulation of the general theory; see Simo [1988a,b].

In addition, in accordance with a standard assumption in metal plasticity, we assume that *plastic flow is isochoric*:

$$\boxed{\det \boldsymbol{F}^p = \det \boldsymbol{C}^p = 1 \Rightarrow J = \det \boldsymbol{F} = \det \boldsymbol{F}^e.} \tag{9.2.1}$$

With these two a priori assumptions, we proceed to outline the governing equations of the model.

9.2.1.1 Stress response. Hyperelastic relationships.

Consistent with the assumption of isotropy and the notion of an intermediate stress-free configuration, we characterize the stress response by a stored-energy function of the form

$$\boxed{\begin{aligned} W &= U(J^e) + \bar{W}(\bar{\boldsymbol{b}}^e), \\ \bar{\boldsymbol{b}}^e &:= J^{e\,-2/3}\boldsymbol{F}^e\boldsymbol{F}^{eT} \equiv J^{e\,-2/3}\boldsymbol{b}^e, \end{aligned}} \tag{9.2.2}$$

where $U : \mathbb{R}_+ \to \mathbb{R}_+ \cup \{0\}$ is a convex function of $J^e := \det[\boldsymbol{F}^e]$. We call $U(J^e)$ and $\bar{W}(\bar{\boldsymbol{b}}^e)$ the *volumetric* and *deviatoric* parts of W, respectively. To make matters as concrete as possible, we consider the following explicit forms

$$\boxed{\begin{aligned} U(J^e) &:= \tfrac{1}{2}\kappa\left[\tfrac{1}{2}(J^{e\,2} - 1) - \ln J^e\right], \\ \bar{W}(\bar{\boldsymbol{b}}^e) &:= \tfrac{1}{2}\mu\left(\mathrm{tr}[\bar{\boldsymbol{b}}^e] - 3\right), \end{aligned}} \tag{9.2.3}$$

where $\mu > 0$ and $\kappa > 0$ are interpreted as the *shear modulus* and the *bulk modulus*, respectively. Note that, in view of (9.2.1), our definition of $\bar{\boldsymbol{b}}^e$ yields $\det\left[\bar{\boldsymbol{b}}^e\right] = 1$, hence, the denomination of the deviatoric part assigned to $\bar{W}$. Also note that

$$\mathrm{tr}[\bar{\boldsymbol{b}}^e] = \mathrm{tr}[\bar{\boldsymbol{C}}^e], \quad \text{where } \bar{\boldsymbol{C}}^e := J^{e\,-2/3}\boldsymbol{F}^{e\,T}\boldsymbol{F}^e. \tag{9.2.4}$$

Now let $W = U(J^e) + \hat{W}(\bar{\boldsymbol{C}}^e)$, and observe that for model (9.2.3) $\bar{W}(\bar{\boldsymbol{b}}^e) = \hat{W}(\bar{\boldsymbol{C}}^e)$. Then the Kirchhoff stress tensor is obtained by the general expression

$$\left.\begin{aligned} \boldsymbol{\tau} &= 2\boldsymbol{F}^e\frac{\partial W}{\partial \boldsymbol{C}^e}\boldsymbol{F}^{e\,T} = J^e U'(J^e)\mathbf{1} + \boldsymbol{s}, \\ \boldsymbol{s} &= 2\,\mathrm{dev}\left[\bar{\boldsymbol{F}}^e\frac{\partial \hat{W}}{\partial \bar{\boldsymbol{C}}^e}\bar{\boldsymbol{F}}^{e\,T}\right], \end{aligned}\right\} \tag{9.2.5}$$

the derivation of which is in Simo, Taylor, and Pister [1985]. Note that the uncoupled, stored-energy function (9.2.2) results in *uncoupled* volumetric-deviatoric stress-strain relationships. From (9.2.4) and (9.2.3), we find that

$$\boxed{\begin{aligned} \boldsymbol{\tau} &= J^e p \mathbf{1} + \boldsymbol{s}, \\ p :&= U'(J^e) = \frac{\kappa}{2}(J^{e2} - 1)/J^e, \\ \boldsymbol{s} :&= \operatorname{dev}[\boldsymbol{\tau}] = \mu \operatorname{dev}[\bar{\boldsymbol{b}}^e]. \end{aligned}} \tag{9.2.6}$$

Observe that $U(J) \rightarrow +\infty$ and $p \rightarrow \pm\infty$, as $J \rightarrow 0$ and $J \rightarrow \infty$. Further, one can easily show that (9.2.6) reduces (for small strains) to the classical isotropic model of the linearized theory.

Remarks 9.2.1.

1. The assumption that the stored-energy function W depends on $\boldsymbol{b}^e = \boldsymbol{F}^e \boldsymbol{F}^{e\,T}$ is consistent with models of the type considered by Lee [1969] and Dafalias [1984]. Alternatively, W is expressed as follows. In view of (9.1.11),

$$\begin{aligned} \operatorname{tr}[\bar{\boldsymbol{b}}^e] &= \mathbf{1} : [J^{e\,-2/3} \boldsymbol{b}^e] \\ &= \mathbf{1} : [J^{-2/3} J^{p\,2/3} \boldsymbol{F} \boldsymbol{C}^{p\,-1} \boldsymbol{F}^T] \\ &= \mathbf{1} : [\bar{\boldsymbol{F}} \bar{\boldsymbol{C}}^{p\,-1} \bar{\boldsymbol{F}}^T] = \bar{\boldsymbol{F}}^T \bar{\boldsymbol{F}} : \bar{\boldsymbol{C}}^{p\,-1}, \end{aligned} \tag{9.2.7}$$

where $\bar{\boldsymbol{C}}^p = J^{p-2/3} \boldsymbol{C}^p$ and $\boldsymbol{C}^p$ is defined by $(9.1.5)_2$. From (9.2.7) and (9.1.19), we conclude that

$$\boxed{\operatorname{tr}[\bar{\boldsymbol{b}}^e] = \bar{\boldsymbol{C}} : \bar{\boldsymbol{C}}^{p\,-1} := \operatorname{tr}[\bar{\boldsymbol{C}} \bar{\boldsymbol{C}}^{p\,-1}].} \tag{9.2.8}$$

Therefore, (9.2.8) and (9.2.3) imply that W has the form

$$\boxed{W = U(J J^{p-1}) + \tfrac{1}{2}\mu \left[\bar{\boldsymbol{C}} : \bar{\boldsymbol{C}}^{p\,-1} - 3\right] =: \tilde{W}\left(\boldsymbol{C}, \boldsymbol{C}^p\right),} \tag{9.2.9}$$

which formally has the same functional form as models considered by Green and Naghdi [1965, 1966]. For a related model, see Simo and Ju [1989]. However, notice that W *does not generally depend on the difference* $\boldsymbol{C} - \boldsymbol{C}^p$, even in the simplest situation afforded by (9.2.3), in contrast with the original proposal of Green and Naghdi [1965].

2. Further insight into the nature of the stored-energy function (9.2.2)–(9.2.3) is gained by examining its time derivative. First, by the well-known formula for the derivative of a determinant,

$$J^e = \det(\boldsymbol{C}^e) \Rightarrow \dot{J}^e = \tfrac{1}{2} J^e \dot{\boldsymbol{C}}^e : \boldsymbol{C}^{e\,-1}. \tag{9.2.10}$$

Second, in view of (9.2.4), from (9.2.3) and (9.2.10),

$$\begin{aligned} \dot{W} &= \left[J^e U'(J^e) \boldsymbol{C}^{e\,-1}\right] : \tfrac{1}{2}\dot{\boldsymbol{C}}^e + \tfrac{1}{2}\mu J^{e\,-2/3}\left[\dot{\boldsymbol{C}}^e - \tfrac{1}{3}\boldsymbol{C}^e(\dot{\boldsymbol{C}}^e : \boldsymbol{C}^{e\,-1})\right] : \mathbf{1} \\ &= \left[J^e p \boldsymbol{C}^{e\,-1} + \mu J^{e\,-2/3}\left(\mathbf{1} - \tfrac{1}{3}\operatorname{tr}[\boldsymbol{C}^e]\boldsymbol{C}^{e\,-1}\right)\right] : \tfrac{1}{2}\dot{\boldsymbol{C}}^e. \end{aligned} \tag{9.2.11}$$

Since $\boldsymbol{C}^e = \boldsymbol{F}^{e\,T}\boldsymbol{F}^e$, a straightforward manipulation of (9.2.11) gives, in view of (9.2.6), the expression

$$\begin{aligned}\dot{W} &= \boldsymbol{F}^{e\,-1}\left[J^e p\mathbf{1} + \mu J^{e\,-2/3}\left(\boldsymbol{b}^e - \tfrac{1}{3}\,\mathrm{tr}[\boldsymbol{b}^e]\mathbf{1}\right)\right]\boldsymbol{F}^{e\,-T} : \tfrac{1}{2}\dot{\boldsymbol{C}}^e \\ &= \left[J^e p\mathbf{1} + \mu\,\mathrm{dev}[\bar{\boldsymbol{b}}^e]\right] : \boldsymbol{F}^{e\,-T}\tfrac{1}{2}\dot{\boldsymbol{C}}^e\boldsymbol{F}^{e\,-1} \\ &= \boldsymbol{\tau} : \tfrac{1}{2}\left[\boldsymbol{F}^{e\,-T}\dot{\boldsymbol{C}}^e\boldsymbol{F}^{e\,-1}\right]. \end{aligned} \tag{9.2.12}$$

Now recall that the total rate of deformation tensor is defined in terms of $\dot{\boldsymbol{C}}$ by formula (9.1.12), which makes intrinsic geometric sense, and is identical in structure to the expression within brackets in (9.2.12). Consequently, by analogy with $\boldsymbol{d} = \frac{1}{2}\boldsymbol{F}^{-T}\dot{\boldsymbol{C}}\boldsymbol{F}^{-1}$, one sets

$$\boxed{\boldsymbol{d}^e := \tfrac{1}{2}\boldsymbol{F}^{e\,-T}\dot{\boldsymbol{C}}^e\boldsymbol{F}^{e\,-1}} \;\Rightarrow\; \boxed{\dot{W} = \boldsymbol{\tau} : \boldsymbol{d}^e.} \tag{9.2.13}$$

Equation (9.2.13) is the natural counterpart of the expression $\dot{W} = \boldsymbol{\sigma} : \dot{\boldsymbol{\varepsilon}}^e$ in the infinitesimal theory.

3. Identical computations are carried out relative to the reference configuration by taking expression (9.2.9) as the point of departure in place of (9.2.3)–(9.2.4). Then an expression is obtained in terms of $\{\boldsymbol{C}, \boldsymbol{C}^p\}$ linear in the strain rates $\{\dot{\boldsymbol{C}}, \dot{\boldsymbol{C}}^p\}$. These observations are consistent with the fact that a form of the constitutive equations in a properly invariant theory should be independent of the description adopted (Lagrangian or Eulerian).

9.2.1.2 Yield condition.

We consider the classical Mises–Huber yield condition formulated in terms of the Kirchhoff stress tensor as

$$\boxed{f(\boldsymbol{\tau}, \alpha) := \|\mathrm{dev}[\boldsymbol{\tau}]\| - \sqrt{\tfrac{2}{3}}\left[\sigma_Y + K\alpha\right] \le 0,} \tag{9.2.14}$$

where, as in Chapter **2**, σ_Y denotes the flow stress, $K > 0$ the isotropic hardening modulus, and α the hardening parameter. Nonlinear hardening laws in which $K(\alpha)$ is a nonlinear function of the hardening parameter are easily accommodated in the formulation, as shown below.

9.2.1.3 The associative-flow rule.

The crucial step in formulating the model lies in developing the corresponding associative flow rule. Remarkably, as in the infinitesimal theory, *given the stored-energy function and the yield condition, the functional form of the corresponding associative flow rule is uniquely determined by the principle of maximum plastic dissipation*.

In the present context, for the Mises-Huber yield condition (9.2.14) and the stored-energy function (9.2.2)–(9.2.3), one can show (Simo [1988a,b]) that the associative flow rule takes the form

$$\boxed{\begin{aligned} \frac{\partial}{\partial t}\{\bar{C}^{p\,-1}\} &= -\tfrac{2}{3}\gamma\,\mathrm{tr}\left[\boldsymbol{b}^e\right]\boldsymbol{F}^{-1}\boldsymbol{n}\boldsymbol{F}^{-T}, \\ \boldsymbol{n} &:= \boldsymbol{s}/\|\boldsymbol{s}\|, \\ \boldsymbol{s} &:= \mathrm{dev}[\boldsymbol{\tau}]. \end{aligned}} \tag{9.2.15}$$

Remarks 9.2.2.

1. From (9.2.8) we observe that $\mathrm{tr}[\boldsymbol{b}^e] = \boldsymbol{C} : \boldsymbol{C}^{p\,-1}$ therefore can be expressed as a function of $\{\boldsymbol{C}, \boldsymbol{C}^p\}$. Further, note that $\boldsymbol{N} := \boldsymbol{F}^{-1}\boldsymbol{n}\boldsymbol{F}^{-T}$ is also a function of $\{\boldsymbol{C}, \boldsymbol{C}^p\}$ as is easily concluded from (9.2.6). Hence, (9.2.15) should be regarded as a *flow rule in strain space giving the evolution of* $\dot{\boldsymbol{C}}^{p\,-1}$ *in terms of* $\{\boldsymbol{C}, \boldsymbol{C}^p\}$.
2. The flow rule (9.2.15) can also be expressed entirely in the spatial description leading to a somewhat surprising result. In fact, from (9.1.15) and (9.2.15), we obtain

$$\boxed{\begin{aligned} L_v\boldsymbol{b}^e &= -\tfrac{2}{3}\gamma\,\mathrm{tr}[\boldsymbol{b}^e]\boldsymbol{n}, \\ \boldsymbol{n} &= \boldsymbol{s}/\|\boldsymbol{s}\|\,. \end{aligned}} \tag{9.2.16}$$

   It should be noted that (9.2.16) defines only the deviatoric part of $L_v\boldsymbol{b}^e$; see Simo [1988a,b] for a detailed elaboration of this point.

9.2.1.4 Isotropic hardening law and loading/unloading conditions.

As in the linear theory, we assume that the evolution of the hardening variable is governed by the rate equation

$$\boxed{\dot{\alpha} = \sqrt{\tfrac{2}{3}}\,\gamma\,,} \tag{9.2.17}$$

where γ is the consistency parameter subject to the standard Kuhn–Tucker loading/unloading conditions

$$\boxed{\gamma \geq 0, \quad f(\boldsymbol{\tau}, \alpha) \leq 0, \quad \gamma f(\boldsymbol{\tau}, \alpha) = 0,} \tag{9.2.18a}$$

which along with the consistency condition

$$\boxed{\gamma \dot{f}(\boldsymbol{\tau}, \alpha) = 0} \tag{9.2.18b}$$

complete the formulation of the model.

9.2.1.5 Kinematic hardening model.

In addition to the isotropic hardening law considered above, other types of hardening response are accommodated in the model by introducing additional internal variables with evolution governed by properly invariant rate equations. In particular, a possible extension to finite strains of the Prager–Ziegler *kinematic hardening* law is constructed as follows.

Let $\bar{\boldsymbol{q}}$ be the back stress interpreted as a spatial second-order (stress-like) contravariant tensor field. Accordingly, we assume that $\bar{\boldsymbol{q}}$ transforms *objectively* under rigid motions superposed on the current configuration, that is

$$\bar{\boldsymbol{q}} \mapsto \bar{\boldsymbol{q}}^{+} := \boldsymbol{Q}\bar{\boldsymbol{q}}\boldsymbol{Q}^{T}, \quad \text{for all } \boldsymbol{Q} \in \mathrm{SO}(3) . \tag{9.2.19}$$

Then consider the following evolution equations for the flow rule and the kinematic hardening law

$$\boxed{\begin{aligned} L_v \boldsymbol{b}^e &= -\gamma \tfrac{2}{3} \operatorname{tr}[\boldsymbol{b}^e]\boldsymbol{n}, \\ L_v \bar{\boldsymbol{q}} &= \gamma \tfrac{2}{3} H \operatorname{tr}[\boldsymbol{b}^e]\boldsymbol{n}, \\ \boldsymbol{n} &:= \left(\boldsymbol{s} - \bar{\boldsymbol{q}}\right) / \|\boldsymbol{s} - \bar{\boldsymbol{q}}\| , \end{aligned}} \tag{9.2.20}$$

where H is the kinematic hardening modulus. As in the linear theory, the Mises yield condition (9.2.14) is modified to accommodate kinematic hardening as follows. Set $q := -\sqrt{\frac{2}{3}}\, K\alpha$, where α is the isotropic hardening variable with evolution equation (9.2.17). Then, in terms of the hardening variables $\boldsymbol{q} = \{q, \bar{\boldsymbol{q}}\}$, the Mises condition (9.2.14) becomes

$$\boxed{\begin{aligned} f(\boldsymbol{\tau}, \boldsymbol{q}) &= \|\operatorname{dev}[\boldsymbol{\tau} - \bar{\boldsymbol{q}}]\| + q - \sqrt{\tfrac{2}{3}}\sigma_Y \leq 0, \\ \dot{q} &= -\tfrac{2}{3} K\gamma, \end{aligned}} \tag{9.2.21}$$

Equations (9.2.20,21) along with the hyperelastic stress-strain relations (9.2.6) and the Kuhn–Tucker complementary conditions (9.2.18) complete the formulation of the model.

9.3 Integration Algorithm for J_2 Flow Theory

In this section we examine in detail the numerical integration of the model problem outlined in Section **2**. It is shown that the resulting integrative algorithm furnishes a canonical extension to finite strains of the classical radial return method of infinitesimal plasticity. Conceptually, the only difference lies in the fact that the elastic predictor (trial elastic state) is evaluated by hyperelastic, finite-strain, stress-strain relationships.

9.3.1 Integration of the Flow Rule and Hardening Law

Proper invariance of the integrated constitutive model under *superposed rigid body motions* on the current configuration is a fundamental restriction placed by the principle of material frame indifference which *must be exactly preserved by the integration algorithm*. To enforce this restriction, we proceed as in Chapter **7**.

9.3.1.1 Outline of the general procedure.

A general procedure that automatically ensures satisfaction of the principle of material frame indifference is as follows:

i. Given a properly invariant evolution equation formulated in the spatial description, we transform the equation to the convected description by using appropriate tensorial transformations involving the deformation gradient (in a more geometric context, one refers to this tensorial transformation as a *pull-back*; see e.g. Marsden and Hughes [1994, Chapter **1**] or Arnold [1978]).

ii. In the convected description, i.e., a description in which $\{\boldsymbol{C}, \boldsymbol{C}^p, \boldsymbol{S}\}$ are the basic variables, we perform the time-stepping algorithm according to any selected integration scheme, typically the generalized midpoint rule. This results in a discrete form of the evolution equation.

iii. We transform the discrete evolution equation back to the spatial description by appropriate tensorial transformations (in a geometric context, we refer to this transformation as a *push-forward*).

In (9.2.16)–(9.2.17), we illustrate in detail the application of this general technique with reference to the flow rule and hardening law.

9.3.1.2 Discrete flow rule and hardening law.

The given equations of evolution in the spatial description are (9.2.16)–(9.2.17). The transformed evolution equations in the convected description are given by (9.2.15) and (9.2.17) (note that α is a scalar), that is,

$$\left.\begin{aligned} \frac{\partial}{\partial t}\{\bar{\boldsymbol{C}}^{p\,-1}\} &= -\tfrac{2}{3}\gamma\left[\boldsymbol{C}^{p\,-1} : \boldsymbol{C}\right]\boldsymbol{F}^{-1}\boldsymbol{n}\boldsymbol{F}^{-T}, \\ \dot{\alpha} &= \sqrt{\tfrac{2}{3}}\,\gamma. \end{aligned}\right\} \tag{9.3.1}$$

Again we remark that $\boldsymbol{F}^{-1}\boldsymbol{n}\boldsymbol{F}^{-T}$ is easily shown to be a function of $\{\boldsymbol{C}, \boldsymbol{C}^p\}$. By applying a backward Euler difference scheme, we obtain the discrete evolution equations as

$$\left.\begin{aligned} \bar{\boldsymbol{C}}^{p-1}_{n+1} - \bar{\boldsymbol{C}}^{p-1}_{n} &= -\tfrac{2}{3}\,\Delta\gamma\left[\boldsymbol{C}^{p\,-1}_{n+1} : \boldsymbol{C}_{n+1}\right]\boldsymbol{F}^{-1}_{n+1}\boldsymbol{n}_{n+1}\boldsymbol{F}^{-T}_{n+1}, \\ \alpha_{n+1} - \alpha_n &= \sqrt{\tfrac{2}{3}}\,\Delta\gamma. \end{aligned}\right\} \tag{9.3.2}$$

Now we complete step **iii** above by transforming (9.3.2) to the current configuration. To this end, a calculation similar to that leading to (9.2.7) now yields

$$\begin{aligned} \operatorname{tr}[\boldsymbol{b}^e_{n+1}] = \boldsymbol{1} : \boldsymbol{b}^e_{n+1} &= \boldsymbol{1} : \boldsymbol{F}_{n+1}\boldsymbol{C}^{p\,-1}_{n+1}\boldsymbol{F}^T_{n+1} \\ &= \boldsymbol{F}^T_{n+1}\boldsymbol{F}_{n+1} : \boldsymbol{C}^{p\,-1}_{n+1} = \boldsymbol{C}_{n+1} : \boldsymbol{C}^{p\,-1}_{n+1}. \end{aligned} \tag{9.3.3}$$

Similarly, since $\bar{\boldsymbol{C}}^p_{n+1} = J^{p-2/3}_{n+1}\boldsymbol{C}^p_{n+1}$, use of (9.1.4) results in the relationship

$$\bar{\boldsymbol{F}}_{n+1}\bar{\boldsymbol{C}}^{p\,-1}_{n+1}\bar{\boldsymbol{F}}^T_{n+1} = J^{p\,2/3}_{n+1}\bar{\boldsymbol{F}}_{n+1}\boldsymbol{C}^{p\,-1}_{n+1}\bar{\boldsymbol{F}}^T_{n+1}$$

$$
\begin{aligned}
&= J_{n+1}^{-2/3} J_{n+1}^{p\,2/3} \boldsymbol{F}_{n+1} \boldsymbol{C}_{n+1}^{p\,-1} \boldsymbol{F}_{n+1}^{T} \\
&= \left(J_{n+1}/J_{n+1}^{p}\right)^{-\frac{2}{3}} \boldsymbol{b}_{n+1}^{e} \\
&= J_{n+1}^{e^{-2/3}} \boldsymbol{b}_{n+1}^{e} =: \bar{\boldsymbol{b}}_{n+1}^{e}.
\end{aligned} \tag{9.3.4}
$$

Now let $\boldsymbol{f}_{n+1}$ and $\bar{\boldsymbol{f}}_{n+1}$ denote the relative deformation gradient and its volume-preserving part, respectively, with respect to configurations $\varphi_n(\mathcal{B})$ and $\varphi_{n+1}(\mathcal{B})$. See Figure **9.4**. By definition,

$$
\boxed{\boldsymbol{f}_{n+1} := \boldsymbol{F}_{n+1} \boldsymbol{F}_n^{-1}}
$$

and

$$
\boxed{\bar{\boldsymbol{f}}_{n+1} = \bar{\boldsymbol{F}}_{n+1} \bar{\boldsymbol{F}}_n^{-1} \equiv \left(J_{n+1}/J_n\right)^{-\frac{1}{3}} \boldsymbol{f}_{n+1}} \tag{9.3.5}
$$

Therefore, proceeding as in the derivation leading to (9.3.4),

$$
\begin{aligned}
\bar{\boldsymbol{F}}_{n+1} \bar{\boldsymbol{C}}_n^{p\,-1} \bar{\boldsymbol{F}}_{n+1}^{T} &= \bar{\boldsymbol{f}}_{n+1} \left[\bar{\boldsymbol{F}}_n \bar{\boldsymbol{C}}_n^{p\,-1} \bar{\boldsymbol{F}}_n^{T}\right] \bar{\boldsymbol{f}}_{n+1}^{T} \\
&= \bar{\boldsymbol{f}}_{n+1} \bar{\boldsymbol{b}}_n^{e} \bar{\boldsymbol{f}}_{n+1}^{T}.
\end{aligned} \tag{9.3.6}
$$

Premultiplying (9.3.2) by $\bar{\boldsymbol{F}}_{n+1}$ and postmultiplying by $\bar{\boldsymbol{F}}_{n+1}^{T}$ along with relationships (9.3.3), (9.3.4), and (9.3.6), yield the spatial, discrete, evolution equations:

$$
\boxed{
\begin{aligned}
\bar{\boldsymbol{b}}_{n+1}^{e} &= \bar{\boldsymbol{f}}_{n+1} \bar{\boldsymbol{b}}_n^{e} \bar{\boldsymbol{f}}_{n+1}^{T} - \tfrac{2}{3} \Delta\gamma \operatorname{tr}[\bar{\boldsymbol{b}}_{n+1}^{e}] \boldsymbol{n}_{n+1}, \\
\boldsymbol{n}_{n+1} &:= \boldsymbol{s}_{n+1}/\|\boldsymbol{s}_{n+1}\|, \\
\boldsymbol{s}_{n+1} &:= \operatorname{dev}[\boldsymbol{\tau}_{n+1}], \\
\alpha_{n+1} &= \alpha_n + \sqrt{\tfrac{2}{3}} \Delta\gamma,
\end{aligned}
} \tag{9.3.7}
$$

where we have used the fact that $\boldsymbol{F}_{n+1}^{-1} = J_{n+1}^{-1/3} \bar{\boldsymbol{F}}_{n+1}^{-1}$. Notice that so far we have

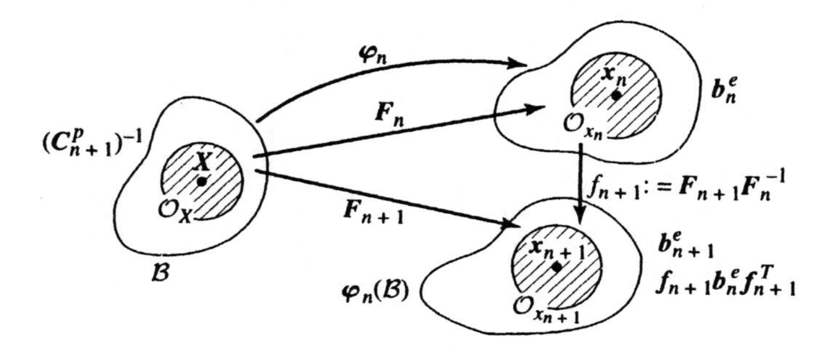

FIGURE 9.4. Total and relative deformation gradients connecting neighborhoods $\mathcal{O}_X$, $\mathcal{O}_{x_n}$, and $\mathcal{O}_{x_{n+1}}$ in $\mathcal{B}$, $\varphi_n(\mathcal{B})$, and $\varphi_{n+1}(\mathcal{B})$, respectively.

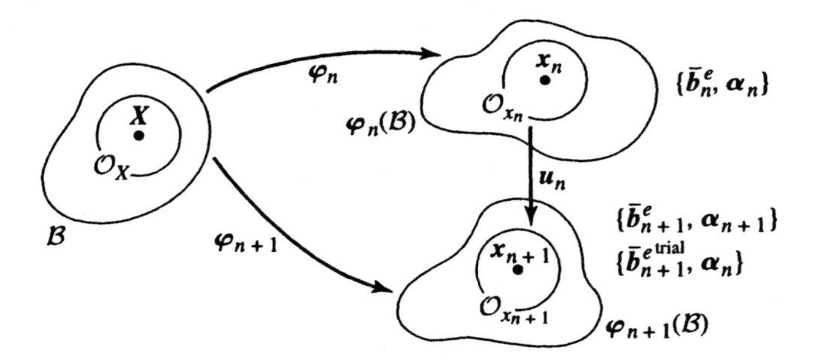

FIGURE 9.5. Update of the current configuration for a given incremental displacement $\boldsymbol{u}_n : \varphi_n(\mathcal{B}) \to \mathbb{R}^3$.

not used the condition $J^p = 1$, which follows from the assumption of isochoric plastic flow.

9.3.2 The Return-Mapping Algorithm

With the preceding developments in hand, we proceed to construct the return-mapping algorithm within the usual computational context which regards the problem essentially as strain-driven.

9.3.2.1 Database and configurational update.

Let $[t_n, t_{n+1}]$ be the time interval of interest. We assume that the following data is known at time t_n:

$$\boxed{\{\varphi_n, \bar{\boldsymbol{b}}^e_n, \alpha_n\}, \quad \boldsymbol{F}_n := D\varphi_n(\boldsymbol{X}).} \tag{9.3.8}$$

Therefore, the Kirchhoff stress tensor $\boldsymbol{\tau}_n$ is also known through the hyperelastic relationships

$$\left.\begin{aligned} \boldsymbol{\tau}_n &= p_n J_n \mathbf{1} + \mu\, \operatorname{dev}[\bar{\boldsymbol{b}}^e_n], \\ p_n &= U'(J_n), \end{aligned}\right\} \tag{9.3.9}$$

where we have enforced the isochoric constraint $J^p = 1 \Longleftrightarrow J^e = J$. Now let

$$\boldsymbol{u}_n : \varphi_n(\mathcal{B}) \to \mathbb{R}^3 \tag{9.3.10}$$

be a *given incremental displacement* field of the configuration $\varphi_n(\mathcal{B})$. Therefore the update of the configuration $\varphi_n(\mathcal{B})$ is immediately obtained simply by setting

$$\boldsymbol{x}_{n+1} = \varphi_{n+1}(\boldsymbol{X}) := \varphi_n(\boldsymbol{X}) + \boldsymbol{u}_n\left[\varphi_n(\boldsymbol{X})\right], \tag{9.3.11a}$$

or in a more compact notation, as

$$\boxed{\varphi_{n+1} = \varphi_n + \boldsymbol{u}_n \circ \varphi_n.} \tag{9.3.11b}$$

The situation is illustrated in Figure **9.5**.

9.3.2.2 The trial elastic state.

Now we consider a state which is obtained by *"freezing" the evolution of plastic flow* on $[t_n, t_{n+1}]$. Consequently, the intermediate configuration remains unchanged, i.e.,

$$\boxed{\begin{aligned} [C_{n+1}^{p\,-1}]^{\text{trial}} &:= C_n^{p\,-1}, \\ \alpha_{n+1}^{\text{trial}} &:= \alpha_n. \end{aligned}} \tag{9.3.12}$$

Premultiplying and postmultiplying (9.3.12) by $\bar{F}_{n+1}$ and $\bar{F}_{n+1}^T$, respectively, where $\bar{F}_{n+1} := \left(D\varphi_{n+1}\right) J_{n+1}^{-\frac{1}{3}}$, yields, in view of (9.1.11) and definition (9.3.5), the relationship

$$\begin{aligned} \bar{b}_{n+1}^{e\,\text{trial}} &:= \bar{F}_{n+1}(\bar{C}_{n+1}^{p\,-1})^{\text{trial}} \bar{F}_{n+1}^T \\ &= \left(\bar{f}_{n+1}\bar{F}_n\right) \bar{C}_n^{p\,-1} \left(\bar{f}_{n+1}\bar{F}_n\right)^T \\ &= \bar{f}_{n+1} \left[\bar{F}_n \bar{C}_n^{p\,-1} \bar{F}_n^T\right] \bar{f}_{n+1}^T \\ &= \bar{f}_{n+1} \bar{b}_n^e \bar{f}_{n+1}^T. \end{aligned} \tag{9.3.13}$$

Thus, with relationships (9.3.12) and (9.3.13) in hand, we define the trial elastic state by the equations

$$\boxed{\begin{aligned} \tau_{n+1}^{\text{trial}} &:= p_{n+1} J_{n+1} \mathbf{1} + s_{n+1}^{\text{trial}}, \\ s_{n+1}^{\text{trial}} &:= \mu \operatorname{dev}[\bar{b}_{n+1}^{e\,\text{trial}}], \\ p_{n+1} &:= U'(J_{n+1}), \\ \bar{b}_{n+1}^{e\,\text{trial}} &:= \bar{f}_{n+1} \bar{b}_n^e \bar{f}_{n+1}^T, \\ \alpha_{n+1}^{\text{trial}} &:= \alpha_n. \end{aligned}} \tag{9.3.14}$$

A schematic illustration of the trial elastic state is given in Figure **9.6**. Note that the intermediate configuration remains unchanged in this trial-state phase of the algorithm.

Remarks 9.3.1.

1. The relative deformation gradient f_{n+1} is computed directly from the update formula (9.3.11) and definition (9.3.5) in terms of the incremental displacement field as follows. Since

$$\begin{aligned} F_{n+1} := D\varphi_{n+1} &= D\varphi_n + \left[\nabla_{x_n} u_n\right] D\varphi_n \\ &= \left[\mathbf{1} + \nabla_{x_n} u_n\right] F_n, \end{aligned} \tag{9.3.15}$$

$$\boxed{f_{n+1} := F_{n+1} F_n^{-1} = \mathbf{1} + \nabla_{x_n} u_n,} \tag{9.3.16}$$

from which we obtain the volume-preserving part as $\bar{f}_{n+1} = \left(\det\left[f_{n+1}\right]\right)^{-\frac{1}{3}} f_{n+1}$.

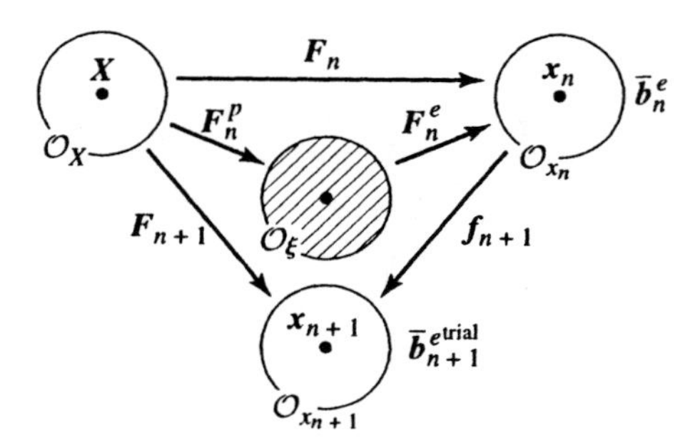

FIGURE 9.6. The trial elastic state kinematic relationships. Note that $f_{n+1} := \mathbf{1} + \nabla_{x_n} u_n$ is given.

2. Expression (9.3.13) is consistent with the following definition of $\bar{b}^{e\,\mathrm{trial}}_{n+1}$:

$$\bar{b}^{e\,\mathrm{trial}}_{n+1} = \bar{F}^{e\,\mathrm{trial}}_{n+1}\left(\bar{F}^{e\,\mathrm{trial}}_{n+1}\right)^T,$$

where (9.3.17)

$$\bar{F}^{e\,\mathrm{trial}}_{n+1} := \bar{F}_{n+1}\bar{F}^{p\,-1}_n,$$

with $\bar{F}_{n+1} := J_{n+1}^{-1/3} F_{n+1}$. A straightforward calculation verifies this result.

9.3.2.3 Discrete governing equations. Loading condition.

The discrete governing equations are conveniently written as follows in terms of the trial elastic state defined above. First, by using (9.3.13) and the fact that plastic flow is isochoric, i.e., $J_{n+1} = J^e_{n+1}$, the discrete evolution equations (9.3.7) are expressed in the equivalent form

$$\boxed{\begin{aligned} \bar{b}^e_{n+1} &= \bar{b}^{e\,\mathrm{trial}}_{n+1} - \tfrac{2}{3}\,\Delta\gamma\ \mathrm{tr}[\bar{b}^e_{n+1}]\,n_{n+1}, \\ \alpha_{n+1} &= \alpha_n + \sqrt{\tfrac{2}{3}}\,\Delta\gamma. \end{aligned}} \tag{9.3.19}$$

On the other hand, the hyperelastic constitutive model (9.2.6) evaluated at time t_{n+1} yields

$$\boxed{\begin{aligned} \tau_{n+1} &= p_{n+1} J_{n+1} \mathbf{1} + s_{n+1}, \\ p_{n+1} &:= U'(J_{n+1}), \\ s_{n+1} &:= \mu\ \mathrm{dev}[\bar{b}^e_{n+1}], \\ n_{n+1} &:= \frac{s_{n+1}}{\|s_{n+1}\|}. \end{aligned}} \tag{9.3.20}$$

Finally, the discrete version of the model is completed by appending the discrete version of the Kuhn–Tucker complementarity conditions (9.2.18) given by

$$\boxed{\Delta\gamma \geq 0, \quad f(\tau_{n+1}, \alpha_{n+1}) \leq 0, \quad \Delta\gamma f(\tau_{n+1}, \alpha_{n+1}) = 0.} \tag{9.3.21}$$

As in the linear theory, the systematic exploitation of these unilateral constraint conditions produces the appropriate return-mapping algorithm. Two alternative situations may arise.

i. First consider the case for which $f_{n+1}^{\text{trial}} \leq 0$, where

$$\boxed{f_{n+1}^{\text{trial}} := f\left(\boldsymbol{\tau}_{n+1}^{\text{trial}}, \alpha_n\right) = \|\boldsymbol{s}_{n+1}^{\text{trial}}\| - \sqrt{\tfrac{2}{3}}\left[\sigma_Y + K\alpha_n\right].} \tag{9.3.22}$$

Then, the trial elastic state with $\Delta\gamma = 0$ satisfies conditions (9.3.21). The remaining equations (9.3.19) and (9.3.20) also hold by construction. Thus, the trial elastic step is the solution at time t_{n+1}.

ii. Alternatively, consider the situation for which $f_{n+1}^{\text{trial}} > 0$. It follows that $\boldsymbol{\tau}_{n+1}^{\text{trial}}$ is nonadmissible and, therefore, cannot be the solution at t_{n+1}. Accordingly, $\boldsymbol{\tau}_{n+1} \neq \boldsymbol{\tau}_{n+1}^{\text{trial}}$ and relationships (9.3.20) imply that $\bar{\boldsymbol{b}}_{n+1}^{e} \neq \bar{\boldsymbol{b}}_{n+1}^{e\ trial}$. Consequently from (9.3.19), we conclude that $\bar{\boldsymbol{b}}_{n+1}^{e} \neq \boldsymbol{b}_{n+1}^{\text{trial}}$ only if $\Delta\gamma > 0$. The preceding discussion shows that whether the point $\boldsymbol{x}_n \in \varphi_n(\mathcal{B})$ experiences loading or unloading during the step $[t_n, t_{n+1}]$ can be concluded solely on the basis of the trial elastic state according to the conditions

$$\boxed{f_{n+1}^{\text{trial}} \begin{cases} \leq 0 & \text{elastic step} \Rightarrow \Delta\gamma = 0\,, \\ > 0 & \text{plastic step} \Rightarrow \Delta\gamma > 0\,. \end{cases}} \tag{9.3.23}$$

The algorithmic procedure is completed by characterizing the solution for $\Delta\gamma > 0$ in terms of the trial step as follows.

9.3.2.4 The radial return algorithm.

Assume that $f_{n+1}^{\text{trial}} > 0 \Longleftrightarrow \Delta\gamma \neq 0$. Since $\text{tr}[\boldsymbol{n}_{n+1}] = 0$, taking the trace in (9.3.19),

$$\boxed{\text{tr}\big[\bar{\boldsymbol{b}}_{n+1}^{e}\big] = \text{tr}\big[\bar{\boldsymbol{b}}_{n+1}^{e\ \text{trial}}\big].} \tag{9.3.24}$$

Then substituting (9.3.24) in (9.3.19) and using the hyperelastic relationships (9.3.20) yields

$$\begin{aligned} \boldsymbol{s}_{n+1} &= \mu\ \text{dev}[\bar{\boldsymbol{b}}_{n+1}^{e\ \text{trial}}] - \tfrac{2}{3}\mu\Delta\gamma\ \text{tr}\left[\bar{\boldsymbol{b}}_{n+1}^{e\ \text{trial}}\right]\boldsymbol{n}_{n+1} \\ &= \boldsymbol{s}_{n+1}^{\text{trial}} - \tfrac{2}{3}\mu\Delta\gamma\ \text{tr}\left[\bar{\boldsymbol{b}}_{n+1}^{e\ \text{trial}}\right]\boldsymbol{n}_{n+1}, \end{aligned} \tag{9.3.25}$$

where we have used definition $(9.3.14)_2$ for $\boldsymbol{s}_{n+1}^{\text{trial}}$. The determination of $\Delta\gamma > 0$ from (9.3.25) now follows the same procedure as in the infinitesimal theory. We set $\boldsymbol{s}_{n+1} = \|\boldsymbol{s}_{n+1}\|\boldsymbol{n}_{n+1}$ and rearrange terms in (9.3.25) to obtain

$$\left.\begin{aligned} \big[\|\boldsymbol{s}_{n+1}\| + 2\bar{\mu}\Delta\gamma\big]\boldsymbol{n}_{n+1} &= \|\boldsymbol{s}_{n+1}^{\text{trial}}\|\boldsymbol{n}_{n+1}^{\text{trial}} \\ \boldsymbol{n}_{n+1}^{\text{trial}} &:= \frac{\boldsymbol{s}_{n+1}^{\text{trial}}}{\|\boldsymbol{s}_{n+1}^{\text{trial}}\|} \\ \bar{\mu} &:= \tfrac{1}{3}\mu\ \text{tr}\left[\bar{\boldsymbol{b}}_{n+1}^{e\ \text{trial}}\right]. \end{aligned}\right\} \tag{9.3.26}$$

Then equation $(9.3.26)_1$ implies

$$\boxed{\boldsymbol{n}_{n+1} \equiv \boldsymbol{n}_{n+1}^{\text{trial}},} \tag{9.3.27}$$

along with the requirement that

$$\boxed{\left[\|\boldsymbol{s}_{n+1}\| + 2\bar{\mu}\Delta\gamma\right] = \|\boldsymbol{s}_{n+1}^{\text{trial}}\|.} \tag{9.3.28}$$

On the other hand, since $\Delta\gamma > 0$, we require that $f(\boldsymbol{\tau}_{n+1}, \alpha_{n+1}) = 0$, and from (9.2.14), $(9.3.19)_2$, (9.3.22), and (9.3.28), we obtain

$$\begin{aligned}
\|\boldsymbol{s}_{n+1}\| - \sqrt{\tfrac{2}{3}}\left[\sigma_Y + K\alpha_{n+1}\right] &= \|\boldsymbol{s}_{n+1}^{\text{trial}}\| - 2\bar{\mu}\Delta\gamma - \sqrt{\tfrac{2}{3}}\left(\sigma_Y + K\alpha_{n+1}\right) \\
&= \|\boldsymbol{s}_{n+1}^{\text{trial}}\| - \sqrt{\tfrac{2}{3}}\left(\sigma_Y + K\alpha_n\right) \\
&\quad - 2\bar{\mu}\Delta\gamma - \sqrt{\tfrac{2}{3}}K\left(\alpha_{n+1} - \alpha_n\right) \\
&= f_{n+1}^{\text{trial}} - 2\bar{\mu}\left[1 + \frac{K}{3\bar{\mu}}\right]\Delta\gamma = 0.
\end{aligned} \tag{9.3.30}$$

Hence

$$\boxed{2\bar{\mu}\Delta\gamma = \frac{f_{n+1}^{\text{trial}}}{1 + \frac{K}{3\bar{\mu}}},}$$

where (9.3.31)

$$\boxed{\bar{\mu} := \tfrac{1}{3}\mu \operatorname{tr}\left[\bar{\boldsymbol{b}}_{n+1}^{e^{\text{trial}}}\right].}$$

Equations (9.3.24), (9.3.27), and (9.3.31) completely determine the discrete governing equations (9.3.19)–(9.3.20) which define the return-mapping algorithm. For the reader's convenience, a detailed step-by-step implementation of the overall algorithmic procedure is given in BOX **9.1**.

Remarks 9.3.2.

1. The update of $\bar{\boldsymbol{b}}_{n+1}^{e}$ is obtained from (9.3.19) and (9.3.24) as

$$\bar{\boldsymbol{b}}_{n+1}^{e} = \bar{\boldsymbol{b}}_{n+1}^{e\ \text{trial}} - \tfrac{2}{3}\Delta\gamma \operatorname{tr}\left(\bar{\boldsymbol{b}}_{n+1}^{e\ \text{trial}}\right)\boldsymbol{n}_{n+1}. \tag{9.3.32}$$

Alternatively, we can employ the following equivalent expression. Solving the elastic constitutive equation for $\operatorname{dev}[\boldsymbol{b}_{n+1}^{e}]$ and using (9.3.24),

$$\bar{\boldsymbol{b}}_{n+1}^{e} = \frac{\boldsymbol{s}_{n+1}}{\mu} + \tfrac{1}{3}\left[\operatorname{tr}\left(\bar{\boldsymbol{b}}_{n+1}^{e\ \text{trial}}\right)\right]\mathbf{1}. \tag{9.3.33}$$

This is the update formula in BOX **9.1**.

BOX 9.1. Return-mapping Algorithm for J_2-Flow Theory. Isotropic Hardening.

1. Update the *current configuration*

$$\boldsymbol{\varphi}_{n+1} = \boldsymbol{\varphi}_n + \boldsymbol{u}_n \circ \boldsymbol{\varphi}_n \qquad \text{(configuration)}$$

$$\boldsymbol{f}_{n+1} = \mathbf{1} + \nabla_{\boldsymbol{x}_n} \boldsymbol{u}_n \qquad \text{(relative deformation gradient)}$$

$$\boldsymbol{F}_{n+1} = \boldsymbol{f}_{n+1} \boldsymbol{F}_n \qquad \text{(total deformation gradient)}$$

2. Compute *elastic predictor*

$$\bar{\boldsymbol{f}}_{n+1} = \left[\det \boldsymbol{f}_{n+1}\right]^{-\frac{1}{3}} \boldsymbol{f}_{n+1}$$

$$\bar{\boldsymbol{b}}^{e\ \text{trial}}_{n+1} = \bar{\boldsymbol{f}}_{n+1} \bar{\boldsymbol{b}}^e_n \bar{\boldsymbol{f}}^T_{n+1}$$

$$\boldsymbol{s}^{\text{trial}}_{n+1} = \mu \ \text{dev}\left[\bar{\boldsymbol{b}}^{e\ \text{trial}}_{n+1}\right]$$

3. Check for *plastic loading*

$$f^{\text{trial}}_{n+1} := \|\boldsymbol{s}^{\text{trial}}_{n+1}\| - \sqrt{\tfrac{2}{3}}\,(K\alpha_n + \sigma_Y)$$

IF $f^{\text{trial}}_{n+1} \le 0$ THEN

Set $(\bullet)_{n+1} = (\bullet)^{\text{trial}}_{n+1}$, *& EXIT*

ELSE

GO TO 4. (Return-mapping)

ENDIF

4. The *return-mapping algorithm*

Set: $\bar{I}^e_{n+1} := \frac{1}{3}\,\text{tr}\left(\bar{\boldsymbol{b}}^{e\ \text{trial}}_{n+1}\right)$

$$\bar{\mu} := \bar{I}^e_{n+1}\mu$$

Compute: $\Delta\gamma := \dfrac{f^{\text{trial}}_{n+1}/2\bar{\mu}}{1 + K/3\bar{\mu}}$

$$\boldsymbol{n} := \boldsymbol{s}^{\text{trial}}_{n+1}/\|\boldsymbol{s}^{\text{trial}}_{n+1}\|$$

Return map:

$$\boxed{\begin{aligned} \boldsymbol{s}_{n+1} &= \boldsymbol{s}^{\text{trial}}_{n+1} - 2\bar{\mu}\Delta\gamma\boldsymbol{n} \\ \alpha_{n+1} &= \alpha_n + \sqrt{\tfrac{2}{3}}\,\Delta\gamma \end{aligned}}$$

5. Addition of the *elastic mean stress*

Mean stress:

$$J_{n+1} := \det\left[\boldsymbol{F}_{n+1}\right]$$

$$p_{n+1} = U'\left(J_{n+1}\right)$$

Stress: $\boxed{\boldsymbol{\tau}_{n+1} = J_{n+1}\, p_{n+1}\mathbf{1} + \boldsymbol{s}_{n+1}}$

6. Update of *intermediate configuration*

$$\boxed{\bar{\boldsymbol{b}}^e_{n+1} = \boldsymbol{s}_{n+1}/\mu + \bar{I}^e_{n+1}\mathbf{1}}$$

2. Note that the intermediate configuration is defined up to a rigid body rotation. In effect, we can compute $\boldsymbol{V}^e_{n+1} := \sqrt{\boldsymbol{b}^e_{n+1}}$ uniquely from $\boldsymbol{b}^e_{n+1}$, so that $\boldsymbol{F}^e_{n+1} = \boldsymbol{V}^e_{n+1}\boldsymbol{R}^e_{n+1}$, where $\boldsymbol{R}^e_{n+1} \in \mathrm{SO}(3)$ is an arbitrary rotation tensor. The arbitrariness in the rotation part of $\boldsymbol{F}^e$ has no effect whatsoever on the computational procedure outlined above, or on the formulation of the model. In this regard, recall that only the symmetric tensor $\boldsymbol{\varepsilon}^p$ is defined in the infinitesimal theory. The plastic spin $\boldsymbol{\omega}^p$ also remains completely arbitrary.
3. The extension of the algorithm outlined above to nonlinear isotropic hardening is straightforward and follows along lines identical to the infinitesimal theory. Assume that

$$f(\boldsymbol{\tau}, \alpha) := \|\boldsymbol{s}\| - \sqrt{\tfrac{2}{3}}\left[\sigma_Y + k(\alpha)\right], \tag{9.3.34}$$

where $k : \mathbb{R} \to \mathbb{R}$ is the non-linear hardening function. Then the counterpart of the consistency equations (9.3.30) becomes

$$\boxed{\begin{aligned} \hat{f}(\Delta\gamma) :&= \|\boldsymbol{s}^{\text{trial}}_{n+1}\| - \sqrt{\tfrac{2}{3}}\sigma_Y \\ &- \left[\sqrt{\tfrac{2}{3}}\,k\big(\alpha_n + \sqrt{\tfrac{2}{3}}\Delta\gamma\big) + 2\bar{\mu}\Delta\gamma\right] = 0\,. \end{aligned}} \tag{9.3.35}$$

This expression furnishes a nonlinear equation for $\Delta\gamma$ which is easily solved by an iterative method. If the derivative $k'(\alpha)$ is easily computed in closed form, a Newton iteration of the form

$$\Delta\gamma^{(k+1)} = \Delta\gamma^{(k)} - \delta^{(k)}\frac{\hat{f}\left[\Delta\gamma^{(k)}\right]}{\hat{f}'\left[\Delta\gamma^{(k)}\right]}, \tag{9.3.36}$$

where $\delta^{(k)} \in (0, 1]$ is the line-search parameter and $\hat{f}'\left[\Delta\gamma^{(k)}\right]$ is given by

$$\hat{f}'(\Delta\gamma^{(k)}) = -2\bar{\mu}\left\{1 + \frac{k'\big(\alpha_n + \sqrt{\tfrac{2}{3}}\Delta\gamma^{(k)}\big)}{3\bar{\mu}}\right\}, \tag{9.3.37}$$

often proves effective. If $-k(\alpha)$ is convex, often encountered in practice, the method is guaranteed to converge at a quadratic rate with $\delta^{(k)} \equiv 1$.
4. The implementation of other types of hardening laws follows the procedure outlined in Section **9.3.1.1**. In particular, this procedure applies to the kinematic hardening model outlined in Section **9.2.1.5**. For further details see Simo [1986, 1988a,b].

9.3.3 Exact Linearization of the Algorithm

As in the linear theory, the algorithm summarized in BOX **9.1** is amenable to exact linearization leading to a closed-form expression for the consistent algorithmic tangent moduli in the finite-strain range. The linearization process is carried out in closed form because of the hyperelastic nature of the stress response. This situation contrasts with the hypoelastic models considered in Chapter **8** for which a closed-form linearization is very difficult to obtain.

Conceptually, the basic step involved in deriving the tangent moduli associated with the algorithm in BOX **9.1** is essentially the same as in the general procedure discussed in Chapter **3** in the context of the infinitesimal theory. However, the actual details of the calculation are far more involved because of the nonlinear nature of the kinematic relationships in the finite-strain theory. Since no new insight is to be gained from this elaborate computation we simply quote the final result, summarized for the reader's convenience in BOX **9.2**; see Simo [1988*a*,*b*] for further details. For extensions to damage, see Simo and Ju [1989].

BOX 9.2. Consistent Elastoplastic Moduli for the Radial Return Algorithm in BOX **9.1**.

1. Spatial elasticity tensor $\mathbf{C}$ for hyperelastic model (9.2.6):

$$\mathbf{C} = (JU')'J\mathbf{1} \otimes \mathbf{1} - 2\,JU'\mathbf{I} + \bar{\mathbf{C}},$$

$$\boxed{\bar{\mathbf{C}} = 2\bar{\mu}[\mathbf{I} - \tfrac{1}{3}\mathbf{1} \otimes \mathbf{1}] - \tfrac{2}{3}\,\|\boldsymbol{s}\|[\boldsymbol{n} \otimes \mathbf{1} + \mathbf{1} \otimes \boldsymbol{n}],}$$

$$\boldsymbol{s} = \mu \operatorname{dev}[\bar{\boldsymbol{b}}^e], \quad \boldsymbol{n} = \boldsymbol{s}/\|\boldsymbol{s}\|,$$

$$\bar{\mu} = \mu\,\tfrac{1}{3}\operatorname{tr}[\bar{\boldsymbol{b}}^e].$$

2. Scaling factors [$k' = K$ for linear hardening]

$$\beta_0 = 1 + \frac{k'}{3\bar{\mu}},$$

$$\beta_1 = \frac{2\bar{\mu}\Delta\gamma}{\|\boldsymbol{s}_{n+1}^{\text{trial}}\|},$$

$$\beta_2 = \left[1 - \frac{1}{\beta_0}\right]\frac{2}{3}\,\frac{\|\boldsymbol{s}_{n+1}^{\text{trial}}\|}{\bar{\mu}}\,\Delta\gamma,$$

$$\beta_3 = \frac{1}{\beta_0} - \beta_1 + \beta_2,$$

$$\beta_4 = \left[\frac{1}{\beta_0} - \beta_1\right]\frac{\|\boldsymbol{s}_{n+1}^{\text{trial}}\|}{\bar{\mu}}.$$

3. Consistent (algorithmic) moduli

$$\mathbf{C}_{n+1}^{\text{ep}} = \mathbf{C}_{n+1}^{\text{trial}} - \beta_1\bar{\mathbf{C}}_{n+1}^{\text{trial}} - 2\bar{\mu}\beta_3\boldsymbol{n} \otimes \boldsymbol{n} - 2\bar{\mu}\beta_4 \operatorname{sym}\left[\boldsymbol{n} \otimes \operatorname{dev}[\boldsymbol{n}^2]\right]^s.$$

9.4 Assessment of the Theory. Numerical Simulations

The objective of this section is to assess the formulation and numerical implementation of the J_2 flow theory presented in the preceding sections. To this end, we consider four representative examples that include one of the few closed-form solutions of plasticity at finite strains, two well documented numerical experiments, and a classical necking bifurcation problem for which extensive experimental data is available. The results below are in excellent agreement with solutions reported in the literature. Two solution strategies are employed: **(a)** classical Newton–Raphson iteration with a linear line search, and **(b)** quasi-Newton iteration employing the BFGS update, a linear line search and periodic refactorization, as advocated in Matthies and Strang [1979]. Procedure **(b)** is extensively employed in the implicit Livermore codes (see Hallquist [1984]). The simulations below were performed with a tolerance value set to 0.6 in the line search, allowing a maximum of 10 BFGS updates before a new factorization is performed. Three remarks should be made about the performance of these solution procedures.

i. The line search is essential for robust performance of Newton's method. (This provides another illustration of a well-established point in nonlinear optimization; see, e.g., Luenberger [1984] or Dennis and Schnabel [1983]).
ii. As in the infinitesimal theory, use of the *consistent tangent moduli* proves crucial in achieving the quadratic rate of asymptotic convergence with Newton's method.
iii. Use of the consistent moduli in the periodic refactorizations of the quasi-Newton method also plays an important role in attaining superlinear rates of asymptotic convergence.

In the numerical simulations that follow, the linear hardening law is replaced by a more general nonlinear hardening law of the form

$$k(\alpha) := \sigma_Y + K\alpha + \big(K_\infty - K_0\big)\big[1 - \exp(-\delta\alpha)\big]\,, \quad \delta \geq 0\,. \tag{9.4.1}$$

Now the consistency parameter $\Delta\gamma$ is obtained from solving the nonlinear equation (9.3.35) by the *local* Newton iteration given by (9.3.36)–(9.3.37). A four-node isoparametric element with bilinear displacement interpolation and constant element volume is employed, and nodal stresses are obtained through a smoothing procedure described in Simo [1988b]. The convergence tolerance is 10^{-20} times the maximum value attained by the energy during the iteration. Nevertheless, this rather severe convergence criterion is easily satisfied.

EXAMPLE: 9.4.1. Expansion of a Thick-Walled Cylinder. This example was considered in Simo and Ortiz [1985] and Simo [1988b]. A thick-walled cylinder with an inner radius of 10 units and an outer radius of 20 units is subjected to internal pressure. The values of the material constants shown are chosen to replicate *rigid plastic behavior* to allow a comparison with the exact solution (Table **9.1A**).

The axisymmetric mesh shown in Figure **9.7a** consists of 20 four-node bilinear isoparametric elements. The inner radius is driven to a value of 85 in 15 equal increments. To provide an idea of the computational effort involved the total num-

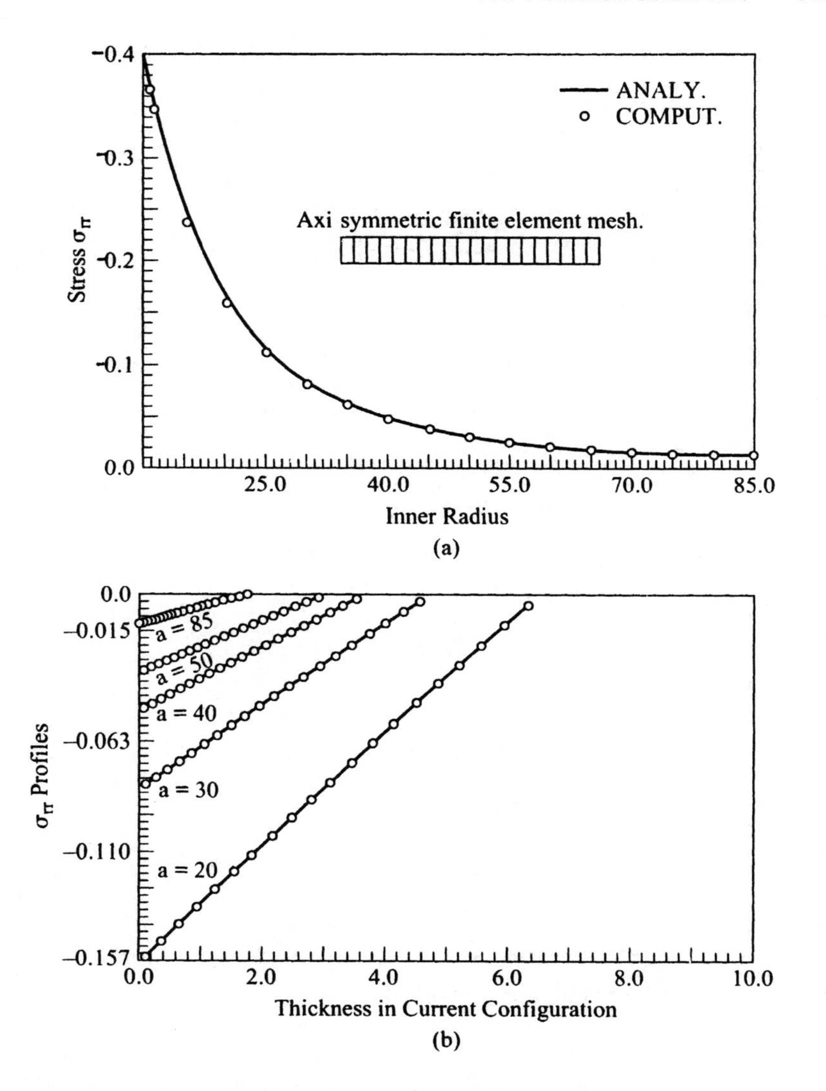

FIGURE 9.7. Expansion of a thick-walled cylinder. **(a)** Inner boundary stress σ_{rr} vs. current inner radius. **(b)** Profiles of σ_{rr} at different inner radii.

TABLE 9.1A. Example **9.4.1**. Material Properties.

Shear modulus	μ	3,800.0 MPa
Bulk modulus	κ	40,000.0 MPa
Flow stress	σ_Y	0.5 MPa
Perfect plasticity		$k = \sigma_Y$
		$H = K = 0$

TABLE 9.1B. Example **9.4.1**. Newton Iterations per Step.

Step	1	2-11	12-15
Iterations	6	5	4

TABLE 9.1C. Example **9.4.1**. Energy norm for steps 8 and 14.

Iteration	**Step 8**	**Step 14**
1	0.41067 E+09	0.11372 E+10
2	0.59505 E+03	0.92209 E+02
3	0.11096 E−01	0.31132 E−03
4	0.46812 E−11	0.36349 E−14
5	0.63522 E−22	

ber of full Newton iterations per step is summarized in Table **9.1B**. Values of the energy norm during the iterations of two typical time steps are given in Table **9.1C**. The quadratic rate of asymptotic convergence is apparent from these results.

A plot of the radial (Cauchy) stress σ_{rr} at the inner boundary vs. current inner radius is given in Figure **9.7a**. Profiles of σ_{rr} corresponding to several values of the inner radius are given in Figure **9.7b**. The computed results are in excellent agreement with the exact solution.

EXAMPLE: 9.4.2. Elastic-Plastic Upsetting of an Axisymmetric Billet. This example is proposed as a severe test problem in Taylor and Becker [1983].

The billet has an initial radius of 10 mm, and its initial height is 30 mm. The calculation is performed with two meshes consisting of 54 and 208 elements, respectively, as shown in Figure **9.8a**. Because of obvious symmetry, only one quarter of the specimen need be considered. The material properties are listed in Table **9.2A**.

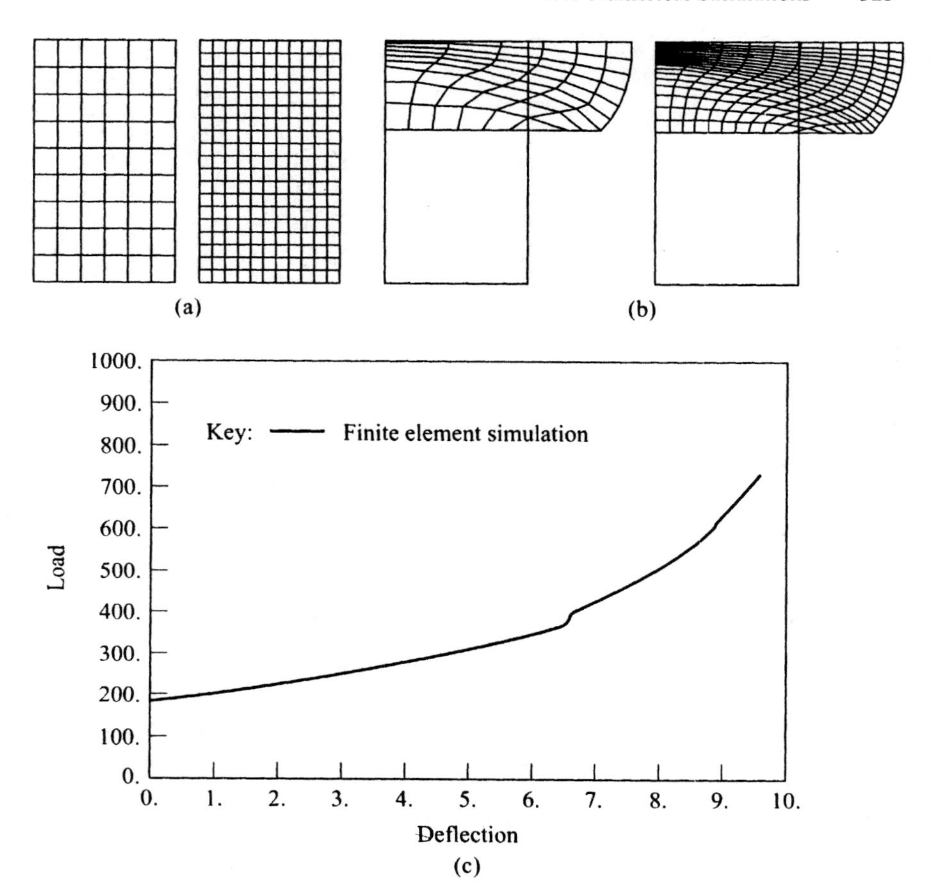

FIGURE 9.8. Example **9.4.2** Elastic-plastic upsetting of an axisymmetric billet. **(a)** Two finite-element meshes. **(b)** Deformed meshes. **(c)** Load-deflection curve.

The deformed meshes corresponding to 64% upsetting are shown in Figure **9.8b**. The final configuration in both calculations is attained in 120 equal steps. To provide an indication of the computational effort involved, values of the energy norm and the more restrictive Euclidean norm of the residual corresponding to two typical steps employing full Newton iterations are shown in Table **9.2C**.

Remarkably, these convergence characteristics are *almost unchanged* for the finer mesh consisting of 208 elements and hence are not reported. Note that an approximate quadratic rate of convergence is exhibited again. A plot of the resulting load-deflection curve is shown in Figure **9.8c**. These results agree with those reported in Taylor and Becker [1983].

TABLE 9.2A. Example **9.4.2**. Material Properties

Shear modulus	μ	384.62 MPa
Bulk modulus	κ	833.33 MPa
Flow stress	σ_Y	1.00 MPa
Linear hardening	K	3.00 MPa

TABLE 9.2B. Example **9.4.2**. Number of Newton Iterations/Step

Step	1	2	3	4-120
Iterations	8	9	10	5

TABLE 9.2C. Example **9.4.2**. Energy and Residual Norms for Typical Steps.

Step 15		**Step 115**	
Energy Norm	**Residual Norm**	**Energy Norm**	**Residual Norm**
0.339 E+02	0.199 E+03	0.128 E+05	0.113 E+06
0.138 E−03	0.164 E+00	0.535 E−02	0.388 E+01
0.865 E−09	0.458 E−03	0.172 E−06	0.140 E−01
0.447 E−14	0.976 E−06	0.963 E−12	0.613 E−04
0.327 E−19	0.329 E−08	0.146 E−16	0.988 E−07

EXAMPLE: 9.4.3. Upsetting of an Axisymmetric Disk. This simulation is considered by Nagtegaal and De Jong [1981] and Taylor and Becker [1983].

The initial finite-element mesh is shown in Figure **9.9a**. The final configuration corresponding to 26.67% upsetting is attained in 100 equal steps and shown in Figure **9.9b**.

To provide a precise indication of the computational effort involved, the number of iterations per time step is summarized in Table **9.3B**. In addition, values of the energy norm during the iterations in a typical load step employing full Newton iterations are shown in Table **9.3C**. The computed load-deflection curve, shown in Figure **9.9c**, is in excellent agreement with the results reported in Nagtegaal and De Jong [1981] for the four-node element. (As one would expect, the results in Taylor and Becker [1983] obtained with cross-triangles are stiffer.)

EXAMPLE: 9.4.4. Necking of a Circular Bar. This experimentally well-documented example is concerned with necking of a circular bar with a radius of 6.413

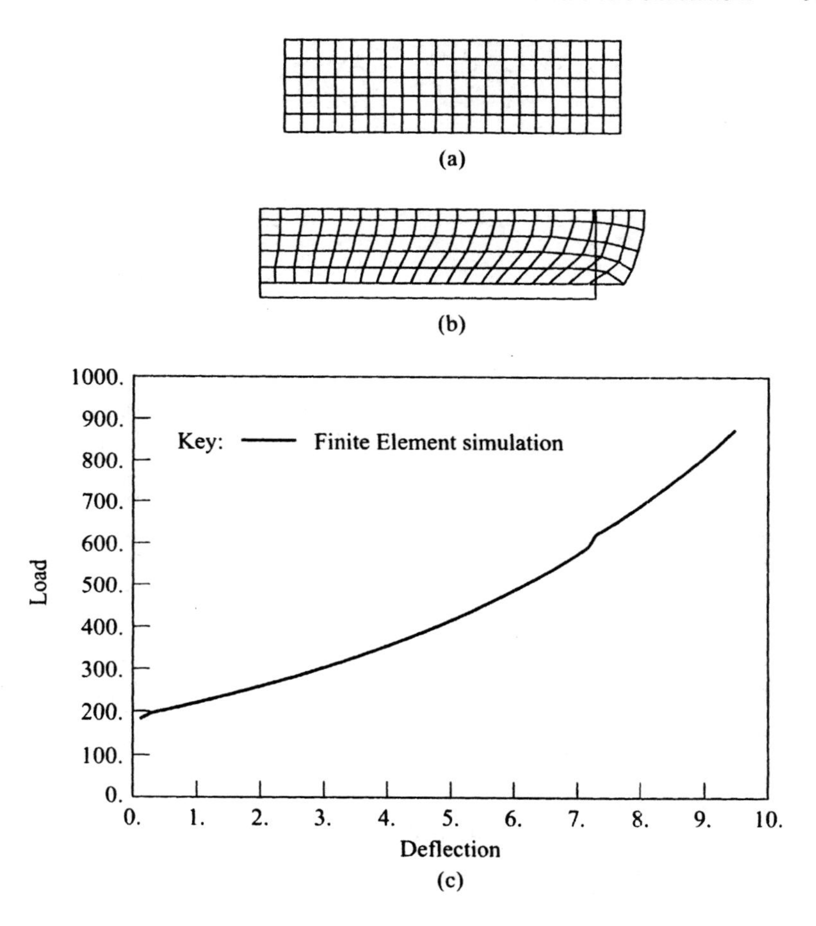

FIGURE 9.9. Example **9.4.3** Upsetting an axisymmetric disk. **(a)** Initial (undeformed) finite-element mesh. **(b)** Final configuration. **(c)** Load-deflection curve.

mm and length 53.334 mm, subjected to uniaxial tension. A fit of the hardening data reported in Hallquist [1984] in terms of equivalent plastic strain with the nonlinear isotropic hardening law (9.4.1) leads to the material properties summarized in Table **9.4A** (see also Figure **9.10a**). Three different meshes consisting of 50, 200, and 400 elements, respectively, are considered to assess the accuracy of discretization. The initial and final meshes after a total axial elongation of 14 mm are shown in Figures **9.10b** and **9.10c**. Note that the results obtained with the coarse meshes (50 and 200 elements) agree well with the results obtained with the finer (400-element) mesh.

The contours of the (Cauchy) stress components σ_{11} and σ_{22} for the 400-element mesh are shown in Figures **9.10d** and **e** and are in excellent agreement with those reported by Hallquist [1984]. Figure **9.10f** shows the ratio of the current to initial

TABLE 9.3A. Example **9.4.3**. Material Properties

Shear modulus	μ	76.92 MPa
Bulk modulus	κ	166.67 MPa
Flow stress	σ_Y	0.30 MPa
Linear hardening	K	0.70 MPa

TABLE 9.3B. Example **9.4.3**. Number of Newton Iterations/Step (Coarse Mesh)

Step	1	2	3	4-100
Iterations	7	6	6	5

TABLE 9.3C. Example **9.4.3**. Energy and Residual Norms for Typical Steps (Coarse Mesh)

Step 15		**Step 95**	
Energy Norm	**Residual Norm**	**Energy Norm**	**Residual Norm**
0.130 E+01	0.848 E+02	0.150 E+01	0.103 E+03
0.175 E−04	0.256 E+00	0.469 E−04	0.353 E+00
0.335 E−09	0.566 E−03	0.217 E−08	0.199 E−02
0.118 E−15	0.843 E−06	0.209 E−14	0.216 E−05
0.682 E−22	0.387 E−09	0.574 E−20	0.381 E−08

radius at the necking section vs. the axial displacement. The results (for the 50-, 200- and 400-element mesh) agree well with experimental and previously reported computational results.

It is of interest to assess the computational effort involved in the calculation. Although 29,000 steps are necessary in an explicit calculation performed with the HEMP code, Giroux [1973], an implicit calculation with NIKE 2D, Hallquist [1984], required only 100 steps. With the present approach, the entire calculation was performed in 15 steps. In addition, the required number of iterations per step necessary to attain the stringent convergence tolerance is quite favorable. Table **9.4B** summarizes the required number of iterations/step for the most demanding calculation employing the 400-element mesh and 15 steps. Table **9.4C** shows values of the energy and residual norms in typical steps of the 15 step simulation. The approximate quadratic rate of convergence is again manifest. The crucial role of the line search during the early stages of the iteration process is emphasized.

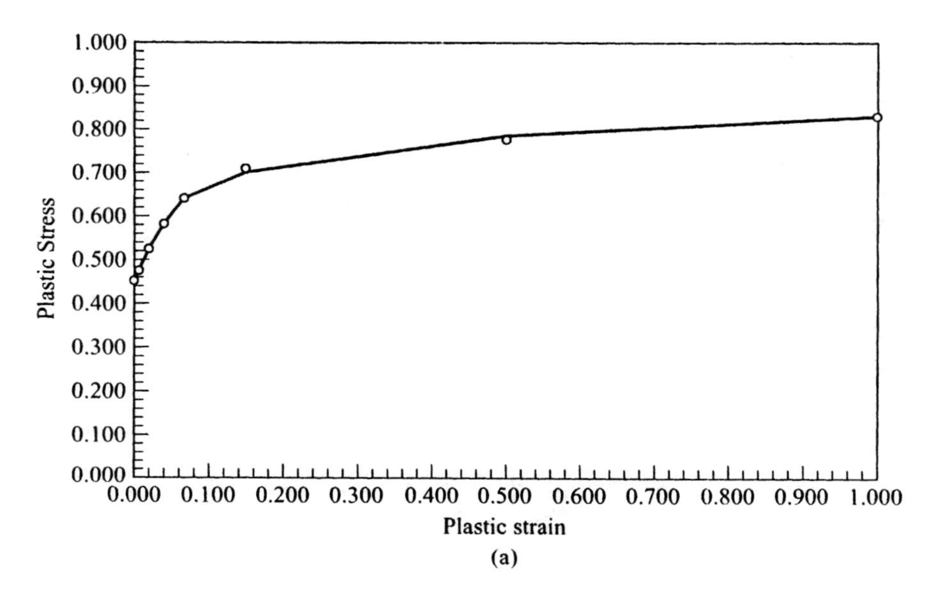

FIGURE 9.10. Example **9.4.4.** Necking of a circular bar. **(a)** Hardening data.

The accuracy of the integration procedure is assessed in Figure **9.10g** where the necking ratio vs. elongation is plotted for three simulations with a total number of 100, 53, and 15 steps, respectively. No dramatic loss of accuracy is observed, even with the largest steps. Finally, we examine the sensitivity of the numerical results to subsequent mesh refinement. For this purpose, we consider additional finite-element meshes consisting of 600 and 1600 elements, as illustrated in Figure **9.10h**, and perform identical numerical simulations as before, leading to a total elongation of 14mm in 100 load steps. The corresponding deformed meshes are shown in Figure **9.10h**. The computed results for the neck radius vs. elongation of the specimen for the 200-, 400-, 600-, and 1600-element meshes are contained in Figure **9.10j**. We observe that the curves corresponding to the calculations with 600- and 1600-element meshes are very close, thus corroborating the insensitivity of the numerical results to mesh refinement.

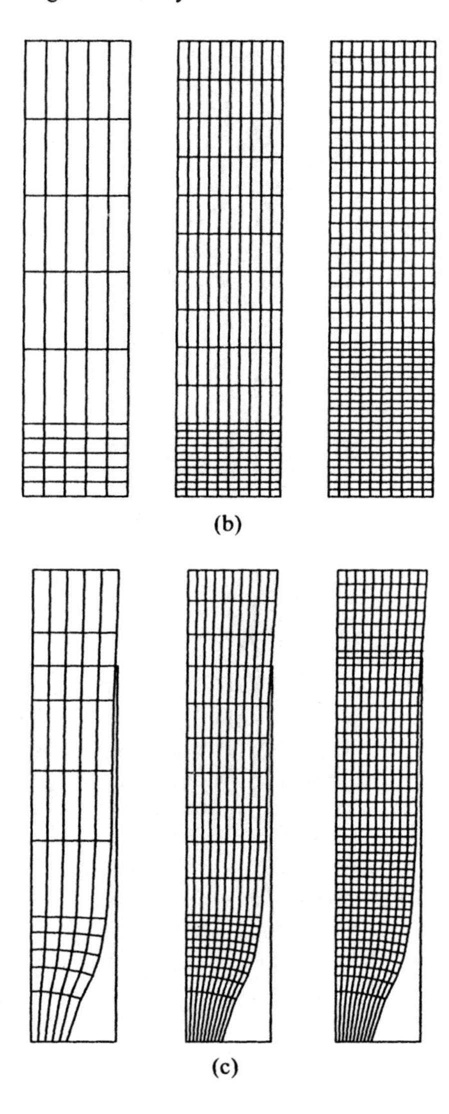

FIGURE 9.10. Example **9.4.4** Necking of a circular bar (continued). Finite-element meshes used in the analysis. 50-, 200-, and 400-element meshes are used with refinement in the necking area. **(b)** Initial configuration. **(c)** Final configuration.

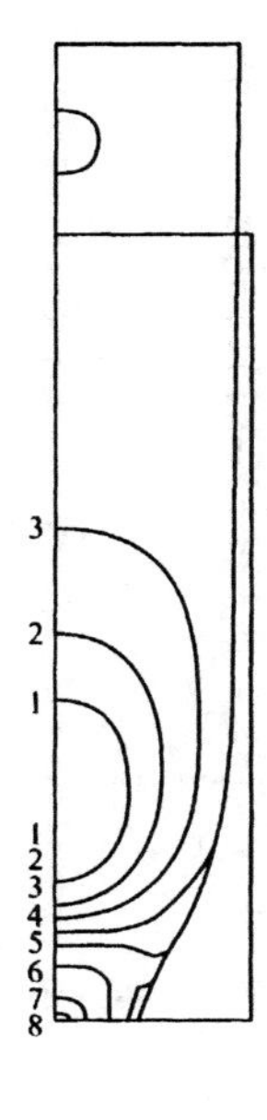

Stress σ_{11}	
Contour	*Value*
1	−0.40
2	−0.25
3	−0.10
4	0.05
5	0.20
6	0.30
7	0.40
8	0.45

(d)

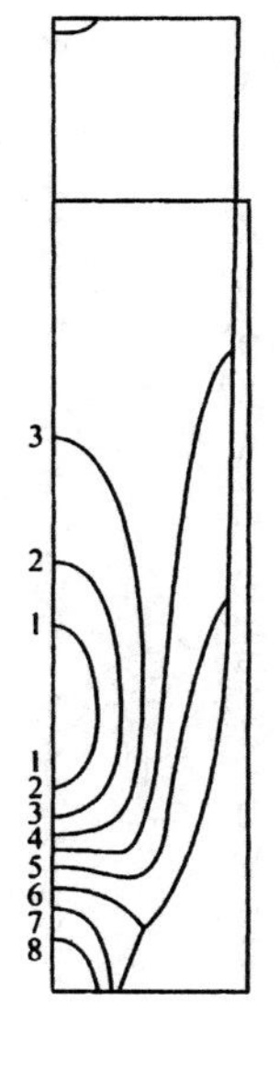

Stress σ_{22}	
Contour	*Value*
1	−0.30
2	−0.10
3	0.10
4	0.30
5	0.50
6	0.80
7	1.10
8	1.30

(e)

FIGURE 9.10. Example **9.4.4** Necking of a circular bar (continued). Selected Cauchy stress contours for the 400-element mesh. **(d)** σ_{11}. **(e)** σ_{22}.

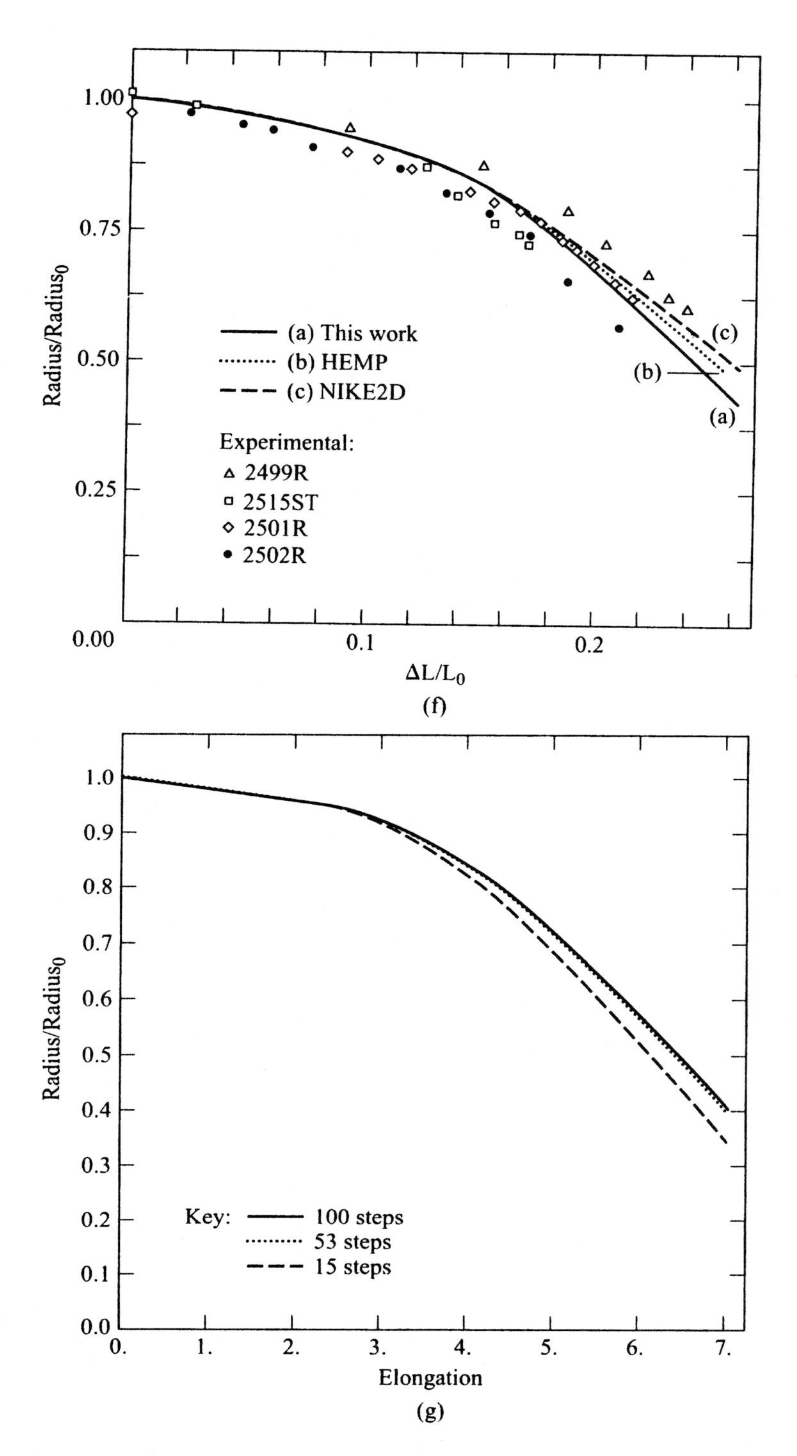

FIGURE 9.10. Example **9.4.4** Necking of a circular bar (continued). Ratio of the current to initial radius at the necking section versus axial displacement. **(f)** Comparison between numerical and experimental results. **(g)** Results for the 400-element mesh using 15, 53, and 100 load steps to arrive at the final configuration.

TABLE 9.4A. Example **9.4.4**. Material Properties

Shear modulus	μ	80.1938	GPa
Bulk modulus	κ	164.206	GPa
Initial flow stress	σ_Y	0.45	GPa
Residual flow stress	Y_∞	0.715	GPa
Linear hardening coefficient	K	0.12924	GPa
Saturation exponent	δ	16.93	

TABLE 9.4B. Example **9.4.4** 400-Element Mesh. Number of Newton Iterations/Step

Step	1	2-4	5	6	7	8-9	10	11-12	13-14	15
Iterations	11	6	7	10	8	9	11	9	8	7

TABLE 9.4C. Example **9.4.4** 400-Element Mesh. Energy and Residual Norms

Step 7		**Step 15**	
Energy Norm	**Residual Norm**	**Energy Norm**	**Residual Norm**
0.191 E+04	0.120 E+04	0.479 E+03	0.604 E+03
0.466 E+00	0.717 E+01	0.174 E+00	0.886 E+01
0.239 E−01	0.177 E+01	0.616 E−03	0.138 E+00
0.214 E−03	0.187 E+00	0.911 E−06	0.165 E−01
0.644 E−06	0.906 E−02	0.202 E−10	0.228 E−04
0.735 E−10	0.871 E−04	0.134 E−14	0.415 E−06
0.149 E−13	0.407 E−06	0.908 E−19	0.918 E−09
0.446 E−17	0.719 E−08		

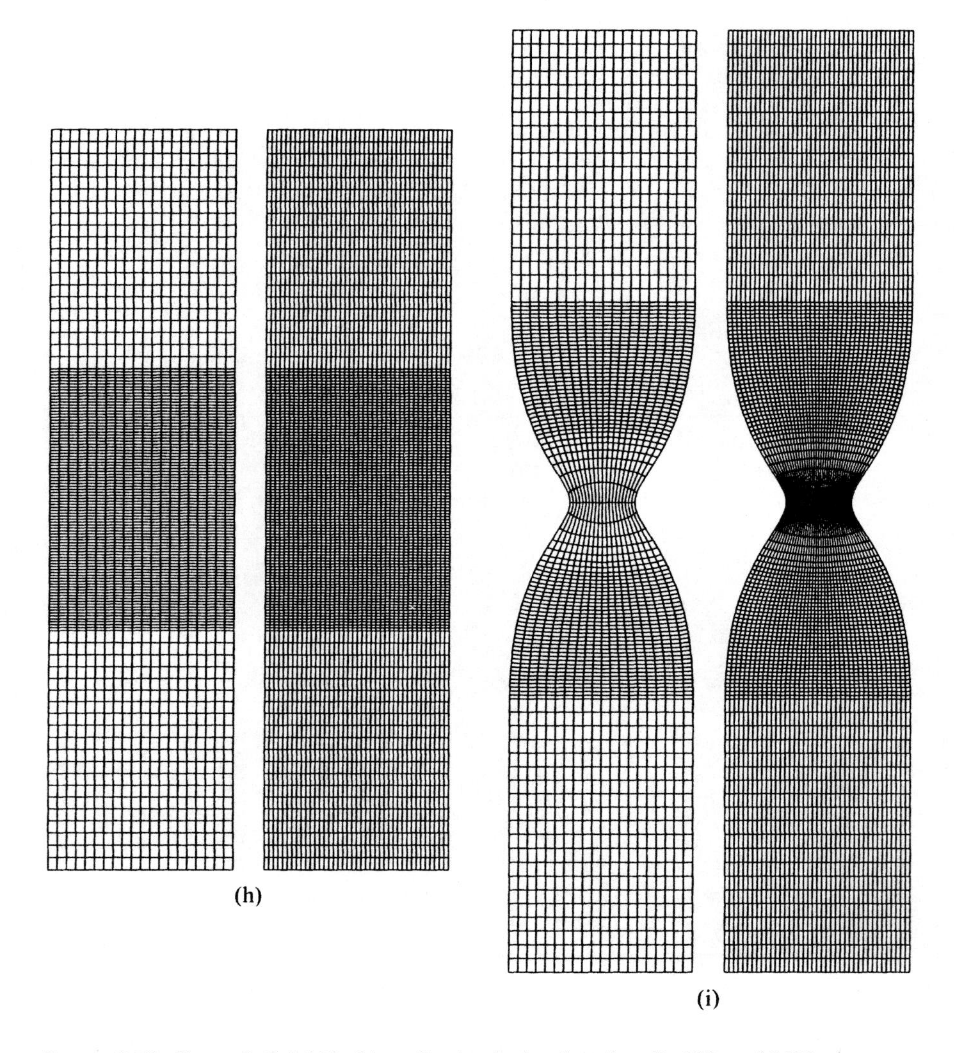

FIGURE 9.10. Example **9.4.4** Necking of a circular bar (continued). 600- and 1600-element meshes used to verify the convergence of the finite-element results. **(h)** Initial configuration. **(i)** Final configuration.

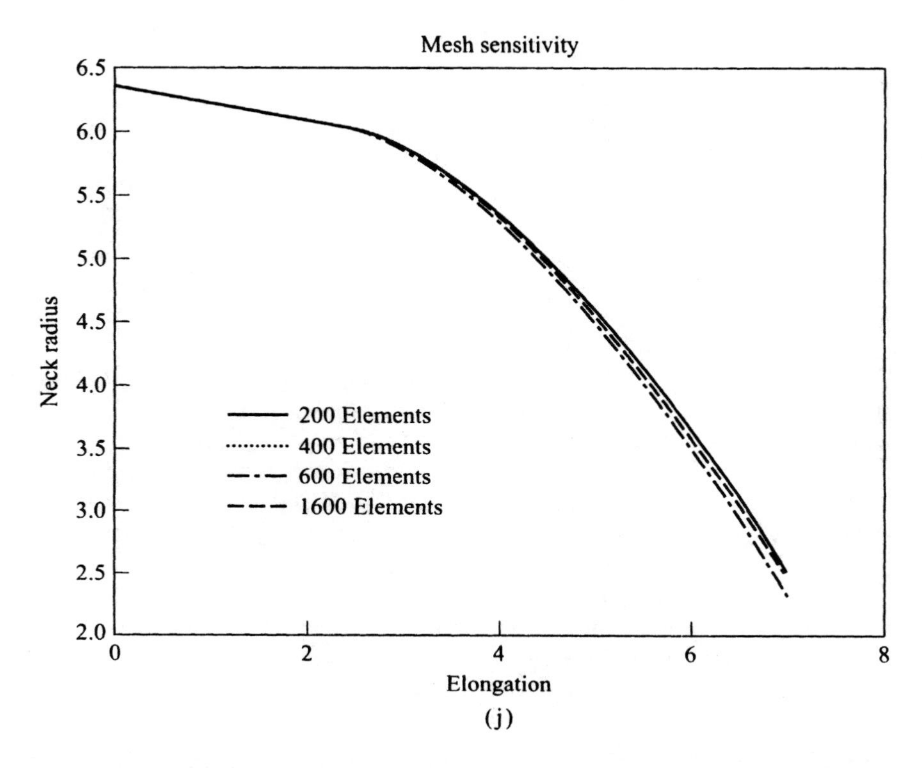

FIGURE 9.10. Example **9.4.4** Necking of a circular bar (continued). **(j)** Necking radius vs. axial displacement for all meshes.

10

Viscoelasticity

In what follows, we present an introduction to linear and nonlinear viscoelasticity. Our objective is to outline the basic mathematical structure of this important class of constitutive models and discuss its algorithmic implementation in detail.

No attempt is made to introduce the foundations of the subject in its full generality. The interested reader should consult standard textbooks; see e.g., Malvern [1969]; Truesdell and Noll [1965] or Christensen [1971] for further information. Despite the rather concrete (and often elementary) framework adopted in our presentation, the formulation discussed below leads to a methodology which, from a computational standpoint, possesses several attractive features. In particular,

1. it is amenable to a rather straightforward numerical implementation in the context of the class of algorithms first suggested in Taylor, Pister, and Goudreau [1970];

2. it has an appealing nonlinear generalization to the nonlinear finite deformation regime which includes anisotropic hyperelasticity as a particular case, as discussed in Simo [1987a]; and

3. an exact separation of volume-preserving and dilatational response is achieved through a multiplicative decomposition of the deformation gradient which goes back to Flory [1961]. This decomposition is first systematically used in Simo, Taylor, and Pister [1985] and Simo [1987a,b] in the formulation and numerical analysis of elasticity, plasticity, and viscoelasticity.

To motivate the structure of the class of viscoelastic constitutive models considered below, first in Section **10.1** we examine the formulation of the simplest possible one-dimensional rheological model. In particular, thermodynamic aspects are introduced in a rather concrete and physically motivated fashion. With this motivation in hand, in Section **10.2** we consider the generalization of these ideas to the three-dimensional physically nonlinear theory. In Section **10.3** we give a rather detailed and complete account of the algorithmic formulation and numerical implementation of the class of viscoelastic models developed in Section **10.2**. The extension of the developments to the nonlinear theory presented in Section **10.2** and **10.3** proceeds in two steps. First, in Section **10.4**, we examine in detail the formulation of nonlinear elasticity theory with uncoupled volumetric/volume-preserving response. As alluded to above, this exact decoupling is achieved by multiplicative decomposition of the deformation gradient into volume-preserving and spherical parts. With this background in hand, in Section **10.5** we consider the formulation of a class of nonlinear viscoelastic constitutive models that generalizes the convolution models of Section **10.2**. Remarkably, in contrast with the classical Coleman–Noll theory (see, e.g., Truesdell and Noll [1965] for an introductory exposition), this model is not restricted to isotropy. Finally, in Section **10.6** we show that the numerical analysis and implementation of the class of models presented in Section **10.5** involves a rather straightforward extension of the ideas presented in Section **10.3**.

10.1 Motivation. One-Dimensional Rheological Models

Consider a one-dimensional mechanical device consisting of two springs and one dashpot, arranged as illustrated in Figure **10.1**. For convenience we assume that the device possesses unit area and unit length so that forces and elongations can be identified with stresses and strains, respectively. Accordingly, we let σ and ε denote the total stress applied to the device, the spring constants are denoted by E_∞ and E, and the viscosity in the (*linear*) dashpot is η, as shown in Figure **10.1**. On physical grounds, we assume that

$$E > 0,$$

$$E_\infty > 0,$$

and

$$\eta > 0. \tag{10.1.1}$$

The response of the device is characterized as an *internal variable model* as follows. Let $[0, T] \subset (\mathbb{R}_+ \cup 0)$ be the time interval of interest. For convenience, we

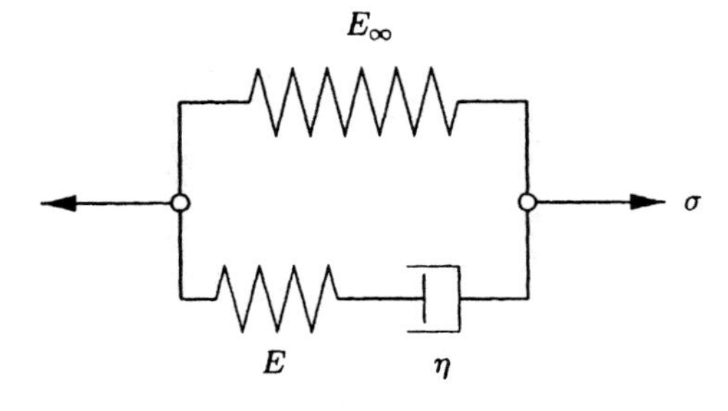

FIGURE 10.1. The one-dimensional standard solid.

introduce the *extended time interval* $(-\infty, T]$, and consider the internal variable

$$\alpha : (-\infty, T] \to \mathbb{R}. \tag{10.1.2}$$

We interpret the internal variable $\alpha(t)$ as the (inelastic) strain in the dashpot. Further, we let

$$\sigma^v : (-\infty, T] \to \mathbb{R}$$

be the stress acting on dashpot, as indicated in Figure **10.2**. We assume the following *linear constitutive relationship* connecting the "viscous" stress σ^v and the *strain rate* $\frac{\partial}{\partial t}\alpha(t)$ in the dashpot:

$$\boxed{\sigma^v(t) = \eta \frac{\partial}{\partial t}\alpha(t)\,.} \tag{10.1.3}$$

This relationship is the constitutive equation for one-dimensional, *linear, viscous flow*. We complete our constitutive hypotheses by assuming a *linear, elastic, stress-strain response in the springs*.

10.1.1 Formulation of the Constitutive Model

Now the governing equations for the model depicted in Figure **10.1** are derived by completely elementary considerations. Inspection of Figure **10.1** leads to the following conclusions:

i. The stress acting on the spring with constant E_∞ equals $(\sigma - \sigma^v)$. Since the strain on this spring is ε,

$$\sigma = E_\infty \varepsilon + \sigma^v. \tag{10.1.4}$$

ii. By equilibrium, the stress in the spring with constant E equals the stress σ^v on the dashpot. Furthermore, the strain on the spring with constant E equals $(\varepsilon - \alpha)$.

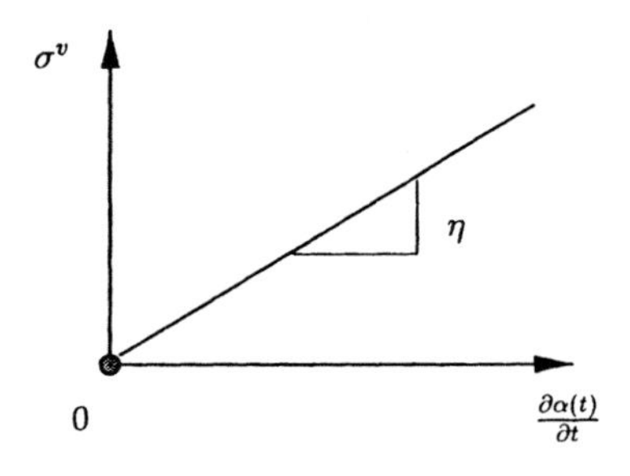

FIGURE 10.2. Viscous stress and strain in the dashpot of a standard solid.

From (10.1.3) and the assumption of linearity on the elastic stress-strain response,

$$\sigma^v = \eta\dot{\alpha} = E(\varepsilon - \alpha), \tag{10.1.5}$$

where $\frac{\partial}{\partial t}(\bullet) = (\dot{\bullet})$. We shall use this convention in what follows.

Upon introducing the constants

$$\left.\begin{aligned} E_0 &:= E_\infty + E \qquad &\text{(initial modulus)},\\ \tau &:= \eta/E \qquad &\text{(relaxation time)}, \end{aligned}\right\} \tag{10.1.6}$$

equations (10.1.4) and (10.1.5) lead to the following constitutive equation for the stress response

$$\boxed{\sigma = E_0\varepsilon - E\alpha,} \tag{10.1.7}$$

where the internal variable α (i.e., *the inelastic strain*) satisfies the evolution equation

$$\boxed{\begin{aligned} &\dot{\alpha} + \frac{1}{\tau}\alpha = \frac{1}{\tau}\varepsilon,\\ &\lim_{t\to-\infty}\alpha(t) = 0. \end{aligned}} \tag{10.1.8}$$

On physical grounds, we have appended the initial condition that $\alpha(t) \to 0$, as $t \to -\infty$.

10.1.2 Convolution Representation

Alternatively, the stress response of the device in Figure **10.1** can be formulated in terms of a convolution integral by eliminating the internal variable $\alpha(t)$ from the constitutive equations as follows.

Equation $(10.1.8)_1$ admits the integration factor $\exp(t/\tau)$, in terms of which $(10.1.8)_1$ becomes

$$\frac{d}{dt}\big[\exp(t/\tau)\alpha(t)\big] = \frac{1}{\tau}\exp(t/\tau)\varepsilon(t)\,. \tag{10.1.9}$$

Integrating this expression and using the boundary condition $(10.1.8)_2$ yields

$$\alpha(t) = \frac{1}{\tau}\int_{-\infty}^{t}\exp[-(t-s)/\tau]\varepsilon(s)\,ds\,, \tag{10.1.10}$$

which, integrated by parts, reduces to

$$\alpha(t) = \varepsilon(t) - \int_{-\infty}^{t}\exp[-(t-s)/\tau]\dot{\varepsilon}(s)\,ds\,. \tag{10.1.11}$$

Note that we have used the condition that $\varepsilon(t) \to 0$, as $t \to -\infty$. Finally, substituting (10.1.11) in (10.1.7) and using (10.1.6) leads to

$$\boxed{\sigma(t) = \int_{-\infty}^{t} G(t-s)\dot{\varepsilon}(s)\,ds\,,} \tag{10.1.12}$$

where $G : \mathbb{R} \to \mathbb{R}_+$ is the function defined by the expression

$$\boxed{G(t) = E_\infty + E \exp(-t/\tau).} \tag{10.1.13}$$

$G(t)$ is called the *relaxation function* associated with the device in Figure **10.1**.

10.1.2.1 Example. Relaxation test.

To gain further physical insight into the response of the model described by (10.1.12) and (10.1.13), consider the strain history

$$\varepsilon(t) = H(t)\varepsilon_0 := \begin{cases} 0 & \text{if } t < 0 \\ \varepsilon_0 & \text{otherwise,} \end{cases} \tag{10.1.14}$$

illustrated in Figure **10.3**.

Then the stress history is calculated by elementary methods using equations (10.1.10) and (10.1.7). Alternatively, using distributional calculus,

$$\begin{aligned} \sigma(t) &= \Big[\int_{-\infty}^{t} G(t-s)\delta(s)\, ds \Big] \varepsilon_0 \\ &= G(t)\varepsilon_0 , \quad t > 0 , \end{aligned} \tag{10.1.15}$$

where $\delta(t)$ is the Dirac delta function (Recall that $d/dt\ H(t) = \delta(t)$, see Stakgold [1978].) The result is illustrated graphically in Figure **10.4**.

Note that the relaxation time $\tau \geq 0$ is computed from the stress history by the relationship

$$\tau = -\frac{\sigma(0)}{\dot{\sigma}(0)}\gamma , \quad \gamma = E/E_0. \tag{10.1.16}$$

$\gamma \leq 1$ is the *stiffness ratio.*

Remarks 10.1.1.

1. The model developed above is known as the *standard solid*. For this model, the stress response can be inverted to express the strain history in terms of the

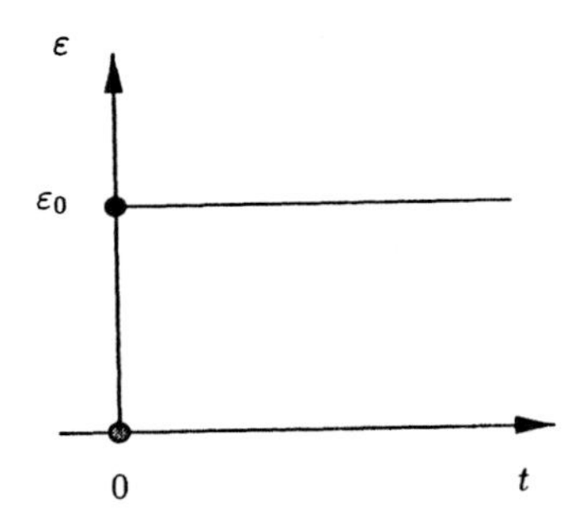

FIGURE 10.3. Strain history in a relaxation test.

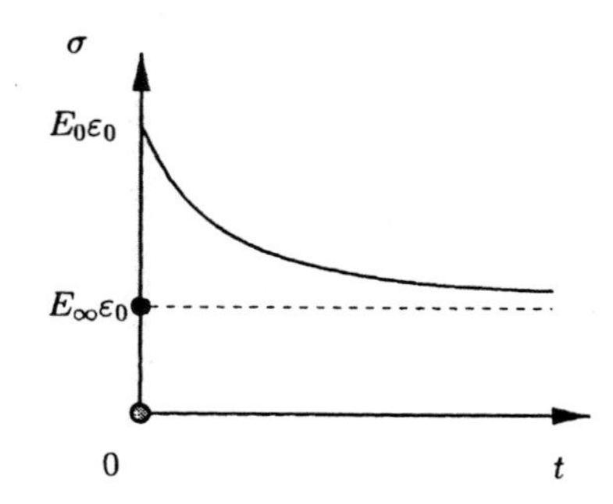

FIGURE 10.4. Stress response in a relaxation test.

stress history through the convolution representation

$$\varepsilon(t) = \int_{-\infty}^{t} J(t - s)\dot{\sigma}(s)\, ds, \tag{10.1.18}$$

where

$$J(t) := \frac{1}{E_\infty}\left[1 - \frac{E}{E_0}\exp\left(-\frac{E_\infty}{\tau E_0} t\right)\right] \tag{10.1.19}$$

is the *creep function*.

2. A special case of the standard solid is the *Maxwell fluid*, which is obtained from the arrangement in Figure **10.1** by setting $E_\infty = 0$. Observe that inversion of the convolution representation (10.1.12) is no longer possible since (10.1.19) is undefined for the Maxwell fluid. See Figure **10.5**. Thus, for constant stress history, the strain response is unbounded. One says that the Maxwell model *exhibits unbounded creep*.
3. By setting $E = 0$ in the arrangement in Figure **10.1**, one obtains the *Kelvin solid*, see Figure **10.6**. Then the stress response is given by

$$\sigma = E_\infty\left[\varepsilon + \frac{1}{\tau}\dot{\varepsilon}\right], \quad \tau := \eta/E_\infty. \tag{10.1.20}$$

In contrast to the Maxwell fluid, only the inverse representation (10.1.18) is defined for the Kelvin solid. Furthermore, the stress response in a relaxation test of the type considered in Example **10.1.2.1** is physically unrealistic, since strain discontinuities lead to unbounded stress response.
4. The algorithmic developments described in the sections that follow rely crucially on the relaxation representation (10.1.12) of the viscoelastic response. Hence, they do not include the Kelvin model as a particular case.

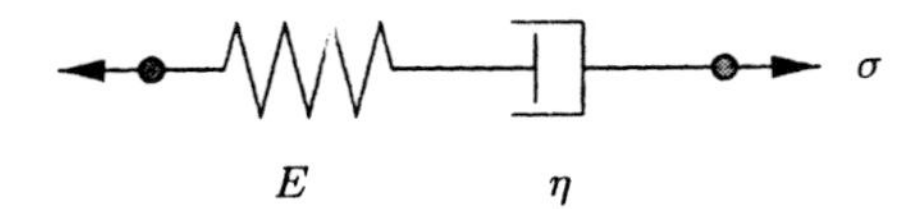

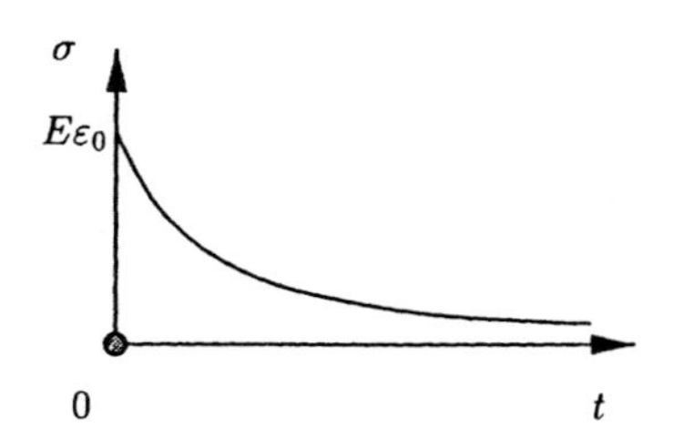

FIGURE 10.5. The Maxwell fluid and its relaxation function.

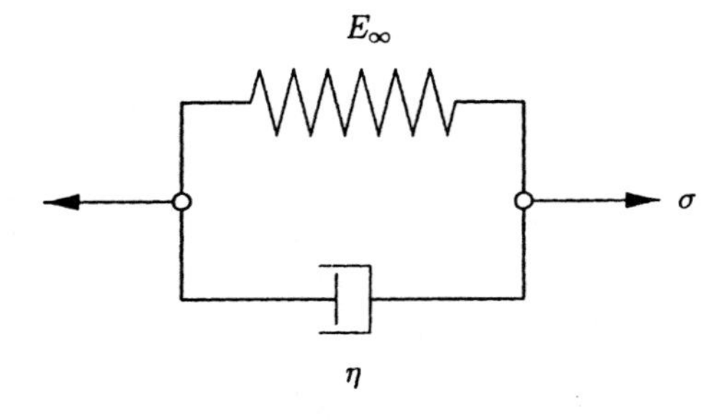

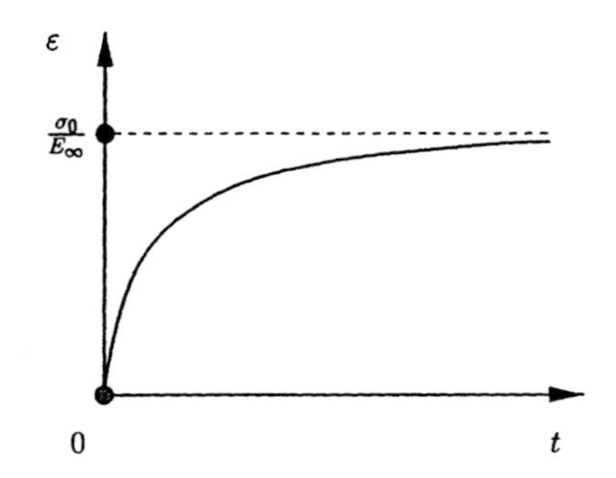

FIGURE 10.6. The Kelvin solid and its creep function.

10.1.3 Generalized Relaxation Models

The elementary model depicted in Figure **10.1** corresponds to a single Maxwell element in parallel with a spring element. This model is easily generalized to include an arbitrary number of Maxwell elements arranged in parallel, as shown in Figure **10.7**.

For this model, the stress response is defined by the relationship

$$\sigma(t) = E_0\varepsilon(t) - \sum_{i=1}^{N} E_i\alpha_i, \tag{10.1.21}$$

where the initial modulus $E_0 > 0$ and the relaxation times $\tau_i \geq 0$ are defined as

$$\left.\begin{aligned} E_0 &:= E_\infty + \sum_{i=1}^{N} E_i, \\ \tau_i &:= \eta_i/E_i\,, \quad i = 1, \ldots, N. \end{aligned}\right\} \tag{10.1.22}$$

Now the internal variables $\alpha_i : (-\infty, T] \to \mathbb{R}$ are governed by the evolution equations

$$\left.\begin{aligned} \dot{\alpha}_i + \frac{\alpha_i}{\tau_i} &= \frac{\varepsilon}{\tau_i}, \\ \lim_{t\to-\infty} \alpha_i(t) &= 0\,. \end{aligned}\right\} \tag{10.1.23}$$

If we define the relaxation function $G\colon \mathbb{R} \to \mathbb{R}_+$ by the expression

$$\boxed{G(t) := E_\infty + \sum_{i=1}^{N} E_i \exp(-t/\tau_i),} \tag{10.1.24}$$

it easily follows that the stress response is expressed as a convolution representation of the form (10.1.12).

10.1.3.1 Elementary thermodynamic considerations.

The discussion that follows provides a concrete motivation for the more general developments to be presented in a more abstract setting in subsequent sections. Here our objective is to provide a physical and rather concrete intuition for the concepts of free energy and dissipation.

i. On purely physical grounds, the *free energy* ψ associated with the arrangement in Figure **10.7** is defined as the elastic stored energy in the springs. Accordingly,

$$\psi(\varepsilon, \boldsymbol{\alpha}) := \tfrac{1}{2} E_\infty\varepsilon^2 + \tfrac{1}{2} \sum_{i=1}^{N} E_i(\varepsilon - \alpha_i)^2\,, \tag{10.1.25}$$

where $\boldsymbol{\alpha} : (-\infty, T] \to \mathbb{R}^N$ is the vector with components $(\alpha_1, \ldots, \alpha_N)$.

ii. Similarly, again on physical grounds, the *dissipation function* associated with the model in Figure **10.7** is defined as the rate of work dissipated in the dashpots.

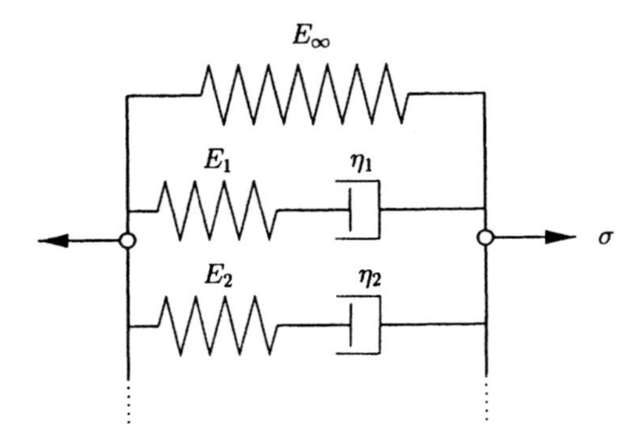

FIGURE 10.7. Generalized relaxation models.

Recall that the force (not necessarily at equilibrium) acting on each dashpot is given by

$$\sigma_i^v := E_i(\varepsilon - \alpha_i) = \eta_i \dot{\alpha}_i \, , \quad i = 1, \ldots, N, \tag{10.1.26}$$

an expression which implies the rate equations (10.1.23). Thus, since $\dot{\alpha}_i$ are the strain rates on the dashpots, the dissipation function, denoted by $\mathcal{D}[\varepsilon, \boldsymbol{\alpha}, \dot{\boldsymbol{\alpha}}]$, takes the form

$$\mathcal{D}[\varepsilon, \boldsymbol{\alpha}, \dot{\boldsymbol{\alpha}}] = \sum_{i=1}^{N} \sigma_i^v \dot{\alpha}_i = \boldsymbol{\sigma}^v \cdot \dot{\boldsymbol{\alpha}} \, . \tag{10.1.27}$$

Here, $\boldsymbol{\sigma}^v$ denotes the vector with components $(\sigma_1^v, \ldots, \sigma_N^v)$.

With these physically motivated notions in hand, now we observe a number of properties which motivate the abstract definitions introduced in Section **10.2.2.**

iii. By substituting the second expression of (10.1.26) in (10.1.27), we conclude that

$$\boxed{\mathcal{D}[\varepsilon, \boldsymbol{\alpha}, \dot{\boldsymbol{\alpha}}] = \sum_{i=1}^{N} \eta_i (\dot{\alpha}_i)^2 \geq 0 \, ,} \tag{10.1.28}$$

i.e., *the dissipation function is nonnegative.*

iv. Differentiating the expression (10.1.25) for free energy with respect to α_i yields, in view of (10.1.26), the relationship

$$\sigma_i^v = E_i(\varepsilon - \alpha_i) = -\frac{\partial \psi(\varepsilon, \boldsymbol{\alpha})}{\partial \alpha_i} \, , \quad \textit{i.e.,} \quad \boldsymbol{\sigma}^v = -\partial_{\boldsymbol{\alpha}} \psi(\epsilon, \boldsymbol{\alpha}) \, . \tag{10.1.29}$$

Thus, (minus) the partial derivative of the free energy relative to the internal variable yields the associated nonequilibrium force on the dashpot; in thermodynamic terms, the flux associated with the internal variable. Furthermore, from (10.1.29) and (10.1.27), we obtain the following *abstract definition of the dissipation function*:

$$\boxed{\mathcal{D}[\varepsilon, \boldsymbol{\alpha}, \dot{\boldsymbol{\alpha}}] = -\sum_{i=1}^{N} \frac{\partial \psi(\epsilon, \boldsymbol{\alpha})}{\partial \alpha_i} \dot{\alpha}_i = -\partial_{\boldsymbol{\alpha}} \psi(\epsilon, \boldsymbol{\alpha}) \cdot \dot{\boldsymbol{\alpha}}.} \tag{10.1.30}$$

Thus, *the dissipation function is the negative rate of change of the free energy with respect to the internal variables.*

v. The derivative of the free energy (10.1.25) with respect to the total strain ε gives

$$\frac{\partial \psi}{\partial \varepsilon}(\varepsilon, \boldsymbol{\alpha}) = E_{\infty}\varepsilon + \sum_{i=1}^{N} E_i(\varepsilon - \alpha_i) \tag{10.1.31}$$

which, in view of (10.1.26) and the arrangement in Figure **10.7**, equals the applied stress on the device. Hence, we conclude that

$$\boxed{\sigma = \frac{\partial \psi}{\partial \varepsilon}(\varepsilon, \boldsymbol{\alpha}).} \tag{10.1.32}$$

To summarize the preceding conclusions, we have seen that the elementary definitions of free energy and dissipation function which lead to expressions (10.1.25) and (10.1.27) are consistent with the abstract definition of dissipation function given by (10.1.30), the non-decreasing property (10.1.28), and the stress-strain relationships (10.1.32). In the following section, we take an expression for the free energy analogous to (10.1.25) as the point of departure and show that property (10.1.28) and the stress-strain relationships (10.1.32) follow from the abstract definition (10.1.30) of dissipation function by a systematic exploitation of the second law of thermodynamics in its classical version known as the *Clausius–Duhem inequality*.

10.1.3.2 Characterization of the equilibrium response.

Intuitively, the device in Figure **10.7**, reaches an equilibrium state under a prescribed strain (or stress) history, when "no further changes" in the dashpots take place. This notion is made precise by placing the following *two conditions on an equilibrium state*:

i. The rate of change of the internal variables is identically zero, i.e.,

$$\boxed{\dot{\boldsymbol{\alpha}}_e = \mathbf{0}.} \tag{10.1.33}$$

This condition simply states that the strain rate in all dashpots is zero at an equilibrium state.

ii. The thermodynamic forces conjugate to the internal variables also vanish at equilibrium. For the model in Figure **10.7**, this condition implies that the (nonequilibrium) viscous forces σ_i^v, acting on the dashpots and given by (10.1.26), vanish at equilibrium. In view of relationship (10.1.29), thus we require that

$$\boxed{-\boldsymbol{\sigma}_e^v := \partial_{\boldsymbol{\alpha}} \psi(\varepsilon, \boldsymbol{\alpha}_e) = \mathbf{0}.} \tag{10.1.34}$$

The preceding conditions along with relationships (10.1.29), (10.1.31), and (10.1.26) imply the following relationships at equilibrium:

$$\left.\begin{aligned} \sigma_e &= E_\infty \varepsilon, \\ \text{and} \\ \alpha_{i_e} &= \varepsilon\,, \quad i = 1, 2, \ldots, N\,. \end{aligned}\right. \tag{10.1.35}$$

In other words, the response at equilibrium is elastic with modulus E_∞. Given the structure of the linear evolutionary equations (10.1.23), it is clear that the equilibrium state is *unique* and is attained in the limit, as $t \to \infty$.

10.1.3.3 Alternative formulation.

We conclude this introductory section by presenting an alternative formulation of the preceding elementary model which is amenable to a straightforward extension to include *nonlinear elastic response.*

The crucial idea is to replace the viscous strains α_i by the following stress-like set of internal variables. Set

$$\boxed{q_i := E_i \alpha_i\,, \quad i = 1, 2, \ldots, N,} \tag{10.1.36}$$

so that, in view of (10.1.31)–(10.1.32), the stress response becomes

$$\boxed{\sigma = E_0 \varepsilon - \sum_{i=1}^{N} q_i\,.} \tag{10.1.37}$$

Next, we define the nondimensional (relative) moduli, $0 \leq \gamma_i < 1$, by the relationships

$$\left.\begin{aligned} \gamma_i &:= E_i / E_0\,, i = 1, 2, \ldots, N, \\ \gamma_\infty &:= E_\infty / E_0, \end{aligned}\right\} \tag{10.1.38}$$

where expression $(10.1.22)_1$ implies the restriction

$$\gamma_\infty + \sum_{i=1}^{N} \gamma_i = 1. \tag{10.1.39}$$

Now let $W^0(\varepsilon)$ be the (initial) stored-energy function defined by the expression

$$W^0(\varepsilon) := \tfrac{1}{2}\,\varepsilon E_0 \varepsilon\,. \tag{10.1.40}$$

With these definitions in hand, in view of (10.1.36), (10.1.37), and (10.1.23), the model problem formulated in the preceding section is recast in the following form:

$$\boxed{\begin{aligned} \sigma &= \frac{\partial}{\partial \varepsilon} W^0(\varepsilon) - \sum_{i=1}^{N} q_i, \\ \dot{q}_i + \frac{1}{\tau_i} q_i &= \frac{\gamma_i}{\tau_i} \frac{\partial}{\partial \varepsilon} W^0(\varepsilon), \\ \lim_{t \to -\infty} q_i(t) &= 0. \end{aligned}} \tag{10.1.41}$$

For the quadratic stored-energy function (10.1.40), the model summarized in (10.1.41) is completely equivalent to that given in Section **10.1.3**. However, for $W^0(\varepsilon)$ an arbitrary convex function of ε, model (10.1.41) remains meaningful and extends our preceding developments to account for nonlinear elastic response. The three-dimensional version of (10.1.41) and its extension to finite deformations is considered subsequently. For further background, see Lubliner [1973].

10.2 Three-Dimensional Models: Formulation Restricted to Linearized Kinematics

In this section, we extend the simple nonlinear models discussed in the preceding section to three-dimensional physically nonlinear elasticity. This extension is patterned after the model presented in Section **10.1.3.2**. First, we discuss the formulation of the general three–dimensional constitutive model. Next, we examine the thermodynamic aspects within the framework of irreversible thermodynamics with internal state variables. Finally, a detailed step-by-step implementation of this class of models in the context of the finite-element method is considered in Section 10.3.

10.2.1 Formulation of the Model

Viscoelastic constitutive models arise, typically, in modeling the response of polymeric materials. The "bulk response" for this class of materials is often elastic and typically much stiffer than the deviatoric response. In fact, in many engineering problems, the assumption of incompressibility holds with a high degree of approximation. Motivated by these considerations, we introduce an additive split of the strain tensor into volumetric and deviatoric parts as follows. Recall that **1** denotes the unit (order two) tensor with components δ_{ij}, the Kronecker delta relative to a Cartesian coordinate system. We set

$$\boldsymbol{\varepsilon} = \boldsymbol{e} + \tfrac{1}{3}\Theta \mathbf{1}, \tag{10.2.1}$$

where $\boldsymbol{e}$ is the deviatoric strain given by

$$\boldsymbol{e} := \operatorname{dev}[\boldsymbol{\varepsilon}] = \boldsymbol{\varepsilon} - \tfrac{1}{3}\operatorname{tr}[\boldsymbol{\varepsilon}]\,\mathbf{1}, \tag{10.2.2}$$

and Θ is the volumetric strain defined as

$$\Theta := \operatorname{tr}[\boldsymbol{\varepsilon}]. \tag{10.2.3}$$

Next, as in Section **10.1.3.2**, we assume an initial stored-energy function of the form

$$\boxed{W^\circ(\boldsymbol{\varepsilon}) = \bar{W}^\circ(\boldsymbol{e}) + U^\circ(\Theta).} \tag{10.2.4}$$

We call the function $U^\circ : \mathbb{R} \to \mathbb{R}_+$ the *elastic volumetric response*. Further, we use the notation

$$\boldsymbol{\sigma}^\circ := \partial_{\boldsymbol{\varepsilon}} W^\circ(\boldsymbol{\varepsilon}) \equiv \frac{\partial W^\circ(\boldsymbol{\varepsilon})}{\partial \boldsymbol{\varepsilon}};$$

i.e.,

$$\sigma^\circ_{ij} = \frac{\partial W^\circ}{\partial \varepsilon_{ij}}. \tag{10.2.5a}$$

Then a straightforward calculation employing the chain rule yields

$$\boxed{\boldsymbol{\sigma}^\circ = \operatorname{dev}\left[\partial_{\boldsymbol{e}} \bar{W}^\circ\right] + U^{\circ\prime}(\Theta)\,\mathbf{1}.} \tag{10.2.5b}$$

In accordance with the model of Section **10.1.3.3**, we assume further that the stress response is given by the expression

$$\boxed{\boldsymbol{\sigma}(t) = \boldsymbol{\sigma}^\circ(t) - \sum_{i=1}^{N} \boldsymbol{q}_i(t),} \tag{10.2.6}$$

where $\{\boldsymbol{q}_i, \cdots, \boldsymbol{q}_N\}$ is a set of internal variables with the equation of evolution defined as follows.

10.2.1.1 Evolution equation for the internal variables.

Once more, motivated by the model in Section **10.1.3.3**, we characterize viscoelastic response by rate equations of the form

$$\left.\begin{aligned} \dot{\boldsymbol{q}}_i + \frac{1}{\tau_i}\boldsymbol{q}_i &= \frac{\gamma_i}{\tau_i} \operatorname{dev}\left[\partial_{\boldsymbol{e}} \bar{W}^\circ(\boldsymbol{e})\right], \\ \lim_{t\to-\infty} \boldsymbol{q}_i &= 0. \end{aligned}\right\} \tag{10.2.7}$$

Here, the material parameters γ_i, $i = 1, \ldots, N$, and γ_∞ are subject to the restrictions

$$\sum_{i=1}^{N} \gamma_i = 1 - \gamma_\infty, \quad 0 \le \gamma_\infty < 1, \tag{10.2.8a}$$

along with the requirement that

$$\gamma_i \geq 0,$$

and

$$\tau_i > 0. \tag{10.2.8b}$$

The evolutionary equations (10.2.7) are *linear* and the solution, therefore, is expressed in closed-form as a convolution representation:

$$\boldsymbol{q}_i = \frac{\gamma_i}{\tau_i} \int_{-\infty}^{t} \exp[-(t-s)/\tau_i] \operatorname{dev}\left\{\partial_{\boldsymbol{e}} \bar{W}^{\circ}\left[\boldsymbol{e}(s)\right]\right\} \mathrm{ds}. \tag{10.2.9}$$

Substitution of (10.2.9) into (10.2.6), integration by parts, and use of the initial condition $(10.2.7)_2$ gives the constitutive equation for the stress tensor as the convolution integral

$$\boxed{\boldsymbol{\sigma}(t) = U^{\circ\prime}(\Theta)\mathbf{1} + \int_{-\infty}^{t} g(t-s) \frac{d}{ds}\left(\operatorname{dev}\left\{\partial_{\boldsymbol{e}} \bar{W}^{\circ}[\boldsymbol{e}(s)]\right\}\right) ds,} \tag{10.2.10}$$

where we have set

$$\boxed{g(t) := \gamma_\infty + \sum_{i=1}^{N} \gamma_i \exp[-t/\tau_i].} \tag{10.2.11}$$

We call the function $g(t)$ the *normalized relaxation function*. This completes the development of the model.

Remarks 10.2.1.

1. One could consider a relaxation function $g(t)$ other than (10.2.11). The functional form (10.2.10) of the model, however, remains unchanged.
2. Similarly, we could consider viscoelastic response in bulk. This generalization involves only a straightforward modification of (10.2.10). We simply need to replace $U^{\circ\prime}(\Theta)$ by the function

$$U^{\circ\prime}\left[\Theta(t)\right] \mapsto \int_{-\infty}^{t} h(t-s) \frac{d}{ds}\{U^{\circ\prime}[\Theta(s)]\} ds, \tag{10.2.12}$$

   where $h(t) : \mathbb{R} \to \mathbb{R}$ is a suitable relaxation function.
3. The crucial advantage of model (10.2.10) should be noted. Essentially, one can accommodate any arbitrary hyperelastic response defined by a convex stored energy $\left[W^{\circ}(\boldsymbol{\varepsilon})\right]$.

10.2.2 Thermodynamic Aspects. Dissipation

With the elementary motivation of Section **10.1.3.1** in hand, now we proceed to examine the thermodynamic foundations of the model (10.2.10) within the framework of irreversible thermodynamics with internal state variables. Throughout our discussion, attention is restricted to the purely mechanical theory.

10.2.2.1 Free energy and the second law.

We start out our development by assuming a free energy function of the form

$$\psi\left(\varepsilon, q_i\right) = W^\circ(\varepsilon) - \sum_{i=1}^{N} q_i \cdot e + \Xi\left(\sum_{i=1}^{N} q_i\right). \tag{10.2.13}$$

The reasons for selecting this functional form for the free energy becomes apparent in the course of the derivation. For simplicity, but without loss of generality, we assume that $N = 1$ in what follows.

Recall that the restriction of the second law of thermodynamics (in the form of the Clausius–Duhem inequality) to the purely mechanical theory leads to the inequality

$$-\dot{\psi}\left(\varepsilon, q\right) + \sigma : \dot{\varepsilon} \geq 0. \tag{10.2.14}$$

We regard this inequality as a constitutive restriction to be satisfied by all admissible states defined by $\{\varepsilon, \alpha, \sigma\}$, and for all rates of deformation $\{\dot{\varepsilon}, \dot{\alpha}\}$. Now, from (10.2.13) (with $N = 1$) and the chain rule,

$$\dot{\psi} = \left(\partial_\varepsilon W^\circ - \text{dev}[q]\right) : \dot{\varepsilon} - \mathcal{D}[\varepsilon, q; \dot{q}], \tag{10.2.15}$$

where

$$\begin{aligned} \mathcal{D}\left[\varepsilon, q; \dot{q}\right] :&= -\partial_q \psi : \dot{q} \\ &= \left[e - \partial_q \Xi(q)\right] : \dot{q} \end{aligned} \tag{10.2.16}$$

is the dissipation function. [Recall our observations in Section 10.1.3.1]. Substituting (10.2.15) and (10.2.16) in (10.2.14) results in the inequality

$$\left\{\sigma - \partial_\varepsilon W^\circ(\varepsilon) + \text{dev}[q]\right\} : \dot{\varepsilon} + \mathcal{D}\left[\varepsilon, q; \dot{q}\right] \geq 0. \tag{10.2.17}$$

Since (10.2.17) must hold for any rates $\{\dot{\varepsilon}, \dot{q}\}$, then a standard argument yields

$$\left.\begin{aligned} \sigma &= \partial_\varepsilon W^\circ(\varepsilon) - \text{dev}[q], \\ \mathcal{D}\left[\varepsilon, q; \dot{q}\right] &:= [e - \partial_q \Xi(q)] : \dot{q} \geq 0. \end{aligned}\right\} \tag{10.2.18}$$

Observe that constitutive equation $(10.2.18)_1$ is the counterpart in three dimensions of equation (10.1.37). Furthermore, the dissipation inequality (10.2.18) is the three-dimensional counterpart of (10.1.30). The function $\Xi(q)$ remains to be determined. We proceed as follows.

10.2.2.2 Conditions for thermodynamic equilibrium.

It is clear that, *given* the rate equations (10.2.7), equilibrium is achieved for

$$\dot{q}_e = \mathbf{0} \quad \Rightarrow \quad q_e = \gamma \ \text{dev}\left[\partial_e \bar{W}^\circ(e_e)\right]. \tag{10.2.19}$$

On the other hand, since no dissipation takes place at equilibrium ($\dot{q}_e = \mathbf{0}$), as in Section **10.1.3.2** we require that the driving conjugate thermodynamic forces be zero, i.e.

$$\partial_q \psi \big|_e = \mathbf{0}. \tag{10.2.20}$$

Thus, from (10.2.18) and (10.2.20), we conclude that

$$\boldsymbol{e}_e = \partial_q \Xi \left(\boldsymbol{q}_e\right). \tag{10.2.21}$$

However, (10.2.19) and (10.2.21) are precisely the relationships that define Ξ as the Legendre transformation of the function $\bar{W}^\circ$ in the sense that

$$\boxed{\Xi(\boldsymbol{q}) = -\gamma \bar{W}^\circ(\boldsymbol{e}) + \boldsymbol{q} : \boldsymbol{e}.} \tag{10.2.22}$$

This relationship completely determines expression (10.2.13) for the free energy (with $N = 1$) associated with our three-dimensional viscoelastic constitutive model. The preceding argument is trivially extended to that case for which $N > 1$.

10.3 Integration Algorithms

The numerical implementation described below is inspired by the algorithmic treatment first suggested in Taylor, Pister, and Goudreau [1970], and Herrmann and Petterson [1968]. The key idea is to transform the convolution representation discussed in the preceding sections into a two-step recurrence formula involving internal variables stored at the quadrature points of a finite-element method. From a computational standpoint, the scheme bypasses the need to store the entire history of strains (at each quadrature point) which would arise if a direct integration of the convolution representation were performed. Unfortunately, the method is restricted to a particular class of relaxation functions consisting of *a linear combination of functions of time which possess the semigroup property*. Extensions of this computationally very attractive scheme to general relaxation functions, such as power laws, not possessing this semigroup property remain an open question.

In what follows, we outline the numerical integration of viscoelasticity with reference to the convolution representation defined by equations (10.2.10) and (10.2.11).

10.3.1 Algorithmic Internal Variables and Finite-Element Database.

Let $[T_0, T] \subset \mathbb{R}$, with $T > 0$ and $T > T_0$, be the time interval of interest. Without loss of generality, we take $T_0 = -\infty$. Further, let

$$[T_0, T] = \bigcup_{n \in \mathbb{I}} [t_n, t_{n+1}], \quad t_{n+1} = t_n + \Delta t_n, \tag{10.3.1}$$

be a partition of time interval $[T_0, T]$ with $\mathbb{I}$ the integers. From an algorithmic standpoint, the problem is cast in the usual *strain-driven format* as follows.

10.3.1.1 Strain-driven algorithmic problem.

Let $\mathcal{B}_e \subset \mathcal{B}$ be a typical finite element of a spatial discretization

$$\mathcal{B} \cong \bigcup_{e=1}^{n_{el}} \mathcal{B}_e. \tag{10.3.2}$$

Then the internal force vector $\boldsymbol{f}_e^{int}(t)$ associated with element $\mathcal{B}_e$ at time $t \in [T_0, T]$ is given by

$$\begin{aligned} \boldsymbol{f}_e^{int}(t) &= \int_{\mathcal{B}_e} \boldsymbol{B}_e^T \boldsymbol{\sigma}(t) dV \\ &= \int_{\square} \left[\boldsymbol{B}_e^T \boldsymbol{\sigma}(t)\right] \circ \phi_e j(\boldsymbol{\xi}) d\boldsymbol{\xi} \\ &\cong \sum_{l=1}^{n_{\text{Gauss}}} \boldsymbol{B}_e^T(\boldsymbol{x}_l) \boldsymbol{\sigma}_l(t) W_l j_l, \end{aligned} \tag{10.3.3}$$

where $\phi_e : \square \to \mathcal{B}_e$ is the standard isoparametric map with Jacobian determinant $j = \det[D\phi_e]$, $\boldsymbol{B}_e$ is the discrete strain-displacement operator, and W_g the quadrature weights. In (10.3.3) a subscript g denotes evaluation at the quadrature point $\boldsymbol{x}_l \in \mathcal{B}_e$ with $l = 1, 2, \cdots, n_{\text{Gauss}}$. The preceding notation is standard; see, e.g., Hughes [1987] or Zienkiewicz and Taylor [1989] for further details and elaboration.

In view of (10.3.3), evaluation of $\boldsymbol{f}_e^{int}(t)$, $(e = 1, \cdots, n_{el})$, requires *knowledge of the stress history* $t \in [T_0, T] \mapsto \boldsymbol{\sigma}_l(t)$ *only at the quadrature points* $\boldsymbol{x}_l \in \mathcal{B}_e$. The discrete algorithmic problem is concerned with determinating this stress history *for a strain history* $t \in [T_0, T] \mapsto \boldsymbol{\varepsilon}_l(t)$ *assumed to be given at the quadrature points*, in the time interval $[T_0, T]$. The relationship between strain and stress histories is defined through the convolution representation (10.2.10):

$$\boldsymbol{\sigma}_l(t) = U^{\circ\prime}(\Theta_l)\mathbf{1} + \int_{T_0}^{t} g(t-s) \frac{d}{ds}\left(\text{dev}\left\{\partial_e \bar{W}^\circ\left[\boldsymbol{e}_l(s)\right]\right\}\right) ds, \tag{10.3.4}$$

where $g(t)$ is given by (10.2.11). Below we show how to transform this convolution integral into a recurrence relationship over the intervals $[t_n, t_{n+1}]$. To simplify our notation, the subscript $l \in \{1, 2, \cdots, n_{\text{Gauss}}\}$ is often omitted in what follows, and all the variable are to be evaluated at the quadrature points.

10.3.1.2 Algorithmic internal variables. Database.

Associated with the relaxation function (10.2.11), we define the following set of N internal variables:

$$\boxed{\boldsymbol{h}^{(i)}(t) := \int_{T_0}^{t} \exp[-(t-s)/\tau_i] \frac{d}{ds} \text{dev}\left\{\partial_e \bar{W}^\circ\left[\boldsymbol{e}(s)\right]\right\} ds.} \tag{10.3.5}$$

Hence, for a typical element $\mathcal{B}_e \subset \mathcal{B}$,

$$\boxed{\begin{array}{l}\text{Number of (algorithmic) internal variables} \\ \quad = N \times n_{\text{Gauss.}}\end{array}} \tag{10.3.6}$$

where $N \geq 1$ is the number of terms in the relaxation function (10.2.11) and n_{Gauss} is the number of quadrature points for element $\mathcal{B}_e$.

Below we show that a second-order accurate, unconditionally stable algorithm, for the integrating the convolution (10.3.4) is constructed by a recurrence scheme based on the following incremental problem:

i. Let the following data be given at time $t_n \in [T_0, T]$:

$$\left\{ s_n^\circ, \boldsymbol{h}_n^{(i)}, \quad i = 1, 2, \cdots, N \right\}. \tag{10.3.7}$$

Here s_n° is an "initial stress" defined by the expressions

$$\left. \begin{aligned} s_n^\circ &:= \operatorname{dev}\left\{ \partial_e \bar{W}^\circ \left[e\left(t_n\right) \right] \right\}, \\ e\left(t_n\right) &:= \operatorname{dev}\left[\varepsilon\left(t_n\right) \right]. \end{aligned} \right\} \tag{10.3.8}$$

ii. Let $\varepsilon_{n+1} = \varepsilon_n + \Delta\varepsilon_n$, where $\Delta\varepsilon_n$ is a strain increment at the quadrature point in question.

iii. **Problem:** Compute the stress σ_{n+1} at time $t_{n+1} = t_n + \Delta t_n$, and update the data base variables in (10.3.7) consistent with the convolution representation (10.3.4).

We remark that a variable with subscript $(\bullet)_n$ denotes the algorithmic approximation of this variable at time t_n. Such an algorithmic approximation is developed next.

10.3.2 One-Step, Unconditionally Stable and Second-Order Accurate Recurrence Formula.

As alluded to above, the crucial property exploited in the recurrence formula for the integration of (10.3.4) is the following.

10.3.2.1 Semigroup property

The following standard property holds for the exponential function $\exp : \mathbb{R} \to \mathbb{R}_+$:

$$\exp[(t + \Delta t)/a] = \exp(\Delta t/a)\exp(t/a), \tag{10.3.9}$$

for any constants Δt and a in $\mathbb{R}$ and $t \in \mathbb{R}$.

10.3.2.2 Recurrence relationship for the algorithmic internal variables.

Let $\boldsymbol{h}^{(i)}(t)$ be defined by formula (10.3.5). Using the semigroup property and the additivity property of the integral with respect to the integration interval, for

$t_{n+1} = t_{n+} \Delta t_{n+1}$,

$$\begin{aligned}
\boldsymbol{h}^{(i)}\left(t_{n+1}\right) :&= \int_{T_0}^{t_n+\Delta t_n} \exp[-(t_n + \Delta t_n - s)/\tau_i] \frac{d}{ds} \boldsymbol{s}^\circ(s) ds \\
&= \int_{T_0}^{t_n} \exp[-\Delta t_{n+1}/\tau_i] \exp[-(t_n - s)/\tau_i] \frac{d}{ds} \boldsymbol{s}^\circ(s) ds \\
&\quad + \int_{t_n}^{t_{n+1}} \exp[-(t_{n+1} - s)/\tau_i] \frac{d}{ds} \boldsymbol{s}^\circ(s) ds \\
&= \exp[-\Delta t_n/\tau_i] \boldsymbol{h}^{(i)}(t_n) \\
&\quad + \int_{t_n}^{t_{n+1}} \exp[-(t_{n+1} - s)/\tau_i] \frac{d}{ds} \boldsymbol{s}^\circ(s) ds. \qquad (10.3.10)
\end{aligned}$$

Thus, $\boldsymbol{h}^{(i)}(t_{n+1})$ is determined by (10.3.10) in terms of $\boldsymbol{h}^{(i)}(t_n)$ and an integral over the time step $\left[t_n, t_{n+1}\right]$. Using the midpoint rule,

$$\begin{aligned}
\int_{t_n}^{t_{n+1}} &\exp[-(t_{n+1} - s)/\tau_i] \frac{d}{ds} \boldsymbol{s}^\circ(s) ds \\
&\cong \exp[-(t_n + \Delta t_n - s)/\tau_i] \frac{d}{ds} \boldsymbol{s}^\circ(s) \Big|_{s=\frac{t_n+t_{n+1}}{2}} \Delta t_n \\
&= \exp(-\Delta t_n/2\tau_i) \frac{d}{ds} \boldsymbol{s}^\circ \left[(t_n + t_{n+1})/2\right] \Delta t_n \\
&= \exp(-\Delta t_n/2\tau_i) \left[\boldsymbol{s}^\circ(t_{n+1}) - \boldsymbol{s}^\circ(t_n)\right]. \qquad (10.3.11)
\end{aligned}$$

This approximation is second-order accurate. Combining (10.3.10)–(10.3.11), we obtain the update formulas

$$\boxed{\begin{aligned}
\boldsymbol{e}_{n+1} &:= \operatorname{dev}\left[\boldsymbol{\varepsilon}_{n+1}\right], \\
\boldsymbol{s}^\circ_{n+1} &:= \operatorname{dev}\left[\partial_{\boldsymbol{e}} \bar{W}^\circ(\boldsymbol{e}_{n+1})\right], \\
\boldsymbol{h}^{(i)}_{n+1} &:= \exp(-\Delta t_n/\tau_i)\boldsymbol{h}^{(i)}_n + \exp(-\Delta t_n/2\tau_i)\left(\boldsymbol{s}^\circ_{n+1} - \boldsymbol{s}^\circ_n\right),
\end{aligned}} \qquad (10.3.12)$$

where $i = 1, 2, \cdots, N$.

10.3.2.3 Computation of the stress tensor.

Expressions (10.3.4) for the stress tensor and (10.2.11) for the relaxation function $g(t)$, along with our definitions (10.3.5) for the algorithmic internal variables, yield

$$\boxed{\boldsymbol{\sigma}_{n+1} = U^{\circ\prime}\left(\Theta_{n+1}\right)\mathbf{1} + \gamma_\infty \boldsymbol{s}^\circ_{n+1} + \sum_{i=1}^{N} \gamma_i \boldsymbol{h}^{(i)}_{n+1},} \qquad (10.3.13)$$

where we have assumed without loss of generality that $\boldsymbol{e}(0) = \operatorname{dev}\left[\boldsymbol{\varepsilon}(0)\right] = \mathbf{0}$.

Remarks 10.3.1.

1. A straightforward argument shows that the algorithm embodied by formulas (10.3.12) and (10.3.13) is *unconditionally stable* and *second-order accurate*.
2. Within the framework of the displacement finite-element method, the *volume variable* Θ_{n+1} is computed simply as

$$\Theta_{n+1} = \operatorname{tr}\left[\varepsilon_{n+1}\right]. \tag{10.3.14}$$

For low-order elements, however, this naive approach leads to well-known locking phenomena in the nearly incompressible limit. Hence, Θ_{n+1} is typically computed via a *mixed, finite-element method.* See Simo, Taylor, and Pister [1985] and Simo [1988] for a detailed description of one possible approach, and Hughes [1987] and Zienkiewicz and Taylor [1989] for general background material on this well-known topic.

3. An alternative update formula for the algorithmic internal variables is obtained by the following argument. Assume that $d/ds[s^{\circ}(s)]$ is constant for $s \in \left[t_n, t_{n+1}\right]$. Then the integral in (10.3.11) is evaluated as follows:

$$\begin{aligned}
&\int_{t_n}^{t_{n+1}} \exp[-(t_{n+1} - s)/\tau_i] \frac{d}{ds} s^{\circ}(s) ds \\
&\quad \cong \frac{d}{ds} s^{\circ}(s)\bigg|_{s=\frac{t_n+t_{n+1}}{2}} \int_{t_n}^{t_{n+1}} \exp[-(t_{n+1} - s)/\tau_i]\, ds \\
&\quad = \left[s^{\circ}\left(t_{n+1}\right) - s^{\circ}\left(t_n\right)\right] \frac{1}{\Delta t_n} \tau_i \exp[-(t_{n+1} - s)/\tau_i]\bigg|_{s=t_n}^{s=t_{n+1}} \\
&\quad = \frac{1 - \exp(-\Delta t_n/\tau_i)}{\Delta t_n/\tau_i} \left(s^{\circ}_{n+1} - s^{\circ}_{n}\right).
\end{aligned} \tag{10.3.15}$$

Combining (10.3.10) and (10.3.15),

$$\boxed{h^{(i)}_{n+1} = \exp(-\Delta t_n/\tau_i) h^{(i)}_n + \frac{1 - \exp(-\Delta t_n/\tau_i)}{\Delta t_n/\tau_i} \left(s^{\circ}_{n+1} - s^{\circ}_{n}\right),} \tag{10.3.16}$$

an update formula often used in place of $(10.3.12)_3$. A straightforward truncation error analysis shows that both $(10.3.12)_3$ and (10.3.16) are in fact second-order accurate. Formula (10.3.16) reduces to that first proposed by Taylor, Goudreau, and Pister [1970].

10.3.3 Linearization. Consistent Tangent Moduli

First we recall that the linearization of the residual force vector f_e^{int} yields the following expression for the (consistent) element tangent matrix:

$$k_{e_{n+1}} := \int_{B_e} B^T \mathbf{C}_{n+1} B\, dV, \tag{10.3.17}$$

where $\mathbf{C}_{n+1}$ is the matrix of tangent moduli obtained by differentiating the stress tensor at t_{n+1} with respect to the strain tensor at t_{n+1}:

$$\boxed{\mathbf{C}_{n+1} := \partial_{\varepsilon_{n+1}} \sigma_{n+1} \equiv \frac{\partial \sigma_{n+1}}{\partial \varepsilon_{n+1}}.} \tag{10.3.18}$$

Observe that σ_{n+1} in this expression is regarded as a function of the strain tensor ε_{n+1} through the algorithmic relationships (10.3.12)–(10.3.13). Of course, (10.3.18) is evaluated at the quadrature points. To obtain an explicit expression for (10.3.18), we differentiate the terms in (10.3.13) as follows.

10.3.3.1 Linearization of the initial stress term s°_{n+1}.

Recall that the deviatoric strain tensor e_{n+1} is a function of ε_{n+1} given by the expression

$$e_{n+1} := \varepsilon_{n+1} - \tfrac{1}{3} \operatorname{tr}\left[\varepsilon_{n+1}\right] \mathbf{1}. \tag{10.3.19}$$

Then differentiating this relationship yields the derivative of the deviatoric strain tensor (relative to the total strain tensor) as

$$\partial_{\varepsilon_{n+1}} e_{n+1} = I - \tfrac{1}{3} \mathbf{1} \otimes \mathbf{1}. \tag{10.3.20}$$

With this expression in hand, we compute the tensor of elastic moduli $\bar{\mathbf{C}}^{\circ}_{n+1} := \partial_{\varepsilon_{n+1}} s^{\circ}_{n+1}$ as follows. Recall that s°_{n+1} is given by the constitutive equation

$$\begin{aligned} s^{\circ}_{n+1} &= \operatorname{dev}\left[\partial_e \bar{W}^{\circ}\left(e_{n+1}\right)\right] \\ &= \partial_e \bar{W}^{\circ}\left(e_{n+1}\right) - \tfrac{1}{3} \operatorname{tr}\left[\partial_e \bar{W}^{\circ}\left(e_{n+1}\right)\right] \mathbf{1}. \end{aligned} \tag{10.3.21}$$

Differentiating (10.3.21) with respect to to ε_{n+1} and using the chain rule yields

$$\begin{aligned} \bar{\mathbf{C}}^{\circ}_{n+1} := \partial_{\varepsilon_{n+1}} s^{\circ}_{n+1} &= \partial_{e_{n+1}} s^{\circ}_{n+1} : \partial_{\varepsilon_{n+1}} e_{n+1} \qquad (10.3.22) \\ &= \left\{\left[\partial^2_{ee} \bar{W}^{\circ}_{n+1}\right] - \tfrac{1}{3} \mathbf{1} \otimes \left[\partial^2_{ee} \bar{W}^{\circ}_{n+1}\right] : \mathbf{1}\right\} : \partial_{\varepsilon_{n+1}} e_{n+1}. \end{aligned}$$

Finally, we combine (10.3.20) and (10.3.22) to obtain the result

$$\boxed{\begin{aligned} \bar{\mathbf{C}}^{\circ}_{n+1} = &\left(\partial^2_{ee} \bar{W}^{\circ}_{n+1}\right) - \tfrac{1}{3} \mathbf{1} \otimes \left[\left(\partial^2_{ee} \bar{W}^{\circ}_{n+1}\right) : \mathbf{1}\right] \\ &- \tfrac{1}{3} \left[\left(\partial^2_{ee} \bar{W}^{\circ}_{n+1}\right) : \mathbf{1}\right] \otimes \mathbf{1} + \tfrac{1}{9} \left(\mathbf{1} : \left[\partial^2_{ee} \bar{W}^{\circ}_{n+1}\right] : \mathbf{1}\right) \mathbf{1} \otimes \mathbf{1}. \end{aligned}} \tag{10.3.23}$$

Observe that this expression depends solely on the assumed form of the deviatoric part of the initial stored-energy function $\bar{W}^{\circ}(e)$.

EXAMPLE: 10.3.3.1. For linear isotropic elasticity,

$$\bar{W}^{\circ}(e) = \mu e : e, \tag{10.3.24}$$

so that

$$\boldsymbol{s}_{n+1}^{\circ} = 2\mu \boldsymbol{e}_{n+1} \quad , \quad \boldsymbol{e}_{n+1} = \operatorname{dev}\left[\boldsymbol{\varepsilon}_{n+1}\right], \tag{10.3.25}$$

and the general result (10.3.33) reduces to

$$\bar{\mathbf{C}}_{n+1}^{\circ} = 2\mu\left(\boldsymbol{I} - \tfrac{1}{3}\mathbf{1}\otimes\mathbf{1}\right), \tag{10.3.26}$$

which is the standard expression for the deviatoric tensor of elastic moduli in linear isotropic elasticity.

10.3.3.2 Linearization of the algorithmic internal variables: Tangent moduli.

Our next objective is to derive the expression for the *algorithmic* tangent moduli consistent with the algorithmic stress update procedure developed above. Recall that these algorithmic tangent moduli are obtained merely by systematically applying the directional derivative relative to a strain increment to the update formulas of the algorithm. In particular, for formula $(10.3.12)_3$,

$$\partial_{\boldsymbol{\varepsilon}_{n+1}} \boldsymbol{h}_{n+1}^{(i)} = \exp(-\Delta t_n/2\tau_i)\partial_{\boldsymbol{\varepsilon}_{n+1}} \boldsymbol{s}_{n+1}^{\circ} \equiv \exp(-\Delta t_n/2\tau_i)\bar{\mathbf{C}}_{n+1}^{\circ}. \tag{10.3.27}$$

Therefore, from the algorithmic expression (10.3.13) and formulas (10.3.23) and (10.3.27), finally we obtain

$$\boxed{\mathbf{C}_{n+1} = U_{n+1}^{\circ\prime\prime}\mathbf{1}\otimes\frac{\partial\Theta_{n+1}}{\partial\boldsymbol{\varepsilon}_{n+1}} + g^*(\Delta t_n)\,\bar{\mathbf{C}}_{n+1}^{\circ},} \tag{10.3.28}$$

where $g^*(\Delta t)$ is the algorithmic expression for the relaxation function (10.2.11) given by the relationship

$$\boxed{g^*(\Delta t_n) = \gamma_\infty + \sum_{i=1}^{N}\gamma_i\exp(-\Delta t_n/2\tau_i).} \tag{10.3.29}$$

Remarks 10.3.2.

1. The explicit expression for the volumetric contribution to the tangent (10.3.28) depends on the type of interpolation employed. For the pure displacement method given by (10.3.14),

$$U_{n+1}^{\circ\prime\prime}\mathbf{1}\otimes\frac{\partial\Theta_{n+1}}{\partial\boldsymbol{\varepsilon}_{n+1}} = U_{n+1}^{0\prime\prime}\mathbf{1}\otimes\mathbf{1}. \tag{10.3.30}$$

2. Expression (10.3.29) for the algorithmic relaxation function $g^*(\Delta t)$ corresponds to the midpoint algorithm. If the update formula $(10.3.12)_3$ is replaced by (10.3.16), the algorithmic relaxation function becomes

$$g^*(\Delta t) = \gamma_\infty + \sum_{i=1}^{N}\gamma_i\,\frac{1-\exp(-\Delta t_n/\tau_i)}{\Delta t_n/\tau_i}. \tag{10.3.30}$$

Other integration algorithms result in different expressions for $g^*(\Delta t)$.

10.4 Finite Elasticity with Uncoupled Volume Response

As a first step toward our development of nonlinear viscoelasticity, we consider the formulation of finite strain elasticity with uncoupled, volumetric/deviatoric response. The crucial idea here, is the introduction of the following volumetric/deviatoric multiplicative split, first suggested in Flory [1961] and systematically exploited in Simo, Taylor, and Pister [1985] among others. For alternative approaches see Glowinski and Le Tallec [1984].

10.4.1 Volumetric/Deviatoric Multiplicative Split

As usual, we let $\mathcal{B} \subset \mathbb{R}^3$ be the reference placement of a continuum body and denote by $\varphi : \mathcal{B} \to \mathbb{R}^3$ the current configuration with deformation gradient $\boldsymbol{F} = D\varphi$ and Jacobian determinant $J = \det\left[\boldsymbol{F}\right]$. Particles in the reference placement are labeled by $\boldsymbol{X} \in \mathcal{B}$, and positions in the current placement $\mathcal{S} = \varphi(\mathcal{B})$ are denoted by $\boldsymbol{x} = \varphi(\boldsymbol{X})$. Let

$$\bar{\boldsymbol{F}} := J^{-\frac{1}{3}}\boldsymbol{F}$$

and (10.4.1)

$$\Theta := \det\left[\boldsymbol{F}\right],$$

and consider the multiplicative decomposition

$$\boxed{\boldsymbol{F} = \Theta^{\frac{1}{3}}\bar{\boldsymbol{F}},}$$

where (10.4.2)

$$\boxed{\det[\bar{\boldsymbol{F}}] = 1.}$$

Although in the present continuum context $\Theta \equiv J$, to construct mixed finite-element approximations, it proves convenient to introduce the preceding notation. $\bar{\boldsymbol{F}}$ and $\Theta\mathbf{1}$ are called the volume-preserving (deviatoric) and spherical parts of the deformation gradient $\boldsymbol{F}$, respectively. The right Cauchy–Green tensors and Lagrangian strain tensors associated with (10.4.1) are given by

$$\left.\begin{aligned} \boldsymbol{C} &:= \boldsymbol{F}^T\boldsymbol{F}, & \boldsymbol{E} &= \tfrac{1}{2}\left(\boldsymbol{C} - \mathbf{1}\right), \\ \bar{\boldsymbol{C}} &:= \bar{\boldsymbol{F}}^T\bar{\boldsymbol{F}}, & \bar{\boldsymbol{E}} &= \tfrac{1}{2}\left(\bar{\boldsymbol{C}} - \mathbf{1}\right). \end{aligned}\right\} \qquad (10.4.3)$$

We record the following relationships which will prove useful in our subsequent developments

Lemma 10.1. *The partial derivatives of* $\bar{\boldsymbol{C}}$ *and* J *with respect to* $\boldsymbol{C}$ *are given by*

$$\boxed{\begin{aligned} \partial_{\boldsymbol{C}}\bar{\boldsymbol{C}} &= J^{-\frac{2}{3}}\left[\boldsymbol{I} - \tfrac{1}{3}\boldsymbol{C} \otimes \boldsymbol{C}^{-1}\right], \\ \partial_{\boldsymbol{C}}J &= \tfrac{1}{2}J\boldsymbol{C}^{-1}. \end{aligned}} \qquad (10.4.4)$$

PROOF. Consider a one-parameter family of right Cauchy–Green tensors of the form

$$C_\varepsilon = C + \varepsilon H, \qquad H = H^T, \quad \text{and} \quad \varepsilon > 0. \tag{10.4.5}$$

(For $\varepsilon > 0$ small enough, $\det\left[C_\varepsilon\right] > 0$). Let

$$J_\varepsilon := \sqrt{\det\left[C_\varepsilon\right]}. \tag{10.4.6}$$

By the formula for the derivative of the determinant,

$$\begin{aligned} \frac{d}{d\varepsilon}\bigg|_{\varepsilon=0} J_\varepsilon &= \frac{1}{2J}\frac{d}{d\varepsilon}\bigg|_{\varepsilon=0}\left[\det\left(C_\varepsilon\right)\right] \\ &= \frac{1}{2J}J^2 \operatorname{tr}\left[C^{-1}\frac{d}{d\varepsilon}\bigg|_{\varepsilon=0} C_\varepsilon\right] \\ &= \tfrac{1}{2}J \operatorname{tr}\left[C^{-1}H\right]. \end{aligned} \tag{10.4.7}$$

Then differentiating the defining relationship $\bar{C}_\varepsilon = J_\varepsilon^{-\frac{2}{3}} C_\varepsilon$ and using (10.4.7) yields

$$\begin{aligned} \frac{d}{d\varepsilon}\bigg|_{\varepsilon=0} \bar{C}_\varepsilon &= J^{-\frac{2}{3}}\left[H - \tfrac{1}{3}\operatorname{tr}\left[C^{-1}H\right]C\right] \\ &= J^{-\frac{2}{3}}\left(I - \tfrac{1}{3}C \otimes C^{-1}\right) : H, \end{aligned} \tag{10.4.8}$$

a relationship which holds for any $H = H^T$. Since, by definition,

$$\frac{d}{d\varepsilon}\bigg|_{\varepsilon=0} \bar{C}_\varepsilon = \partial_C \bar{C} : \frac{d}{d\varepsilon}\bigg|_{\varepsilon=0} C_\varepsilon = \partial_C \bar{C} : H, \tag{10.4.9}$$

the result follows from (10.4.7), (10.4.8), and (10.4.9). □

With the preceding kinematic decomposition at hand, next we consider the formulation of finite strain nonlinear constitutive relationships that generalize our developments in Section 10.2.1.

10.4.2 Stored-Energy Function and Stress Response

Motivated by expression (10.2.4), now we consider a stored-energy function of the following form:

$$W^\circ(C) = U^\circ(\Theta) + \bar{W}^\circ\left(\bar{C}\right). \tag{10.4.10}$$

As in the geometrically linear theory, the functions U° and $\bar{W}^\circ$ define the volumetric and volume-preserving contributions to the stored-energy function, respectively. Then the stress response is obtained from the derivatives of the stored-energy function (10.4.10), as follows.

i. *Stress response in the material description.* Differentiating (10.4.10) relative to C and using the chain rule along with Lemma **10.1** yields (recall that $\Theta \equiv J$):

$$\begin{aligned} 2\partial_C W^\circ(C) &= 2U^{\circ\prime}(\Theta)\partial_C\Theta + 2\partial_{\bar{C}}\bar{W}^\circ(\bar{C}) : \partial_C\bar{C} \\ &= JU^{\circ\prime}(\Theta)C^{-1} \\ &\quad + 2J^{-\frac{2}{3}}\left[\partial_{\bar{C}}\bar{W}^\circ(\bar{C}) - \tfrac{1}{3}\left(\partial_{\bar{C}}\bar{W}^\circ(\bar{C}) : C\right)C^{-1}\right] \end{aligned} \tag{10.4.11}$$

To simplify the expressions in our subsequent developments, we introduce the notation

$$\boxed{\begin{aligned} \bar{S}^\circ &:= J^{-\frac{2}{3}}\,\mathrm{DEV}\left[2\partial_{\bar{C}}\bar{W}^\circ(\bar{C})\right] \\ &:= J^{-\frac{2}{3}}\left\{2\partial_{\bar{C}}\bar{W}^\circ(\bar{C}) - \tfrac{1}{3}[2\partial_{\bar{C}}\bar{W}^\circ(\bar{C}) : C]\,C^{-1}\right\}. \end{aligned}} \tag{10.4.12}$$

Thus, denoting the second Piola-Kirchhoff tensor by S°, with the notation in (10.4.12), expression (10.4.11) yields the constitutive equation

$$\boxed{S^\circ := 2\partial_C W^\circ\,(C) = JU^{\circ\prime}\,(\Theta)\,C^{-1} + \bar{S}^\circ.} \tag{10.4.13}$$

It should be noted that DEV [•] defined by (10.4.12) gives the correct notion of "deviatoric" stress tensor in the convected or material representation (i.e., in terms of X and C) since

$$\begin{aligned} C : \mathrm{DEV}\,[\bullet] &= C : (\bullet) - \tfrac{1}{3}\left[(\bullet) : C\right]C^{-1} : C \\ &= C : (\bullet) - (\bullet) : C \equiv 0. \end{aligned} \tag{10.4.14}$$

ii. *Stress response in the spatial description.* The expression for the Kirchhoff stress tensor, denoted by $\boldsymbol{\tau}^\circ$, follows at once from from (10.4.13) and the relationship $\boldsymbol{\tau}^\circ = FS^\circ F^T$. We obtain

$$\begin{aligned} \boldsymbol{\tau}^\circ = JU^{\circ\prime}(\Theta)FC^{-1}F^T + J^{-\frac{2}{3}}&\left[2\,F\partial_{\bar{C}}\bar{W}^\circ(\bar{C})F^T\right. \\ &\left. -\tfrac{1}{3}\left\{[2\,F\partial_{\bar{C}}\bar{W}^\circ(\bar{C})F^T] : \mathbf{1}\right\}FC^{-1}F^T\right], \end{aligned} \tag{10.4.15}$$

where we have used the following relationship

$$(\bullet) : C = (\bullet) : F^T\mathbf{1}F = F\,(\bullet)\,F^T : \mathbf{1}. \tag{10.4.16}$$

Since $\bar{F} = J^{-\frac{1}{3}}F$, expression (10.4.15) is written as

$$\boxed{\boldsymbol{\tau}^\circ = JU^{\circ\prime}(\Theta)\mathbf{1} + \mathrm{dev}\left[2\bar{F}\partial_{\bar{C}}\bar{W}^\circ(\bar{C})\bar{F}^T\right],} \tag{10.4.17}$$

where

$$\mathrm{dev}\,[\bullet] = (\bullet) - \tfrac{1}{3}\left[(\bullet) : \mathbf{1}\right]\mathbf{1}, \tag{10.4.18}$$

is the deviator operator in the spatial description. Relations (10.4.12)–(10.4.13) and (10.4.17) define the stress response in the convected and spatial descriptions, respectively.

EXAMPLE: 10.4.1. A simple example of a stored-energy function of the form (10.4.10) is given by the relationships

$$\left.\begin{aligned} U^\circ(J) &= \tfrac{1}{2}\kappa(\tfrac{1}{2}(J^2-1)-\ln J), \\ \bar{W}^\circ(\bar{C}) &= \tfrac{1}{2}\mu(\bar{C}:\mathbf{1}-3) = \tfrac{1}{2}\mu(\mathrm{tr}[\bar{C}]-3). \end{aligned}\right\} \tag{10.4.19}$$

Then from (10.4.17),

$$\tau^\circ = \frac{\kappa}{2}(J^2-1)\mathbf{1} + \mu\,\mathrm{dev}(\bar{F}\bar{F}^T). \tag{10.4.20}$$

Note that $U^{\circ\prime\prime}(J) = \frac{1}{2}\kappa(1+\frac{1}{J^2}) > 0$ for all $J \in \mathbb{R}$. It follows that $U^\circ : \mathbb{R}_+ \to \mathbb{R}_+$ is a convex function.

10.4.3 Elastic Tangent Moduli

We conclude our development of uncoupled volumetric/deviatoric finite deformation elasticity by recording the explicit expressions for the elastic tangent moduli. The following notation will be used throughout.

i. *Material tangent moduli.* We let

$$\mathbf{C}^\circ = 2\partial_C S^\circ,$$

i.e.,

$$\mathbf{C}^\circ_{IJKL} := 2\frac{\partial S^\circ_{IJ}}{\partial C_{KL}}. \tag{10.4.21}$$

Furthermore,

$$S^\circ = JpC^{-1} + \bar{S}^\circ, \tag{10.4.22}$$

where $p := U^{\circ\prime}(\Theta)$ is the mean stress and $\bar{S}^\circ$ is the deviatoric part of S° defined by (10.4.12). In addition, we set

$$\bar{\mathbf{C}}^\circ := 2\partial_C\bar{S}^\circ,$$

i.e.,

$$\bar{\mathbf{C}}^\circ_{IJKL} := 2\frac{\partial \bar{S}^\circ_{IJ}}{\partial C_{KL}}. \tag{10.4.23}$$

ii. *Spatial tangent moduli.* These elastic moduli are defined through the transformation (push-forward)

$$\mathbf{c}^\circ_{ijkl} := F_{iI}F_{jJ}F_{kK}F_{lL}\mathbf{C}^\circ_{IJKL}, \tag{10.4.24a}$$

and similarly

$$\bar{\mathbf{c}}^\circ_{ijkl} := F_{iI}F_{jJ}F_{kK}F_{lL}\bar{\mathbf{C}}^\circ_{IJKL}. \tag{10.4.24b}$$

For hyperelasticity, the property that the stress response derives from an elastic potential (the stored-energy function) implies the major symmetry. In summary, we have

$$\mathbf{C}^{\circ}_{IJKL} = \mathbf{C}^{\circ}_{KLIJ} = \mathbf{C}^{\circ}_{IJLK} = \mathbf{C}^{\circ}_{JIKL} \tag{10.4.25}$$

and the analogous relationships for the spatial elastic moduli $\mathbf{c}^{\circ}_{ijkl}$. In particular, in view of expressions (10.4.10), (10.4.12), and (10.4.13)

$$\mathbf{C}^{\circ}_{IJKL} = 4\frac{\partial^2 W^{\circ}}{\partial C_{IJ}\partial C_{KL}},$$

and (10.4.26)

$$\bar{\mathbf{C}}^{\circ}_{IJKL} = 4\frac{\partial^2 \bar{W}^{\circ}}{\partial C_{IJ}\partial C_{KL}}.$$

However, since $\bar{W}^{\circ}$ is a function of $\bar{C} := J^{-\frac{2}{3}}C$, it is necessary to express relationship $(10.4.26)_2$ in terms of the derivatives of $\bar{W}^{\circ}\left(\bar{C}\right)$ with respect to $\bar{C}$. This involves a somewhat involved application of the chain rule employing relationships (10.4.4). For the reader's convenience, we outline the procedure and record the basic results below.

10.4.3.1 The material tangent moduli $\bar{\mathbf{C}}^{\circ}$.

Using the relationships (10.4.4) and the chain rule, from definition (10.4.23), we find that

$$\boxed{\begin{aligned}\bar{\mathbf{C}}^{\circ} &= \bar{\mathbf{C}}^{\circ}_{DEV} - \tfrac{2}{3}\bar{S}^{\circ}\otimes C^{-1} - \tfrac{2}{3}C^{-1}\otimes\bar{S}^{\circ} \\ &\quad + \tfrac{2}{3}J^{-\frac{2}{3}}\left(2\partial_{\bar{C}}\bar{W}^{\circ} : C\right)\left[I_{C^{-1}} - \tfrac{1}{3}C^{-1}\otimes C^{-1}\right],\end{aligned}} \tag{10.4.27}$$

where $I_{C^{-1}}$ is the fourth-order tensor with components

$$\left(I_{C^{-1}}\right)_{IJKL} := \tfrac{1}{2}\left[C^{-1}_{IK}C^{-1}_{JL} + C^{-1}_{IL}C^{-1}_{JK}\right] \tag{10.4.28}$$

and $\bar{\mathbf{C}}^{\circ}_{DEV}$ is the fourth-order tensor of deviatoric tangent moduli defined by the expression

$$\boxed{\begin{aligned}\bar{\mathbf{C}}^{\circ}_{DEV} :&= 4J^{-\frac{4}{3}}\left[\partial^2_{\bar{C}\bar{C}}\bar{W}^{\circ} + \tfrac{1}{9}(C : \partial^2_{\bar{C}\bar{C}}\bar{W}^{\circ} : C)C^{-1}\otimes C^{-1}\right. \\ &\quad \left. - \tfrac{1}{3}C^{-1}\otimes(\partial^2_{\bar{C}\bar{C}}\bar{W}^{\circ} : C) - \tfrac{1}{3}(\partial^2_{\bar{C}\bar{C}}\bar{W}^{\circ} : C)\otimes C^{-1}\right].\end{aligned}} \tag{10.4.29}$$

The preceding notation is motivated by the following property which is an easy consequence of (10.4.29):

$$C : \bar{\mathbf{C}}^{\circ}_{DEV} = \bar{\mathbf{C}}^{\circ}_{DEV} : C = \mathbf{0}, \tag{10.4.30}$$

and shows that $\bar{\mathbf{C}}^{\circ}_{DEV}$ is a fourth-order *deviatoric tensor*.

10.4.3.2 The total tangent moduli $\bar{\mathbf{C}}^{\circ}$.

By the chain rule, we find that

$$\mathbf{C}^{\circ} = \bar{\mathbf{C}}^{\circ} + 4\left(U^{\circ\prime\prime}(\Theta)\frac{\partial\Theta}{\partial \boldsymbol{C}} \otimes \frac{\partial\Theta}{\partial \boldsymbol{C}} + U^{\circ\prime}(\Theta)\frac{\partial^2\Theta}{\partial \boldsymbol{C}\partial \boldsymbol{C}}\right). \tag{10.4.31}$$

As in the geometrically linear theory, in a finite-element context, the explicit expression for the term $\partial\Theta/\partial \boldsymbol{C}$ depends on the specific form of the approximation for Θ; for instance,

$$\Theta = J \Longrightarrow 2\partial_{\boldsymbol{C}}\Theta = J\boldsymbol{C}^{-1}, \text{ etc.} \tag{10.4.32}$$

For bilinear isoparametric quadrilateral elements, a widely used approximation is given by the formula

$$\Theta = \frac{\int_{\mathcal{B}_e} J dV}{\int_{\mathcal{B}_e} dV}, \tag{10.4.33}$$

where $\mathcal{B}_e$ is a typical finite element. Expression (10.4.33) is known as the *mean dilatation approximation*. Then the term $\partial_{\boldsymbol{C}}\Theta$ must be computed from (10.4.33); see Simo, Taylor, and Pister [1985] for further details and references to the literature.

10.4.3.3 The spatial versions of $\bar{\mathbf{C}}^{\circ}$ and $\mathbf{C}^{\circ}$.

Using the transformation rule (10.4.24b) along with the relationships

$$\left.\begin{aligned} \boldsymbol{F}\boldsymbol{C}^{-1}\boldsymbol{F}^T &= \mathbf{1}, \\ \operatorname{dev}\left[\bar{\boldsymbol{F}}(\bullet)\bar{\boldsymbol{F}}^T\right] &= J^{-\frac{2}{3}}\boldsymbol{F}\mathrm{DEV}[\bullet]\boldsymbol{F}^T, \end{aligned}\right\} \tag{10.4.34}$$

from (10.4.27), we obtain

$$\begin{aligned} \bar{\mathbf{c}}^{\circ} = \bar{\mathbf{c}}^{\circ}_{dev} &- \tfrac{2}{3}\operatorname{dev}(\boldsymbol{\tau}^{\circ}) \otimes \mathbf{1} - \tfrac{2}{3}\mathbf{1} \otimes \operatorname{dev}(\boldsymbol{\tau}^{\circ}) \\ &+ \tfrac{2}{3}\operatorname{tr}(2\bar{\boldsymbol{F}}\,\partial_{\bar{\boldsymbol{C}}}\bar{W}^{\circ}\,\bar{\boldsymbol{F}}^T)\left(\boldsymbol{I} - \tfrac{1}{3}\mathbf{1} \otimes \mathbf{1}\right), \end{aligned} \tag{10.4.35}$$

where

$$\left(\bar{\mathbf{c}}^{\circ}_{dev}\right)_{ijkl} \cdot = F_{iI}F_{jJ}F_{kK}F_{lL}\left(\bar{\mathbf{C}}^{\circ}_{DEV}\right)_{IJKL}. \tag{10.4.36}$$

Similarly, from (10.4.31) and relationships (10.4.34), we readily obtain (for the case $\Theta = J$)

$$\boxed{\mathbf{c}^\circ = J^2 U^{\circ\prime\prime}(\Theta)\,\mathbf{1}\otimes\mathbf{1} + J U^{\circ\prime}(\Theta)\big(\mathbf{1}\otimes\mathbf{1} - 2I\big) + \bar{\mathbf{c}}^\circ.} \tag{10.4.37}$$

Equations (10.4.35) and (10.4.37) define the tangent elastic moduli of the the material in the spatial representation.

10.5 A Class of Nonlinear, Viscoelastic, Constitutive Models

In this section, we consider the extension of the viscoelastic constitutive models, motivated in Section **10.1.3** and formulated in Section **10.2** in the context of the infinitesimal theory, to finite deformations. Although of a particular functional form, the resulting class of nonlinear viscoelastic models possesses a number of attractive features.

i. For very fast or very slow processes (in the sense precisely described below), the model reduces to finite deformation anisotropic elasticity.
ii. The proposed class of models incorporates viscoelastic versions of the well-known Ogden elastic materials which are particularly well suited for phenomenological modeling of rubber elasticity; see Ogden [1984] and references therein.
iii. Computationally, the algorithmic implementation presented below involves a straightforward extension of the schemes developed in Section **10.3** in the context of the infinitesimal theory. The resulting class of algorithms is second-order accurate, unconditionally stable and satisfies, by construction, the condition of incremental objectivity. Furthermore, the algorithms are amenable to exact linearization leading to a closed-form expression for the algorithmic tangent moduli.
iv. As in our formulation of finite-strain plasticity, volumetric and deviatoric response are exactly decoupled via a multiplicative decomposition of the deformation gradient into spherical and volume-preserving parts.

Nonlinear viscoelasticity has received considerable attention in recent years, particularly within the context of rubber elasticity and, nowadays, constitutes an active topic of research. The objective of this chapter is to present an introduction to the topic from the perspectives of continuum mechanics and numerical analysis.

10.5.1 Formulation of the Nonlinear Viscoelastic Constitutive Model

The formulation of the proposed class of viscoelastic constitutive relations is patterned after our developments in Section **10.2.1**. Accordingly, we consider the following steps.

10.5.1.1 Stress response.

The generalization of relationship (10.2.6) to the finite deformation regime is accomplished by the expression

$$\boxed{S(t) = S^{\circ}(t) - J^{-\frac{2}{3}}\,\text{DEV}\left[\sum_{i=1}^{N} Q_i(t)\right],} \tag{10.5.1}$$

where $S^{\circ}(t)$ is defined by (10.4.12)–(10.4.13) with $C(t)$ now a function of time. Here, $Q_i(t)$, $i = 1, 2, \cdots, N$, are internal variables, which *remain unaltered under superposed spatial rigid body motions*. This fundamental requirement is the same invariance property classically placed on the second Piola–Kirchhoff tensor $S(t)$ and automatically ensures frame indifference of the constitutive relationship (10.5.1).

10.5.1.2 Evolution equations for the interval variables $Q_i(t)$.

Motivated by our developments in Section **10.2.1.1**, we consider the following set of rate equations governing the evolution of the interval variables $Q_i(t)$:

$$\boxed{\begin{aligned} &\dot{Q}_i(t) + \frac{1}{\tau_i} Q_i(t) = \frac{\gamma_i}{\tau_i}\,\text{DEV}\left\{2\partial_{\bar{C}}\bar{W}^{\circ}\left[\bar{C}(t)\right]\right\}, \\ &\lim_{t\to-\infty} Q_i(t) = \mathbf{0}, \end{aligned}} \tag{10.5.2}$$

where $\gamma_i \in [0, 1]$ and $\tau_i > 0$ are subject to restrictions (10.2.8), that is,

$$\sum_{i=1}^{N} \gamma_i = 1 - \gamma_{\infty},$$

with (10.5.3)

$$\gamma_{\infty} \in [0, 1).$$

The evolution equations (10.5.2) are *linear* and, therefore, explicitly lead to the following convolution representation:

$$Q_i(t) = \frac{\gamma_i}{\tau_i}\int_{-\infty}^{t} \exp[-(t-s)/\tau_i]\,\text{DEV}\left\{2\partial_{\bar{C}}\bar{W}^{\circ}\left[\bar{C}(s)\right]\right\} ds. \tag{10.5.4}$$

10.5.1.3 Convolution representation.

Now the stress response function is obtained by substituting (10.5.4) in (10.5.1) and integrating by parts along with use of relationships (10.4.12)–(10.4.13). The following expression is obtained:

$$S(t) = JU^{\circ\prime}(\Theta)C^{-1}(t) + \gamma_{\infty}J^{-\frac{2}{3}}\,\text{DEV}\left\{2\partial_{\bar{C}}\bar{W}^{\circ}[\bar{C}(t)]\right\}$$

$$+\sum_{i=1}^{N} J^{-\frac{2}{3}} \gamma_i \int_{-\infty}^{t} \exp[-(t-s)/\tau_i] \frac{d}{ds}$$

$$\cdot \left(\mathrm{DEV}\left\{2\partial_{\bar{C}} \bar{W}^{\circ}[\boldsymbol{C}(s)]\right\}\right) ds. \tag{10.5.5}$$

Now, define the relaxation function for the viscoelastic model with internal variables governed by the evolutionary equations (10.5.2) with the same expression as in the physically linear theory, namely,

$$\boxed{g(t) = \gamma_\infty + \sum_{i=1}^{N} \gamma_i \exp(-t/\tau_i).} \tag{10.5.6}$$

Combining (10.5.4) and (10.5.5) and integrating by parts along with the initial condition $(10.5.2)_2$, we obtain, exactly as in the linear theory, the following convolution representation for the second Piola–Kirchhoff stress tensor (keep in mind, we assume $\Theta = J$):

$$\boxed{\begin{aligned} \boldsymbol{S}(t) &= J U^{\circ\prime}(\Theta) \boldsymbol{C}^{-1}(t) \\ &+ J^{-\frac{2}{3}}(t) \int_{-\infty}^{t} g(t-s) \frac{d}{ds} \left(\mathrm{DEV}\left\{2\partial_{\bar{C}} \bar{W}^{\circ}[\bar{\boldsymbol{C}}(s)]\right\}\right) ds. \end{aligned}} \tag{10.5.7}$$

The counterpart in the spatial description of expression (10.5.7) is easily derived using the following relationship between the operator DEV[•], defined in the material description, and the operator dev[•] defined in the spatial description:

$$J^{-\frac{2}{3}} \boldsymbol{F} (\mathrm{DEV}\,[\bullet])\, \boldsymbol{F}^T = \mathrm{dev}\left[\bar{\boldsymbol{F}}\,(\bullet)\,\bar{\boldsymbol{F}}^T\right]. \tag{10.5.8}$$

Then the convolution representation (10.5.7) in terms of the Kirchhoff stress takes the form

$$\boxed{\begin{aligned} \boldsymbol{\tau}(t) &= J U^{\circ\prime}(\Theta)\, \mathbf{1} \\ &+ \int_{-\infty}^{t} g(t-s) \frac{d}{ds} \left(\mathrm{dev}\left\{2\bar{\boldsymbol{F}} \partial_{\bar{C}} \bar{W}^{\circ}[\bar{\boldsymbol{C}}(s)] \bar{\boldsymbol{F}}^T\right\}\right) ds. \end{aligned}} \tag{10.5.9}$$

Remarks 10.5.1.

1. The final convolution model, given either by (10.5.7) or (10.5.9), exactly decouples volumetric effects, which are assumed elastic, from the viscoelastic volume-preserving response.
2. As in our discussion of the geometrically linear theory, viscoelastic effects on the bulk response are easily accommodated by a procedure similar to that outlined in Remark **10.2.1**.
3. We emphasize that the form of stored-energy function $\bar{W}^{\circ}\left(\bar{\boldsymbol{C}}\right)$ in the constitutive equation (10.5.7) (or (10.5.9)) is completely arbitrary and **not** restricted to isotropic response.

4. The preceding model is derived by a systematic exploitation of the second law of thermodynamics (in the form of the Claussius–Duhem inequality) by postulating the following free energy function:

$$\psi\left(\boldsymbol{C}, \boldsymbol{Q}_i\right) = U^{\circ}(J) + \bar{W}^{\circ}\left(\bar{\boldsymbol{C}}\right) - \sum_{i=1}^{N} \tfrac{1}{2}\bar{\boldsymbol{C}} : \boldsymbol{Q}_i + \Xi\left(\sum_{i=1}^{N} \boldsymbol{Q}_i\right), \tag{10.5.10}$$

where Ξ is a certain function of the internal variables. For additional background on thermodynamic considerations see Coleman and Gurtin [1967] and Coleman and Noll [1963].

10.6 Implementation of Integration Algorithms for Nonlinear Viscoelasticity

In this section we consider numerically integrating the class of viscoelastic constitutive models developed in the preceding section. We show that the development of stable and accurate integration algorithms for this class of models involves only a straightforward modification of the results presented in Section **10.3**.

10.6.1 One-Step, Second-Order Accurate Recurrence Formula.

The steps involved in implementing the recurrence formula presented below are essentially identical to those discussed in detail in Section **10.2** and proceed as follows.

10.6.1.1 Algorithmic internal variables and recurrence relationship.

Define internal (algorithmic) variables $\boldsymbol{H}^{(i)}(t), i = 1, 2, \cdots, N$, by the expression

$$\boldsymbol{H}^{(i)}(t) := \int_{-\infty}^{t} \exp[-(t-s)/\tau_i] \frac{d}{ds}\left(\mathrm{DEV}\left\{2\partial_{\bar{C}}\bar{W}^{\circ}\left[\bar{\boldsymbol{C}}(s)\right]\right\}\right) ds. \tag{10.6.1}$$

Now use the semigroup property (10.3.9) and the property of additivity of the integral over the interval of integration to arrive at the following recurrence relationship, identical to (10.3.10):

$$\boxed{\begin{aligned}\boldsymbol{H}^{(i)}\left(t_{n+1}\right) &= \exp[-\Delta t_n/\tau_i]\boldsymbol{H}^{(i)}(t_n) \\ &\quad + \int_{t_n}^{t_{n+1}} \exp[-(t_{n+1}-s)/\tau_i] \frac{d}{ds}\left(\mathrm{DEV}\left\{2\partial_{\bar{C}}\bar{W}^{\circ}\left[\bar{\boldsymbol{C}}(s)\right]\right\}\right) ds.\end{aligned}} \tag{10.6.2}$$

Finally, we use the midpoint rule to approximate the integral over $[t_n, t_{n+1}]$, with $t_{n+1} := t_n + \Delta t_n$, to arrive at the update formula:

$$\boxed{\begin{aligned} \boldsymbol{H}_{n+1}^{(i)} &= \exp(-\Delta t_n/\tau_i)\boldsymbol{H}_n^{(i)} + \exp(-\Delta t_n/2\tau_i)\left(\tilde{\boldsymbol{S}}_{n+1}^{\circ} - \tilde{\boldsymbol{S}}_n^{\circ}\right), \\ \tilde{\boldsymbol{S}}_{n+1}^{\circ} &:= \mathrm{DEV}_{n+1}\left[2\partial_{\bar{\boldsymbol{C}}}\bar{W}^{\circ}(\bar{\boldsymbol{C}}_n)\right], \\ \tilde{\boldsymbol{S}}_n^{\circ} &:= \mathrm{DEV}_n\left[2\partial_{\bar{\boldsymbol{C}}}\bar{W}^{\circ}(\bar{\boldsymbol{C}}_n)\right]. \end{aligned}} \tag{10.6.3}$$

Note carefully that $\mathrm{DEV}_{n+1}[\bullet]$ and $\mathrm{DEV}_n[\bullet]$ are computed according to (10.4.12) with $\boldsymbol{C}_{n+1}$ and $\boldsymbol{C}_n$, respectively, i.e.,

$$\mathrm{DEV}_{n+1}[\bullet] = (\bullet) - \tfrac{1}{3}\left[(\bullet) : \boldsymbol{C}_{n+1}\right]\boldsymbol{C}_{n+1}^{-1}, \tag{10.6.4}$$

and an analogous expression for $\mathrm{DEV}_n[\bullet]$. The explicit computation of the volume-preserving tensors $\bar{\boldsymbol{C}}_{n+1}$ and $\bar{\boldsymbol{C}}_n$, and the overall implementation of the algorithm is considered in Section **10.5.3**.

10.6.1.2 Computation of stress tensors.

From the convolution representation (10.5.5), it readily follows that the algorithmic approximation for the second Piola–Kirchhoff stress takes the form (for $\Theta = J$):

$$\boxed{\begin{aligned} \boldsymbol{S}_{n+1} &= J_{n+1}U^{\circ\prime}\left(\Theta_{n+1}\right)\boldsymbol{C}_{n+1}^{-1} \\ &\quad + \gamma_{\infty}\bar{\boldsymbol{S}}_{n+1}^{\circ} + \sum_{i=1}^{N}\gamma_i J_{n+1}^{-\frac{2}{3}}\,\mathrm{DEV}_{n+1}[\boldsymbol{H}_{n+1}^{(i)}], \end{aligned}} \tag{10.6.5}$$

where we recall that

$$\bar{\boldsymbol{S}}_{n+1}^{\circ} := J_{n+1}^{-\frac{2}{3}}\tilde{\boldsymbol{S}}_{n+1}^{\circ} = J_{n+1}^{-\frac{2}{3}}\,\mathrm{DEV}_{n+1}\left[2\partial_{\bar{\boldsymbol{C}}}\bar{W}^{\circ}\left(\bar{\boldsymbol{C}}_{n+1}\right)\right] \tag{10.6.6}$$

and $\boldsymbol{H}_{n+1}^{(i)}$ is given by the recurrence formula $(10.6.3)_1$. The (spatial) Kirchhoff stress tensor is computed via the standard transformation (push-forward)

$$\boldsymbol{\tau}_{n+1} = \boldsymbol{F}_{n+1}\boldsymbol{S}_{n+1}\boldsymbol{F}_{n+1}^{T}. \tag{10.6.7}$$

Using the fact that $\boldsymbol{F}_{n+1}\boldsymbol{C}_{n+1}^{-1}\boldsymbol{F}_{n+1}^{T} = \mathbf{1}$ and the relationship

$$\mathrm{dev}[\bar{\boldsymbol{F}}_{n+1}(\bullet)\bar{\boldsymbol{F}}_{n+1}^{T}] = J_{n+1}^{-\frac{2}{3}}\boldsymbol{F}_{n+1}\left[\mathrm{DEV}_{n+1}(\bullet)\right]\boldsymbol{F}_{n+1}^{T}, \tag{10.6.8}$$

which easily follows from (10.6.4) and the fact that $\bar{\boldsymbol{F}}_{n+1} := J_{n+1}^{-\frac{1}{3}}\boldsymbol{F}_{n+1}$, we find that

$$\boxed{\begin{aligned} \boldsymbol{\tau}_{n+1} &= J_{n+1}U^{\circ\prime}\left(\Theta_{n+1}\right)\mathbf{1} + \gamma_{\infty}\mathrm{dev}\left[2\bar{\boldsymbol{F}}_{n+1}\partial_{\bar{\boldsymbol{C}}}\bar{W}^{\circ}(\bar{\boldsymbol{C}}_{n+1})\bar{\boldsymbol{F}}_{n+1}^{T}\right] \\ &\quad + \sum_{i=1}^{N}\gamma_i\mathrm{dev}[\bar{\boldsymbol{F}}_{n+1}\boldsymbol{H}_{n+1}^{(i)}\bar{\boldsymbol{F}}_{n+1}^{T}]. \end{aligned}} \tag{10.6.9}$$

This expression gives the update formula for the Kirchhoff stress and constitutes the counterpart as the spatial description of the updated formula (10.6.5).

Remarks 10.6.1.

1. It is apparent from expression (10.6.9) that the present formulation and algorithmic treatment completely decouples the bulk response, which is assumed elastic, from viscoelastic effects, which influence only the deviatoric part of the stress tensor.
2. Once more, we recall that, in the context of a standard displacement-like finite-element method, the variable Θ_{n+1} is simply given by

$$\Theta_{n+1} = J_{n+1}. \tag{10.6.10}$$

For low-order elements, however, such an approximation leads to a locking response in the incompressible limit. For bilinear isoparametric interpolations, a more suitable approximation is obtained via the formula (10.4.33).

10.6.2 Configuration Update Procedure

The update formulas (10.6.3) and (10.6.4) involve the kinematic variables $\boldsymbol{C}_{n+1}$ and $\bar{\boldsymbol{C}}_{n+1}$. These variables are easily computed within the context of a *strain-driven* type of algorithm as follows.

Let $\mathcal{S}_n = \varphi_n(\mathcal{B})$ be the current placement of the body at time t_n, defined by the configuration $\varphi_n : \mathcal{B} \to \mathbb{R}^3$, which is *assumed to be given*. Let Δt_n be the time step size, and assume that an incremental displacement field, denoted by

$$\boldsymbol{U} : \mathcal{B} \to \mathbb{R}^3, \tag{10.6.11}$$

is *given*. Then we compute the updated placement $\mathcal{S}_{n+1} = \varphi_{n+1}(\mathcal{B})$ simply by setting

$$\varphi_{n+1}(\boldsymbol{X}) = \varphi_n(\boldsymbol{X}) + \boldsymbol{U}(\boldsymbol{X}), \tag{10.6.12}$$

which defines the configuration $\varphi_{n+1} : \mathcal{B} \to \mathbb{R}^3$ at time $t_{n+1} := t_n + \Delta t_n$. Then the deformation gradient and Jacobian determinant are computed via the standard expressions as

$$\boldsymbol{F}_{n+1} = D\varphi_{n+1}(\boldsymbol{X}),$$

and (10.6.13)

$$J_{n+1} = \det\left[\boldsymbol{F}_{n+1}\right].$$

Finally, the total and volume-preserving right Cauchy–Green tensors are determined by the formulas

$$\boldsymbol{C}_{n+1} = \boldsymbol{F}_{n+1}^T \boldsymbol{F}_{n+1}$$

and (10.6.14)

$$\bar{\boldsymbol{C}}_{n+1} = J_{n+1}^{-\frac{2}{3}} \boldsymbol{C}_{n+1}.$$

By substituting (10.6.14) into the formulas (10.6.5) and (10.6.9), the stress-tensors S_{n+1} and τ_{n+1} become nonlinear functions of the updated configuration φ_{n+1}.

10.6.3 Consistent (Algorithmic) Tangent Moduli

The preceding algorithm defines the second Piola-Kirchhoff stress tensor S_{n+1} at the configuration φ_{n+1} by formula (10.6.5). Therefore the associated material tangent moduli are obtained by differentiating (10.6.5) with respect to C_{n+1}, that is, we define *algorithmic moduli in the material description* by the formula

$$\mathbf{C}_{n+1} := 2\partial_{C_{n+1}} S_{n+1}. \tag{10.6.15}$$

Similarly, we set

$$\bar{\mathbf{C}}_{n+1} := 2\partial_{C_{n+1}} \bar{S}_{n+1}, \tag{10.6.16}$$

where $\bar{S}_{n+1}$ is the deviatoric part of S_{n+1}. Here S_{n+1} and $\bar{S}_{n+1}$ are understood to be the functions of C_{n+1} defined by the algorithm (10.6.5). We call (10.6.15) the *algorithmic tangent moduli*. These moduli appear naturally in linearizing the weak forms of the equilibrium equations discretized by finite-element methods. Recall that this linearized weak form gives rise to the *tangent stiffness matrix* associated with the discrete boundary-value problem, which becomes completely determined by the algorithmic tangent moduli and the stress tensor.

Since the volumetric response is elastic, exactly as in Section **10.4.3.2**, we compute the expression

$$\boxed{\mathbf{C}^{\circ}_{n+1} = 4(U^{\circ\prime\prime}(\Theta_{n+1})\frac{\partial \Theta_{n+1}}{\partial C_{n+1}} \otimes \frac{\partial \Theta_{n+1}}{\partial C_{n+1}} + U^{\circ\prime}(\Theta_{n+1})\frac{\partial^2 \Theta_{n+1}}{\partial C_{n+1}\partial C_{n+1}} + \bar{\mathbf{C}}^{\circ}_{n+1}.} \tag{10.6.17}$$

Therefore, only $\bar{\mathbf{C}}^{\circ}_{n+1}$ remains to be determined. Note, if $\Theta = J$, (10.6.17) simplifies.

10.6.3.1 The algorithmic tangent moduli $\bar{\mathbf{C}}^{\circ}_{n+1}$.

To differentiate the algorithm, we rewrite the update formula (10.6.3), as follows

$$\left.\begin{aligned} \tilde{H}^{(i)}_{n} &:= \exp(-\Delta t_n/\tau_i)H^{(i)}_{n} - \exp(-\Delta t_n/2\tau_i)\tilde{S}^{\circ}_{n}, \\ H^{(i)}_{n+1} &:= \tilde{H}^{(i)}_{n} + \exp(-\Delta t_n/2\tau_i)\tilde{S}^{\circ}_{n+1}. \end{aligned}\right\} \tag{10.6.18}$$

With this notation, the deviatoric part of (10.6.5) becomes

$$\bar{S}_{n+1} = g^*(\Delta t_n)\bar{S}^{\circ}_{n+1} + \sum_{i=1}^{N} \gamma_i J_{n+1}^{-\frac{2}{3}} \mathrm{DEV}_{n+1}\left[\tilde{H}^{(i)}_{n}\right], \tag{10.6.19}$$

where $g^*(\Delta t_n)$ is the algorithmic relaxation function given by

$$g^*(\Delta t_n) = \gamma_\infty + \sum_{i=1}^{N} \gamma_i \exp(-\Delta t_n/2\tau_i). \tag{10.6.20}$$

Observe carefully that $\tilde{\boldsymbol{H}}_n^{(i)}$ is a constant as far as the linearization process is concerned.

With expression (10.6.19) in hand, we can immediately computate (10.6.16) by exploiting our results in Section **10.4.1**. In fact, using (10.4.4), from (10.6.19), we obtain

$$\boxed{\begin{aligned}\bar{\mathbf{C}}_{n+1} := {} & g^*(\Delta t_n)\,\bar{\mathbf{C}}^{\circ}_{n+1} \\ & + \sum_{i=1}^{N} \gamma_i J_{n+1}^{-\frac{2}{3}} \Big\{ -\tfrac{2}{3}\,\mathrm{DEV}_{n+1}[\tilde{\boldsymbol{H}}_n^{(i)}] \otimes \boldsymbol{C}_{n+1}^{-1} \\ & - \tfrac{2}{3}\boldsymbol{C}_{n+1}^{-1} \otimes \mathrm{DEV}_{n+1}[\tilde{\boldsymbol{H}}_n^{(i)}] \\ & + \tfrac{2}{3}\,(\tilde{\boldsymbol{H}}_n^{(i)} : \boldsymbol{C}_{n+1}) \Big(\boldsymbol{I}_{\boldsymbol{C}_{n+1}^{-1}} - \tfrac{1}{3}\boldsymbol{C}_{n+1}^{-1} \otimes \boldsymbol{C}_{n+1}^{-1} \Big) \Big\},\end{aligned}} \tag{10.6.21}$$

where $\bar{\mathbf{C}}^{\circ}_{n+1}$ is given by (10.4.27). Observe that (10.6.21) *is symmetric*.

10.6.3.2 The spatial algorithmic tangent moduli.

The relationship between material and spatial algorithmic tangent moduli is also given by (10.4.24*a*,*b*). Thus, as in (10.4.37), we find that (again for $\Theta = J$)

$$\mathbf{c}_{n+1} = J_{n+1}^2 U^{\circ\prime\prime}(\Theta_{n+1})\,\mathbf{1}\otimes\mathbf{1} + J_{n+1} U^{\circ\prime}(\Theta_{n+1})\,(\mathbf{1}\otimes\mathbf{1} - 2\boldsymbol{I}) + \bar{\mathbf{c}}_{n+1}, \tag{10.6.22}$$

where $\bar{\mathbf{c}}_{n+1}$ is computed from (10.6.21) and relationships (10.4.24*b*) as

$$\boxed{\begin{aligned}\bar{\mathbf{c}}_{n+1} := {} & g^*(\Delta t_n)\bar{\mathbf{c}}^{\circ}_{n+1} \\ & + \sum_{i=1}^{N} \gamma_i \Big\{ -\tfrac{2}{3}\,\mathrm{dev}[\bar{\boldsymbol{F}}_{n+1}\tilde{\boldsymbol{H}}_n^{(i)}\bar{\boldsymbol{F}}_{n+1}^T] \otimes \mathbf{1} \\ & - \tfrac{2}{3}\mathbf{1} \otimes \mathrm{dev}[\bar{\boldsymbol{F}}_{n+1}\tilde{\boldsymbol{H}}_n^{(i)}\bar{\boldsymbol{F}}_{n+1}^T] \\ & + \tfrac{2}{3}\,\mathrm{tr}[\bar{\boldsymbol{F}}_{n+1}\tilde{\boldsymbol{H}}_n^{(i)}\bar{\boldsymbol{F}}_{n+1}^T] \Big(\boldsymbol{I} - \tfrac{1}{3}\mathbf{1}\otimes\mathbf{1} \Big) \Big\}.\end{aligned}} \tag{10.6.23}$$

For the reader's convenience and easy reference, we have summarized the overall implementation of the algorithm developed above in Box **10.1** and Box **10.2**.

BOX 10.1. Recursive Update Procedure

1. Database at each Gaussian point

$$\left\{\tilde{\boldsymbol{S}}_n^\circ, \boldsymbol{H}_n^{(i)}, \quad i = 1, 2, \cdots, N\right\}.$$

2. Given $\boldsymbol{U} : \mathcal{B} \to \mathbb{R}^3$, update configurations

$$\varphi_{n+1} = \varphi_n + \boldsymbol{U}, \quad \boldsymbol{F}_{n+1} = D\varphi_{n+1}, \quad J_{n+1} := \det\left[\boldsymbol{F}_{n+1}\right]$$

$$\boldsymbol{C}_{n+1} = \boldsymbol{F}_{n+1}^T \boldsymbol{F}_{n+1}, \quad \bar{\boldsymbol{F}}_{n+1} := J_{n+1}^{-1/3} \boldsymbol{F}_{n+1}, \quad \bar{\boldsymbol{C}}_{n+1} := J_{n+1}^{-2/3} \boldsymbol{C}_{n+1}.$$

3. Compute initial elastic stress Kirchhoff tensor

$$\bar{\boldsymbol{\tau}}_{n+1}^\circ = \operatorname{dev}\left[2\bar{\boldsymbol{F}}_{n+1} \partial_{\bar{\boldsymbol{C}}} \bar{\boldsymbol{W}}^\circ(\bar{\boldsymbol{C}}_{n+1}) \bar{\boldsymbol{F}}_{n+1}^T\right].$$

4. Update algorithmic internal variables

$$\tilde{\boldsymbol{H}}_n^{(i)} := \exp(-\Delta t_n/\tau_i)\boldsymbol{H}_n^{(i)} - \exp(-\Delta t_n/2\tau_i)\tilde{\boldsymbol{S}}_n^\circ$$

$$\tilde{\boldsymbol{S}}_{n+1}^\circ = \bar{\boldsymbol{F}}_{n+1}^{-1} \bar{\boldsymbol{\tau}}_{n+1}^\circ \bar{\boldsymbol{F}}_{n+1}^{-T}$$

$$\boldsymbol{H}_{n+1}^{(i)} = \tilde{\boldsymbol{H}}_n^{(i)} + \exp(-\Delta t_n/2\tau_i)\tilde{\boldsymbol{S}}_{n+1}^\circ.$$

5. Compute Kirchhoff stress tensor

$$p_{n+1} := U^{\circ\prime}\left(\Theta_{n+1}\right)$$

$$\bar{\boldsymbol{h}}_n := \sum_{i=1}^N \gamma_i \operatorname{dev}\left[\bar{\boldsymbol{F}}_{n+1} \tilde{\boldsymbol{H}}_n^{(i)} \bar{\boldsymbol{F}}_{n+1}^T\right]$$

$$\bar{h}_n := \sum_{i=1}^N \gamma_i \operatorname{tr}\left[\bar{\boldsymbol{F}}_{n+1} \tilde{\boldsymbol{H}}_n^{(i)} \bar{\boldsymbol{F}}_{n+1}^T\right]$$

$$\boldsymbol{\tau}_{n+1} = J_{n+1} p_{n+1} \mathbf{1} + g^*(\Delta t_n)\, \bar{\boldsymbol{\tau}}_{n+1}^\circ + \bar{\boldsymbol{h}}_n.$$

BOX 10.2. Algorithmic Spatial Tangent Moduli

1. Compute "initial" elastic moduli

$$\bar{\mathbf{c}}^{\circ}_{n+1} = \bar{\mathbf{c}}^{\circ}_{dev_{n+1}} - \tfrac{2}{3}\bar{\boldsymbol{\tau}}^{\circ}_{n+1} \otimes \mathbf{1} - \tfrac{2}{3}\mathbf{1} \otimes \bar{\boldsymbol{\tau}}^{\circ}_{n+1}$$
$$+ \tfrac{2}{3}\operatorname{tr}\left(2\bar{\boldsymbol{F}}_{n+1}\partial_{\bar{C}}\bar{W}^{\circ}_{n+1}\bar{\boldsymbol{F}}^{T}\right)\left(\boldsymbol{I} - \tfrac{1}{3}\mathbf{1}\otimes\mathbf{1}\right),$$

where

$$\left(\bar{\mathbf{c}}^{\circ}_{dev}\right)_{ijkl}\Big|_{n+1} = \left[F_{iI}F_{jJ}F_{kK}F_{lL}(\bar{\mathbf{C}}^{\circ}_{DEV})_{IJKL}\right]_{n+1}$$

and

$$\bar{\mathbf{C}}^{\circ}_{DEV} := 4J^{-\frac{4}{3}}\left[\partial^2_{\bar{C}\bar{C}}\bar{W}^{\circ} + \tfrac{1}{9}(\boldsymbol{C} : \partial^2_{\bar{C}\bar{C}}\bar{W}^{\circ} : \boldsymbol{C})\boldsymbol{C}^{-1}\otimes\boldsymbol{C}^{-1}\right.$$
$$\left. - \tfrac{1}{3}\boldsymbol{C}^{-1}\otimes(\partial^2_{\bar{C}\bar{C}}\bar{W}^{\circ} : \boldsymbol{C}) - \tfrac{1}{3}(\partial^2_{\bar{C}\bar{C}}\bar{W}^{\circ} : \boldsymbol{C})\otimes\boldsymbol{C}^{-1}\right].$$

2. Introduce viscoelastic effects

$$\bar{\mathbf{c}}_{n+1} := g^{*}(\Delta t_n)\bar{\mathbf{c}}^{\circ}_{n+1} - \tfrac{2}{3}\bar{\boldsymbol{h}}_n \otimes \mathbf{1} - \tfrac{2}{3}\mathbf{1}\otimes\bar{\boldsymbol{h}}_n + \tfrac{2}{3}\bar{h}_n\left(\boldsymbol{I} - \tfrac{1}{3}\mathbf{1}\otimes\mathbf{1}\right).$$

3. Add mean stress contribution (for the case $\Theta = J$)

$$\mathbf{c}_{n+1} = \bar{\mathbf{c}}_{n+1} + J_{n+1}p_{n+1}\left(\mathbf{1}\otimes\mathbf{1} - 2\boldsymbol{I}\right) + J^2_{n+1}U^{\circ\prime\prime}_{n+1}\mathbf{1}\otimes\mathbf{1}.$$

References

Abraham, R., and J.E. Marsden, [1978], *Foundations of Mechanics*, The Benjamin/Cummings Publishing Company, Reading, Mass.

Abraham, R., J.E. Marsden, and T. Ratiu, [1983], *Manifolds, Tensor Analysis and Applications*, Addison–Wesley, Reading, Mass.

Agah–Tehrani, A., E.H. Lee, R.L. Mallet, and E.T. Oñate, [1986], "The Theory of Elastic-Plastic Deformation at Finite Strain with Induced Anisotropy Modeled as Combined Isotropic-Kinematic Hardening," *Metal Forming Report*, Rensselaer Polytechnic Institute, Troy, N.Y., June 1986.

Aravas, N., [1987], "On the Numerical Integration of a Class of Pressure-Dependent Plasticity Models," *International Journal for Numerical Methods in Engineering*, **24**, 1395–1416.

Argyris, J.H., [1965], "Elasto-Plastic Matrix Analysis of Three Dimensional Continua," *Journal of the Royal Aeronautical Society* **69**, 633–635.

Argyris, J.H., and J.St. Doltsinis, [1979], "On the Large Strain Inelastic Analysis in Natural Formulation. Part I: Quasi-Static Problems," *Computer Methods in Applied Mechanics and Engineering* **20**, 213–251.

Argyris, J.H., and J.St. Doltsinis, [1980], "On the Large Strain Inelastic Analysis in Natural Formulation. Part II: Dynamic Problems," *Computer Methods in Applied Mechanics and Engineering* **21**, 213–251.

Argyris J.H., J.St. Doltsinis, W.C. Knudson, L.E. Vaz, and K.S. Willam, [1979], "Numerical Solution of Transient Nonlinear Problems," *Computer Methods in Applied Mechanics and Engineering* **17/18**, 341–409.

Arnold, V.I., [1978], *Mathematical Methods of Classical Mechanics*, Springer, Berlin.

Asaro, R.J., [1979], "Geometrical Effects in the Inhomogeneous Deformation of Ductile Single Crystals," *Acta Metallurgica* **27**, 445–453.

Asaro, R.J., [1983], Micromechanics of Crystals and Polycrystals, in *Advances in Applied Mechanics*, Volume **23**, Academic Press, New York.

Asaro, R.J., and J.R. Rice, [1977], "Strain Localization in Ductile Single Crystals," *Journal of Mechanics and Physics of Solids* **25**, 309–338.

Atluri, S.N., and E. Reissner, [1988], "On the Formulation of Variational Theorems Involving Volume Constraints," in "Computational Mechanics '88,"

Proceedings of the International Conference on Computational Engineering Science, Springer-Verlag, Berlin.

Ball, J.M., [1977], "Convexity Conditions and Existence Theorems in Nonlinear Elasticity," *Archive for Rational Mechanics and Analysis* **63**, 337–403.

Bertsekas, D.P., [1982], *Constrained Optimization and Lagrange Multiplier Methods*, Academic Press, New York.

Budiansky, B., N.J. Dow, R.W. Peters, and R.P. Sheperd, [1951], "Experimental Studies of Polyaxial Stress-Strain Laws of Plasticity," *Proceedings of the 1st U.S. National Congress on Applied Mechanics*, ASME, New York, 503–512.

Burrage, K., and J.C. Butcher, [1979], "Stability Criteria for Implicit Runge-Kutta Methods," *SIAM Journal of Numerical Analysis* **16** (1), 46–57.

Burrage, K., and J.C. Butcher, [1980], "Nonlinear Stability of a General Class of Differential Equation Methods," *BIT* **20**, 185–203.

Butcher, J.C., [1975], "A Stability Property of Implicit Runge-Kutta Methods," *BIT* **15**, 358–361.

Carey, G, and J.T. Oden, [1983], *Finite Elements. Volume II: A Second Course*, Prentice–Hall Inc., Englewood Cliffs, N.J.

Casey, J., and P.M. Naghdi, [1981], "On the Characterization of Strain in Plasticity" *Journal of Applied Mechanics* **48**, 285–295.

Casey, J., and P.M. Naghdi, [1983a], "On the Nonequivalence of the Stress Space and Strain Space Formulations of Plasticity Theory," *Journal of Applied Mechanics* **50**, 350–354.

Casey, J., and P.M. Naghdi, [1983b], "On the Use of Invariance Requirements for Intermediate Configurations Associated with the Polar Decomposition of a Deformation Gradient," *Quarterly of Applied Mathematics*, 339–342.

Chadwick, P., [1976], *Continuum Mechanics*, John Wiley and Sons, Inc., New York.

Chen, W.F., [1984], *Plasticity in Reinforced Concrete*, McGraw–Hill, New York.

Chorin, A., T.J.R. Hughes, M.F. McCracken, and J.E. Marsden, [1978], "Product Formulas and Numerical Algorithms," *Communications on Pure and Applied Mathematics* **31**, 205–256.

Christoffersen, J., and J.W. Hutchinson, [1979], "A Class of Phenomenological Theories of Plasticity," *Journal of Mechanics and Physics of Solids* **27**, 465–487.

Christensen, R.M., [1971], *Theory of Viscoelasticity: An Introduction*, Academic Press, New York.

Ciarlet, P., [1978], *The Finite Element Method for Elliptic Problems*, North-Holland Publishing Company, Amsterdam.

Ciarlet, P., [1988], *Mathematical Elasticity. Volume I: Three-dimensional Elasticity*, North-Holland Publishing Company, Amsterdam.

Coleman, B.D., and W. Noll, [1963], "The Thermodynamics of Elastic Materials with Heat Conduction and Viscosity," *Archive for Rational Mechanics and Analysis* **13**, 167–178.

Coleman, B.D., and M. Gurtin, [1967], "Thermodynamics with Internal Variables," *Journal of Chemistry and Physics* **47**, 597–613.

Cormeau, I.C., [1975], "Numerical Stability in Quasi-Static Elasto/Visco- plasticity," *International Journal for Numerical Methods in Engineering* **9**, 109–127.

Dafalias, Y.F., [1984], "A Missing Link in the Formulation and Numerical Implementation of Finite Transformation Elastoplasticity," in *Constitutive Equations: Macro and Computational Aspects*, ed., K.J. Willam, ASME, New York.

Dahlquist, G., [1963], "A Special Stability Problem for Linear Multistep Methods," *BIT* **3**, 27–43.

Dahlquist, G., [1975], "Error Analysis of a Class of Methods of Stiff Non-linear Initial Value Problems," *Numerical Analysis, Dundee* 1975, Springer Lecture Notes in Mathematics, No. 506, 60–74.

Dahlquist, G., [1978], "G–Stability is Equivalent to A–Stability," *BIT* **18**, 384–401.

Dahlquist, G., and R. Jeltsch, [1979], "Generalized Disks of Contractivity and Implicit Runge–Kutta Methods," *Inst. for Numerisk Analys*, Report No. TRITA–NA–7906, Stockholm.

Demengel, F., [1989], "Compactness Theorems for Spaces of Functions with Bounded Derivatives and Applications to Limit Analysis Problems in Plasticity," *Archive for Rational Mechanics and Analysis* **105**, No.2, pp.123–161.

Dennis, J.E., and R.B. Schnabel, [1983], *Numerical Methods for Unconstrained Optimization*, Prentice–Hall, Inc., Englewood Cliffs, N.J.

Desai, C.S., and H.J. Siriwardane, [1984], *Constitutive Laws for Engineering Materials, with emphasis on geologic materials*, Prentice-Hall. Englewood Cliffs, N.J.

Dienes, J.K., [1979], "On the Analysis of Rotation and Stress Rate in Deforming Bodies," *Acta Mechanica* **32**, 217–232.

DiMaggio, F.L., and I.S. Sandler, [1971], "Material Models for Granular Soils," *Journal of Engineering Mechanics* **97**, 935–950.

Doherty, W.P., E.L. Wilson, and R.L. Taylor, [1969], "Stress Analysis of Axisymmetric Solids Utilizing Higher Order Quadrilateral Finite Elements," SESM Report No. 69-3, Department of Civil Engineering, University of California, Berkeley.

Duvaut, G., and J.L. Lions, [1972], *Les Inequations en Mecanique et en Physique*, Dunod, Paris.

Flory, R.J., [1961], "Thermodynamic Relations for Highly Elastic Materials," *Transactions of the Faraday Society*, **57**, 829–838.

Fortin, M., and R. Glowinski, [1983], *Augmented Lagrangian Methods: Applications to the Numerical Solution of Boundary-Value Problems*, North-Holland Publishing Company, Amsterdam.

Franca, L. P., [1989], "Algorithm to Compute the Square Root of a 3 × 3 Positive Definite Matrix", *Computers and Mathematics with Applications* **18**, 459–466.

Fung, Y.C., [1965], *Foundations of Solid Mechanics*, Prentice–Hall International Series in Dynamics, Prentice–Hall, Englewood Cliffs, N.J.

Gear, C.W., [1971], *Numerical Initial Value Problems in Ordinary Differential Equations*, Prentice–Hall, Englewood Cliffs, N.J.

Gelfand, I.M., and S.V. Fomin, [1963], *Calculus of Variations*, Prentice–Hall, Englewood Cliffs, N.J.

Girault, V., and P.A. Raviart, [1986], *Finite Element Methods for Navier-Stokes Equations. Theory and Algorithms*, Springer-Verlag, Berlin.

Giroux, E.D., [1973], "HEMP User's Manual," Report UCRL-51079, Lawrence Livermore National Laboratory, Livermore, Calif.

Glowinski, R., and P. Le Tallec, [1984], "Finite Element Analysis in Nonlinear Incompressible Elasticity," in *Finite Elements, Vol V: Special Problems in Solid Mechanics*, eds., J.T. Oden and G.F. Carey, Prentice–Hall, Englewood Cliffs, N.J.

Glowinski, R., and P. Le Tallec, [1989], *Augmented Lagrangian and Operator-splitting Methods in Nonlinear Mechanics*, Society for Industrial and Applied Mathematics, Philadelphia, Volume **9**.

Glowinski, R., [1984], *Numerical Methods for Nonlinear Variational Problems*, Springer-Verlag, New York.

Goldstein, H., [1981], *Classical Mechanics*, 2nd ed., Addison–Wesley, Reading, Mass.

Goudreau, G.L., and J.O. Hallquist, [1982], "Recent Developments in Large-Scale Finite Element Lagrangian Hydrocode Technology," *Computer Methods in Applied Mechanics and Engineering* **33**, 725–757.

Green, A.E., and P.M. Naghdi, [1965], "A General Theory of an Elastic-Plastic Continuum," *Archive for Rational Mechanics and Analysis*, **18**, 251–281.

Green, A.E., and P.M. Naghdi, [1966], "A Thermodynamic Development of Elastic-Plastic Continua," *Proceedings of IUTAM Symposium*, Vienna, June 22–28, 1966.

Green, A.E., and W. Zerna, [1960], *Theoretical Elasticity*, 2nd ed., Clarendon Press, Oxford.

Gurtin, M.E., [1972], "The Linear Theory of Elasticity," in *Handbuch der Physik*, Vol. VIa/2, Mechanics of Solids II, ed., C. Truesdell, Springer-Verlag, Berlin.

Gurtin, M.E., [1981], *An Introduction to Continuum Mechanics*, Academic Press, Orlando, Fl.

Hairer, E., S.P. Norsett, and G. Wanner, [1987], *Solving Ordinary Differential Equations I, Nonstiff Problems*, Springer-Verlag, Berlin.

Hallquist, J.O., [1979], "NIKE 2D: An Implicit, Finite Deformation, Finite Element Code for Analyzing the Static and Dynamic Response of Two-Dimensional Solids," Lawrence Livermore National Laboratory, Report UCRL-52678, University of California, Livermore.

Hallquist, J.O., [1988], "DYNA 3D: An Explicit, Finite Element Code for Dynamic Analysis in Three Dimensions," Lawrence Livermore National Laboratory, University of California, Livermore.

Halphen, B. and Q.S. Nguyen, [1975], "Sur les Materiaux Standards Generalises," *Journal de Mecanique*, **14**, 39–63.

Harrier, E., and G. Wanner, [1991], *Solving Ordinary Differential Equations II, Stiff and Differential-Algebraic Problems*, Springer-Verlag, Berlin.

Herrmann, L.R., [1965], "Elasticity Equations for Incompressible and Nearly Incompressible Materials by a Variational Theorem," *AIAA Journal* **3** 1896-1900.

Herrmann, L.R., and F.E. Peterson, [1968], "A Numerical Procedure for Viscoelastic Stress Analysis," in Proceedings 7th Meeting of ICRPG Mechanical Behavior Working Group, Orlando, Fl.

Herstein, J.N., [1964], *Topics in Algebra*, Ginn and Company, Waltham, Mass.

Hestenes, M., [1969], "Multiplier and Gradient Methods," *Journal of Optimization Theory and Applications* **4**, 303–320.

Hill, R., [1950], *The Mathematical Theory of Plasticity*, Oxford University Press, Oxford, U.K.

Hill, R., [1958], "A General Theory of Uniqueness and Stability in Elasto-Plastic Solids," *Journal of Mechanics and Physics of Solids* **6**, 236–249.

Hill, R., [1966], "Generalized Constitutive Relations for Incremental Deformation of Metal Crystals by Multislip," *Journal of Mechanics and Physics of Solids* **14**, 95–102.

Hill, R., [1978], "Aspects of Invariance in Solid Mechanics," *Advances in Applied Mechanics* **18**, 1–75.

Hill, R., and J.R. Rice, [1972], "Constitutive Analysis of Elastic-Plastic Crystals at Arbitrary Strain," *Journal of Mechanics and Physics of Solids* **20**, 401–413.

Hinton, E., and D.R.J. Owen, [1980], *Finite Elements in Plasticity: Theory and Practice*, Pineridge Press, Swansea, Wales.

Hoger, A., and D.E. Carlson, [1984a], "Determination of the Stretch and Rotation in the Polar Decomposition of the Deformation Gradient," *Quarterly of Applied Mathematics* **42**, 113–117.

Hoger, A., and D.E. Carlson, [1984b], "On the Derivative of the Square Root of a Tensor and Guo's Rate Theorems," *Journal of Elasticity* **14**, 329–336.

Hughes, T.J.R., [1980], "Generalization of Selective Integration Procedures to Anisotropic and Nonlinear Media," *International Journal for Numerical Methods in Engineering* **15**, 1413–1418.

Hughes, T.J.R., [1983], "Analysis of Transient Algorithms with Particular Emphasis on Stability Behavior," in *Computational Methods for Transient Analysis*, eds., T. Belytschko and T.J.R. Hughes, North-Holland Publishing Company, Amsterdam 67–155.

Hughes, T.J.R., [1984], "Numerical Implementation of Constitutive Models: Rate-Independent Deviatoric Plasticity," in *Theoretical Foundations for Large Scale Computations of Nonlinear Material Behavior*, eds., S. Nemat-Nasser, R. Asaro, and G. Hegemier, Martinus Nijhoff Publishers, Dordrecht, The Netherlands 29–57.

Hughes, T.J.R., [1987], *The Finite Element Method*, Prentice–Hall, Englewood Cliffs, N.J.

Hughes, T.J.R, and K.S. Pister, [1978], "Consistent Linearization in Mechanics of Solids and Structures," *Computers and Structures* **9**, 391–397.

Hughes, T.J.R., and F. Shakib, [1986], "Pseudo-Corner Theory: A Simple Enhancement of J_2-Flow Theory for Applications Involving Non-Proportional Loading," *Engineering Computations* **3** (2), 116–120.

Hughes, T.J.R., and R.L. Taylor, [1978], "Unconditionally Stable Algorithms for Quasi-Static Elasto/Viscoplastic Finite Element Analysis," *Computers and Structures* **8**, 169–173.

Hughes, T.J.R., R.L. Taylor, and J.L. Sackman [1975], "Finite Element Formulation and Solution of Contact-Impact Problems in Continuum Mechanics–III," SESM Report No. 75-3, Department of Civil Engineering, University of California, Berkeley.

Hughes, T.J.R., and J. Winget, [1980], "Finite Rotation Effects in Numerical Integration of Rate Constitutive Equations Arising in Large-Deformation Analysis," *International Journal for Numerical Methods in Engineering* **15**, 1862–1867.

Hundsdorfer, W.H., [1985], *The Numerical Solution of Nonlinear Stiff Initial Value Problems: An Analysis of One Step Methods*, CWI Tract, Center for Mathematics and Computer Science, Amsterdam, The Netherlands.

Hungerford, T.W., [1974], *Algebra*, Springer Verlag, New York.

Iwan, D.W., and P.J. Yoder, [1983], "Computational Aspects of Strain-Space Plasticity," *Journal of Engineering Mechanics*, **109**, 231–243.

Johnson, C., [1976a], "Existency Theorems for Plasticity Problems," *Journal de Mathematiques Pures et Appliques* **55**, 431–444.

Johnson, C., [1976b], "On Finite Element Methods for Plasticity Problems," *Numerische Mathematik* **26**, 79–84.

Johnson, C., [1977], "A Mixed Finite Element for Plasticity," *SIAM Journal of Numerical Analysis* **14**, 575–583.

Johnson, C., [1978], "On Plasticity with Hardening," *Journal of Applied Mathematical Analysis* **62**, 325–336.

Johnson, C., [1987], *Numerical Solution of Partial Differential Equations by the Finite Element Method*, Cambridge University Press, Cambridge, U.K.

Johnson, G., and D.J. Bammann, [1984], "A Discussion of Stress Rates in Finite Deformation Problems," *International Journal of Solids and Structures* **20**, 725–737.

Kachanov, L.M., [1974], *Fundamentals of the Theory of Plasticity*, MIR Publishers, Moscow.

Karmanov, V., [1977], *Programmation Mathematique*, MIR Publishers, Moscow.

Key, S.W., [1969], "A Variational Principle for Incompressible and Nearly Incompressible Anisotropic Elasticity," *International Journal of Solids Structures* **5**, 951–964.

Key, S.W., [1974], "HONDO - A Finite Element Computer Program for the Large Deformation Response of Axisymmetric Solids," Report 74–0039, Sandia National Laboratories, Albuquerque, N.M.

Key, S.W., C.M. Stone, and R.D. Krieg, [1981], "Dynamic Relaxation Applied to the Quasi-Static, Large Deformation, Inelastic Response of Axisymmetric

Solids," in *Nonlinear Finite Element Analysis in Structural Mechanics*, eds., W. Wunderlich et al., Springer, Berlin.

Key, S.W., and R.D. Krieg, [1982], "On the Numerical Implementation of Inelastic Time Dependent and Time Independent, Finite Strain Constitutive Equations in Structural Mechanics," *Computer Methods in Applied Mechanics and Engineering* **33**, 439–452.

Koiter, W.T., [1953], "Stress-Strain Relations, Uniqueness and Variational Theorems for Elastic-Plastic Materials with a Singular Yield Surface," *Quarterly of Applied Mathematics* **11**, 350–354.

Koiter, W.T., [1960], "General Theorems for Elastic-plastic Solids," in *Progress in Solid Mechanics* **6**, eds. I.N. Sneddon and R. Hill, North-Holland Publishing Company, Amsterdam, 167–221.

Kratochvil, J., [1973], "On a Finite Strain Theory of Elastic-Inelastic Materials," *Acta Mechanica* **16**, 127–142.

Krieg, R.D., and S.W. Key, [1976], "Implementation of a Time Dependent Plasticity Theory into Structural Computer Programs," *Constitutive Equations in Viscoplasticity: Computational and Engineering Aspects*, eds., J.A. Stricklin and K.J. Saczalski, AMD-20, ASME, New York.

Krieg, R.D. and D.B. Krieg, [1977], "Accuracies of Numerical Solution Methods for the Elastic-Perfectly Plastic Model," *Journal of Pressure Vessel Technology* **99**, 510–515.

Kroner, E. and C. Teodosiu, [1972], "Lattice Defect Approach to Plasticity and Viscoplasticity," in *Problems of Plasticity*, ed., A. Sawczuk, Noordhoff.

Lang, S., [1983], *Real Analysis*, 2nd ed., Addison–Wesley Publishing Company, Inc., Reading, Mass.

Lee, E.H., [1969], "Elastic-Plastic Deformations at Finite Strains," *Journal of Applied Mechanics* **36**, 1–6.

Lee, E.H., and D.T. Liu, [1967], "Finite Strain Elastic-Plastic Theory Particularly for Plane Wave Analysis," *Journal of Applied Physics*, **38**, 19–27.

Lee, R.L., P.M. Gresho, and R.L. Sani, [1979], "Smoothing Techniques for Certain Primitive Variable Solutions of the Navier-Stokes Equations," *International Journal for Numerical Methods in Engineering* **14**, (12), 1785–1804.

Lemaitre, J., and J.-L. Chaboche, [1990], *Mechanics of Solid Materials*, Cambridge University Press, Cambridge, U.K.

Loret, B., and J.H. Prevost, [1986], "Accurate Numerical Solutions for Drucker-Prager Elastic-Plastic Models," *Computer Methods in Applied Mechanics and Engineering* **54**, 259–277.

Lubliner, J., [1972], "On the Thermodynamic Foundations of Non-Linear Solid Mechanics," *International Journal of Non-Linear Mechanics* **7**, 237–254.

Lubliner, J., [1973], "On the Structure of the Rate Equations of Materials with Internal Variables," *Acta Mechanica* **17**, 109–119.

Lubliner, J., [1984], "A Maximum-Dissipation Principle in Generalized Plasticity," *Acta Mechanica* **52**, 225–237.

Lubliner, J., [1986], "Normality Rules in Large-Deformation Plasticity," *Mechanics of Materials* **5**, 29–34.

Luenberger, D.G., [1972], *Optimization by Vector Space Methods*, John Wiley & Sons Inc., New York.

Luenberger, D.G., [1984], *Linear and Nonlinear Programming*, Addison–Wesley Publishing Company, Reading, Mass.

Maier, G., [1970], "A Matrix Structural Theory of Piecewise Linear Elastoplasticity with Interacting Yield Planes," *Meccanica* **5**, 54–66.

Maier, G., and D. Grierson, [1979], *Engineering Plasticity by Mathematical Programming*, Pergamon Press, New York.

Malkus, D.S. and T.J.R. Hughes, [1978], "Mixed Finite Element Methods — Reduced and Selective Integration Techniques: A Unification of Concepts," *Computer Methods in Applied Mechanics and Engineering* **15** (1), 68–81.

Malvern, L., [1969], *An Introduction to the Mechanics of a Continuous Medium*, Prentice–Hall, Englewood Cliffs, N.J.

Mandel, J., [1964], "Contribution Theorique a l'Etude de l'Ecrouissage et des Lois de l'Ecoulement Plastique," *Proceedings of the 11th International Congress on Applied Mechanics*, 502–509.

Mandel, J., [1965], "Generalisation de la Theorie de la Plasticite de W.T. Koiter," *International Journal of Solids and Structures* **1**, 273–295.

Mandel, J., [1974], "Thermodynamics and Plasticity," in *Foundations of Continuum Thermodynamics*, eds., J.J. Delgado et al., Macmillan, New York.

Marcal, P.V., and I.P. King, [1967], "Elastoplastic Analysis of Two-Dimensional Stress Systems by the Finite Element Method," *International Journal of Mechanical Science* **9**, 143–155.

Marsden J.E. and T.J.R Hughes, [1994], *Mathematical Foundations of Elasticity*, Dover, New York.

Matthies, H., [1978], "Problems in Plasticity and their Finite Element Approximation", Ph.D. Thesis, Department of Mathematics, Massachusetts Institute of Technology, Cambridge, Mass.

Matthies, H. [1979], "Existence theorems in thermo–plasticity," *Journal de Mecanique* **18** (4), 695–711.

Matthies, H., and G. Strang, [1979], "The Solution of Nonlinear Finite Element Equations," *International Journal for Numerical Methods in Engineering* **14** (11), 1613–1626.

Matthies, H., G. Strang, and E. Christiansen, [1979], "The saddle point of a differential," in *Energy Methods in Finite Element Analysis*, R. Glowinski, E.Y. Rodin, and O.C. Zienkiewicz, eds., John Wiley & Sons, New York.

McMeeking, R.M., and J.R. Rice, [1975], "Finite-Element Formulations for Problems of Large Elastoplastic Deformations," *International Journal of Solids and Structures* **11**, 601–616.

Mehrabadi, M.M., and S. Nemat–Nasser, [1987], "Some Basic Kinematical Relations for Finite Deformations of Continua," *Mechanics of Materials* **6**, 127–138.

Mikhlin, S.G., [1970], *An Advanced Course of Mathematical Physics*, American Elsevier, New York.

Moreau, J.J., [1976], "Application of Convex Analysis to the Treatment of Elastoplastic Systems," in *Applications of Methods of Functional Analysis to Problems in Mechanics*, eds., P. Germain and B. Nayroles, Springer-Verlag, Berlin.

Moreau, J.J., [1977], "Evolution Problem Associated with a Moving Convex Set in a Hilbert Space," *Journal of Differential Equations* **26**, 347.

Morman, K.N., [1987], "The Generalized Strain Measure with Application to Non-Homogeneous Deformations in Rubber-Like Solids," *Journal of Applied Mechanics* **53**, 726–728.

Naghdi, P.M., [1960], "Stress-Strain Relations in Plasticity and Thermoplasticity," in Proceedings of the 2nd Symposium on Naval Structural Mechanics, Pergamon Press, London.

Naghdi, P.M., and J.A. Trapp, [1975], "The Significance of Formulating Plasticity Theory with Reference to Loading Surfaces in Strain Space," *International Journal of Engineering Science*, **13**, 785–797.

Nagtegaal, J.C., [1982], "On the Implementation of Inelastic Constitutive Equations with Special Reference to Large Deformation Problems," *Computer Methods in Applied Mechanics and Engineering* **33**, 469–484.

Nagtegaal, J.C., and J.E. De Jong, [1981] , "Some Computational Aspects of Elastic-Plastic Large Strain Analysis," *International Journal for Numerical Methods in Engineering*, **17**, 15–41.

Nagtegaal, J.C., D.M. Parks, and J.R. Rice, [1974], "On Numerically Accurate Finite Element Solutions in the Fully Plastic Range," *Computer Methods in Applied Mechanics and Engineering*, **4**, 153–177.

Nagtegaal, J.C., and F.E. Veldpaus, [1984], "On the Implementation of Finite Strain Plasticity Equations in a Numerical Model," in *Numerical Analysis of Forming Processes*, eds., J.F.T. Pittman, O.C. Zienkiewicz, R.D. Wood, and J.M. Alexander, John Wiley & Sons Ltd., Chicherster, U.K.

Nayak, G.C., and O.C. Zienkiewicz, [1972], "Elastoplastic Stress Analysis. Generalization of Various Constitutive Equations Including Stress Softening," *International Journal for Numerical Methods in Engineering* **5**, 113–135.

Neal, B.G., [1961], "The Effect of Shear and Normal Forces on the Fully Plastic Moment of a Beam of Rectangular Cross Section," *Journal of Applied Mechanics* **28**, 269–274.

Needleman, A., [1982], "Finite Elements for Finite Strain Plasticity Problems," *Plasticity of Metals at Finite Strains: Theory, Computation, and Experiment*, eds., E.H. Lee and R.L. Mallet, Division of Applied Mechanics, Stanford University, Stanford, California, 387–436.

Needleman, A., and V. Tvergaard, [1984], "Finite Element Analysis of Localization Plasticity," in *Finite elements, Vol V: Special Problems in Solid Mechanics*, eds., J.T. Oden and G.F. Carey, Prentice–Hall, Englewood Cliffs, N.J.

Nemat–Nasser, S., [1982], "On Finite Deformation Elasto-Plasticity," *International Journal of Solids Structures* **18**, 857–872.

Nemat–Nasser, S., [1983], "On Finite Plastic Flow of Crystalline Solids and Geomaterials" *Journal of Applied Mechanics*, **50**, 1114–1126.

Nguyen, Q.S., [1977], "On the elastic-plastic initial boundary value problem and its numerical integration," *International Journal for Numerical Methods in Engineering* **11**, 817–832.

Oden, J.T., and G. Carey, [1983], *Finite Elements. Volume IV. Mathematical aspects,* Prentice–Hall, Englewood Cliffs, N.J.

Oden, J.T., and J.N. Reddy, [1976], *Variational Methods for Theoretical Mechanics*, Springer-Verlag, Berlin.

Ogden, R.W., [1982], "Elastic Deformations in Rubberlike Solids," in *Mechanics of Solids, the Rodney Hill 60th Anniversary Volume*, (eds., H.G. Hopkins and M.J. Sewell), Pergamon Press, Oxford, 499–537.

Ogden, R.W., [1984], *Non-Linear Elastic Deformations*, Ellis Horwood Ltd., West Sussex, England.

Ortiz, M., and J.C. Simo, [1986], "Analysis of a New Class of Integration Algorithms for Elastoplastic Constitutive Relations," *International Journal for Numerical Methods in Engineering* **23**, 353–366.

Ortiz, M., and E.P. Popov, [1985], "Accuracy and Stability of Integration Algorithms for Elastoplastic Constitutive Equations," *International Journal for Numerical Methods in Engineering* **21**, 1561–1576.

Ortiz, M., and J.E. Martin, [1989], "Symmetry-Preserving Return Mapping Algorithms and Incrementally Extremal Paths: A Unification of Concepts," *International Journal for Numerical Methods in Engineering* **28**, 1839–1853.

Pazy, A., [1983], *Semigroups of Linear Operators and Applications to Partial Differential Equations*, Springer-Verlag, New York.

Perzyna, P., [1971], "Thermodynamic Theory of Viscoplasticity," in *Advances in Applied Mechanics*, (ed., Chia-Shun Yih), Academic Press, New York, **11**, 313–354.

Pinsky, P., M. Ortiz, and K.S. Pister, [1983], "Numerical Integration of Rate Constitutive Equations in Finite Deformation Analysis," *Computer Methods in Applied Mechanics and Engineering* **40**, 137–158.

Pitkaranta, J., and R. Sternberg, [1984], "Error Bounds for the Approximation of the Stokes Problem Using Bilinear/Constant Elements on Irregular Quadrilateral Meshes," Report MAT-A222, Helsinki University of Technology, Institute of Mathematics, Helsinki.

Powell, M.J.D., [1969], "A Method for Nonlinear Constraints in Minimization Problems," in *Optimization*, ed., R. Fletcher, Academic Press, London.

Prager, W., [1956], "A New Method of Analyzing Stress and Strains in Work-Hardening Plastic Solids," *Journal of Applied Mechanics* **23**, 493–496.

Pschenichny, B.N., and Y.M. Danilin, [1978], *Numerical Methods in Extremal Problems*, MIR Publishers, Moscow.

Resende, L., and J.B. Martin, [1986], "Formulation of Drucker-Prager Cap Model," *Journal of Engineering Mechanics* **111**(7), 855–865.

Rice J.R. and D.M. Tracey, [1973], "Computational Fracture Mechanics," in *Proceedings of the Symposium on Numerical Methods in Structural Mechanics*, ed., S.J. Fenves, Urbana, Illinois, Academic Press.

Richtmyer, R.D., and K.W. Morton [1967], *Difference Methods for Initial Value Problems*, 2nd ed., Interscience, New York.

Rolph, W.D., and K.J. Bathe, [1984], "On a Large Strain Finite-Element Formulation for Elasto-Plastic Analysis," in *Constitutive Equations, Macro and Computational Aspects*, ed., K.J. Willam, Winter Annual Meeting, 1984, ASME, New York.

Rubinstein R., and S.N. Atluri, [1983], "Objectivity of Incremental Constitutive Relations over Finite Time Steps in Computational Finite Deformation Analyses," *Computer Methods in Applied Mechanics and Engineering* **36**, 277–290.

Sandler, I.S., DiMaggio, F.L., and Baladi, G.Y., [1976], "Generalized CAP Model for Geological Materials," *Journal of the Geotechnical Engineering Division*, **102** (GT7), July, 683–699.

Sandler, I.S., and D. Rubin, [1979], "An Algorithm and a Modular Subroutine for the Cap Model," *International Journal for Numerical and Analytical Methods in Geomechanics* **3**, 173–186.

Schreyer, H.L., R.L. Kulak, and J.M. Kramer, [1979], "Accurate Numerical Solutions for Elastic-Plastic Models," *Journal of Pressure Vessel Technology*, **101**, 226–334.

Sidoroff, F., [1974], "Un Modele Viscoelastique Non Lineaire Avec Configuration Intermediaire," *Journal de Mecanique* **13**, 679–713.

Simo, J.C., [1986], "On the Computational Significance of the Intermediate Configuration and Hyperelastic Relations in Finite Deformation Elastoplasticity," *Mechanics of Materials* **4**, 439–451.

Simo, J.C., [1987a], "On a Fully Three-Dimensional Finite-Strain Viscoelastic Damage Model: Formulation and Computational Aspects," *Computer Methods in Applied Mechanics and Engineering* **60**, 153–173.

Simo, J.C., [1987b], "A J_2-Flow Theory Exhibiting a Corner-Like Effect and Suitable for Large-Scale Computation," *Computer Methods in Applied Mechanics and Engineering* **62**, 169–194.

Simo, J.C., [1988a], "A Framework for Finite Strain Elastoplasticity Based on Maximum Plastic Dissipation and the Multiplicative Decomposition: Part I. Continuum Formulation," *Computer Methods in Applied Mechanics and Engineering* **66**, 199–219.

Simo, J.C., [1988b], "A Framework for Finite Strain Elastoplasticity Based on Maximum Plastic Dissipation and the Multiplicative Decomposition: Part II. Computational Aspects," *Computer Methods in Applied Mechanics and Engineering* **68**, 1–31.

Simo, J.C., [1991], "Nonlinear Stability of the Time-Discrete Variational Problem of Evolution in Nonlinear Heat Conduction and Elastoplasticity," *Computer Methods in Applied Mechanics and Engineering* **88**, 111–131.

Simo, J.C., and S. Govindjee, [1988], "Exact Closed-Form Solution of the Return Mapping Algorithm for Plane Stress Elasto-Viscoplasticity," *Engineering Computations* **3**, 254–258.

Simo, J.C., and S. Govindjee, [1991], "Nonlinear B–Stability and Symmetry Preserving Return Mapping Algorithms for Plasticity and Viscoplasticity" *International Journal for Numerical Methods in Engineering* **31**, 151–176.

Simo, J. C., K. D. Hjelmstad, and R. L. Taylor, [1984], "Numerical Formulations for Finite Deformation Problems of Beams Accounting for the Effect of Transverse Shear," *Computer Methods in Applied Mechanics and Engineering* **42**, 301–330.

Simo, J.C., and T.J.R. Hughes, [1986], "On the Variational Foundations of Assumed Strain Methods," *Journal of Applied Mechanics* **53**, 51–54.

Simo, J.C., and T.J.R. Hughes, [1987], "General Return Mapping Algorithms for Rate Independent Plasticity," in *Constitutive Laws for Engineering Materials*, (ed., C.S. Desai), Elsevier, N.Y.

Simo, J.C., and J.W. Ju, [1989], "Finite Deformation Damage-Elastoplasticity: A Non-Conventional Framework," *International Journal of Computational Mechanics* **5**, 375–400.

Simo, J.C., Ju, J. W., Pister, K. S., and Taylor, R. L., [1988], "An Assessment of the Cap Model: Consistent Return Algorithms and Rate-Dependent Extension," *Journal of Engineering Mechanics* **114**, 191–218.

Simo, J.C., J.G. Kennedy, and S. Govindjee, [1988], "Non-Smooth Multisurface Plasticity and Viscoplasticity. Loading/Unloading Conditions and Numerical Algorithms," *International Journal for Numerical Methods in Engineering* **26**, 2161–2185.

Simo, J.C., J.G. Kennedy, and R.L. Taylor, [1988], "Complementary Mixed Finite Element Formulations of Elastoplasticity," *Computer Methods in Applied Mechanics and Engineering* **74**, 177–206.

Simo J.C., and J.E. Marsden, [1984], "On the Rotated Stress Tensor and the Material Version of the Doyle-Ericksen Formula," *Archive for Rational Mechanics and Analysis* **86**, 213–231.

Simo, J.C., J.E. Marsden, and P.S. Krishnaprasad, [1988], "The Hamiltonian Structure of Elasticity. The Convective Representation of Solids, Rods and Plates," *Archive for Rational Mechanics and Analysis* **104**, 125–183.

Simo J.C., and M. Ortiz, [1985], "A Unified Approach to Finite Deformation Elastoplasticity Based on the use of Hyperelastic Constitutive Equations," *Computer Methods in Applied Mechanics and Engineering* **49**, 221–245.

Simo J.C., and K.S. Pister, [1984], "Remarks on Rate Constitutive Equations for Finite Deformation Problems: Computational Implications," *Computer Methods in Applied Mechanics and Engineering* **46**, 201–215.

Simo, J.C., and L.V. Quoc, [1986], "A 3-Dimensional Finite Strain Rod Model. Part II: Geometric and Computational Aspects," *Computer Methods in Applied Mechanics and Engineering* **58**, 79–116.

Simo J.C., and R.L. Taylor, [1985], "Consistent Tangent Operators for Rate Independent Elasto-Plasticity," *Computer Methods in Applied Mechanics and Engineering* **48**, 101–118.

Simo, J.C, and R.L. Taylor, [1986], "Return Mapping Algorithm for Plane Stress Elastoplasticity," *International Journal for Numerical Methods in Engineering* **22**, 649–670.

Simo, J.C., R.L. Taylor, and K.S. Pister, [1985], "Variational and Projection Methods for the Volume Constraint in Finite Deformation Elastoplasticity," *Computer Methods in Applied Mechanics and Engineering* **51**, 177–208.

Simo, J.C., and R.L. Taylor, [1991], "Quasi-Incompressible Finite Elasticity in Principal Stretches. Continuum Basis and Numerical Algorithms," *Computer Methods in Applied Mechanics and Engineering* **85**, 273–310.

Simo, J.C., and K. Wong, [1991], "Unconditionally Stable Algorithms for the Orthogonal Group That Exactly Preserve Energy and Momentum," *International Journal for Numerical Methods in Engineering* **31**, 19–52

Sokolnikoff, I.S., [1956], *Mathematical Theory of Elasticity*, 2nd ed., McGraw Hill, New York.

Stakgold, [1979], *Green's Functions and Boundary Value Problems*, John Wiley & Sons, New York.

Strang, G., [1986], *Introduction to Applied Mathematics*, Wellesley-Cambridge Press, Wellesley, Mass.

Strang, G., H. Matthies, and R. Temam, [1980], "Mathematical and Computational Methods in Plasticity," in *Variational Methods in the Mechanics of Solids*, ed., S. Nemat–Nasser, Pergamon Press, Oxford.

Suquet, P., [1979], "Sur les Eq́uations de la Plasticite," *Ann. Fac. Sciences Toulouse* **1**, 77–87.

Suquet, P.M., [1981], "Sur les Equations de la Plasticite: Existence et Regularite des Solutions," *Journal de Mecanique* **20**, 3–39.

Taylor, G.I., [1938], "Analysis of Plastic Strain in a Cubic Crystal," in *Stephen Timoshenko 60th Anniversary Volume*, ed., J.M. Lessels, Macmillan, New York.

Taylor, G.I., and C.F. Elam, [1923], "Bakerian Lecture: The Distortion of an Aluminum Crystal During a Tensile Test," *Proceedings of the Royal Society of London* **A102**, 643–667.

Taylor, G.I., and C.F. Elam, [1925], "The Plastic Extension and Fracture of Aluminum Crystals," *Proceedings of the Royal Society of London* **A108**, 28–51.

Taylor, L.M. and E.B. Becker, [1983], "Some Computational Aspects of Large Deformation, Rate-Dependent Plasticity Problems," *Computer Methods in Applied Mechanics and Engineering* **41**, No. 3, 251–278.

Taylor, R.L., K.S. Pister, and G.L. Goudreau, [1970], "Thermomechanical Analysis of Viscoelastic Solids," *International Journal for Numerical Methods in Engineering* **2**, 45–79.

Taylor, R.L., K.S. Pister, and L.R. Herrmann, [1968], "On a Variational Theorem for Incompressible and Nearly-Incompressible Elasticity," *International Journal of Solids and Structures* **4**, 875–883.

Taylor, R. L., J.C. Simo, O. C. Zienkiewicz, and A.C. Chan, [1986], "The Patch Test: A Condition for Assessing Finite Element Convergence," *International Journal for Numerical Methods in Engineering* **22**, 39–62.

Temam, R., and G. Strang, [1980], "Functions of Bounded Deformation," *Archive for Rational Mechanics and Analysis* **75**, 7–21.

Temam, R., [1985], *Mathematical Problems in Plasticity*, Gauthier-Villars, Paris (Translation of original 1983 French edition).

Ting, T.T., [1985], "Determination of $C^{1/2}$, $C^{-1/2}$ and More General Isotropic Tensor Functions of C," *Journal of Elasticity* **15**, 319–323.

Triantafyllidis, N., A. Needleman, and V. Tvergaard, [1982], "On the Development of Shear Bands in Pure Bending," *International Journal of Solids and Structures* **18**, 121–138.

Troutman, J.L., [1983], *Variational Calculus with Elementary Convexity*, Springer-Verlag, New York.

Truesdell C., and W. Noll, [1965], "The Nonlinear Field Theories," *Handbuch der Physik*, Band III/3, Springer-Verlag, Berlin.

Truesdell C., and R.A. Toupin, [1960], "The Classical Field Theories," in *Handbook der Physik*, III/1, Springer-Verlag, Berlin.

Tsai, S.W. and E.M. Wu [1971], "A General Theory of Strength for Anisotropic Materials," *Journal of Composite Materials* **5**, 58.

Tvergaard, V., A. Needleman, and K.K. Lo, [1981], "Flow Localization in the Plane Strain Tensile Test," *Journal of Mechanics and Physics of Solids* **29**, 115–142.

Vainberg, M.M., [1964], *Variational Methods for the Study of Nonlinear Operators*, Holden–Day, Inc., San Francisco.

Voce, E., [1955], "A Practical Strain Hardening Function," *Metallurgica* **51**, 219–226.

Wanner, G., [1976], "A Short Proof of Nonlinear A–Stability," *BIT* **16**, 226–227.

Whittaker, E.T., [1944], *A Treatise on the Analytical Dynamics of Particles and Rigid Bodies*, Dover, New York.

Wilkins, M.L., [1964], "Calculation of Elastic-Plastic Flow," in *Methods of Computational Physics 3*, eds., B. Alder et. al., Academic Press, New York.

Yoder, P.J., and R.L. Whirley, [1984], "On the Numerical Implementation of Elastoplastic Models," *Journal of Applied Mechanics* **51**, (2), 283–287.

Zaremba, S., [1903], "Sur une Forme Perfectionée de la Théorie de la Relaxation," *Bul. Int. Acad. Sci. Cracovie* **8**, 594–616.

Zeidler, E. [1985], *Nonlinear Functional Analysis and its Applications III: Variational Methods and Optimization*, Springer–Verlag, Berlin.

Ziegler, H., [1959], "A Modification of Prager's Hardening Rule," *Quarterly of Applied Mathematics* **17**, 55–65.

Zienkiewicz, O.C., and I.C. Cormeau, [1974], "Viscoplasticity – Plasticity and Creep in Elastic Solids – a Unified Numerical Solution Approach," *International Journal for Numerical Methods in Engineering* **8**, 821–845.

Zienkiewicz, O.C., [1977], *The Finite Element Method, 3rd ed.*, McGraw–Hill, London.

Zienkiewicz, O.C., and R.L. Taylor [1989], *The Finite Element Method, 4th ed.*, McGraw–Hill, London, Vol. I.

Zienkiewicz, O.C., R.L. Taylor, and J.M. Too, [1971], "Reduced Integration Technique in General Analysis of Plates and Shells," *International Journal for Numerical Methods in Engineering* **3**, 275–290.

Index

Printed in the United Kingdom
by Lightning Source UK Ltd.
111940UKS00005B/137

9 780387 975207